# Preface

Twenty years ago, recognizing the need for fostering understanding of the new technology in space travel, the staff of McLaughlin Research Corporation compiled the *Space Age Dictionary* under the editorial guidance of Charles McLaughlin, then President of the Company. The success of that publication and the recognition of a similar need for basic information in an old but newly stimulated technology, led us to the decision to compile this *Energy Dictionary*.

The *Energy Dictionary* has been prepared to meet the need for an up-to-date, authoritative yet concise compilation of the salient terms associated with the broad field of energy. This book covers both the conventional sources of fossil energy, and the nuclear and advanced energy systems—solar, geothermal, ocean, and wind. It also includes environmental and conservation areas, and the major scientific physical concepts, processes, and technological advances.

All definitions have been expressed as clearly and simply as possible without altering their acknowledged definition by the source from which they were selected. Many of the terms have multiple meanings, and all the interpretations are not always listed. In general, the selected definition was determined by its direct relationship to some facet of the energy picture. The aim throughout this book has been to make it simple, but sufficiently complete to be a valuable reference source for anyone desiring to enlarge his knowledge of the world of energy.

In addition to more than 4000 entries, this book includes some 320 charts, graphs, process diagrams, and latest available photographs to assist the reader in quickly grasping the key points of each definition. An overview of the status of the current technology and its impact on the nation's energy situation is also provided. For further reading and convenience, a comprehensive bibliography has been organized by specific energy source areas. The asterisked references delineate the sources used. The Appendix includes conversion factors and a glossary of the acronyms that appear in the book.

BRUCE MCLAUGHLIN
President
McLaughlin Research Corporation
New York City

V. DANIEL HUNT
Manager, Energy Systems Division
McLaughlin Research Corporation
Burke, Virginia

# *Acknowledgments*

We wish to acknowledge the contribution of those members of the McLaughlin Research staff who were assigned key roles in this work. First of all, V. Daniel Hunt was assigned the principal role of editorial guidance and acquisition of source material. David Whelchel provided able assistance in the research effort. Helen Mack organized the selection of terms and took charge of technical editing. Allen Higgs and his staff prepared the technical graphics which are an essential complement to the text.

We are indebted to the following organizations whose cooperation in providing source material was indispensable in preparing this book.

- The Department of Energy
- The Energy Research and Development Agency
- The American Gas Association
- The National Coal Association
- The American Nuclear Society
- The American Association for the Advancement of Science
- The Electric Power Research Institute
- The National Solar Heating and Cooling Information Center
- The American Petroleum Institute
- The National Academy of Sciences
- The National Aeronautics and Space Administration
- The Environmental Protection Agency
- The Library of Congress

Finally, I wish to credit specifically all of the sources reviewed for the selection of the terms and definitions used in this Energy Dictionary:

*A Dictionary of Mining, Mineral, and Related Terms*, Paul W. Thrush and Staff of Bureau of Mines (eds.).

*A Guide to Federal Power Commission Public Information*, Federal Power Commission.

*American National Standard, Glossary of Terms in Nuclear Science and Technology*, American Nuclear Standards Committee.

*A National Plan for Energy Research, Development, and Demonstration*, U.S. Energy Research and Development Administration.

*Annual Review of Energy*, Jack M. Hollander (ed.).

*Basic Nuclear Engineering*, Arthur R. Foster and Robert L. Wright, Jr.

*Breeder Backgrounder*, U.S. Energy Research and Development Administration and Breeder Reactor Corporation.

*Coal, Conversion and Utilization*, U.S. Energy Research and Development Administration.

*Coal Mining Safety Manual No. 1*, M. G. Zabetakis and L. D. Phillips.

*Coal Power and Combustion, Quarterly Report Oct.–Dec., 1975*, U.S. Energy Research & Development Administration, Office of Fossil Energy.

*Common Environmental Terms, A Glossary*, U.S. Department of Commerce.

*Criteria for Energy Storage R & D*, The National Research Council, National Academy of Sciences.

*Direct Conversion of Energy*, William R. Corliss.

*Direct Use of the Sun's Energy*, Farrington Daniels.

*Dynasurge*, Catalyst Research Corporation.

*Efficient Electricity Use*, Craig B. Smith (ed.).

*Energy Alternatives, A Comparative Analysis*, U.S. Council on Environmental Quality.

*Energy and the Future*, Allen L. Hammond, William D. Metz, and Thomas H. Maugh II.

*Energy Environment Source Book*, John M. Fowler.

*Energy Facts II*, Science Policy Research Division, U.S. House of Representatives, 94th Congress, First Session.

*Energy For Man—From Windmills to Nuclear Power*, Hans Thirring.

*Energy for Survival—The Alternative to Extinction*, Wilson Clark.

*Energy From Coal*, U.S. Energy Research and Development Administration, Division of Coal Conversion and Utilization.

*Energy Intelligence & Analysis for Energy Consumers*, Phillip J. Berardelli (ed.).

*Energy Microthesaurus*, U.S. Department of Commerce.

*Energy Planning Report*, Resources News Service, Inc.

*Environmental Information Summaries C-3*, U.S. Department of Commerce, National Oceanic and Atmospheric Administration.

Geothermal Energy—The Hot Prospect, *EPRI Journal*, Electric Power Research Institute.

*Glossary for the Gas Industry*, American Gas Association.

*Glossary of Chemical Terms*, Clifford A. Hampel and Gessner G. Hawley.

*Glossary of Electric Utility Terms*, Edison Electric Institute.

*Glossary of Surface Mining and Reclamation Technology*, Council for Surface Mining and Reclamation Research in Appalachia (ed.), National Coal Association.

*Help: The Useful Almanac: 1976–1977*, Arthur E. Rowse (ed.).

*How To Build a Solar Heater*, Ted Lucas.

*Information from ERDA Reference Information: Special Issue*, U.S. Energy Research and Development Administration.

*Information from ERDA Weekly Announcements*, U.S. Energy Research and Development Administration.

*Living with Natural Energy, Design for a Limited Planet*, Norma Skurka and Jon Naar.

*Master Plan: ERDA-76/122*, U.S. Energy Research and Development Administration, Division of Safeguards and Security.

*Minimum Energy Dwelling*, U.S. Energy Research and Development Administration, Office of Conservation, Division of Buildings and Community Systems.

*Naval Petroleum Reserves*, Committee on Armed Services, House of Representatives, 93rd Congress, First Session.

*News Photo Catalog for the Trans Alaska Pipeline*, Alyeska Pipeline Service Company.

*New Water*, U.S. Department of the Interior, Office of Water Research and Technology.

*NRDC Newsletter*, Marc Reisner (ed.).

*Nuclear Terms, A Glossary*, U.S. Atomic Energy Commission.

*Petropolitics and the American Energy Shortage*, Committee on Government Operations, United States Senate, 93rd Congress, First Session.

*Popular Science*, Times Mirror Magazine, Inc.

*Solar Energy*, William W. Eaton.

*Solar Energy for Space Heating & Hot Water*, U.S. Energy Research and Development Administration, Division of Solar Energy.

*Solar Energy Update*, U.S. Energy Research and Development Administration.

*Summary Project Description of the Trans Alaska Pipeline System*, Alyeska Pipeline Service Company.

*Sun Language*, Solar Energy Institute of America.

*The Clinch River Breeder Reactor Plant and its Impact on the Environment*, Breeder Reactor Corporation.

*The Coming Age of Solar Energy*, D. S. Halacy.

*The Dependable Life Support Cell*, Catalyst Research Corporation.

*The Environmental Impact of Electrical Power Generation: Nuclear and Fossil: ERDA-69*, U.S. Energy Research and Development Administration, Division of Biomedical and Environmental Research.

*The Seven Sisters*, Anthony Sampson.

*The Wonderful World of Energy*, Lancelot Hogben.

*Van Nostrand's Scientific Encyclopedia*, Douglas M. Considine (ed.).

*Wind Machines*, Frank R. Eldridge.

*Windpower—A Handbook on Wind Energy Conversion Systems*, V. D. Hunt, D. R. Scherr (Draft).

Credit for the photographs and illustrations selected from these and other sources is indicated where appropriate throughout the book.

# Contents

PREFACE iii
ACKNOWLEDGMENTS v
ENERGY OVERVIEW 1
Conservation 1
Transportation Sector 2
Residential/Commercial Sector 3
Industrial Sector 4
Expansion of Existing Fuel Sources 4
Produce Additional Petroleum and Natural Gas 5
Direct Use of Coal 5
Light Wave Reactors 7
Technologies That Use Fuels 9
Shale Oil 10
Geothermal 10
Solar Heating and Cooling 12
Energy from Wastes and Biomass 13
Solar Electric Systems 13
ENERGY DEFINITIONS 17
CONVERSION FACTORS 500
GLOSSARY 505
BIBLIOGRAPHY 509

# *Energy Overview*

The demand for energy is increasing while the supplies of oil and natural gas are diminishing. Unless the U.S. makes a timely adjustment before world oil becomes very scarce and very expensive in the 1985's, the nation's economic security and the American way of life may be gravely endangered.

Tc help solve the energy problem, new technology must be developed to (1) increase conservation of energy, (2) expand the use of existing fuels, and (3) make the transition to new fuels.

Increased conservation can have the greatest immediate impact on the nation's energy system between now and the year 2000. This impact can be obtained through a coordinated set of actions involving voluntary efforts, economic incentives, regulatory actions, and development of more efficient technologies to use and produce energy.

The expansion in the production and use of existing fuels (such as oil and natural gas, coal, and uranium (using a once-through fuel cycle in light water reactors) can also contribute substantially to solving the Nation's energy problem. Expansion of existing fuels combined with increased energy efficiency will provide most of the impetus in meeting the nation's energy goals.

Because of the long development times required for difficult new technologies such as solar electric, hot dry rock geothermal, breeder concepts, and fusion, their major energy contributions are expected to occur after the turn of the century. But the research and development needed must be pursued today if the potentials of these technologies are to be realized in time. Meanwhile, the new fuel technologies likely to expand energy supplies significantly prior to year 2000 include solar heating and cooling, hydrothermal-geothermal, geopressured geothermal, biomass, and shale oil. Although the combined contribution from these new fuel technologies is likely to be smaller at the end of the century than the contribution from either conservation or existing fuels technologies, new fuels will continue to grow in importance as the transition to renewable and essentially inexhaustible fuels continues into the 21st century. Prior to that time, energy contributions from these sources can and will ease the burden on existing depleting fuels and on imports, and thus play a key role in the energy transition process.

The following material provides a brief overview of the energy technology, which is in a rapid state of change in response to our critical energy needs.

## CONSERVATION

The number of individual technologies required for effective conservation in transportation, residential/commercial, and industrial sectors is very large; the following discussion highlights only a few of the significant options.

## Transportation Sector

The transportation sector embraces many different energy-consuming systems and accounted in 1976 for 20 quads (or 26 percent) of total domestic energy consumption. The passenger vehicle, accounting for more than nine quads of the total, is by far the largest energy user in this sector. Consequently, it is a prime target for conservation measures.

In the near term, conservation will be achieved through evolutionary modifications of existing types of vehicles, engines, and components. Smaller, lighter cars with smaller engines (and less power per unit weight) will provide some of the mandated improvements in miles per gallon. Other easily adopted modifications of the present internal combustion engine, such as computer-controlled ignition and fuel/air ratios, and lock-up torque converters, will contribute substantially—eventually perhaps as much as reductions in vehicle size and lowered performance—to conservation.

Two alternative engine systems, the turbine and the Stirling, offer significant advantages in meeting the national objective of improved fuel mileage and clear air. Both engines involve continuous combustion processes which reduce air pollutant emissions and improve fuel economy.

The turbine is further advanced than the Stirling engine, since the former has drawn heavily from aircraft jet-engine technology. While difficult problems remain in the development program, the turbine offers efficiencies as much as 50 percent greater than those of conventional engines when ceramic components capable of withstanding the necessary high temperatures can be inexpensively produced. An infinitely variable transmission with a more conventional turbine offers an alternative approach.

The Stirling cycle has been limited mainly to fixed installations at near-constant loads. An automotive version has been under development for only a few years. To meet space, weight, and performance objectives, the automotive Stirling must operate at pressures of about 3,000 pounds per square inch, which are difficult to maintain. The heat exchanger must operate at high temperatures and very high pressure, while the piston rod seal must have very low leakage in a difficult environment. In addition, a control system must be developed that meets automotive requirements without reducing efficiency. The successful resolution of these problems could improve fuel efficiency by 50 to 60 percent.

Neither of these systems could have a significant impact on conservation for about 10 years. At least five years of development remain before production and production-line engineering can begin in earnest. Even after a commitment to either or both of the new engines is made, production lines cannot be converted to the new systems in less than a decade. However, if the new engines are successfully developed, they will add important savings to those achieved by other automotive improvements and act directly to further reduce the levels of petroleum consumption in that critical time period.

The development and introduction of electric and hybrid automobiles is also being pursued. These would provide a substitution capability for many gasoline- and diesel-powered vehicles currently used in routine, short-haul, low-load applications in commercial use in urban as well as rural areas. The Congress enacted the Electric and Hybrid Vehicle Research, Development, and Demonstration Act of 1976 (Public Law 94-413) to require (a) the demonstration of the technical and economic practicability of such vehicles, (b) the establishment of performance standards, and (c) the implementation of a loan guarantee program to encourage the commercial production of such vehicles.

Nonautomotive transportation—trucks, airplanes, railroads, and ships—shows potential for significant energy savings by the year 2000. Some of these changes depend on major capital investments such as the purchase of new fleets of commercial aircraft in which improved energy efficiency is but one of several decision criteria.

## Residential/Commercial Sector

The residential/commercial sector accounted, in 1976, for 37 percent of domestic energy consumption. Some 63 percent of this use was for space conditioning. Because of the characteristically long life of buildings, conservation systems are required that can be retrofitted into the existing stock of buildings, as well as those that can be incorporated into new buildings at the design stage.

One important class of energy systems, for new or retrofit application, is the heat pump. Attention is being given to increasing their variety, improving their economics, and extending their application. Improved heat pumps are being developed in larger sizes (for multiple-dwelling units or commercial buildings), for broader temperature ranges, for economic application in different temperature zones of the country, and with alternative power drives such as gas engines. Currently, one problem with heat pumps in many areas is that, to be economic, they must be used in conjunction with central air conditioning.

Other developments, such as improved insulating materials and installation practices, are applicable to both new and retrofit markets. Improved space conditioning equipment may also be suitable for both markets. More elaborate energy-conserving space conditioning systems such as integrated energy systems for single dwellings, apartments, or entire communities, are appropriate only for the new building market.

The simpler systems, applicable to existing buildings, can have significant near-term impacts, but the more elaborate systems will probably require 10 to 15 years for development and significant market penetration.

### Industrial Sector

In 1976, the industrial sector consumed 37 percent of all energy used in the United States. Efficiencies in many industrial processes are low because the industrial complex evolved over a period of abundant and low-cost energy. Consequently, significant opportunities exist for development of new energy-efficient processes.

Industrial energy-conservation programs have two thrusts: (1) toward the most energy-intensive processes of the six most energy-intensive industries—steel, chemicals/petroleum, glass, pulp and paper, cement, and food processing—which together use 70 percent of all industrial energy; and (2) toward the energy processes that are used across a wide spectrum of industries.

An important area under development and generally applicable to most industrial energy processes is waste–heat recovery.

For example, the organic Rankine bottoming cycle converts some of the waste heat from diesel- or gas-turbine power plants to mechanical and electric energy. A Rankine system using a pressurized organic working fluid, such as toluene, in a closed system will be demonstrated during the next five years. The hot exhaust (260° C to 538° C) of a diesel- or gas-turbine engine converts the fluid to a vapor, which drives a turbine, thereby converting some of the waste-heat energy into mechanical and, possibly, electrical energy. The combination of diesel engine and Rankine bottoming cycle may recover an additional 8 percent of the initial fuel energy; a gas turbine bottoming cycle combination may save 15 percent.

Five 600-kilowatt organic Rankine bottoming cycle units will be field-tested in selected utility and industrial plants by 1980. Other bottoming cycles have been developed to increase the efficiency of diesels and gas turbines used for industrial processes, for pumping gas and petroleum, and for propelling ships. Since they can be installed on existing diesels and gas turbines, these bottoming units may have a significant near-term impact.

One specific industrial activity not generally considered part of the conservation program is uranium enrichment. Introduction of the new gas centrifuge technology will save over 90 percent of the electric energy presently used in gaseous diffusion and enrich the uranium ore more efficiently. This will contribute significantly to energy savings through the end of this century.

## EXPANSION OF EXISTING FUEL SOURCES

Although energy conservation is a critical element in the National Energy Plan, additional energy supplies will also be required. Because long lead times are needed to develop other sources and introduce new technologies, existing or imminent technologies that use present fuel sources must be stressed in the early 1980's. The nation must concentrate on technologies that:

- Produce additional petroleum and natural gas through enhanced recovery techniques;
- Expand the direct use of coal in both the utility and the industrial sectors;
- Expand the use of light water reactors in a once-through fuel cycle for electric power production.

In the late 1980's and beyond, these efforts can be augmented by:

- Expanding the industrial use of coal through improved technologies including conversion to synthetic fuel;
- Converting coal to electric power in an environmentally sound and more efficient manner;
- Developing advanced nuclear technologies consistent with nonproliferation objectives.

Descriptions of selected and illustrative technology approaches follow:

## Produce Additional Petroleum and Natural Gas

Although the demand for petroleum and natural gas has risen, the domestic capacity to meet that demand has not kept pace. Domestic petroleum unrecovered from oil fields after conventional production, and natural gas held in geologic formations considered currently unexploitable, could supply substantial additional quantities of energy. For example, it is estimated that approximately 300 billion barrels of petroleum resources remain in existing, developed fields after conventional primary and secondary production. Obtaining these untapped supplies requires enhanced recovery techniques for both petroleum and natural gas.

## Direct Use of Coal

Coal will meet the greatest portion of increased U.S. energy needs. A comprehensive coal program is, therefore, a high priority. Such a program would focus on more effectively and economically meeting environmental requirements and seek to expand the substitution of coal for petroleum and natural gas products.

Most of coal's impact between now and 1985 should result from the expansion of the existing technology used to burn coal directly. A major effort to develope economically attractive and environmentally sound techniques for burning coal, particularly higher sulfur coal, is underway. Success of this effort will accomplish two important objectives: it will increase the magnitude of the usable coal reserve, and it will reduce the costs involved (in many instances) by permitting greater use of the eastern coals which are closer to the larger demand centers in the east.

The current coal program has three generic approaches. One is to remove the sulfur and other impurities before combustion. Another is to develop new combustion techniques that will burn higher sulfur coals in an environmentally sound manner. The third approach is to convert coal of varying quality to a clean, synthetic fuel. The illustrative technologies needed to increase the use of coal by the industrial and utility sector are:

Advanced cleaning of beneficiation methods (example of first approach)

- Atmospheric fluidized bed combustion (example of second approach)
- Low-Btu gasification (example of third approach).

Beneficiation, or cleaning, improves the environmental quality of mined coal to broaden its range of applications. The process involves the reduction of "free" sulfur and of rock, shales, and other "impurities" from the coal through grinding, washing, or floating. Beneficiated coal would be suitable for use in existing and in future direct coal combustion technology.

Atmospheric fluidized bed combustion is a new system for burning coal efficiently, even in relatively small boilers, while controlling air pollutants. In this system, the coal is mixed with limestone and burned in a fluid bed. The limestone reacts with sulfur in the coal, effectively eliminating sulfur from the exhaust gas. Problems with respect to coal and limestone feed, ash disposal, and sorbent regeneration systems for eliminating the large quantities of limestone or dolomite required for combustion are among those under investigation.

More advanced fluidized bed systems, commercially applicable in the mid-term, are being developed. These advanced units will be capable of operating in a combined cycle mode as part of a total community energy system. One such mode uses a high-pressure fluidized bed system. The combustion products pass first through an advanced gas turbine, generating electricity, and then through a conventional heat exchanger/boiler system, generating steam to run a turbine to generate more electricity. A final step would use the steam turbine exhaust to supply residential and commercial space heating services to the surrounding community.

In low-BTU coal gasification, coal is burned in a limited amount of air to produce a fuel gas with an energy content of 10 to 50 percent of natural gas. Because of this low-energy content, the gas cannot be economically stored or transported over long distances. Rather, it is intended for direct on-site use, either as a clean burning fuel or as a chemical feedstock.

Although low Btu coal gasification is potentially an environmentally acceptable use of coal, there are still some unresolved questions associated with the actual coal processing and conversion, particularly with respect to the sulfur and trace elements in the coal and the compounds produced during gasification.

Several demonstration projects, from power generation in small generating units to the manufacture of hydrogen for chemical production, are planned or under way. Since this technology evolves from existing systems and can be used in small-sized applications, it can be implemented in the near term.

In addition to the illustrative near-term technologies discussed above, there are other technologies with potential long-term applications. Among these are magnetohydrodynamics, high-Btu gasification, and technologies that produce synthetic crude oil from coal.

High-Btu synthetic gas produced from coal may provide a substitute for declining natural gas supplies. The development program for these technologies focuses on using advanced technologies. The technology for producing synthetic crude oil from coal is not as well developed as synthetic gas technologies.

### Light Water Reactors

The light water reactors (LWR's) now in operation are generating about 10 percent of present electricity needs with an in-place capacity of about 46,000 megawatts electric (MWe). An additional 185,000 MWe are under construction or planned for introduction through the early 1990's. Thus, LWR technology with a once-through fuel cycle represents a current energy production capability that can play a significant part in the nation's energy future by easing the demand to expand other fuel systems and by contributing to the reduction of U.S. petroleum imports.

Even though the necessary power plant technology is currently available, the streamlining of regulatory requirements would undoubtedly enhance the energy contribution of nuclear power while protecting the public interests of health, safety, and the environment. Moreover, the full potential of nuclear energy will not be realized until the problems associated with three closely related aspects of the fuel cycle are fully resolved. Namely:

- Assuring that peaceful uses of nuclear power will not contribute to the proliferation of nuclear weapons;
- Finding and providing sufficient fuel to expand nuclear capacity in an economically and environmentally acceptable manner;
- Disposing of waste from nuclear facilities.

These problems, which are directly attacked by elements of the president's comprehensive National Energy Plan, are discussed below.

Effective safeguards systems require a balance of physical protection, material control, and material accountability. Developmental and operational activities are under way in each of these areas to ensure that safeguards and security systems are technically and economically viable.

To promote measures at the international level, to prevent the proliferation of nuclear weapons, the U.S. has:

- Taken a lead role in promoting the International Atomic Energy Agency (IAEA) safeguards and adherence to the nuclear safeguards systems administered by that body.
- Agreed to place its facilities with no direct national security significance under IAEA safeguards;
- Strongly endorsed widespread adherence to the nuclear nonproliferation supplier state policies aimed at the development of common and prudent nuclear export policies;
- Judged that its nonproliferation influence would best be fostered through a program of carefully controlled nuclear cooperation rather than through a posture of strict embargo. To this end, the U.S. has concluded approximately 33 nuclear cooperation agreements with 30 nations and international organizations designed to permit the export of U.S. reactors and fuels under effective bilateral and IAEA controls.

There are still some differences of view among the industrialized nations as to how nonproliferation goals can best be fostered. In particular, the United States is pressing for a new international perspective on how the LWR fuel cycle should be approached to assure that nonproliferation considerations receive highest priority.

In this regard, the nation has opposed the spread of nationally controlled reprocessing and enrichment facilities, favoring instead a moratorium on the export of such technologies. Moreover, this country has raised fundamental new questions about the value of reprocessing and plutonium recycling in LWR's, and has argued that neither is necessary or inevitable for the U.S.

LWR's, as designed and operated today, require about 5,300 tons of uranium oxide ($U_3O_8$—commonly called "yellowcake") for each 1,000 MWe of capacity over a normal plant lifetime. Domestic uranium reserves recoverable at a forward production of \$30 per pound $U_3O_8$ are 680,000 tons $U_3O_8$ with another 140,000 tons $U_3O_8$ estimated to be available as a byproduct from phostate and copper production. In addition to the reserves, there may be additional resources of 2,700,000 tons $U_3O_8$ potentially available. Of this potential, 1,090,000 tons are estimated to be in the probable category which, when combined with the reserves, form a base of over 1,800,000 tons $U_3O_8$—sufficient to support all LWR's now operating or planned for the U.S. over their full operating lifetime. If the total combined reserves and potential resources are realized, they could support the lifetime requirement of over 600 nuclear reactors.

The extent to which nuclear power can contribute in the future will therefore be strongly affected by the rate at which potential resources are discovered and translated into reserves, and by the ability of the uranium mining and milling industry to expand to meet anticipated uranium production demands. It is currently estimated that a production rate of up to 60,000 tons $U_3O_8$ per year could be attained by the early 1990's, provided that industry expands steadily in the interim period.

The National Uranium Assessment Program (NURE) will provide a more reliable estimate of the total uranium resource base and lead eventually to increased reserves. NURE is being redesigned to give greater attention to resource assessment and expanded to acquire data on the thorium resource base.

In addition to expanding the resource base, energy production capability can be expanded in several ways. Alternate reactor concepts or LWR's which utilize less uranium per kilowatt hour of electricity produced can be developed. Therefore, emphasis is being given to advanced concepts other than the plutonium breeder. A second approach is to improve LWR technology for increasing capacity factors and decreasing plant construction time and costs through standardization of designs. A third approach is to operate on the front end of the fuel cycle. If greater amounts of uranium-235 were extracted in the enrichment process, less uranium ore would be required for each kilowatt hour of electricity produced in an LWR. All three approaches are being pursued.

In order to meet its domestic and worldwide obligations, the three U.S. uranium enrichment plants presently operating are being upgraded, and U.S. enrichment capacity will be further expanded to provide an additional 8.8 million separative work units per year using a vastly improved and energy-efficient technology—centrifuge enrichment. This technology uses less than one-tenth the electric power needed for gaseous diffusion.

Government involvement in high-level waste management is unique in that it is required by law to accept and safely store high-level radioactive wastes generated by commercial operations. High-level radioactive wastes produced in the fission process present potential long-term hazards and need to be isolated from man's environment for extremely long periods of time. These wastes are contained in the spent fuel assemblies and can be stored or disposed of in that form. The basic approach to waste management is to place multiple barriers between radioactive wastes and man's environment. The current program objective is to develop, construct, and operate the first of a number of terminal storage facilities for the long-term safe storage of radioactive wastes.

## TECHNOLOGIES THAT USE NEW FUELS

During the late 1980's and 1990's, conservation and expanded use of existing supply sources will be the major complements to diminishing sources of petroleum and natural gas, but new fuels can and must be introduced into the market. Several technologies that use new fuels—shale oil, synthetic fuels from wastes and biomass, geothermal heat, and solar energy—can have substantial impacts on the energy situation between 1985 and 2000.

Beyond 2000, the nation will have to rely increasingly on energy systems based on "essentially inexhaustible" or renewable resources (i.e. solar electric, hot dry rock geothermal, or fusion). The status of some of the technologies that use new fuels is discussed below.

## Shale Oil

Billions of barrels of oil may someday be recovered from shale deposits in the western states if environmental and economic problems can be overcome. Optimistic economic projections indicate that oil-from-shale could be produced profitably if priced at today's foreign crude petroleum prices. Major uncertainties, however, need to be resolved. Among them are: (1) the environmental impacts of shale oil production; (2) the availability of water (a scarce resource with other competing needs in the oil shale regions); (3) the actual economics of shale oil production (the most developed technology, which involves mining shale and using heat to separate oil from crushed shale in a processing plant, appears to be the most costly); (4) inadequate knowledge of shale geology and characteristics; and (5) the attractiveness of the substantial initial investments required, since new technologies generally require higher investments per unit of net energy output. In spite of these uncertainties, shale oil could become an attractive new energy source for the U.S. and could help to fill the liquid fuels demand-supply gap that may exist in the latter half of the century.

Past research by both government and industry has advanced surface retorting technology through the pilot plant stage. Pilot-plant testing of several alternative approaches is continuing, using both surface and in-situ (or underground) processing techniques. Processes involving development of eastern shales are less well developed than those for western shales, but one major in-situ gasification project involving eastern shale is under way. If the economic problems can be resolved, and the attendant environmental questions on air pollution and water availability can be answered, shale oil technologies could begin to have an impact in the late 1980's.

## Geothermal

Geothermal energy exists everywhere beneath the earth's crust, but in most places the heat is too diffused or too deep to be a potentially usable energy resource. National and regional surveys by the U.S. Geological Survey show that in the U.S. the potentially exploitable hydrothermal geothermal resources suitable for both electricity production and thermal applications are largely confined to the Rocky Mountain and western states. Recent data indicate that areas of hydrothermal resources suitable for moderate thermal application may also exist on the eastern seaboard. Geopressured geothermal resources are located principally in deep sediments along the Gulf Coast, and hot dry rock resources may underlie most of the country. Hydrothermal resources are most readily accessible and nearer development than the geopressured and hot dry rock resource bases. However, the latter resource bases are estimated to be much larger.

Current domestic efforts are focused on (1) exploring and assessing resources, (2) improving energy extraction and utilization technology and its economics, (3) resolving environmental concerns involving the release of toxic materials into the environment, subsidence, and waste water, and (4) reducing those institutional barriers at all levels of government which tend to inhibit the use of geothermal energy. A federal loan guarantee program has also been implemented to make capital more available to the geothermal industry.

To stimulate the development of geothermal resources, the National Energy Plan proposes a tax reduction for intangible drilling costs, additional funds to evaluate the geopressured and liquid-dominated hydrothermal resources, and measures to streamline the leasing and environmental review procedures.

The geopressured resource base consists principally of hot brine and associated dissolved methane confined under pressure at considerable depths (5 to 15,000 feet) by deep impermeable rocks. Serious environmental problems must be resolved to tap this energy resource. First, the withdrawal of the hot brine may produce subsidence as reservoir pressure is decreased unless appropriate control measures, such as brine reinjection, are taken. Second, the quantities of air pollutants that may be released are not known. Third, where a fresh water aquifer occurs above a geothermal reservoir, the fresh water could be contaminated by the brine released. Nevertheless, geopressured resources are of particular interest because of the magnitude of the potentially recoverable energy (about 2,500 quads), which is thought to include significant amounts of dissolved methane gas. Such a potential warrants major efforts to refine the knowledge of the actual magnitude and the content of the reservoirs, to develop the technology to drill economically at great depths, and to reduce the technological and economical uncertainties concerning reservoir producibility and longevity.

For the longer term, hot dry rock offers an even larger potential renewable resource option to provide substantial quantities of high grade energy. Rock deep in the earth contains vast quantities of heat, but little or no fluid to bring the heat to the surface. Therefore, introduction and circulation of a heat transfer fluid, such as water, are required to extract usable energy. However, the economic costs and technical difficulties of drilling to such depths and recovering such highly heated fluids or other transfer mechanism without undesirable environmental impact suggest that this resource is unlikely to enter the market in a significant way until the next century. Determination of the characteristics of the resource and development of techniques for heat extraction and reservoir stimulation are in process. Experiments pursued at Fenton Hill, New Mexio, are evaluating heat extraction and reservoir stimulation concepts.

## Solar Heating and Cooling

The sun is an inexhaustible source of energy. Growing public interest is evidenced by some 200 installations of solar heating systems in 1975 and an estimated 1,000 or more systems in 1976. The Federal Government subsidized about one of every seven of these installations. There are now an estimated 500 companies offering solar systems and components on a commercial basis. The president's National Energy Plan calls for tax credits for solar installations and the use of solar heating in 2.5 million homes by 1985.

The primary objective is to develop and demonstrate economically competitive systems with a wide range of applications. Current efforts are aimed at perfecting key components—collectors, storage and heat exchanger units, heat pumps and air conditioners, and appropriate controls for the systems. A large-scale program to demonstrate residential and commercial solar heating systems is also under way. These efforts should permit widespread introduction of solar hot water systems and solar space heating in the late 1970's and early 1980's. Increased market penetration depends on further reducing per-unit costs, improving system performance, and removing institutional barriers.

Similar efforts are under way to apply solar heating in agricultural and industrial situations where large quantities of low-temperature heat are required. It is estimated that over 40 percent of the industrial and nearly all of the agricultural heat needed at the point of application and for preheating is below 204° C. Solar energy technology is already available for producing hot water, hot air, or saturated steam to meet these requirements. A number of projects are under construction to demonstrate processes such as crop and grain drying, heating of animal shelters and greenhouses, production of hot water for commercial textile dyeing operations, and curing of concrete blocks. Solar production of agricultural and industrial process heat should be competitive with conventional energy sources in some areas of the country in the near future.

## Energy From Waste and Biomass

A variety of readily available resources—municipal waste, sewage sludge, agricultural and forest products, and animal residues—can be burned directly or converted to synthetic liquid and gaseous fuels. In addition, some of the resulting synthetics may be valuable as petrochemical feedstocks such as methane, hydrogen, and ammonia. The technologies used for conversion can therefore be viewed as a means of controlling pollution and conserving energy as well as a means of producing new fuels.

Biomass, which includes animal manure, agricultural and forest crops and their residues, and aquatic plants such as algae and kelp, can be converted by a series of processes into clean fuels and other energy-intensive products. Methane, for example, can be produced from animal manure by anaerobic digestion and then upgraded to pipeline-quality Substitute Natural Gas (SNG). Cellulosic biomass such as wood or sugar cane can be broken down into sugars which, when fermented, yield ethanol, a liquid which can be used as a gasoline extender in an unmodified internal combustion engine. Gasification of biomass in a manner analogous to the thermal reactions of petroleum refining can produce gaseous fuels or a synthesis gas which can be transformed into methanol, SNG, hydrogen, or ammonia. Some forms of biomass can also be burned directly to produce heat for a variety of uses.

In the long run, both terrestrial and aquatic biomass may be purposefully grown (energy farming) for conversion to fuels, with wood a large source of the biomass. Questions concerning the economics of production, collection, and transportation of biomass to a conversion facility, biomass availability, competing uses of land, and potential environmental impacts must be answered before fuels from biomass can make a major impact on the national energy problem.

## Solar Electric Systems

Various technologies—photovoltaics, thermal electric, wind, and ocean thermal—are part of the technology category known as solar electric. Solar power is a renewable energy source relatively free from the pollution concerns that face other energy supply systems.

Although most of these technologies have already been demonstrated, it is difficult to predict when these systems will become marketable. The problem is one of engineering design and development to provide economically competitive systems. Some of these systems may find initial application in small, specialized markets where higher costs are acceptable. Such systems might also find long-term applications as decentralized energy sources.

Photovoltaic or solar cells have been used in the space program for some time and are commercially available. A federal program is also under way to help reduce the current cost of solar arrays by a factor of 30 by the mid-1980's—from about $15 per peak watt of output at present to about 50 cents per peak watt by 1986. Initial domestic residential/commercial use of such solar systems is expected to occur in the mid-1980's, but widespread use will depend on additional cost reductions (i.e. to 20 cents per peak watt by the 1990's). This latter advance may require the introduction of new cell materials or other configurations that offer higher cost reductions. There seem to be a variety of uses abroad that may be economically competitive now.

Thermal electric conversion systems use concentrated solar energy to heat water or other working fluids to power turbines which will drive electric generators. Such systems can also be incorporated into total energy systems that supply heat for industrial processes or space heating and cooling.

Large tracking mirrors (heliostats) can focus large quantities of solar energy on central receiver systems—boiling working fluids and producing steam to power electric generating equipment. Commercialization depends on the development of an economical system for concentrating the sun's energy. System costs must be reduced by a factor of 5 to 10 if such systems are to be used widely in intermediate-load situations.

Local heat rejection, potential negative effects of shadowing large areas of land, and competing alternate uses for land areas are other issues needing resolution prior to large-scale commercialization.

Wind energy conversion systems are more modern and economic versions of the old-fashioned windmill. Several projects are under way to encourage the development of more cost competitive wind energy systems. A 100-kilowatt system has been constructed and tested for over a year as a cooperative ERDA/NASA project. The first of two 200-kilowatt experimental machines is being fabricated. One of these wind energy conversion systems has been installed and is providing power for Clayton, New Mexico. A 1.5-megawatt experimental system design is under way. In addition, smaller wind systems under development for decentralized applications have the potential for early commercialization of wind energy technologies. To achieve market penetration for this technology, the economics must be improved by a factor of 2 to 4, and questions concerning controls, structural dynamics, site characteristics, television interference, and the service life of large rotors must be resolved.

Ocean thermal electric conversion systems utilize the enormous but diffuse quantities of heat collected and stored in the oceans to generate electricity. Ocean thermal systems use warm surface water to heat a secondary system liquid such as ammonia, causing it to vaporize and turn a turbine connected to a generator. Cold water from the ocean depths condenses the ammonia vapor, and the cycle is repeated.

Such systems might provide electricity for mainland distribution or for energy-intensive processes such as manufacturing ammonia or processing aluminum ore at sea or on island sites. Although small units have been operated, major cost reductions and performance improvements are required before commercialization can be contemplated. Technical problems include producing an efficient heat exchanger, controlling biofouling and corrosion, and constructing and positioning large, seagoing platforms. These factors are being studied by TRW Inc. for the Department of Energy. In addition, a variety of socio-economic, institutional, and environmental questions have to be answered since these systems are still in the early developmental stages.

Experimental studies for heat exchangers and biofouling control technologies are under way, and critical component testing has begun.

## Advanced Nuclear Reactors

Due to concern with the proliferation dangers associated with the plutonium fuel cycle, the United States is currently reorienting its advanced nuclear reactor research and development program. The president has proposed to defer efforts to commercialize the Liquid Metal Fast Breeder Reactor (LMFBR). He has proposed that the systems design for the Clinch River Breeder Reactor Demonstration (CRBR) plant be completed, but that construction and operation be cancelled. However, an extended discussion by Congress appears to favor continued development and implementation of the CRBR. The Fast Flux Test reactor facility under construction at Hanford will be completed and operable by 1980.

Alternative reactor systems, including breeders and advanced converters, will be investigated with emphasis on nonproliferation and safety factors. Spectral shift and tandem cycle techniques are being considered as methods to improve the performance of converter reactors. Coprocessing of spent fuel from converter reactors is being examined as a possible method for increasing fuel supply to converter reactors or breeder reactors while reducing proliferation dangers. A variety of thorium breeders as well as converter reactors are under consideration as alternatives to the LMFBR. The fuel cycle alternatives studies will be completed within about two years.

## Fusion

Although fusion technology is still decades away from commercial market introduction, a strong incentive for development is that the fuel is available in virtually unlimited amounts and at negligible costs.

The present program for developing fusion involves two very different approaches. The first, magnetic confinement, involves the confinement and heating of a "plasma" consisting of deuterium and tritium to a point at which the high velocity nuclei fuse on collision. The first experimental test facility designed to produce significant thermonuclear energy is under construction at Princeton. Development programs are now in place in all fusion technology problem areas: materials, plasma heating, fueling, magnets, impurities, vacuums, tritium handling, maintenance, energy storage and transfer, and power density.

The second approach to developing fusion—inertial confinement—involves laser, electron-beam, and ion-beam sources to implode pellets of deuterium and tritium, resulting in fusion of the nuclei and release of energy. A major laser facility is expected to demonstrate high energy gain from pellets in the 1980's and thus establish the scientific feasibility of net energy production from an operating device.

Although the major objective of both fusion approaches is to develop commercially viable electric power reactors, and although the inertial confinement fusion program has significant weapons technology applications, other possible applications are also being considered. Among these are direct production of hydrogen gas and/or synthetic fuels, chemical and material processing, fissile fuel production, fusion/fission hybrid reactors, and auxiliary use of reactor heat.

This overview reflects President Carter's ideals in the development of a comprehensive and effective national energy plan.

# A

**abatement.** Reducing the effects of pollution; the reduction of the pollution effects of mine drainage.

**abrasion drilling.** An oil-drilling technique using abrasive material under pressure instead of the conventional drill steam and bit.

**ABS.** Acrylonitrile-butadiene-styrene.

**absolute pressure.** Pressure measured with respect to a vacuum (zero pressure); absolute pressure can be zero only in a perfect vacuum.

**absolute temperature.** Temperature measured with respect to absolute zero on a thermodynamic or Kelvin scale. The zero of the Kelvin scale is −273.16°C.

**absolute zero.** The temperature at which a perfect gas kept at constant volume would exhibit no pressure. It is equal to −273.16°C.

**absorbed dose.** When ionizing radiation passes through matter, some of its energy is imparted to the matter. The amount absorbed per unit mass of irradiated material is called the absorbed dose and is measured in rems and rads.

**absorber.** In the nuclear field, any material that absorbs or diminishes the intensity of ionizing radiation. Neutron absorbers—like boron, hafnium, and cadmium—are used in control rods for reactors. Concrete and steel absorb gamma rays and neutrons in reactor shields. A thin sheet of paper or metal will absorb or attenuate the alpha particles and all except the most energetic beta particles. The term also applies to a surface, usually blackened metal, in a solar collector, which absorbs solar radiation. (See Fig. p. 18)

**absorber plate.** The surface in a flat-plate collector upon which incident solar radiation is absorbed.

**Absorbite.** Trade name for activated charcoal.

**absorptance.** A ratio of the radiation absorbed by a body of material to the radiation incident upon it; in a solar collector, the soaking up of heat measured as a percent of the total radiation available.

**absorption.** Penetration of a substance into the body of another; the extraction of one or more components from a mixture of gases when gases and liquids are brought into contact.

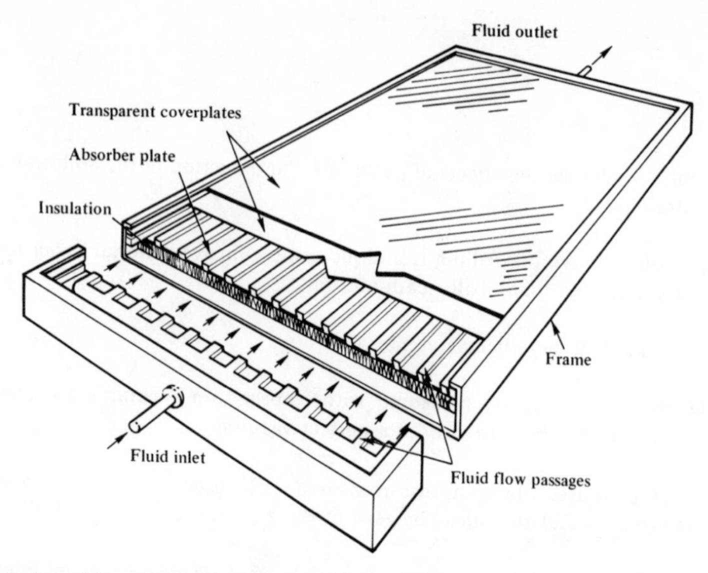


Detailed view of a typical solar collector absorber. (*Courtesy Energy Research Development Administration*)

**absorption air conditioner.** Designed to use solar hot water, the process starts with a generator filled with a solution of lithium bromide (the absorber) and water (the refrigerant). Solar-heated hot water flows through the heating coil which causes the lithium bromide solution in the generator to boil violently. Water vapor shoots up through the tube carrying the lithium bromide solution in the form of spray and droplets. This mixture is sorted out by baffles in the separator and the lithium bromide solution flows, by gravity, back down the tube, while the water vapor flows to the condenser. There, cooling water passing through coils condenses it to liquid water, which flows through the restriction and into the cooling coil, where the sudden drop in pressure makes the water boil, producing refrigeration. Some of the vapor in the cooling coil recondenses and collects in the reservoir, but most flows to the absorber. There, the concentrated lithium bromide solution absorbs it. The lithium bromide solution used at this stage was separated from water vapor in the separator, flowed through the heat exchanger, where it was partially cooled, and then on to the absorber. After absorbing water vapor, the cool lithium bromide flows down the tube through the heat exchanger, where it is partially reheated, and back to the generator to start the cycle again. The condensing water that flows through the absorber and condensing coils comes from an outside source—usually a cooling tower. The solar hot water comes from a system of solar collectors on the roof and suitable insulated storage tanks so that cooling can be provided on sunless days and at night. (See Fig. p. 19)

**absorption control.** In a nuclear reactor, adjusting the quantity or properties of neutron-absorbing material so as to change the reactivity.

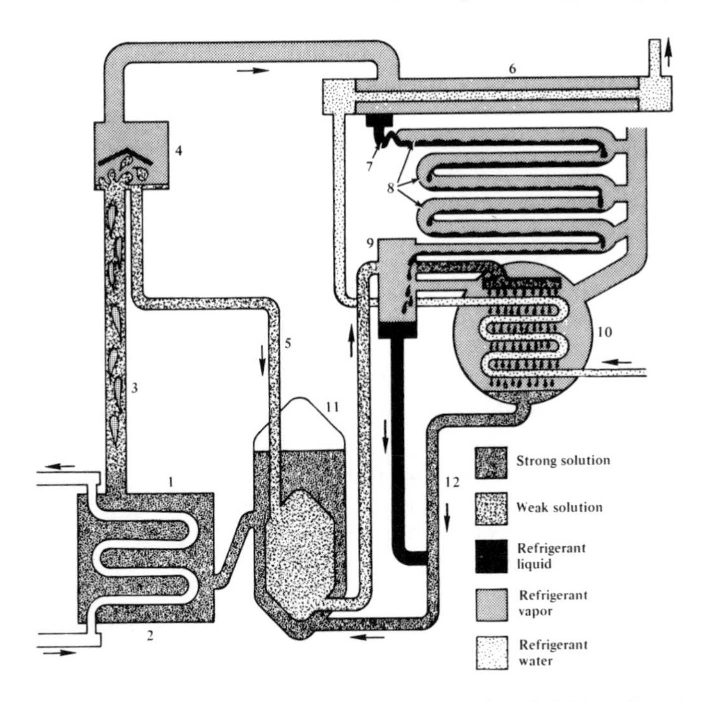


Absorption air conditioner: (1) generator, (2) heating coil, (3) lithium bromide tube, (4) separator, (5) gravity flow tube, (6) condenser, (7) liquid water flow, (8) cooling coil, (9) reservoir, (10) absorber, (11) heat exchanger, (12) cooled lithium bromide tube (*Courtesy Popular Science, September 1977*)

**absorption curve.** A plot of the relationship between thickness of absorbing material and intensity of transmitted radiation.

**absorption gasoline.** Gasoline extracted from natural or refinery gas.

**absorption loss.** The loss of water occurring during initial filling of a reservoir in wetting rocks and soil; that part of the transmission loss which is due to the dissipation or the conversion of sound energy into some other form of energy, usually heat.

**absorption oil.** An oil used to separate the heavier components from a vapor mixture in recovering natural gasoline from wet gas; oil containing little or no gasoline.

**absorption plant.** A device that removes hydrocarbon compounds from natural gas, especially casinghead gas.

**absorption refrigerating system.** A system in which a secondary fluid absorbs the refrigerant, thus giving up heat, and then releases the refrigerant and allows the heat to be reabsorbed. The absorption cycle consists of the expansion of a vapor, absorbing heat; the absorption of that vapor by a liquid or solid; the desorption of the vapor from the liquid or solid by the application of heat sometimes from a solar collector, a flame, or steam; the condensation of the vapor, giving off heat; the expansion that starts the cycle again. Ammonia or water is used as the vapor in commercial absorption cycle refrigerators and water or lithium bromide is the absorber.

**absorptivity.** The ratio of the radiant energy number absorbed by a body to that falling upon it. It is equal to the emissivity for radiation of the same wavelength.

**abundance ratio.** The proportions of the various isotopes making up a particular specimen of an element.

**abutment.** A point or surface provided for withstanding forces; in coal mining, the point of contact between the ends of an embankment and the natural ground material.

**accelerated supply strategy.** Actions taken by the federal government to increase the domestic supply of energy including accelerated federal leasing of the Outer Continental Shelf, opening the Naval Petroleum Reserves to commercial exploration and development, and removing regulatory constraints.

**accelerated weathering test.** A test to indicate the effect of weather on coal, in which the coal is alternately exposed to freezing, wetting, warming, and light. The alternation may be varied to suit. This test may be applied to other bituminous material.

**acceleration.** The rate of change of velocity, measured as a change in velocity in unit time. It is expressed as feet per square second.

**accelerator.** A device for increasing the velocity and energy of charged elementary particles, for example, electrons or protons, through application of electrical and/or magnetic forces. Accelerators have made particles move at velocities approaching the speed of light.

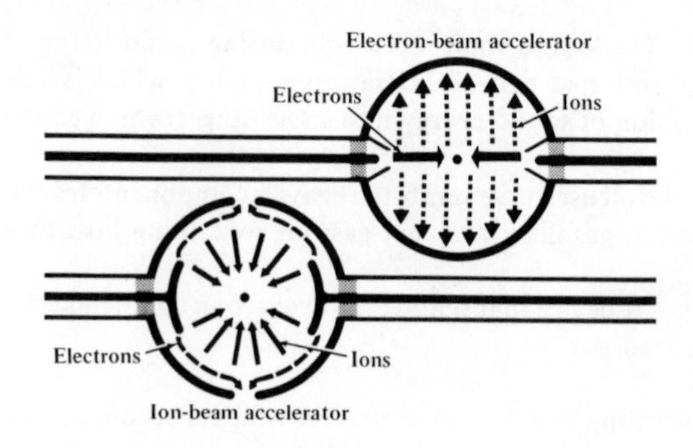


Ion-beam accelerator for laser fusion. (*Courtesy Popular Science, 1976*)

**accelerometer.** An instrument used to measure the acceleration of the system with which it moves; for example, seismograph designed to measure earth particle accelerations.

**acceptor levels.** Energy levels formed within the energy gap by a deficiency of electrons.

**acceptors.** In the coal field, calcined carbonates that absorb carbon dioxide evolved during gasification.

**accountability.** The quantitative accounting for material inventories through a system of measurements, records, and reports.

**accumulated dose.** The total dose absorbed by the system considered resulting from exposure to radiation.

**accumulated provision for depletion.** The net accumulated credit resulting from offsetting charges to income for the prorata cost of extracted depletable natural resources such as coal, gas, and oil.

**accumulator.** A cylinder containing water or oil under pressure of a weighted piston for hydraulic presses and hoists.

**ACES heating cycle.** The ACES (annual cycle energy system) heating cycle consists of a methanol-water loop, a refrigerant loop, and a loop for domestic hot water. Heat is removed from a large tank of water for winter heating; this operation turns the water into ice, which is stored for cooling the house in summer.

**Acheson furnace.** A resistance-type furnace for the production of silicon carbide and synthetic graphite.

**acid.** A compound that dissociates in a water solution to furnish hydrogen ions, having acid-forming constituents present in excess of the proportion required to form a neutral or normal compound. In geology, the term refers to a test for composition of rocks.

**acid clay.** A naturally occurring clay which, after evaluation, usually with acid, is used mainly as a decolorant or refining agent, and sometimes as a desulfurizer, coagulant, or catalyst; a clay which yields hydrogen ions in a water suspension.

**acid-gas removal.** The removal of hydrogen sulfide and carbon dioxide from the gas stream.

**acidizing.** The practice of applying acids to the walls of oil and gas wells to remove any material which obstructs the entrance of fluids; also the process of forcing acid into limestone, dolomite, or sandstones in order to increase permeability and porosity.

**acid mine drainage.** Acidic drainage from bituminous coal mines containing a high concentration of acidic sulfates, especially ferrous sulfate.

**acid producing materials.** In coal mining, rock strata containing significant pyrite which, if exposed, will cause acids to form when acted upon by air and water.

**acid soil.** A soil deficient in available bases, particularly calcium; a soil having a preponderance of hydrogen over hydroxyl ions in the soil solution; generally, a soil that is acid throughout most or all of the parts of it that plant roots occupy.

**acid spoil.** Spoil material which contains sufficient pyrite so that weathering produces acid water.

**acre feet of water.** The volume of water that would cover one acre to a depth of one foot. An acre-foot contains 43,560 cubic feet or 325,851.4 gallons.

**acre foot of sand.** A unit of measurement applied to petroleum and natural gas reservoirs. It is an acre of producing formation one-foot thick.

**acrometer.** An instrument for determining the density of gases.

**acrylonitrile-butadiene-styrene (ABS).** The forming of ABS resins through the polymerization of acrylonitrile and styrene liquids and butadiene gas.

**actinide series.** The series of chemical elements of increasing atomic number beginning with actinium (atomic number 89) and continuing through lawrencium (atomic number 103), which together occupy one position in the Periodic Table. The series includes uranium (atomic number 92) and all the man-made transuranic elements. The group is also referred to as the "Actinides."

**activated carbon.** Carbon obtained by carbonization in the absence of air, preferably in a vacuum. Has the property of absorbing large quantities of gases, solvent vapors; used also for clarifying liquids.

**activation.** The process of making a material radioactive by bombardment with neutrons, protons, or other nuclear particles. Also called radioactivation.

**activation analysis.** A method for identifying and measuring chemical elements in a sample of material. The sample is first made radioactive by bombardment with neutrons, charged particles, or gamma rays. The newly formed radioactive atoms in the sample then give off characteristic nuclear radiations (such as gamma rays) that can identify the kinds of atoms present and indicate their quantity. Activation analysis is usually more sensitive than chemical analysis.

**activation detector.** A device which is used to determine particle flux density. Particles or radiation creating radioactive isotopes in a thin film of material is measured at the end of a fixed period of time as a measure of the number of particles of the amount of radiation that passed through the film.

**activation energy.** The amount of energy that must be applied to initiate a reaction per molecule of a chemical reaction; or per atom of a nuclear reaction.

**activation foil.** An activation detector.

**active residential solar heating system.** A solar heating system that is combined with the primary heating system in the house.

**active solar system.** Any solar system that needs mechanical means such as motors, pumps, valves, etc., to operate.

**activity.** In nuclear physics, the rate of decay of atoms by radioactivity. It is the number of spontaneous nuclear disintegrations occurring in a given quantity of material during a suitably small interval of time divided by that interval of time. It is usually measured in curies.

**activity concentration.** The activity of a material divided by its volume.

**acute radiation sickness syndrome.** An acute organic disorder following exposure to relatively severe doses of ionizing radiation.

**additive.** Any materials incorporated in finished petroleum products for the purpose of improving their performance in existing applications or for broadening the areas of their utility.

**add-on systems.** Storage components that may be added optionally to basic space-conditioning systems.

**adiabatic.** An action or a process during which no heat is added or subtracted.

**adiabatic compression.** Compression in which no heat is added to or subtracted from the air and the internal energy of the air is increased by an amount equivalent to the external work done by the air. The increase in temperature of the air during adiabatic compression tends to increase the pressure on account of the decrease in volume alone; therefore, the pressure during adiabatic compression rises faster than the volume diminishes.

**adsorption.** Physical adhesion of molecules to the surfaces of solids without chemical reaction; also the process by which some substances, notably those of a carbonaceous nature, are able to compress and hold on their surface large quantities of gas.

**advanced batteries.** Large battery storage systems for harnessing solar or wind energy or storing excess electricity during low demand periods for use during periods of higher demand.

**Advanced Gasification process.** Coal gasification process in which two pressurized, fluidized-bed vessels are used. Air, steam, and char react in the gasifier. Resulting hot gases provide heat for the devolatilizer/desulfurizer where dolomite is added to remove sulfur. Low-Btu product gases may be used as a fuel gas or in a combined power system. Process conditions in the reactors are 10 to 20 atms with temperatures at 1038°C in the gasifier and 871°C in the devolatilizer.

**adverse hydro.** Water conditions such as to limit the production of hydroelectric power.

**aerodynamics.** The branch of dynamics that deals with the motion of air and other gases, and the forces acting upon bodies passing through them—especially aircraft, rockets, and missiles.

**aerosol.** Solid or liquid particles of microscopic size which are suspended in air or another gas. Colloidal suspensions of coal dust or other particles (smoke) and of fine water droplets (fog) are instances of natural aerosols.

**aerospere.** The atmosphere considered as a spherical shell of gases surrounding the earth.

**afforestation.** Forest crops established artifically by planting or sowing on land which has not previously grown tree crops.

**afterburner.** A device that removes undesirable organic gases through incineration.

**aftercooling.** The cooling of a reactor after it has been shut down.

**afterheat.** The heat produced by the continuing decay of radioactive atoms in a nuclear reactor after fission has stopped. Most of the afterheat is due to the radioactive decay of fission products.

**afterpower.** The power corresponding to the afterheat.

**age.** In nuclear physics, one-sixth of the normalized second spatial moment of the neutron flux density (flux age) at energy E, or of the neutron slowing-down density past energy E (slowing-down age), for a point isotropic neutron source.

**agglomerating burner gasification process.** A process using a self-agglomerating fluidized bed coal burner to produce synthesis gas by steam gasification of coal.

**agglomeration.** The bringing together of small particles of a solid into larger clumps. For example, when coal is heated at low temperatures, the sticky, tar-like material that forms makes the small coal particles gather into larger lumps.

**aggregate.** Sand, gravel, or any clastic material in a bedded iron ore, sometimes so abundant as to make it resemble a puddingstone; uncrushed or crushed gravel, crushed stone or rock, sand, or artificially produced inorganic materials which form the major part of concrete. The term also means to bring together, or collect, or to gather into a mass.

**agricultural limestone.** A limestone whose calcium and magnesium carbonate content must be equivalent to not less than 80 percent calcium carbonate and must be fine enough so that not less than 90 percent will pass through a U.S. Standard No. 10 sieve and not less than 35 percent will pass through a U.S. Standard No. 50 sieve.

**air.** The mixture of gases enveloping the earth and forming its atmosphere. It is composed, by volume, of 21 percent oxygen and 78 percent nitrogen; by weight, about 23 percent oxygen and 77 percent nitrogen. It also contains about 0.03 percent carbon dioxide, some aqueous vapor, and some argon.

**air batteries.** Advanced battery systems which use air for one electrode; iron-air and lithium-air systems. They are potential candidates for use in electric vehicles.

**airborne pollutants.** Include aerosols—such as dusts, smoke, and fumes—and gases or organic vapors—such as sulfur dioxide, chlorine, oxides of nitrogen, aldehydes, carbon monoxide, etc.

**air change.** The quantity of air infiltration and/or ventilation in cubic feet per hour or per minute divided by the volume of the room gives the number of so-called air changes during that interval of time.

**air classification.** The separation of light materials from heavier materials by means of forcing air up a cylindrical container at a controlled velocity.

**air cleaning.** A coal cleaning method that utilizes air tables to remove the dust and waste from coal. Air cleaning requires that the coal contain less than 5 percent of surface moisture as a rule. It is effective only in the coarsest sizes (plus 10 to 28 mesh) and is best suited to coals having a sharply defined line between coal and refuse material. Predrying to reduce the moisture content of the coal head of the air table treatment is not uncommon. It is a less expensive and a less accurate method of cleaning coal than the wet cleaning method.

**air conditioning.** The control of the quality, quantity, and temperature—humidity of the air in a room or building within prescribed limits.

**air curtain.** A method for mechanically containing oil spills. It is also used as a barrier by blowing a stream of air across the door of a heated or air-conditioned building to reduce heat loss or gain.

**air diffuser.** A device, such as a louver, which allows mixing of the air, delivered by it into a room, with the room air.

**Airdox.** A system for breaking down coal by which compressed air, generated locally by a portable compressor at 10,000 pounds per square inch, is used in a releasing cylinder which is placed in a hole drilled in the coal. Slow breaking results in producing, with no flame, an amount of lump coal larger than that produced by explosives. Its principal advantage is that it may be used safely in gaseous and dusty mines.

**air-gas ratio.** The ratio of the air volume to the gas volume in order to achieve a desired character of combustion.

**air heating system.** A solar heating system in which air is heated in the solar collector and used as the energy transfer medium to the rest of the system.

**air horsepower.** The rate at which energy is used in horsepower units, in moving air between two points.

**air mass.** A body of air extending as high as the stratosphere; also the path length of solar radiation through the earth's atmosphere.

**air pollution.** Air contaminants such as sulfur dioxide, automotive exhaust products, toxic dusts from coal smoke, and other particulates and radioactive emanations.

Air pollution in downtown area of Chicago. (*Courtesy Argonne National Laboratory*)

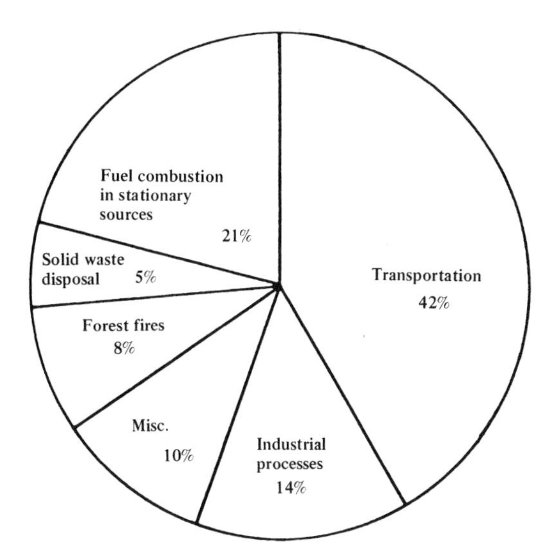


Sources of air pollution. (*Courtesy National Science Teachers Association*)

**air quality control region.** Inter- or intrastate areas designated by the federal government which are treated as a single unit for the purpose of air pollution control.

**air quality criteria.** The criteria for establishing the basis for air quality standards, as determined from scientific compendiums documenting the relationship between various concentrations of air pollutants and their adverse effects.

**air quality standards.** The prescribed level of air quality which protects the public health or welfare.

**air sampling.** The collection and analysis of samples of air to measure its radioactivity or to detect the presence of radioactive substances or chemical pollutants.

**air shutter.** Device for varying the size of air inlet(s).

**air-to-rock solar heating system.** A solar heating system of flat plate collectors which circulate sun-heated air into a bin of rocks where its heat is transferred to the rocks to be stored for later use.

**air wave.** The acoustic energy pulse transmitted through the air as a result of the detonation of a seismic shot.

**Alcator.** A fusion device, a tokamak, developed by the Massachusetts Institute of Technology, which uses powerful magnetic fields instead of a physical container to bottle the plasma.

**alcohol gasoline.** The use of alcohol as a blending agent in gasoline.

**aldehyde.** Any of various organic compounds derived from a hydrocarbon by oxidation.

**algal coal.** Coal composed mainly of algal remains, such as Pila, Reinschia, etc.

**aliphatics.** A class of organic compounds characterized by an open chain structure and consisting of the paraffin, olefin, acetylene hydrocarbons and their derivatives (such as the fatty acids).

**alkali.** Soluable salts, usually of sodium, potassium, magnesium, or calcium, which combine with acids to form neutral salts. They are often used in water and wastewater treatment plants to control acidity.

**alkylation.** The process of introducing one or more alkyl groups into the structure of hydrocarbons to form high octane fuels.

**allobar.** Form of an element having a different atomic weight due to a different isotopic composition.

**allocated pool.** A pool in which the total oil or natural gas production is restricted and allocated to various wells therein in accordance with proration schedules.

**allotropic.** Those substances which exist in two or more forms, as diamond and graphite.

**allowables.** The rate of production from a well that is allowed by a particular state or governing body.

**alluvial.** Adjective used to identify particular types of, or minerals found associated with, deposits made by flowing water.

**alpha decay.** Radioactive decay in which an alpha particle is emitted.

**alpha particle.** A positively charged particle emitted by certain radioactive materials. It is composed of two neutrons and two protons, and, hence, is identical to the nucleus of a helium atom. It can also be produced through nuclear reactions. It is the least penetrating of the three common types of radiation (alpha, beta, and gamma).

**alpha ray.** A stream of alpha particles.

**alternating current.** Electrical current in which the flow of electrons reverses at regular intervals, usually 120 reversals per second or 60 cycles per second.
(See Fig. p. 29)

**alternative combustion engines.** Alternatives to the current spark ignition and

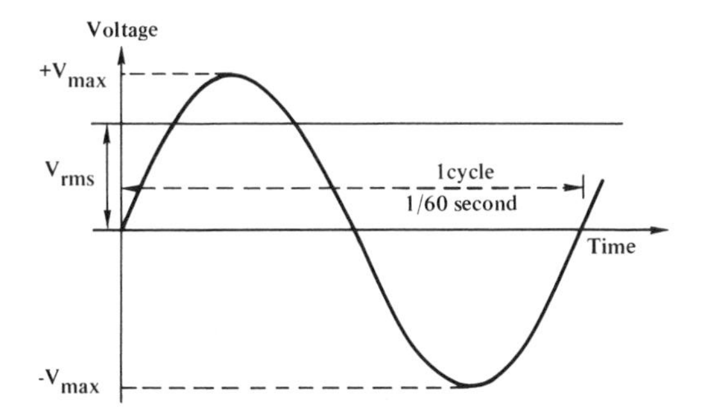


Alternating current voltage. (*Courtesy National Science Teachers Association*)

diesel combustion engines, emphasizing the gas turbine and Stirling engines which can burn a variety of liquid fuels, including kerosene, diesel fuels, methanol, and coal-derived liquids.

**alternator.** A generator producing alternating current by the rotation of its rotor, which is driven by a steam or water turbine. A gas turbine or a diesel engine can be used as a prime mover in certain cases.

**ambient.** The surrounding atmosphere; encompassing on all sides; the environment surrounding a body but undisturbed or unaffected by it.

**ambient air temperature.** Ambient air temperature is the temperature of the air surrounding solar collectors being tested.

**ambient temperature.** Temperature of the atmosphere surrounding a particular location.

**amendment.** Material, such as lime and gypsum, added to the soil to make it more productive.

**amine.** Any of various organic compounds containing the chemical group $NH_2$, NH or N, and a hydrocarbon group such as $CH_3$.

**ampere.** A unit of measure for an electric current; the current produced by 1 volt acting through a resistance of 1 ohm.

**amplitude.** The maximum displacement from the mean position in connection with vibration.

**anabatic wind.** An upslope, rising wind.

**anaerobic digestion.** The process of using microorganisms that live only in the absence of oxygen to decompose organic materials into more valuable or more easily disposed of components. The chief use of anaerobic digestion is reduction of the volume of sewage sludge. It can also be used for the production of synthetic natural gas from organic wastes or from algae.

**ancillary energy.** A measure of the external energy required for an energy process, such as energy for process heat, electricity for pumps, and fuel for truck, train, or barge transportation.

**anemometer.** An instrument for measuring air velocity.

**ANFO.** A blasting agent which is composed of ammonium nitrate and fuel oil.

**angle of dip.** The angle at which strata or mineral deposits are inclined to the horizontal plane.

**angle of polarization.** That angle, the tangent of which is the index of refraction of a reflecting substance.

**angle of repose.** The maximum slope at which a heap of any loose or fragmented solid material will stand without sliding or come to rest when poured or dumped in a pile or on a slope.

**angle of slide.** The slope measured in degrees of deviation from the horizontal, on which loose or fragmented solid materials will start to slide. It is a slightly greater angle than the angle of repose.

**angstrom.** A unit of length equal to $10^{-10}$ meters or $10^{-4}$ microns, and used to measure the spacing of atoms in crystals, the size of atoms and molecules, and the length of light waves. Named after Anders Jons Angstrom (1814–1874), a Swedish spectroscopist.

**animal waste conversion.** The process for obtaining oil from animal wastes. A Bureau of Mines experiment has obtained 80 gallons of oil per ton from cow manure. In comparison, average oil shale yields 25 gallons of oil per ton of ore.

**annealing.** A process involving controlled heating and subsequent controlled cooling for such purposes as inducing ductility in metals, producing a desired microstructure, or obtaining desired mechanical, physical, or other properties.

**Annual Cycle Energy System (ACES).** The ACES (annual cycle energy system) heating cycle consists of a methanol-water loop, a refrigerant loop, and a loop for domestic hot water. Heat is removed from a large tank of water for winter heating; this operation turns the water into ice, which is stored for cooling the house in summer. (See Fig. p. 31)

**annual load fraction.** A fraction of the annual heating needs supplied by solar energy.

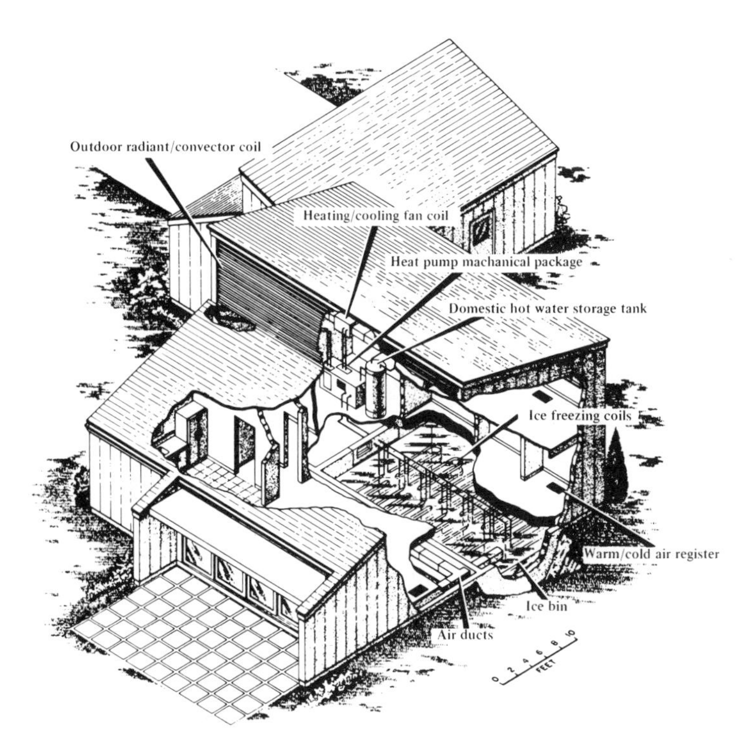


A 1,500 square foot, three-bedroom house built on University of Tennessee property outside Knoxville, demonstrates the ACES. The system is designed to save at least 50 percent of the electrical energy now used to heat and cool a home. (*Courtesy Oak Ridge National Laboratory*)

**annual maximum demand.** The greatest of all demands of the electrical load which occurred during a prescribed interval in a calendar year.

**annual plant.** A plant that completes its life cycle in 1 year or less.

**annual throughput.** The average yearly amount of fresh fuel, introduced as a replacement or spent fuel.

**anode.** The positive terminal of an electrolytic cell; the negative terminal of a primary cell or of a storage battery that is delivering current; the electrode at which oxidation or corrosion occurs, or from which the current is transmitted to the electrolyte.

**anode carbon.** Carbon of high purity, widely used in Leclanche cells, in electric arcs, and in nuclear reactors.

**anthracite.** A hard, black lustrous coal containing a high percentage (between 92%

and 98%) of fixed carbon and a low percentage (between 2% and 8%) of volatile matter and, generally, a heating value of 12,000–15,000 British thermal units per pound.

**anthracite silt.** Minute particles of anthracite too fine to be used in ordinary combustion.

**anthracite stove.** A closed-in type of domestic stove specially designed to burn anthracite and used mainly for heating purposes.

**anthracitization.** The process of transformation of bituminous coal into anthracite.

**anthracosis.** A deposition of coal dust within the lungs caused from the inhalation of sooty air.

**anticline.** The upward geological fold in rock strata in which the beds of layers dip in opposite direction from the crest, permitting possible entrapment of oil and gas.

**antimatter.** Matter in which the ordinary nuclear particles (neutrons, protons, electrons, etc.) are conceived of as being replaced by their corrersponding antiparticles (antineutrons, antiprotons, positrons, etc.). An antihydrogen atom, for example, would consist of a negatively charged antiproton with an orbital positron. Normal matter and antimatter would mutually annihilate each other upon contact, converting totally into energy.

**antitrade winds.** The prevailing westerly winds frequently present above the trade winds.

**API engine service classification system.** Classifications and designations for automotive engine lubricating oils approved by the American Petroleum Institute (API) Lubrication Committee and adopted by the API Division of Marketing.

**API gravity**. A measure of the mass of a fluid relative to water; the standard American Petroleum Institute (API) method for specifying the density of crude petroleum. The density in degrees API equals (141.5/P)-131.5 where P is the specific gravity of the particular oil at 16°C.

**apparent day.** Solar day; interval between successive transits of the sun's center across the observer's meridian. The time thus measured is not uniform or clock time.

**appliance or customer saturation.** The number of specified appliances, or users, divided by the basic units or total potential of the whole body of things involved, i.e., gas heating saturation related to customers is the total number of customers with space heating divided by the total number of customers.

**aquifer.** An underground bed or stratum of earth, gravel, or porous stone that contains water.

**aquifer storage.** The storage of gas underground in porous and permeable rock stratum, the pore space of which was originally filled with water.

**aquitard.** A rock unit with relatively low permeability that retards the flow of water.

**area mining.** Surface mining carried on in flat to gentle rolling regions.

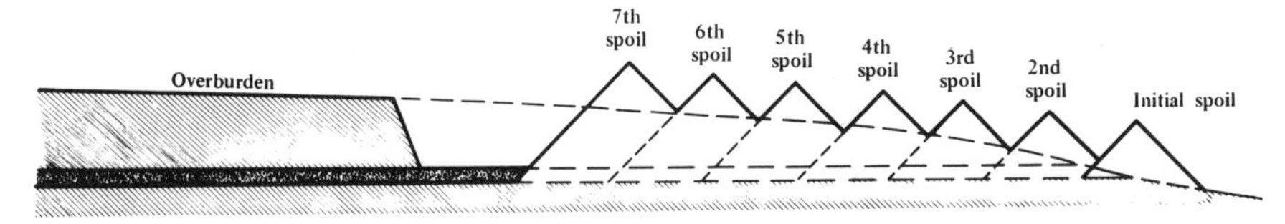


The production of coal by area mining. (*Courtesy U.S. Energy Research and Development Administration, Division of Coal Conversion and Utilization*)

**area source.** In air pollution, any small individual fuel combustion source, including any transportation sources.

**arenaceous.** Applies to rocks that have been derived from sand or that contain sand.

**argillaceous.** Rocks or substances having a high clay content.

**aromatic hydrocarbon.** An unsaturated cyclic hydrocarbon containing one or more six-carbon rings.

**aromatics.** Cyclic hydrocarbons and their derivatives; characterized by the presence of at least one benzene ring. The principal aromatics are benzene, toluene, and xylene (commonly referred to as the BTX group).

**artesian.** Ground water under sufficient pressure to cause the water level in a drilled hole to rise above the top of the rock unit.

**artificial liquid fuels.** Fuels created by the hydrogenation of coal, the destructive distillation of coal, lignite, or shale at low temperature, and by a recombination of the constituents of water gas in the presence of a suitable catalyst.

**artificial satellites.** Man-made vehicles designed to orbit the earth, moon, or any other heavenly body. Satellites are used to obtain and radio back to earth information about conditions in the upper atmosphere, the ionosphere, and outer space. Communication satellites are used for relaying radio and television signals around the curved surface of the earth. Their instruments are powered by solar cell batteries or SNAP atomic batteries. Satellites are being considered as platforms for solar power stations.

Four AEC-developed nuclear generators (atomic batteries) provide electric power for NASA's Pioneer 10 spacecraft. (*Courtesy National Aeronautics and Space Administration*)

**asbestos.** A mineral fiber with countless industrial uses; a hazardous air pollutant when inhaled. Its effects as a water pollutant are under scrutiny.

**ash.** Noncombustible mineral matter contained in coal. These minerals are generally similar to ordinary sand, silt, and clay in chemical and physical properties.

**ash content.** The percentage of incombustible material in a fuel; that portion of a laboratory sample remaining after heating until all the combustible matter has been burned away.

**aspect.** The direction toward which a slope faces.

**asphalt.** A bitumen of variable hardness comparatively nonvolatile, composed principally of hydrocarbons containing little or no crystallizable paraffins. It occurs in nature, but can also be obtained as the residue from the refining of certain petroleums and is then known as artificial asphalt.

**asphalt base petroleum.** Crude oils which, upon processing, yield large amounts of asphaltic residues.

**asphalt cement.** A fluxed or unfluxed asphalt specially prepared as to quality and consistency for direct use in the manufacture of bituminous pavements.

**aspirator.** An appliance, such as a suction pump, exhaust fan, or the friction of a water jet, for causing a movement of gases or liquids by suction.

**associated-dissolved gas.** Associated gas is free natural gas in immediate contact, but not in solution, with crude oil in the reservoir; dissolved gas is natural gas in solution in crude oil in the reservoir.

**associated natural gas.** Free natural gas in immediate contact, but not in solution, with crude oil in the reservoir.

**ASTM coal classification.** A system based on proximate analysis in which coals containing less than 31 percent volatile matter on the mineral matter free basis (Parr formula) are classified only on the basis of fixed carbon (that is, 100 percent volatile matter). They are divided into five groups, the first three of which are called anthracites and the last two bituminous coals: above 98 percent fixed carbon; 98 to 92 percent fixed carbon; 92 to 86 percent fixed carbon; 86 to 78 percent fixed carbon; and 78 to 69 percent fixed carbon. The remaining bituminous coals, the subbituminous coals and the lignites, are then classified into groups as determined by the calorific value of the coals containing their natural bed moisture, that is, the coals as mined but free from any moisture on the surface of the lumps.

**ATGAS.** A process for coal gasification being developed where the primary feature of the process is the dissolving of coal in a bath of molten iron.

**atmosphere.** The gaseous envelope surrounding the earth; the outdoor air in general; also a mixture of gases within any specific chamber such as a heat-treating furnace.

**atmospheric pollution.** Degradation of the quality of the atmosphere due to combustion products from industrial plants, power plants, or vehicular engines.

**atmospheric pressure.** The pressure of the air at sea level. The standard atmospheric pressure will support a column of mercury 760 millimeters high, or 29.9 inches high, in a barometer.

**atmospheric pressure fluidized beds combustion.** Involves using a cushion of air to suspend particles of burning coal and dolomite/limestone. The dolomite/limestone is used to absorb sulfur oxides and thereby clean the combustion gases prior to release through a stack.

**atmospheric water.** Water which exists in the atmosphere in the gaseous, liquid, or solid state.

**atom.** The smallest particle of an element that can exist either alone or in combination with similar particles of the same element or of a different element. According to present theory, the atom consists of a nucleus of protons and neutrons positively charged, surrounded by negatively charged particles called electrons.

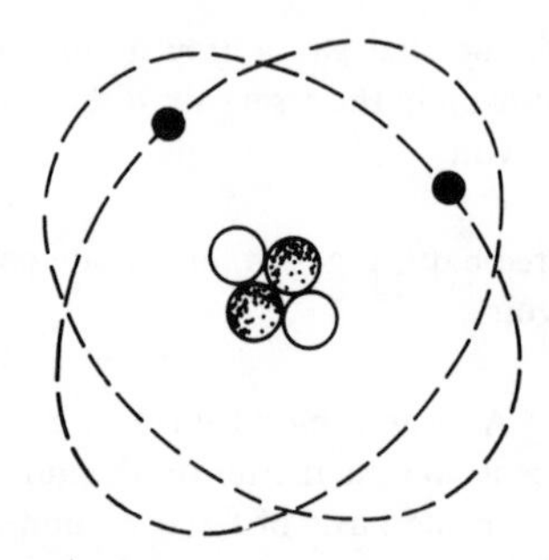

Helium atom containing two neutrons and two protons in the nucleus; two electrons are in orbit outside the nucleus. (*Courtesy Basic Nuclear Engineering*)

**atomic battery.** A radioisotopic generator.

**atomic bomb.** A bomb whose energy comes from the fission of heavy elements, such as uranium or plutonium.

**atomic charge.** Electrical charge density due to gain or loss of one or more electrons.

**atomic clock.** A device that uses the extremely fast vibrations of molecules or atomic nuclei to measure time. These vibrations remain constant with time; consequently, short intervals can be measured with much higher precision than by mechanical or electrical clocks.

**atomic cloud.** The cloud of hot gases, smoke, dust, and other matter that is carried aloft after the explosion of a nuclear weapon in the air or near the surface. The cloud frequently has a mushroom shape.

**atomic disintegration.** Conversion of the nucleus of an atom of one element into that of some other element.

**atomic energy.** Nuclear energy; the energy released by a nuclear reaction or by radioactive decay. Nuclear energy is the preferred term.

**Atomic Energy Commission.** The independent civilian agency of the federal government with statutory responsibility to supervise and promote use of nuclear energy. Functions were taken over in 1974 by the Energy Research and Development Administration and Nuclear Regulatory Commission.

**atomic mass.** The mass of any species of atom, usually expressed in atomic mass units.

**atomic mass units.** A unit of mass used for expressing the masses of different isotopes of elements; one-twelfth the mass of a neutral atom of the most abundant isotope of carbon-12, or 1/16 of the atomic mass of the most abundant oxygen isotope, which is about $1.66035 \times 10^{-24}$ gram or in terms of equivalent energy to about 931 electron volts.

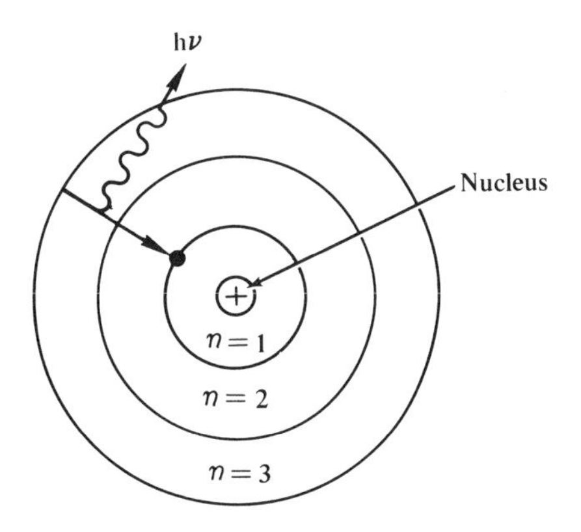


Bohr's model of the hydrogen atom showing allowable electron orbits. (*Courtesy Basic Nuclear Engineering, 3rd Edition, Arthur R. Foster and Robert L. Wright, Jr., Allyn and Bacon, Inc., Boston, Mass.*)

**atomic number.** The number of protons in the nucleus of an atom, and also its positive charge. This number represents the location of an element in the Periodic Table; it is always the same as the number of negatively charged electrons in the shells. Thus an atom is electrically neutral, except in an ionized state. Atomic numbers range from 1, for hydrogen, to 106, for the most recently discovered element which has not been named, as yet.

**atomic pile.** A nuclear reactor.

**atomic reactor.** A nuclear reactor.

**atomic weapon.** An explosive weapon in which the energy is produced by nuclear fission or fusion.

**atomic weight.** The mass of an atom relative to other atoms. The present-day basis of the scale of atomic weights is carbon; the commonest isotope of this element has arbitrarily been assigned an atomic weight of 12. The unit of the scale is 1/12 the weight of the carbon-12 atom, or roughly the mass of one proton or one neutron. The atomic weight of any element is approximately equal to the number of protons and neutrons in its nucleus.

**atomize.** To divide a liquid into extremely minute particles.

**atom smasher.** An accelerator.

**attenuation.** The reduction of a radiation quantity upon passage of radiation through matter resulting from all types of interaction with that matter. Attenuation usually does not include geometric attenuation.

**attenuation factor.** For a given attenuating body in a given configuration, the factor by which a radiation quantity at some point of interest is reduced owing to the interposition of the body between the source of radiation and the point of interest.

**attic oil.** Preferred term is crude oil.

**attitude of bedrock.** A general term describing the relation of some directional features to a rock in a horizontal surface.

**auger.** A rotary drilling device used to drill shotholes or geophone holes in which the cuttings are removed by the device itself without the use of fluids.

**auger mining.** A method of extracting coal using instruments to bore horizontally into coal seams; generally practiced in but not restricted to hilly regions.

**augmentors.** Devices that increase the speed of air flow through wind machines.

**autoclave.** A vessel constructed of thick-walled steel (alloy steel or nickel alloys) for carrying out chemical reactions under pressure and at high temperatures.

**autofining.** A fixed-bed catalytic process for desulfurizing distillates.

**automatic controller.** A device which operates automatically to regulate a controlled variable in response to a command and a feedback signal.

**automatic damper.** A device which is used to maintain automatically the flow of hot or cold air to or from a room.

**automatic ignition.** A means which provides for the automatic lighting of gas at the burner when the gas valve controlling flow is turned on.

**automatic input flow control valve.** A device for controlling the gas supply to the main burner without manual attention.

**automatic shutoff valve.** A device designed to shut off gas flow upon flame failure without manual attention.

**automotive engine oil.** Preferred term is motor oil.

**autoradiograph.** A photographic record of radiation from radioactive material in an object, made by placing the object very close to a photographic film or emulsion. The process is called autoradiography. It is used, for example, to locate radioactive atoms or tracers in metallic or biological samples.

**auxillary.** The component physically connected as part of a solar system and that provides less than half of the annual energy supplied by that solar system.

**auxiliary heat source.** The provision of supplementary heat when the primary source is insufficient.

**available energy.** That part of the total energy which can be usefully employed; in a perfect engine, that part which is converted to work.

**available heat.** The amount of energy that may be converted into useful heat from the chemical energy in a unit of fuel.

**available nutrient.** Consists of ions that are dissolved in the soil moisture or are absorbed in the clay minerals of the soil. They constitute the mineral content of the soil that is immediately available for uptake by the plant.

**available power loss.** The available power loss of a transducer connecting an energy source and an energy load is the transmission loss measured by the ratio of the source power to the output power of the transducer.

**average demand.** The demand on a system or any of its parts over an interval of time, determined by dividing the total energy supplied by the number of units of time in the interval.

**average life.** The average of the individual lives of all the atoms of a particular radioactive substance. It is 1.443 times the radioactive half-life of the substance.

**average wind speed.** The mean wind speed over a specified period of time.

**Avogadro's Law.** Principle that equal volumes of different gases at the same temperature and pressure contain the same number of molecules.

**axial flow compressor.** One in which air is compressed in a series of stages as it flows axially through a decreasing tubular area.

# B

**back.** The roof or upper part in any underground mining cavity; the ore body between a level and the surface, or between two levels.

**back arch.** A concealed arch carrying the backing or inner part of a wall in a mine where the exterior facing material is carried by a lintel.

**back blade.** In regrading, to drag the blade of a bulldozer or grader as the machine moves backward, as opposed to pushing the blade forward.

**backcast stripping.** A strip mining method using two draglines, one of which strips and casts the overburden while the other recasts a portion of the overburden.

**backfill.** The material excavated from a site and reused for filling.

**background radiation.** The radiation in man's natural environment, including cosmic rays and radiation from the naturally radioactive elements, both outside and inside the bodies of men and animals. It is also called natural radiation.

**back pressure.** Resistance transferred from rock into drill stem when the bit is being fed at a faster rate than the bit can cut.

**back pressure valve.** A valve built to maintain a given pressure in a piping system by remaining in a closed position until the given pressure is reached, at which time it opens to permit flow until the pressure falls below the specified pressure.

**backscatter.** When radiation of any kind strikes matter (gaseous, solid, or liquid), some of it may be reflected or scattered back in a general direction of the source. An understanding or exact measurement of the amount of backscatter is important when beta particles are being counted in an ionization chamber, in medical treatment with radiation, or in the use of industrial radioisotope thickness gauges.

**backup.** The reserve generating capacity of a power system.

**bad ground.** Soft, highly fractured, or cavernous rock formations in which drilling a borehole is a slow procedure involving time-consuming cementing or casing operations.

**baffles.** Devices such as steel plates, louvers, or screens used to retard the flow of materials.

**bagasse.** The fibrous material remaining after the extraction of the juice from sugarcane. It is used as a fuel and as a mix in making lightweight refractories; the dried and pulverized or shredded sugarcane fibers sometimes are added to a drilling fluid to plug crevices in, and prevent loss of circulation liquid from a borehole.

**bag filter.** An apparatus for removing dust from dust-laden air, employing cylinders of closely woven material which permit passage of air but retain solid particles.

**baghouse.** Chamber in which exit gases from roasting, smelting, calcining are filtered through membranes (bags) which arrest solids; a fabric filter used to separate particulates from an airstream; an air pollution abatement device used to trap particulates by filtering gas streams through large fabric bags, usually made of glass fibers.

**balanced winding.** The conventional method of winding in a mine shaft. As the cage containing the loaded cars ascends, the other cage containing the empties descends and, thus, the cages and cars are balanced, implies the use of a balance rope.

**baling.** A means of reducing the volume of solid waste by compaction.

**balloon framing.** The system of building currently used for most wood frame house construction.

**band.** Slate or other rock interstratified with coal. It is commonly called middle band in Arkansas; also, dirt band, sulfur band, or slate band.

**banded coal types.** Banded bituminous coal consists of bands made from various types of coal; formerly known as bright coal, dull coal, and mother of coal.

**bank.** The top of the shaft (mine shaft), or out of the shaft; a mound of mine refuse.

**Banka method.** A manual method of boring used for sampling alluvial deposits.

**bare.** To cut coal by hand; to hole by hand; to remove overburden. In the nuclear field, the term means not having a reflector.

**bar hole.** Small hole made in the ground in the vicinity of gas piping for the purpose of extracting a sample of the ground atmosphere for analysis, such as when searching for leaks.

**bar hole leak survey.** Leakage surveys made by driving or boring holes at regular intervals along the route of an underground gas pipe and testing the atmosphere in the holes with a combustible gas detector or other suitable device.

**barn.** A unit area used in expressing the cross sections of atoms, nuclei, electrons, and other particles. One barn is equal to $10^{-24}$ square centimeter.

**barometer.** Instrument used for measuring atmospheric pressure.

**barometer holiday.** In mining, any day on which no work is carried on underground, because of the very low state of the barometer (for instance, when it drops below 29 inches), as much firedamp may be expected to be given off in the mine.

**barrel.** A liquid measure of oil, usually crude oil, equal to 42 American gallons or about 306 pounds. One barrel equals 5.6 cubic feet or 0.159 cubic meters. For crude oil, one barrel is about 0.136 metric tons, 0.134 long tons, and 0.150 short tons. The energy values in millions of British thermal units per barrel of petroleum products are: crude petroleum—5.6; residual fuel oil—6.29; distillate fuel oil—5.83; gasoline—5.25; jet fuel (kerosine type)—5.67; jet fuel (naphtha type)—5.36; kerosine—5.67; petroleum coke—6.02; and asphalt—6.64.

**barricade shield.** A type of movable shield for protection from radiation.

**barrier.** Blocks of coal left between the workings of different mine owners and/or within the workings of a particular mine for reasons of safety and cost reduction. Barriers help to prevent disasters of inundation, explosions, and/or fires involving an adjacent mine or another part of a mine and prevent water running from one mine to another, or from one section to another of the same mine.

**barrier shield.** A wall or enclosure shielding the operator from an area where radioactive material is being used or processed by remote control equipment.

**barring.** The end and side timber bars used for supporting a rectangular shaft; also, using an iron bar to remove loose rocks after blasting.

**bar screen.** A stationary inclined screen, comprising longitudinal bars, spaced at intervals onto which the material to be screened is fed at the upper end; in waste water treatment, a screen that removes large floating and suspended solids.

**baryon.** One of a class of heavy elementary particles that includes hyperons, neutrons, and protons.

**barytes concrete.** A type of heavy concrete containing barytes added to improve its shielding characteristics.

**basal metabolism.** The amount of heat liberated by a person at rest in a comfortable environment (about 40 British thermal units per hour).

**baseball.** A plasma confinement machine used in fusion research in which the plasma is held in a linear magnetic bottle sealed at both ends by an arrangement of magnetic fields that deflects escaping ions back into the container.

**baseboard radiator.** A heat disseminating unit located at the lower perimeter of a room.

**baseload plant.** An electrical generation facility which is designed primarily to satisfy a continuous demand. Generally, capacity factors are 0.6 to 0.9.

**base gas.** The total volume of gas which will maintain the required rate of delivery during an output cycle.

**base load.** The minimum load of an average utility (gas or electric).

**base load station.** In the gas field, a station which is normally operated to take all or part of the base load of a system and which, consequently, operates essentially at a high load factor.

**base of highwall.** The point of intersection between the highwall and the plane forward at the base of the excavated material.

**base oil.** A refined or untreated oil used in combination with other oils and additives to produce lubricants.

**base pressure.** A standard to which measurements of a volume of gas are referred. The American Gas Association and the Federal Power Commission have adopted 14.73 pounds per square inch gage as the standard pressure base.

**base pressure index.** A device which automatically compensates to correct gas volume at base pressure without regard for any correction for temperature.

**base production control level.** The total number of barrels of domestic crude petroleum produced from a particular property in the corresponding month of 1972.

**base volume index.** A device which automatically compensates to correct gas volumes measured at operating temperature and pressure to volume at a specified base temperature and pressure.

**basic energy science.** A broad-based program of scientific investigation into the fundamental nature of the universe to develop greater understanding of the nature and behavior of matter. The program includes research in the molecular, material, nuclear, and biological sciences.

**basicity.** Of an acid, the number of hydrogen atoms per molecule which can be replaced by a metal.

**basin.** The lowest part of a mine or area of coal lands; a general region with an overall history of subsidence and thick sedimentary section. Geologic basins may be filled with sediment and thus be invisible from the surface.

**batch oil.** A pale, lemon-colored, neutral oil having a viscosity of about 80 Saybolt at 21°C.

**batch process.** A process in which the feed is introduced as discrete charges, each of which is processed to completion separately.

**bathyclinograph.** In oceanography, an instrument for measuring vertical currents in the deep sea.

**bathymeter.** An instrument that measures temperature, pressure, and sound velocity to depths up to 7 miles. The devices are completely transistorized and use frequency modulation for telemetering.

**battery.** A device which converts energy from chemical compounds (called reactants) into electricity. A battery consists of one or more units (called cells) which are enclosed in an insulated casing. Each cell contains a positive and negative electrode separated by an electrolyte.

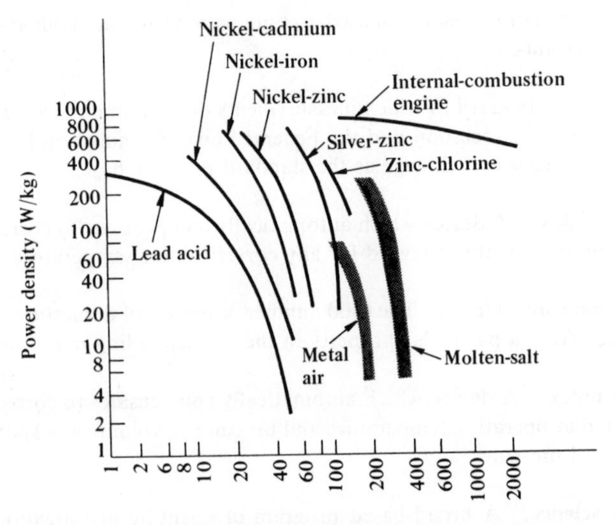


Power and energy capabilities of various batteries. (*Courtesy Electric Power Research Institute and the Applied Nucleonics Co., Inc.*)

**Battery Energy Storage Test (BEST) Facility.** It will provide the cabability for testing battery installations up to 10,000 kilowatt-hours in an actual utility environment.

**Baume scale.** A device for determining the specific gravity of liquids, particularly petroleum products. It has been superseded to a considerable extent by the American Petroleum Institute scale (°API, instead of °B or Be).

**beam hole.** An opening through a reactor shield and, generally, through the reactor reflector, which permits a beam of radioactive particles or radiation to be used for experiments outside the reactor.

**beam radiation.** Solar radiation which is not scattered by dust or water droplets. It is capable of being focused and casts shadows.

**beam reactor.** A reactor designed to produce beams of neutrons to be used for research outside the reactor.

**bearing.** The practice of underholing or undermining.

**bearing wall.** A wall which supports a vertical load in addition to its own weight.

**Beaufort scale.** A numerical scale, graded from 0 to 12, devised by Admiral Beaufort in the 19th century to indicate wind speeds. This scale has been adopted internationally.

| Code Number | Wind Velocity (mph) | Description |
|---|---|---|
| 0 | less than 1 | calm |
| 1 | 1-3 | light air |
| 2 | 4-7 | light breeze |
| 3 | 8-12 | gentle breeze |
| 4 | 13-18 | moderate breeze |
| 5 | 19-24 | fresh breeze |
| 6 | 25-31 | strong breeze |
| 7 | 32-38 | moderate gale (or near gale) |
| 8 | 39-46 | fresh gale (or gale) |
| 9 | 47-54 | strong gale |
| 10 | 55-63 | whole gale (or storm) |
| 11 | 64-74 | storm (or violent storm) |
| 12 | over 74 | hurricane |

Beaufort wind scale.

**Becke test method.** In optical mineralogy, a test method for determining relative indices of refraction. The method involves determining microscopically the index of refraction of a mineral compared with that of an oil or another substance, such as Canada balsam, in which it is immersed, or of two adjacent minerals in a microscopic thin section.

**bed.** A deposit, parallel to the stratification, later in origin of coal than the rock below and older than the rock above, and thus constituting a regular member of the series of formation and not an intrusion.

**bedplate.** An iron plate forming the bottom for a furnace.

**bedrock.** Any solid rock exposed at the surface of the earth or overlain by unconsolidated material.

**Belanger's critical velocity.** That condition in open channels for which the velocity head equals one-half the mean depth.

**Belknap process.** Old method of coal cleaning in a bath of heavy liquid produced by dissolving calcium chloride in water. The shale sinks and the coal floats.

**bell-and-spigot pipe.** Pipe made with an enlarged diameter or bell at one end into which the plan or spigot end is inserted. The joint is then sealed with a suitable substance such as solvent cement, oakum, lead, or rubber, which is calked into the bell and around the spigot.

**bell hole.** Hole dug or excavation made at the section joints of a pipeline for the purpose of repairs.

**bell joint clamp.** A sealing device to prevent leakage, which is attached at the joint of a bell-and-spigot pipe.

**belt conveyor.** A moving endless belt that rides on rollers and on which coal or other materials can be carried for various distances.

**bench.** One of two or more divisions of a coal seam, separated by slate, etc., or simply separated by the process of cutting the coal, one bench or layer being cut before the adjacent one; also to cut the coal in benches.

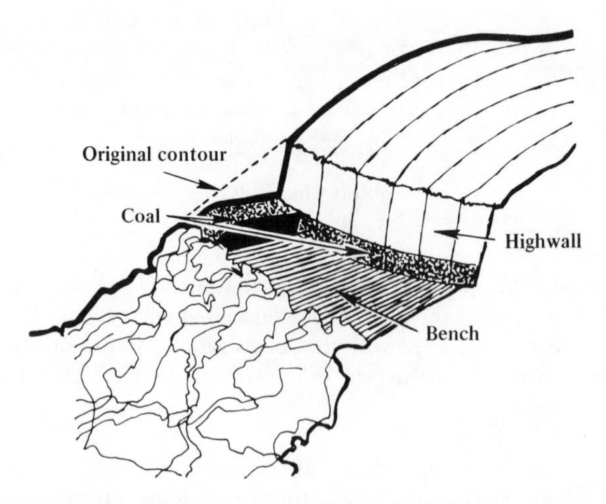


Cutaway showing divisions of a coal seam. (*Courtesy Mining Enforcement and Safety Administration*)

**bench-scale unit.** A small-scale laboratory unit for testing process concepts.

**Bender process.** A continuous, fixed-bed chemical treating process, using a lead sulfide catalyst for sweetening light distillates. The process converts mercaptans to disulfides by oxidation.

**bending moment.** The algebraic sum of the couples or the moments of the external forces, or both, to the left or to the right of any section on a member subjected to bending by couples or transverse forces, or both.

**beneficiation.** A coal cleaning process, which improves the environmental quality of mined coal to broaden its range of application. The process involves the reduction of "free" sulfur and of rock, shales, and other impurities from the coal through grinding, washing, or floating.

**benzine.** An obsolete term for light petroleum distillates covering the gasoline and naphtha range.

**Bergius Process Hydrogenation.** Coal is pulverized and made into paste with heavy oil derived from the process. Powdered catalyst and hydrogen are mixed with the paste and the liquid phase hydrogenation is carried out in a series of converters. Gasoline and middle oil are eventually distilled from liquid products. The catalyst is composed of $FeSO_4$–$H_2O$ and sodium sulfide. Reactor process conditions are 427° to 482°C and 250 to 700 atms. The yield range is 50 to 67 percent of coal fed by weight.

**berm.** A horizontal shelf or ledge built into an embankment or sloping wall of an open pit or quarry to break the continuity of an otherwise long slope for the purpose of strengthening and increasing the stability of the slope or to catch or arrest slope slough material. A berm may also serve as a bench above which material is excavated from a bench face.

**berm interval.** Vertical distance from crest of berm to its underlying toe, as in a bank or bench.

**beta decay.** Radioactive decay in which a beta particle is emitted or in which orbital electron capture occurs.

**beta particles.** An electron, positive or negative. Beta rays are high-velocity electrons, usually negative, originating in radioactive atoms or in particle accelerators.

**beta treatment.** The process of heating followed by rapid cooling to form beta-phase uranium.

**betatron.** A doughnut-shaped accelerator in which electrons are accelerated by a changing magnetic field. Energies as high as 340 million electron volts have been attained.

**Betz law.** The theoretical maximum amount of the wind energy that can be extracted by a machine was calculated to be 59.3 percent by A. Betz. In practice, the efficiency limit is approximately 40 percent.

**BIGAS process.** A process for producing pipeline quality gas. The gasifier is a two-stage entrained-flow reactor. Coal fed into the top stage of the reactor is entrained and devolatilized by hot synthesis gas rising from the lower stage. Unreacted char from the top stage is gasified in the lower stage with oxygen and steam under slagging conditions. The partially methanated product gases in the top stage are cleaned and further methanated into pipeline quality gas. Process conditions are 50 to 100 atmospheres with temperatures of 1482 to 1538°C in the lower stage and 760 to 927°C in the top stage.

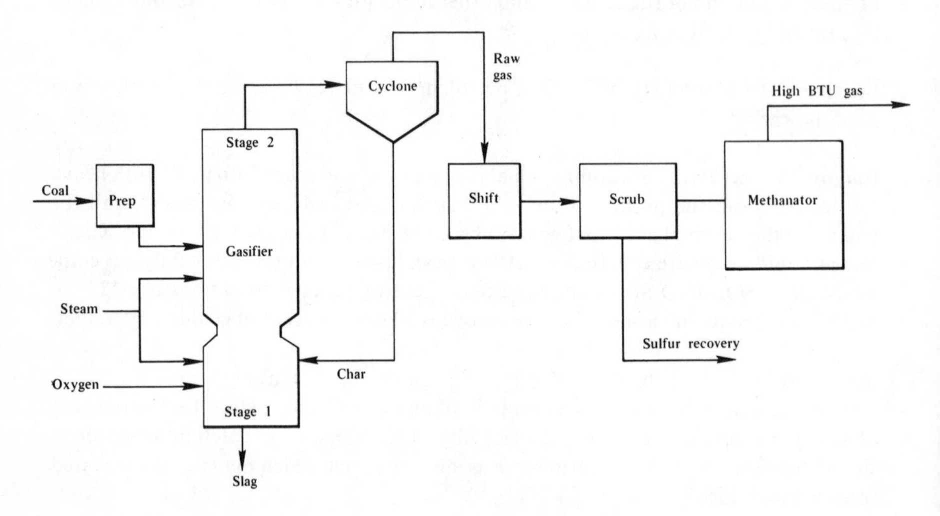


The basic steps involved in the BIGAS coal gasification process. (*Courtesy U.S. Council on Environmental Quality*)

**binary cycle.** Combination of two turbine cycles utilizing two different working fluids in electrical generation plants. The waste heat from the first turbine cycle provides the heat energy for the second turbine cycle. The purpose of a binary cycle is to obtain higher efficiencies from the energy source.

**binary-fluid systems.** A system in which hot fluid is passed through a heat exchanger to transfer heat to a low boiling point fluid which is then used as the working fluid. (See Fig., p. 49)

**binder.** Any substance, such as carbon products, tars, etc., used to impart cohesion to the body to be formed; a streak of impurity in a coal seam, usually difficult to remove.

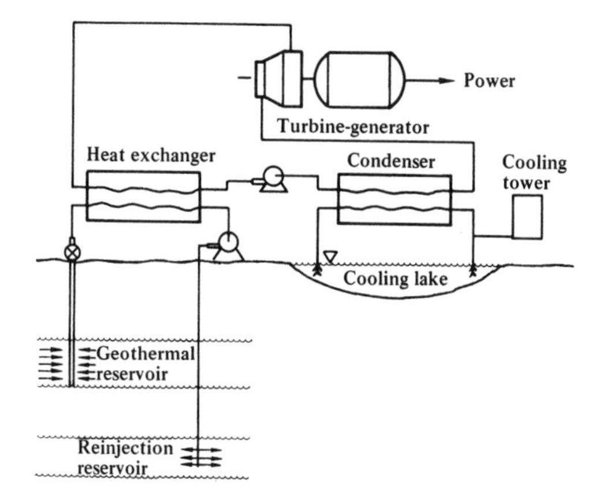


Binary-fluid system for electric power generation from geothermal fluids. (*Courtesy Annual Review of Energy, Jack M. Hollander* (ed.), *Annual Reviews, Inc., Palo Alto, CA 94306*)

**binding energy.** The binding energy of a nucleus is the minimum energy required to dissociate it completely into its component neutrons and protons. Neutron or proton binding energies are those required to remove a neutron or a proton, respectively, from a nucleus. Electron binding energy is that required to remove an electron completely from an atom or a molecule.

**binding energy per nucleon.** The total binding energy of a nuclide divided by its mass number.

**biochemical oxygen demand (BOD).** The amount of oxygen required by bacteria to convert organic material into stable compounds.

**bioconversion.** A general term describing the conversion of one form of energy into another by plants or microorganisms. Synthesis of organic compounds from carbon dioxide by plants is bioconversion of solar energy into stored chemical energy. Similarly, digestion of solid wastes or sewage by microorganisms to form methane is bioconversion of one form of stored chemical energy into another, more useful form. (See Fig., p. 50)

**biodegradable.** Used in sewage disposal and water pollution to describe those substances that can be quickly broken down by the bacteria used for this purpose at sewage disposal plants.

**biofouling.** The growth of small marine organisms which can decrease the heat transfer efficiency of heat exchangers used in the ocean in ocean thermal energy conversion systems.

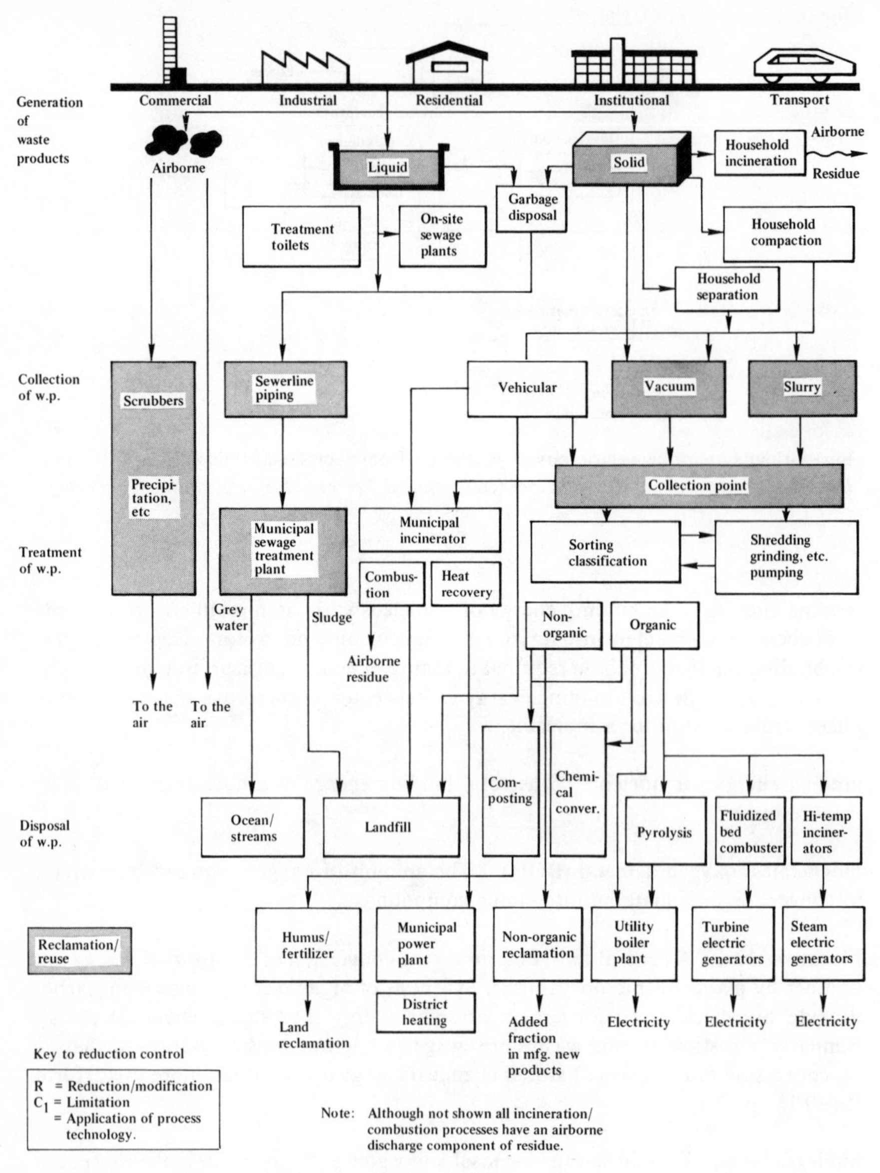


Urban waste products generation-disposal/rinse cycles. (*Courtesy Electric Power Research Institute and Applied Nucleonics Co., Inc.*)

**Biogas process.** An anaerobic digestion process for converting solid municipal waste and sewage into pipeline quality fuel gas and an odor free, stable solid.

**biological dose.** The radiation dose absorbed in biological material. Measured in rems.

**biological energy conversion.** Using a living plant or animal to convert one form of energy to another; for example, trees use solar radiation as an energy source to produce cellulose which can then be used as a fuel.

**biological half-life.** The time required for a biological system, such as a man or an animal, to eliminate, by natural processes, one-half the amount of a substance (such as a radioactive material) that has entered it.

**biological shield.** The outer portion of shielding around a reactor designed to reduce the neutron and gamma fluxes to safe working levels.

**biomass.** The amount of living matter in a unit area or volume; the living weight. Usually limited to plant material; trees and other living plants, crop residues, wood and bark residues, and animal manures.

**biomass fuels.** The production of synthetic fuels, such as methane hydrogen, from terrestrial and sea plants and from animal wastes.

**biomedical environmental research.** The scientific investigation of the health and environmental effects of radiation and other pollutants on the environment and its inhabitants. This program includes the study of ecological relationships and the development of systems and methods to measure the release of noxious or harmful substances.

**biomedical irradiation reactor.** A reactor used for the primary purpose of affecting biological processes by utilization of the reactor-generated ionizing radiation.

**biomonitoring.** The use of living organisms to test the suitability of effluent for discharge into receiving waters and to test the quality of such waters downstream from a discharge.

**biosphere.** Zone at and adjacent to the earth's surface where all life exists; all living organisms of the earth.

**biostabilizer.** A machine used to convert solid waste into compost.

**bird cage.** In the nuclear field, a container and attached cagelike structure for maintaining a safe distance between a body of fissile material and other objects which, if brought too close, might give rise to criticality.

**bit.** Any device that may be attached to, or is an integral part of, a drill string and is used as a cutting tool to bore into or penetrate rock or other materials by utilizing power applied to the bit percussively or by rotation.

**bituminous.** Containing much organic, or carbonaceous matter, mostly in the form of the tarry hydrocarbons which are usually described as bitumen.

**bituminous coal.** A coal which is high in carbonaceous matter and gaseous constituents, having between 15 and 50 percent volatile matter; soft coal; coal other than anthracite and low-volatile coal and lignite. It is dark brown to black in color and burns with a smoky, luminous flame. When volatile matter is removed from bituminous coal by heating in the absence of air, the coal becomes coke.

**bituminous sand.** A sand naturally impregnated with bitumen or petroleum residue.

**bituminous shale.** A shale containing hydrocarbons or bituminous material; when rich in such substances, it yields oil or gas on distillation (i.e., oil shale).

**blackbody.** A body that completely absorbs all wavelengths of incident radiant energy and reflects or transmits none. The surface in question emits radiant energy at each wavelength at the maximum rate possible for the temperature of the surface.

**black light.** Electromagnetic radiation not visible to the human eye. The portion of the spectrum generally used in fluorescent inspection, such as in examining well cuttings for oil, falls in the ultraviolet region.

**black lung.** A respiratory ailment, similar to emphysema, which is caused by inhalation of coal dust.

**black oil.** A residue from petroleum or from its distillates. It is used as a cheap lubricant.

**black shale.** Usually a very thin bedded shale, rich in sulfides (especially pyrite which may have replaced fossils) and organic material, deposited under barred basin conditions causing anaerobic accumulation; a highly carbonaceous shale.

**blanket.** The area immediately surrounding the reactor core is a liquid metal fast breeder reactor. Its major function is to produce plutonium-239 from uranium-238.

**blast furnace gas.** A low-grade producer gas, low in heat content.

**blasting.** The operation of breaking coal, ore, or rock by boring a hole in it, inserting an explosive charge, and detonating or firing it. It is also called shot firing.

**blast wave.** A pulse of air, propagaged from an explosion, in which the pressure increases sharply at the front of a moving air mass, accompanied by strong, transient winds.

**blending.** The process of mixing two or more oils having different properties to obtain a final blend having the desired characteristics. This can be accomplished by off-line batch processes or by in-line operations as part of continuous-flow operations.

**blending naphtha.** A distillate used to thin heavy stocks to facilitate processing (i.e., to thin lubricating oil in dewaxing processes).

**blending stock.** Any of the stock used to make commercial gasoline, including natural gasoline, straight-run gasoline, cracked gasoline, polymer gasoline, alkylate, and aromatics.

**blind drain.** A trench filled with stones selected so as to fill a trench, yet to allow the flow of water through it. It is also called rubble drain.

**blind flange.** A steel plate inserted between flanges of a pipeline, thus cutting off the line.

**blinding.** In uranium leaching, reduced permeability of ion-exchange resins due to adherent slimes; in sieving, blocking of screen apertures by particles.

**block coal.** A bituminous coal that breaks into large lumps or cubical blocks, also, coal passing over certain-sized screens instead of through them.

**block cut method.** Method of surface mining that removes overburden and places it around the periphery of a box-shaped cut. After coal is removed, the spoil is pushed back into the cut and the surface is blended into the topography.

**blowout preventer.** Equipment installed at the wellhead for the purpose of controlling pressures in the annular space between the casing and drill pipe or in an open hole during drilling and completion operations.

**body of coal.** A term frequently used to indicate the "fatty," flammable property in coal, which is the basis of the phenomenon called combustion.

**boiler.** A closed vessel in which water is converted to pressurized steam.

**boiler efficiency.** Ratio of heat absorbed by the water in the boiler to the total heat supplied to the boiler.

**boiler feed.** The water supplied to a boiler.

**boiler fuel.** Natural gas used as a fuel for the generation of steam.

**boiler horsepower.** Represents the conversion of 34.5 pounds of water per hour to steam at a pressure of 14.7 pounds per square inch and a temperature of 100°C.

**boiler pressure.** The pressure of the steam or water in a boiler. It is generally expressed in pounds per square inch gauge.

**boiler rating.** The heating capacity of a steam boiler expressed in British thermal units per hour. Sometimes expressed in horsepower or pounds of steam per hour.

**boiling point.** The temperature at which a substance changes its state from liquid to gas. The boiling point of water is 100°C. The term also refers to the temperature at which crude oil, on being heated, begins to give forth its different distillates.

**boiling water reactor.** A nuclear reactor in which water, used as both coolant and moderator, is allowed to boil in the nuclear core. The resulting steam can be used directly to drive a turbine.

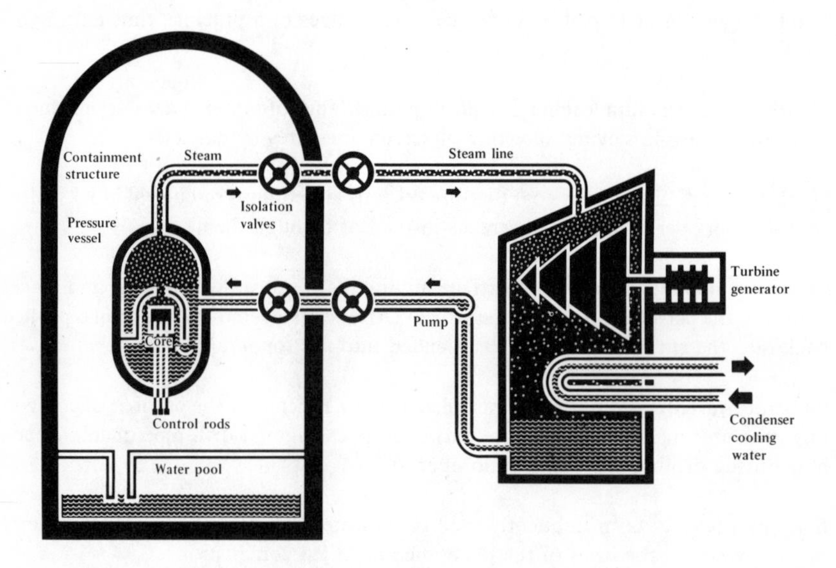


Basic design of a boiling water reactor. (*Courtesy U.S. Council on Environmental Quality*)

**bond.** The cohesion or adhesion that develops between particles; also the link between two atoms due to an electron pair resonating or rotating between them. If each atom contributes an electron, the bond is atomic and non polar; if held unequally, it is polar. If one atom contributes both electrons, the bond is molecular.

**bone.** A hard coal-like substance high in noncombustible mineral matter. It is often found above or below, or in partings between layers of relatively pure coal.

**bone seeker.** A radioisotope that tends to accumulate in the bones when it is introduced into the body. An example is strontium-90, which behaves chemically like calcium.

**boom.** Any beam attached to lifting or excavating equipment; also a floating device that is used to contain oil on a body of water.

**booster.** Any device or substance to augment or improve performance, volume, or force, such as a compressor used to raise pressure in a gas or oil pipeline.

**booster station.** A facility containing equipment which increases pressure on oil or gas in a pipeline.

**boral.** A sandwich of boron carbide crystals in aluminum, used as a shielding material against the passage of thermal neutrons.

**borehole.** A hole with a drill, auger, or other tools for exploring strata in search of minerals, for water supply, for blasting purposes, for proving the position of old workings, faults, and for releasing accumulations of gas or water.

**boron chamber.** An ionization chamber which utilizes a nuclear reaction with boron to detect or measure slow neutrons.

**boron counter tube.** A radiation counter tube which utilizes a nuclear reaction with boron to detect slow neutrons.

**borrow pit.** A type of excavation which involves taking earth from some bank or pit for use in filling or embanking.

**bottle.** A gas-tight container used for storing or transporting gas.

**bottled gas.** The liquefied petroleum gases propane and butane, contained under moderate pressure (about 125 pounds per square inch and 30 pounds per square inch respectively), in cylinders.

**bottom hole contract.** A contract providing for the drilling of a well to a specified depth.

**bottom hole pressure.** The pressure, expressed in pounds per square inch, at the bottom of a closed-in well.

**bottoming cycle.** A means to increase the thermal efficiency of a steam electric generating system by converting some waste heat from the condenser into electricity rather than discharging all of it to the environment.

**bottom water.** In oil wells, the water that lies below the productive sand and is separated from it.

**bowing.** Bending due to a nonuniform temperature distribution, excess moisture, or lack of moisture.

**box cut.** Initial excavation in a mine that penetrates a hill resulting in walls on three sides, with spoils dumped over the slope.

**Boyle's law.** Principle that at a constant temperature, the volume of a gas varies inversely as the absolute pressure while the density varies directly as the pressure.

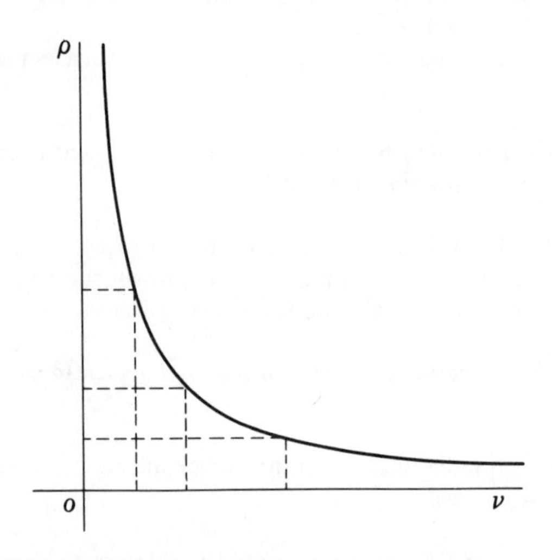


Equilateral hyperbola representing Boyle's law. The rectangular areas (pv) are all equal. (*Courtesy Van Nostrand's Scientific Encyclopedia, 5th Edition, Douglas M. Considine (ed.), Van Nostrand Reinhold Company, New York, 10001*)

**brachytherapy.** Radiation treatment using a solid or enclosed radioisotopic source on the surface of the body or at a short distance from the area to be treated.

**brackish water.** A mixture of fresh and salt water.

**brandenhead.** In oil well drilling, an iron or steel head screwed into the top of the casing that enables the use of one size pipe inside another for the subsequent control of products being delivered from either one of the two pipes.

**Bradford breaker.** A machine which combines coal crushing and screening. It consists of a revolving cylindrical screen.

**Bragg-Grey principle.** A relationship used as the basis of many measurements of ionizing radiation. It states that the ionization produced in a small gas-filled cavity in a homogeneous medium by a uniform field of ionizing radiation is proportional to the absorbed dose in that medium.

**brake horsepower.** The power output delivered by the crankshaft of an engine.

**branching decay.** Radioactive decay of a nuclide that can proceed in more than one way.

**branching fraction.** In branching decay, the fraction of nuclei which disintegrates in a specified way, usually expressed as a percentage.

**brattice.** A board of plank lining, or other partition, in any mine passage to confine the air and force it into the working places. Its object is to keep the intake air from finding its way by a short route into the return airway. Temporary brattices are often made of cloth.

**Brayton cycle.** The Brayton cycle has potential and is a practical alternative to the steam Rankine cycle for solar power and for high-temperature gas-cooled nuclear reactors. The Brayton cycle is most familiar in its open form as used in aircraft gas turbines. The open Brayton cycle cannot compete with steam-Rankine in efficiency. In a power-generation application, cycle efficiencies on the order of 20 percent would be expected. However, the Brayton cycle can achieve higher efficiency through recuperation, sometimes called regeneration. The working fluid is an inert gas, typically helium. Inert gas mixtures, such as helium-xenon, have been studied and have potential advantages. The recuperated Brayton cycle approaches Carnot efficiency in the ideal limit. As compressor and turbine work are reduced, the average temperatures for heat addition and rejection approach the cycle limit temperature. The limit is reached as compressor and turbine work (and cycle pressure ratio) approach zero and fluid mass flow per unit power output approaches infinity. It can be expected from this that practical, recuperated Brayton cycles would operate at relatively low pressure ratios, but be very sensitive to pressure drop.

**Brayton cycle engine.** Turbine cycle engines using internal heat sources, usually from the burning of fossil fuels.

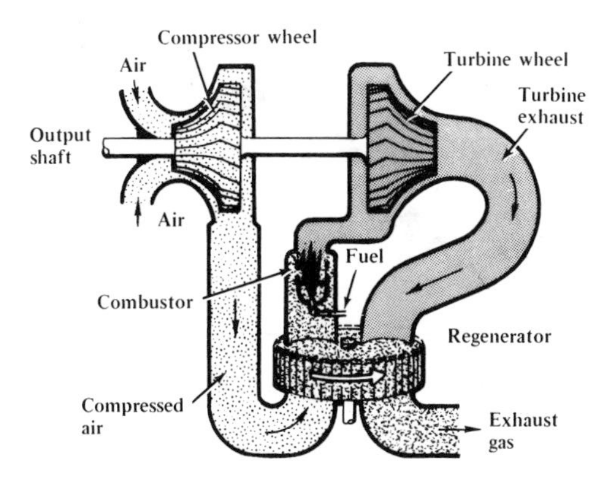


Brayton cycle engine. (*Courtesy Popular Science, April 1976, p. 84*)

**Brayton engine.** Gas turbine engine employing a split shaft, with the compressor turbine physically separated from the power turbine connected to the transmission. With single-shaft design, the compressor and power turbines turn together. Air enters the compressor and is ducted through two heat exchangers heated by exhaust gases. The heated, compressed air enters the combustor, where fuel is continuously injected. The mixture is ignited and expands. Expanding gas turns both the compressor and power turbines.

**breaker.** In anthracite mining, the structure in which the coal is broken, sized, and cleaned for market.

**break-in oil.** Oil for lubricating new engines.

**breakthrough.** Interception of an underground mine by surface or auger mining; also, the point at which a drill bit leaves the rock and enters a natural or a man-made opening; also, in leaching, the arrival of traces of uranium in the final ion-exchange column during the adsorption cycle.

**breeching.** A passageway, usually constructed of sheet metal, for conveying the smoke from the flue gases to the smokestack.

**breeder.** A nuclear reactor that produces more fuel than it consumes. Breeding is possible because of two facts of nuclear physics: (1) Fission of atomic nuclei produces on the average more than two neutrons for each nucleus undergoing reaction. In simplified terms, then, one neutron can be used to sustain the fission chain reaction and the excess neutrons can be used to create more fuel. (2) Some nonfissionable nuclei can be converted into fissionable nuclei by capture of a neutron of proper energy. Nonfissionable uranium-238, for example, can thus be bred into fissionable plutonium-239 upon irradiation with high-speed neutrons.

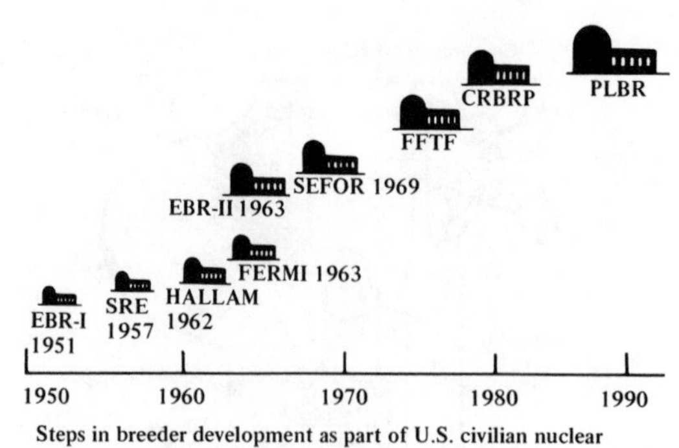

Steps in breeder development as part of U.S. civilian nuclear power program.

Breeder reactor development. (*Courtesy Breeder Reactor Corporation*)

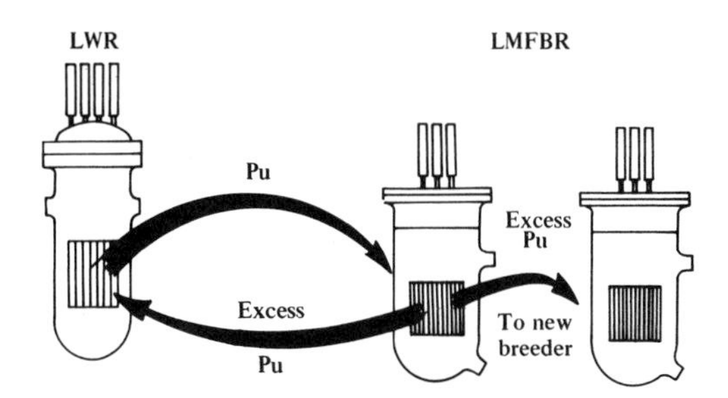


Breeder reactors will create enough new nuclear fuel to supply themselves and new LMFBR's coming on line as well as present light water nuclear power plants. (*Courtesy Breeder Reactor Corporation*)

**breeding ratio.** The conversion ratio when it is greater than the unity; the ratio of the number of fissionable atoms produced in a breeder reactor to the number of fissionable atoms consumed in the reactor.

**breeze.** In general, any light or moderate wind blowing from land to sea or from sea to land. In the coal industry, the term refers to coke of small size or the dust from coke or coal.

**bremsstrahlung.** Electromagnetic radiation emitted (as photons) when a fast-moving charged particle (usually an electron) loses energy upon being accelerated and deflected by the electric field surrounding a positively charged atom nucleus. X-rays produced in ordinary x-ray machines are bremsstrahlung.

**bridge.** A device to measure the resistance of wire or other conductor forming a part of an electric circuit.

**bright annealing.** Annealing in a protective medium to prevent discoloration of the bright surface.

**bright coal.** The constituent of banded coal which is of a jet black, pitchy appearance. It is more compact than dull coal, and breaks with a conchoidal fracture when viewed macroscopically. In thin section, bright coal always shows preserved cell structure of woody plant tissue, either of stem branch or root.

**bright stock.** High-viscosity, fully-refined and dewaxed lubricating oils produced by the treatment of residual stocks and used to compound motor oils.

**brine.** A strong saline solution, such as calcium chloride, used in refrigeration.

**briquet.** A block of compressed coal dust used as fuel.

**British thermal unit (Btu).** The amount of heat required to raise the temperature of one pound of water one degree Fahrenheit under stated conditions of pressure and temperature (equal to 252 calories, 778 foot-pounds, 1055 joules, and 0.293 watt-hours). It is the standard unit for measuring quantity of heat energy.

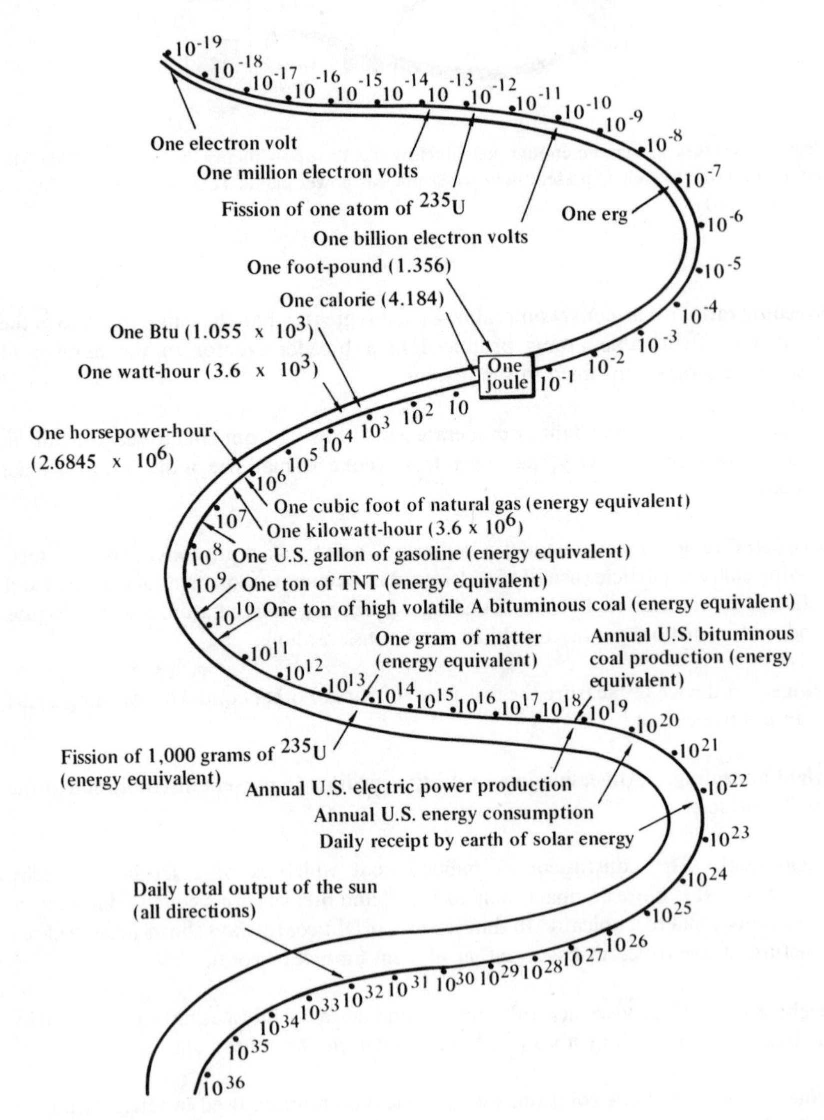


Spectrums of various energy quantities. (*Courtesy Van Nostrand's Scientific Encyclopedia*)

**broad-beam attenuation.** The portion of the unscattered and some of the scattered radiation that reaches the detector.

**brown coal.** A low-rank coal which is brown, brownish-black, but rarely black. It commonly retains the structures of the original wood. It is high in moisture, low in heat value, and checks badly upon drying.

**brownout.** The lowering of voltage by electric utility companies when demand for power exceeds generating capacity.

**brushes.** In a generator, spring-loaded flanges made of carbon that brush against the rotor, thereby collecting the electricity produced.

**BTX.** Benzene, and its derivatives, toluene (methyl benzene) and xylene (dimethyl benzene); aromatic hydrocarbons.

**bubble chamber.** A device used for detection and study of elementary particles and nuclear reactions. Charged particles from an accelerator are introduced into a superheated liquid so that each forms a trail of bubbles along its path. The trails are photographed, and by studying the photograph scientists can identify the particles and analyze the nuclear events in which they originate.

**bubble tower.** A closed cylindrical tower arranged with shelves on which is absorbing oil. The distilled gas is caused to bubble, and the heavier fractions of gas are absorbed.

**bucket-wheel excavator.** A continuous mining machine which uses scoops mounted in a circular rotating frame to remove overburden and deposits.

**buckling.** Producing a bulge, bend, or other wavy condition in sheets or plates by compressive stresses.

**building energy ratio.** The space-conditioning load of a building.

**building overall energy loss coefficient—area product.** The factor which when multiplied by the monthly degree-days yields the monthly space heating load.

**bulk density.** The ratio of the weight of a collection of discrete particles to the volume which it occupies.

**bulkhead.** In mines, a tight partition of wood, rock, and mud or concrete for protection against gas, fire, and water.

**bulk mining.** A method of mining in which large quantities of low-grade ore are mined without attempt to segregate the high-grade portions.

**bulk plant.** A wholesale distributing unit for petroleum products.

**bull plug.** A plug that is inserted into the end of an unfinished pipeline to keep out dirt and moisture.

**bunker oil.** A heavy fuel oil used by ships, industry, and large scale heating installations, formed by stabilization of the residual oil remaining after the cracking of crude petroleum.

**burden.** All types of rock or earthy materials overlying bedrock; also overburden.

**Bureau of Mines.** A government agency in the U.S. Department of the Interior concerned with the conservation and utilization of mineral resources and with the health and safety regulations in the mining industry.

**burnable poison.** A neutron absorber (or poison), such as boron, which, when purposely incorporated in the fuel or fuel cladding of a nuclear reactor, gradually "burns up" (is changed into nonabsorbing material) under neutron irradiation. This process compensates for the loss of reactivity that occurs as fuel is consumed and fission-product poisons accumulate, and keeps the overall characteristics of the reactor nearly constant during its use.

**burner capacity.** The maximum British thermal units per hour that can be released by a burner with a stable flame and satisfactory combustion.

**burner reactor.** A converter reactor; a nuclear reactor that produces some fissionable fuel, but less than it consumes; also a nuclear reactor that produces fissionable material different from the fuel burned, regardless of the ratio.

**burner unit.** An assembly of one or more burner heads receiving gas through a single set of control valves.

**burning oil.** A common name for kerosine.

**burning point.** The temperature at which a volatile oil in an open vessel will ignite from a match held close to its surface.

**burnout point.** For a liquid cooled reactor, any combination of values of heat-transfer parameters which results in fuel burnout.

**burnout ratio.** A computed quantity used to establish safe design limits.

**burnup.** A measure of nuclear reactor fuel consumption. It can be expressed as either the percentage of fuel atoms that have undergone fission or the amount of energy per unit weight of fuel fissioned.

**burnup figure of merit.** For a reactor, this may be expressed either as a percent of the fuel that is consumed before fuel elements have to be replaced; or in terms of the megawatt-days of energy obtained per unit mass of fuel in a charge.

**burst cartridge detection (BCD).** System used in gas cooled reactors to sense gaseous fission products which may have escaped from exposed uranium.

**busbar.** A heavy metal conductor for high-amperage electricity.

**bushings.** A fitting for the purpose of connecting dissimilar-size pipes; also a metal cylinder between a shaft and a support or a wheel that serves to reduce rotating friction and to protect the parts.

**butane.** Bottle gas; a compound generally stored and delivered in liquefied form and used as a fuel in gaseous form, obtained either by processing natural gas as produced or from a process in petroleum refining. Chemical formula $C_4H_{10}$.

**butane-air plant.** A gasification plant for preparing liquid butane for customer use.

**butt.** Opposite of face; coal exposed at right angles to the face, and in contrast to the face, generally having a rough surface; also to bring two flat surfaces together.

**Buys Ballot's law.** Statement of the principle of winds; the law that when the observer has his back to the wind, lower barometric pressure is to his left in the northern hemisphere, and to his right in the southern hemisphere. This is due to the earth's rotation.

**bypass.** Usually refers to an extra gas pipe connection around a valve or other control mechanism to prevent a complete stoppage of the flow of gas when the valve or mechanism is closed while adjustments or repairs are made on the control which is bypassed.

**byproduct material.** Any radioactive material (except source material or fissionable material) obtained during the production or use of source material or fissionable material. It includes fission products and many other radioisotopes produced in nuclear reactors.

**byproducts (residuals).** Secondary products, such as coke, tar, and ammonia, or ammonium sulphate.

# C

**cable system.** One of the well-known drilling systems, sometimes designated as the American or rope system. The drilling is performed by a heavy string of tools suspended from a flexible manila or steel cable to which a reciprocating motion is imparted by an oscillating "walking beam" through the suspension rope or cable.

**cable tool drilling.** One of two principal methods of drilling for gas and oil; the other is rotary. Cable tool, the older method, consists of raising and dropping a heavy drill bit, suspended from the end of a cable, so that it pounds and pulverizes its way through the subsurface structures. Water in the hole keeps the cuttings in suspension for removal at intervals by bailing.

**cadmium ratio.** The ratio of the neutron-induced saturated activity in an unshielded foil to the saturated activity of the same foil when it is covered with cadmium.

**cage.** Mining term for elevator.

**caisson.** A cylindrical steel section of shaft, used for sinking through running or waterlogged ground; a chamber used for the purpose of gaining access to the bed of a stream.

**calorie.** Originally, the amount of heat energy required to raise the temperature ot 1 gram of water 1 degree Centigrade. Because this quantity varies with the temperature of the water, the calorie has been redefined in terms of other energy units. One calorie is equal to 4.2 joules. (When capitalized, Calorie means 1000 calories (or 1 kcal). Therefore, one Calorie is equal to 4184 joules.

**calorific intensity.** The temperature of a fuel attained by its combustion.

**calorific power.** The quantity of heat liberated when a unit weight or a unit volume of a fuel is completely burned.

**calorifics.** The science of heating.

**calorific value.** The amount of heat liberated by the combustion of a unit quantity of a fuel (solid or liquid).

**calorimeter.** Any apparatus for measuring the amount of heat generated in a body, or emitted by it.

**cam.** A rotating piece, either noncircular or eccentric, used to convert rotary into reciprocating motion; often of irregular outline and giving motion that is irregular in direction, rate, or time.

**can.** A sealed container for nuclear fuel; in a nuclear reactor, the container in which fuel rods are sealed to protect the fuel from corrosion and to prevent gaseous diffusion products from escaping into the coolant.

**candlepower.** The international standard unit of luminous intensity, defined as 1/60 of the intensity of 1 square centimeter of a blackbody radiator at the temperature of solidification of platinum. It is used as a measure of illuminants.

**candu reactor.** Canadian deuterium moderated nuclear reactor capable of more efficient fuel utilization than a light water reactor.

**cannel coal.** Old term for a coal burning with a steady luminous flame. The term is now used for sapropelic coal containing spores. When viewed microscopically, it shows no stratification. It is generally dull and has a more or less pronounced waxy luster.

**canyon.** In nuclear technology, a space enclosed with heavy shields constituting a part of a building used for fuel reprocessing.

**capability margin.** The difference between net electrical system capability and system maximum (peak) load requirements.

**capacitance rate.** Mass flow rate times specific heat of the fluid flowing through a component such as a heat exchanger.

**capacitor.** Electrical appliance working on the condensor principle. Two conducting plates are separated by an insulating layer. When alternating current is applied, the capacitor is adjusted so that its leading current balances the lag of the circuit giving a highpower factor.

**capacity.** The maximum power output or load for which a machine, apparatus, station, or system is rated. (See Fig. p. 66)

**capacity factor.** The ratio of the amount of electricity produced by a plant or system to its maximum theoretical productive capacity. A measure of utilization of capacity defined in the electric power industry to be the ratio of net kWh production for the year divided by 8760 to the design capacity in kWh.

**capacity of the wind.** The total amount of detrital material of a given kind that can be sustained (per unit volume of air) by a wind of a given velocity. In the aggregate, wind transports more material than water, although water at the same speed of flow is capable of transporting much larger particles. During a dust storm, the wind may carry from 160 short tons up to 126,000 short tons per cubic mile of air.

**capillary water.** Water held above the water table in soil by capillary force.

**capital intensive.** Requiring heavy capital investment. The energy industry, for

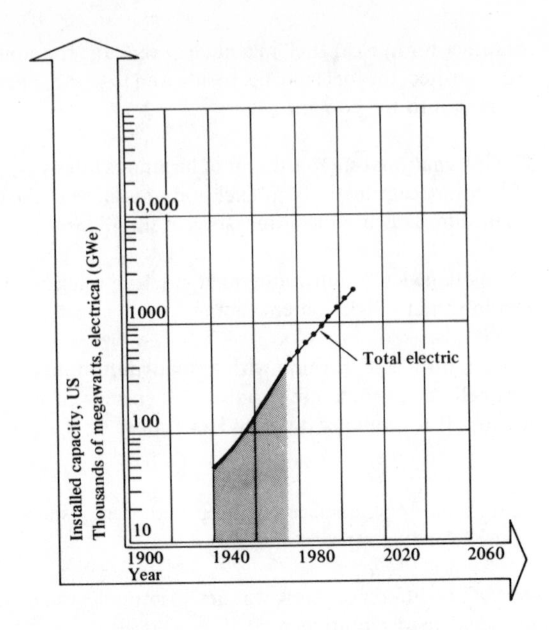


Projected electrical generating capacity. (*Courtesy Electric Power Research Institute and the Applied Nucleonics Co., Inc.*)

example, is said to be capital intensive rather than labor intensive because it employs relatively more dollars than people.

**caprock.** An impervious geological stratum immediately overlying an oil- or gas-bearing rock.

**capture.** A process in which an atomic or a nuclear system acquires an additional particle; for example, the capture of electrons by positive ions, or capture of electrons or neutrons by nuclei.

**carbide miner.** A pushbutton mining machine with a potential range of 1000 feet into the seam from the highwall, a maximum production of some 600 tons per shift, and a recovery of 65 to 75 percent of the coal within the reach of the machine.

**carbon.** A nonmetallic element existing in diamonds, graphite, coal, petroleum, asphalt, limestone, and other carbonates, and in all organic compounds. It is also obtained artifically in varying degrees of purity.

**carbonaceous.** Containing carbon or coal, especially shale or other rock containing some particles of carbon distributed evenly throughout the whole mass.

**carbonate.** Rocks composed predominantly of carbon dioxide, such as limestone, dolomite, etc.

**carbon black.** Almost pure amorphous carbon consisting of extremely fine particles, usually produced from burning hydrocarbons such as mineral oils in conditions where combustion is incomplete.

**carbon dioxide.** A compound of carbon and oxygen formed whenever carbon is burned. Chemical formula $CO_2$.

**carbonization.** The destructive distillation of coal accompanied by the formation of char (coke), liquid (tar), and gaseous products.

**carbon mass transfer.** The transport of carbon by a fluid from one point of a circuit to another.

**carbon monoxide.** A compound of carbon and oxygen produced by the incomplete combustion of carbon. It is emitted by automobiles and is the major air pollutant on the basis of weight. Chemical formula CO.

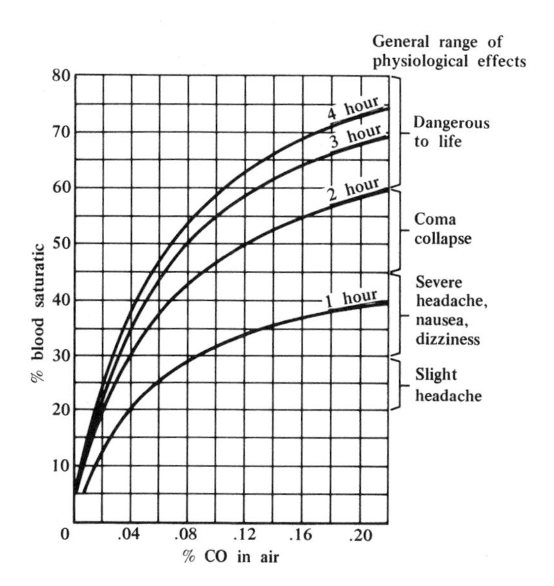


Carbon monoxide effects on human beings. (*Courtesy Van Nostrand's Scientific Encyclopedia*)

**carbon oxides.** Compounds of carbon and oxygen produced when the carbon of fossil fuels combines with oxygen during burning. The two most common such oxides are carbon monoxide, a very poisonous gas, and carbon dioxide.

**carbon zinc cell battery.** A cell for the production of electric energy by galvanic oxidation of carbon. It is commonly used in household appliances and radios and has a relatively short life. (See Fig., p. 68)

**carcinogen.** A cancer-producing substance or agent.

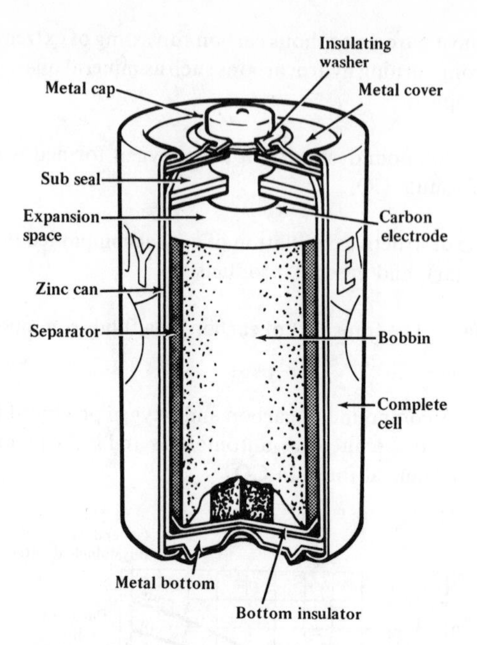


Round carbon-zinc cell battery. (*Courtesy Van Nostrand's Scientific Encyclopedia*)

**Carnot cycle.** An ideal heat engine cycle, conceived by Sadi Carnot, in which the sequence of operations forming the working cycle consists of isothermal expansion, adiabatic expansion, isothermal Compression, and adiabatic compression back to its initial state. An ideal Carnot cycle engine converts heat into work with the maximum theoretical efficiency.

**Carnot efficiency.** The maximum efficiency with which work can be produced from heat in ideal processes. Carnot efficiency is only dependent upon the maximum and minimum temperatures available.

**carrier.** A stable isotope, or a normal element, to which radioactive atoms of the same element can be added to obtain a quantity of radioactive mixture sufficient for handling, or to produce a radioactive mixture that will undergo the same chemical or biological reaction as the stable isotope; a substance in weighable amount which, when associated with a trace of another substance, will carry the trace through a chemical, physical, or biological process.

**cascade.** A connected arrangement of units of equipment for separation of isotopes. A single device or process usually can produce only a small amount of isotopic separation, but, if a number of these are connected together, the effect can be multiplied and a significant amount of separation achieved. An example is a cascade of barriers for the gaseous diffusion process; also a nuclear cascade of short lived isotopes.

**Cascade Improvement Program (CIP).** A program initiated by the Atomic Energy Commission (now DOE) in an effort to expand gaseous diffusion plant production to increase the supply of enriched uranium.

**cascade tails assay.** The concentration of one or more isotopes in the waste product of a cascade.

**case.** A small fissure, admitting water into the mine working.

**casing.** A zone of material altered by vein action and lying between the unaltered country rock and the vein; special steel tubing lowered into a borehole to prevent entry of loose rock, gas, or liquid, or to prevent loss of circulation liquid; also piping used to support the sides of a borehole.

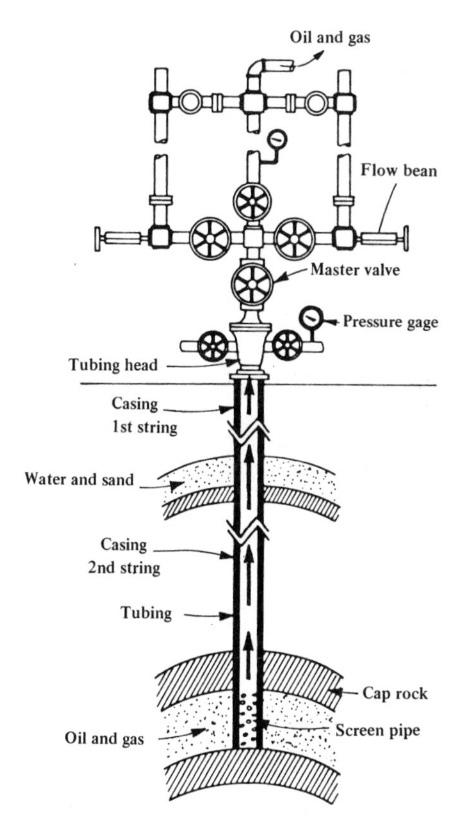


Casing enclosing the drill pipe in a master valve and pipe assembly at top of well. (*Courtesy Van Nostrand's Scientific Encyclopedia*)

**casinghead gas.** Natural gas, rich in oil vapors, usually collected or separated from the oil at the casinghead. It is also called combination gas, wet gas, or Bradenhead gas.

**cask.** A heavily shielded container used to store and/or ship radioactive materials.

**catabolism.** Chemical and physical processes within living organisms involving the release of energy.

**catalysis.** Modification (especially an increase in the rate) of a chemical reaction induced by material unchanged chemically at the end of the reaction; any reaction brought about by a separate agent.

**catalyst.** A substance that changes the rate of a reaction without itself undergoing any net change; a substance that induces catalysis.

**catalytic converter.** An air pollution abatement device used to reduce nitrogen oxide emissions from motor vehicles.

**catalytic cracking.** A refinery process that converts a high-boiling range fraction of petroleum (gas oil) to gasoline, olefin feed for alkylation, distillate, fuel oil, and fuel gas by use of a catalyst and heat.

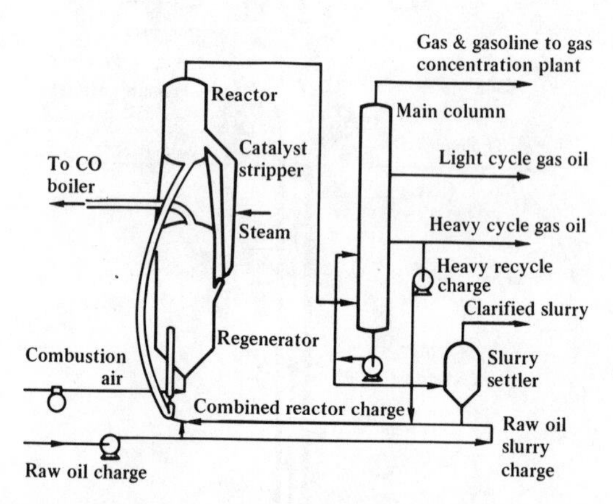


Catalytic cracking process. (*Courtesy U.S. Council on Environmental Quality*)

**catalytic hydrogenation of coal tar.** The process of converting sulfur-bearing coal into nonpolluting fuel.

**catalytic reforming.** A catalytic process to improve the anti-knock quality of low-grade naphthas and virgin gasolines by the conversion of naphthenes (such as cyclohexane) and paraffins into higher-octane aromatics (such as benzene, toluene, and xylenes). There are approximately ten commercially licensed catalytic reforming processes, including fully regenerative and continuously regenerative designs.

**cathode.** The electrode where electrons enter (current leaves) an operating system such as a battery, an electrolytic cell, an x-ray or a vacuum tube. In a battery or electrolytic cell, it is the electrode where reduction occurs.

**cathode protection.** A method of controlling the corrosion of steel pipe and connected metallic equipment through the use of electrolysis.

**cathode rays.** A stream of electrons emitted by the cathode or negative electrode of a gas-discharge tube, or by a hot filament in a vacuum tube such as a television tube.

**caulking.** Latex or elastomeric materials used to reduce air leaks by packing cracks or air-space openings. Caulking is commonly used wherever two different materials or parts of the house meet, such as around windows, door frames, and siding joints.

**caustic.** Capable of destroying texture or eating away substance by chemical action; burning; corrosive.

**cave.** A chamber beneath the surface of the earth; in nuclear technology, a shielded cavity for storing radioactive materials; also, a hot cell.

**ceiling panel heating.** A system using ceiling panels as heating surfaces.

**cell.** A battery unit consisting of two electrodes separately contacting an electrolyte so that there is a potential difference between them; with respect to solid waste disposal, earthen compartments in which solid wastes are dumped.

**cellulosic fiber.** A form of loose home insulation that is blown in for installation. It is more effective than mineral fiber insulations such as rock wool and glass fiber.

**Centigrade (or Celsius) scale.** A temperature scale that takes the melting point of ice as 0° C and the boiling point of water as 100° C. It is used by scientists throughout the world and by laymen as well in those countries where the metric system is used. To convert Centigrade to Fahrenheit, multiply the reading by 9, divide by 5, and add 32.

**Central Receiver Power Plant.** Pilot power plant to be constructed under DOE sponsorship. The objective of the Central Receiver Power Plant project is to make possible by the 1990's the use of solar power plants to produce supplementary electric power to meet utility systems requirements for peak and intermediate-load demands.

**central-receiver systems.** The central receiver system focuses all of the incoming direct solar radiation on a single point rather than along the length of an absorber pipe. Consequently, significantly higher temperatures are attainable with this system than with line-focusing cylindrical distributed collectors. These higher temperatures, currently upwards of 540° C, have resulted in greater overall conversion efficiencies of approximately 24 percent. The basic system configuration uses arrays of tracking mirrors called heliostats, which direct solar energy to a single large receiver mounted atop a tower. The working fluid contained in the receiver is heated by the focused solar energy and then piped to the turbogenerator unit (usually located at the base of the tower) where it powers, either directly or indirectly, a conventional turbine and thus generates electricity.

**centrifugal collector.** Mechanical system employing centrifugal force for removing aerosols from a gas stream.

**central station power.** Production of power—usually electrical—in large quantities at a generation plant as opposed to production at the point of consumption.

**centrifugal separator.** A device which separates two fluids, or a fluid and a solid of different density, by rotating them rapidly and forcing the denser material to the outside.

**centrifuge.** A high speed rotating device for separating liquids of different specific gravities or for separating suspended colloidal particles according to particle-size fractions by centrifugal force.

**ceramic fuel.** Fuel consisting of refractory compounds.

**ceramic radiants.** Baked clay devices which radiate heat.

**Cerenkov radiation.** Light emitted when charged particles pass through a transparent material at a velocity greater than that of light in that material. It can be seen, for example, as a blue glow in the water around the fuel elements of pool reactors.

**cermet.** A combination of the words ceramic and metal to identify mixtures of these materials for use at high temperatures. Used in fabricating ceramic fuels, such as uranium-plutonium cermets, which offer the dual advantages of high-temperature use and resistance to radiation damage.

**cermet fuel.** Fuel consisting of a mixture of metallic materials and refractory compounds.

**chain fission yield.** Fissions giving rise to nuclei of a particular mass number.

**chain reaction.** Any process of molecular or nuclear reaction in which one nuclear transformation triggers off a whole series.

**Chance process.** A method of cleaning coal by using a fluid mixture of sand and water which floats off a clean coal product but allows slate and other impurities to sink.

**Chapman engine.** Chapman engines have developed a new positive displacement, vane-type orbital bottoming cycle engine. After flowing through the engine, hot coolant enters the heat exchanger. Hot exhausts and oil also flow through this heat exchanger. Heat from these sources, normally wasted in conventional automobiles, heat Freon flowing through the exchanger. Freon passes through the throttle valve, expanding in the orbital engine and turning the engine's main shaft. Used Freon condenses in heat exchanger, giving up its heat to cooled coolant coming from the radiator.

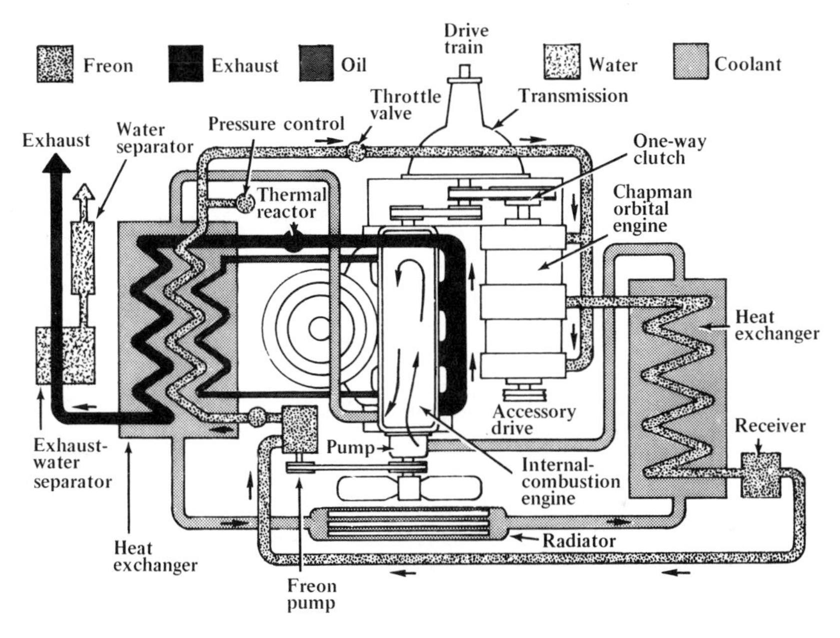


Chapman's orbital engine. (*Courtesy Popular Science, January, 1976, p. 83*)

**char.** A porous, solid residue resulting from the incomplete combustion of organic material. If produced from coal, it is called coke; if produced from wood or bone, it is called charcoal. It is closer to pure carbon than the coal, wood, or bone from which it is produced.

**charcoal.** A dark-colored or black porous form of carbon made from vegetable or animal substances (i.e., by charring wood in a kiln or by excluding air from retort) and used for fuel.

**charge.** The fuel placed in a reactor; in refrigeration, the quantity of refrigerant in a system; also the liquid or solid materials fed into a furnace for its operation.

**charged particle.** An ion; an elementary particle that carries a positive or negative electric charge.

**Charles's law.** The volume of a gas, kept at constant pressure, varies directly as the absolute temperature.

**Char-Oil Energy Development (COED) process.** A process being developed for low-temperature distillation of coal carbonization products. The process is designed to produce clean liquids, gases, and char for fuel, with the product balance depending upon economic factors.

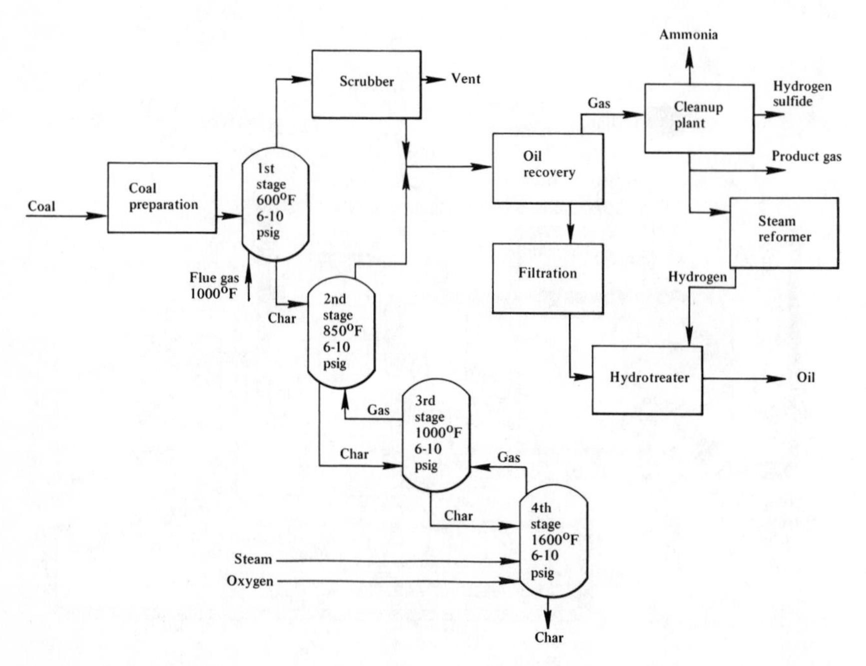


Char-Oil Energy Development (COED) coal liquefaction process. (*Courtesy U.S. Council on Environmental Quality*)

**check dam.** A dam that divides a drainage way into two sections to decrease the streamflow velocity and promote deposition of sediment.

**checks.** Numerous very small cracks in metal or other material caused in processing.

**check valve.** A ball-type valve placed in core barrels, soil samples, or drill rods to control the directional flow of liquids.

**chemical compatibility.** The ability of materials and components in contact with each other to resist mutual chemical degradation, such as that caused by electrolytic action or plasticizer migration.

**chemical decladding.** The use of chemical means to remove cladding.

**chemical dosimeter.** A detector for indirect measurement of radiation by indicating the extent to which the radiation causes a definite chemical change to take place.

**chemical oxygen demand (COD).** The amount of oxygen required to convert (oxidize) organic compounds into stable forms—usually carbon dioxide and water; includes all compounds requiring oxidation.

**chemical shim.** A chemical, usually boric acid, placed in the coolant system of a nuclear reactor to serve as a neutron absorber that compensates for fuel burnup during normal operation. A chemical shim can also compensate for temperature changes in the coolant, for buildup and decay of various elements, and for depletion of fissionable material.

**chemonuclear reactor.** A reactor designed as a radiation source for making chemical transformations on an industrial scale. Also called chemical processing reactor.

**cherry coal.** A soft noncaking coal which burns readily with a yellowish flame; a deep black, dull or lustrous bituminous coal, with a somewhat conchoidal fracture, readily breaking up into cuboidal fragments.

**chilling effect.** The cooling of the earth's temperature because of the increase of atmospheric particulates which prevent penetration of the sun's energy.

**chimney effect.** The tendency of heated air or gas to rise in a duct or other vertical passage, as in a chimney, a small enclosure, or building, due to its lower density compared to the surrounding air or gas.

**chimney lid.** Device used on chimney, which operates by remote control to open or close and prevent heat loss through some opening such as a fireplace.

**Chinook.** A warm, dry wind that descends the eastern side of the Rocky Mountains, and generally blows from the southwest.

**choke.** A device designed to reduce pressure and/or control production from a gas well or an oil well.

**chokedamp.** A mine atmosphere that causes choking, or suffocation, because of insufficient oxygen.

**chopper.** A rotating shutter for interrupting an otherwise continuous stream of particles. Choppers can release short bursts of neutrons with known energies and are used to measure nuclear cross sections.

**chord.** The distance from the leading to the trailing edge of an airfoil.

**Christmas Tree.** The assembly of valves, pipes, and fittings used to control the flow of oil and gas from a well. (See Fig. p. 76)

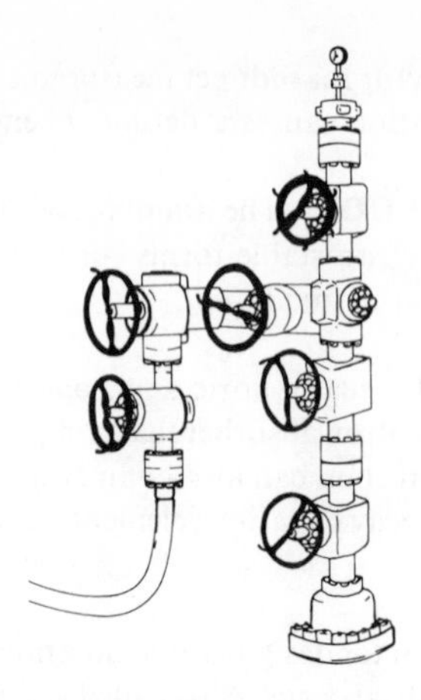

Christmas Tree of control valves at the wellhead. (*Courtesy U.S. Council on Environmental Quality*)

**chromosome.** The determiner of heredity within a cell.

**chute.** A channel or shaft underground, or an inclined trough above ground, through which ore falls or is shot by gravity from a higher level to a lower level.

**circuit.** A conducting part or a system of conducting parts through which an electric current flows or is intended to flow.

**circuit breaker.** An overload protective device to interrupt the flow of electric current when it becomes excessive or exceeds a predetermined value.

**circulated gas-oil ratio.** The number of cubic feet of gas introduced into the well for gas-lift operations per barrel of oil lifted.

**circulating-fuel reactor.** A reactor design in which the fuel circulates through the core.

**cladding.** The outer jacket of nuclear fuel elements which prevents corrosion of the fuel and the release of fission products into the coolant. Aluminum or its alloys, stainless steel and zirconium alloys are common cladding materials.

**clarifier.** A centrifuge or other device for separating suspended solid matter from a liquid. It is used in wastewater treatment to clean the water of some suspended solids. Rotary scrapers in square or circular tanks are used to move the sludge in the water toward the center of the tank where it is removed by pumping.

**classifications of fuel oil.** The grading of fuel oils referred to as light, medium, and heavy, or domestic and industrial.

**Claus process.** A process for recovering elemental sulfur from hydrogen sulfide gas at high temperatures with oxygen reacting with the hydrogen sulfide to yield dry sulfur and steam.

**Claus recovery plant.** A Claus plant takes emission gas streams containing 10 percent or more hydrogen sulfide and oxidizes the hydrogen sulfide, producing elemental sulfur of high purity.

**clean.** In reactor technology, the absence of induced radioactivity and fission products.

**clean bomb.** A nuclear bomb that produces relatively little radioactive fallout. A fusion bomb.

**Clean Coke Plus Liquids process.** A process wherein crushed coal is split into two fractions. One fraction is processed in a carbonization unit where it is devolatilized and partially desulfurized to produce char which is further processed to metallurgical coke. The second portion is slurried with recycled oil and processed in a hydrogenation unit which produces liquid and gas products. Liquid products from both the carbonization and the hydrogenation unit are treated in a liquid processing unit to get liquid fuels and chemical feedstocks. Operating conditions are 649°–760°C and 9 to 100 pounds per square inch in the carbonizer. Operating conditions are 482°C and 3000 to 4000 pounds per square inch in the hydrogenation unit.

**Clean Coke process.** Process that combines carbonization and hydrogenation reactions to convert nonmetallurgical-grade coals to low-sulfur metallurgical coke, chemical feedstocks, and liquid and gaseous fuels.

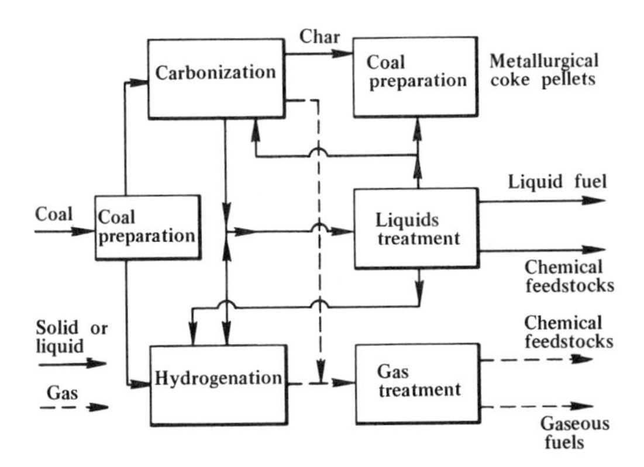


Simplified diagram of Clean Coke process. (*Courtesy Van Nostrand's Scientific Encyclopedia*)

**cleaned coal.** Coal produced by a cleaning process (wet or dry).

**clean fuel.** Usually means fuel in which there is very little sulfur.

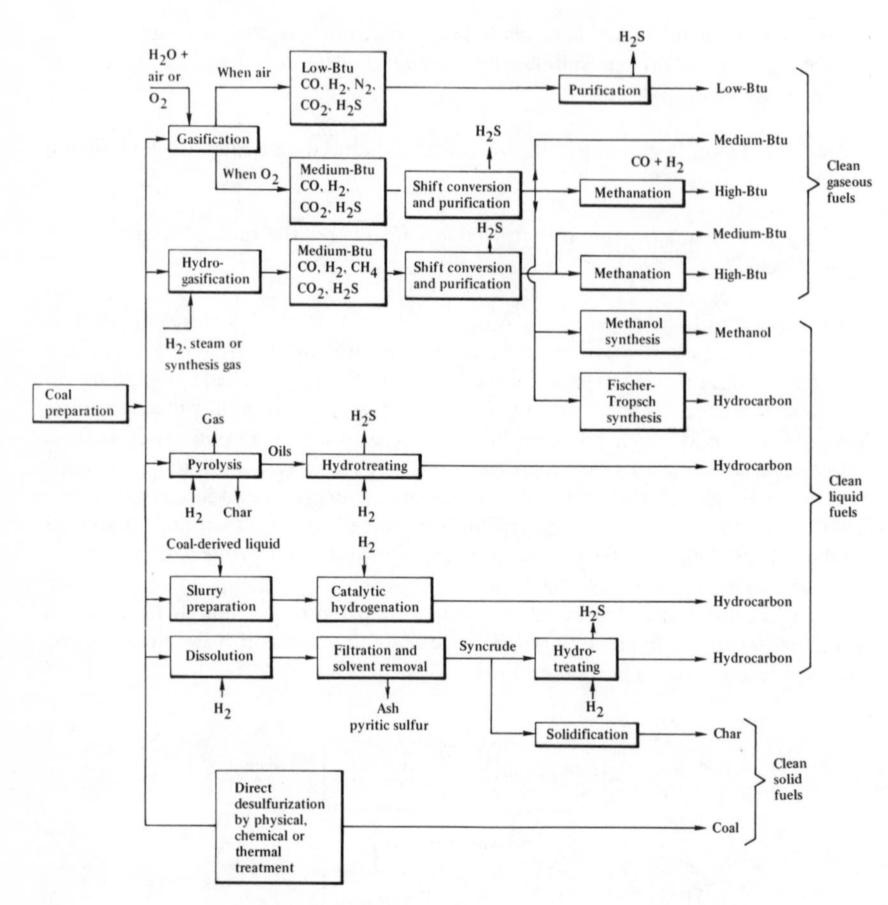


Clean fuels from coal. (*Courtesy U.S. Energy Research and Development Administration, Division of Coal Conversion and utilization*)

**cleaning coal.** Coal may be cleaned in any of a number of ways. Originally, it was cleaned by hand (hand picking). However, a modern cleaning plant may use a variety of methods depending on the size and ultimate use of the coal. For example, coarse coal may be washed in special units designed to separate the coal from the heavier rock and other refuse. This coal is passed on to a dryer and the refuse discarded. Fine coal and refuse may be separated by uniform or pulsating air currents (dry process) or by the use of special frothing agents in a wet separator. (See Fig. p. 79)

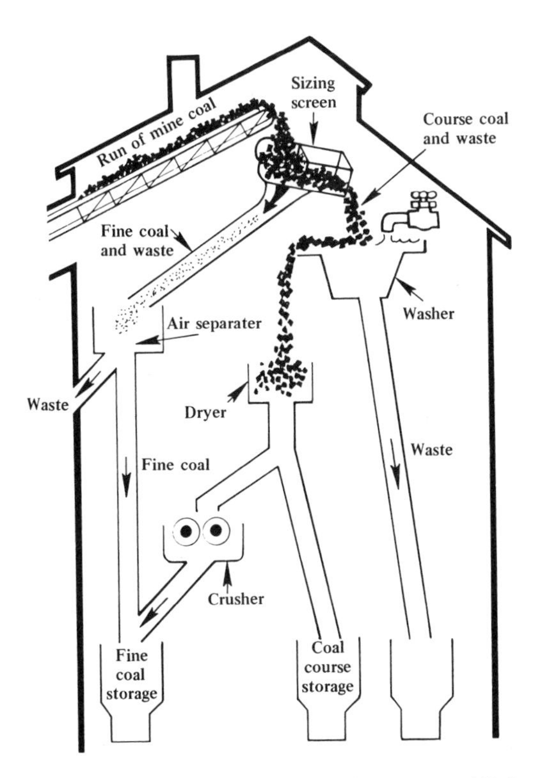


Diagram of a coal cleaning plant. (*Courtesy Mining Enforcement and Safety Administration*)

**cleaning plant refuse.** The difference in volume between total raw coal sent to the preparation plant and clean coal received from the plant.

**clear octane.** The octane number of a gasoline before the addition of antiknock additives.

**cleat.** Main joint in a coal seam along which it breaks most easily. It runs in two directions, along and across the seam.

**Clinch River Breeder Reactor Plant.** The Clinch River Breeder Reactor Plant (CRBRP) project is the joint government and industry effort to build the Nation's first large-scale (350–400 megawatt electrical) demonstration breeder nuclear power plant. CRBRP is designed to demonstrate the commercial potential and enviromental advantages of a large-scale liquid metal fast breeder reactor (LMFBR) as a source of electrical generation in a utility environment. (See Fig., p. 80)

Photograph of the Clinch River breeder reactor plant. (*Courtesy L. McCord for Burns & Roe, Inc.*)

**closed cycle.** A thermodynamic power cycle in which the working fluid is recycled; also applicable to a cooling system in which the coolant is cycled repeatedly through the source of heat, itself being cooled in another part of the cycle.

**closed-cycle reactor system.** A reactor design in which the primary heat of fission is transferred outside the reactor core to do useful work by means of a coolant circulating in a completely closed system that includes a heat exchanger.

**closed-loop solar heating system.** A solar heating system wherein the pipes, circulating liquid from the collectors to the storage vehicle, form a closed, continuous loop.

**closed water piping system.** A heating system which utilizes an air tank as a means of pressurizing the system, and circulating water as a heat medium.

**cloud chamber.** A device in which the tracks of charged atomic particles, such as cosmic rays or accelerator beams, are displayed. It consists of a glass-walled chamber filled with a supersaturated vapor, such as wet air. When charged particles pass through the chamber, they trigger a process of condensation, and so produce a track of tiny liquid droplets, much like the vapor trail of a jet plane. This tract permits scientists to study the particles' motions and interactions.

**$CO_2$ Acceptor process.** Two fluidized-bed reactors are used to convert highly reactive coals, such as lignite and subbituminous coal, into a medium-Btu gas which can be upgraded to pipeline quality gas. Coal fed into the gasifier is devolatilized and then gasified with steam. Heated calcium oxide provides reaction heat and combines with (acceptor) $CO_2$ from reaction products. Char and calcium carbonate products from the gasifier are fed to the regenerator where char supported combustion reverses the acceptor reaction, thereby recycling the calcium oxide acceptor to the gasifier. The process operates at a pressure of 150 psi with a gasifier temperature of 1500° F and a regenerator temperature of 1870° F.

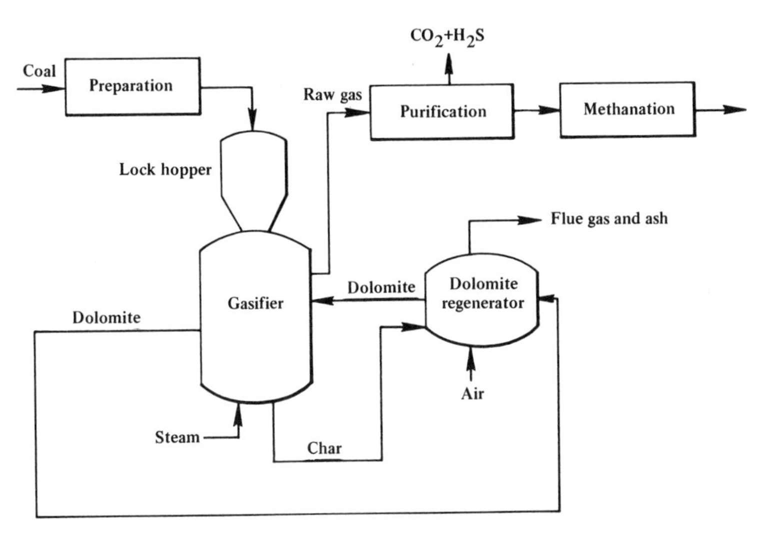


$CO_2$ acceptor coal gasification process. (*Courtesy U.S. Council on Environmental Quality*)

**coagulation.** The union of fine particles to form larger particles; also the clumping of particles in order to settle out impurities.

**coal.** A solid, brittle, stratified, combustible carbonaceous rock formed by the decomposition of vegetation; varies in color from dark brown to black; very insoluble. Chemically, coal is composed chiefly of condensed aromatic ring structures of high molecular weight. It is made up of varying quantities of the elements carbon, hydrogen, oxygen, and nitrogen. The amount of each material depends on the depth of the coal below the surface and on other factors. (See Fig., p. 82)

**coal alkylation.** A process to convert sulfur-bearing coal into a non-polluting fuel.

**coal analysis.** The determination, by chemical methods, of the proportionate amounts of various constituents of coal. Two kinds of coal analyses are ordinarily made: (10 proximate analysis, which divides the coal into moisture, volatile matter, fixed carbon, and ash; and (2) ultimate analysis, which determines the percentages of the chemical elements carbon, hydrogen, oxygen, nitrogen, and sulfur.

**coal ash.** Noncombustible matter in coal.

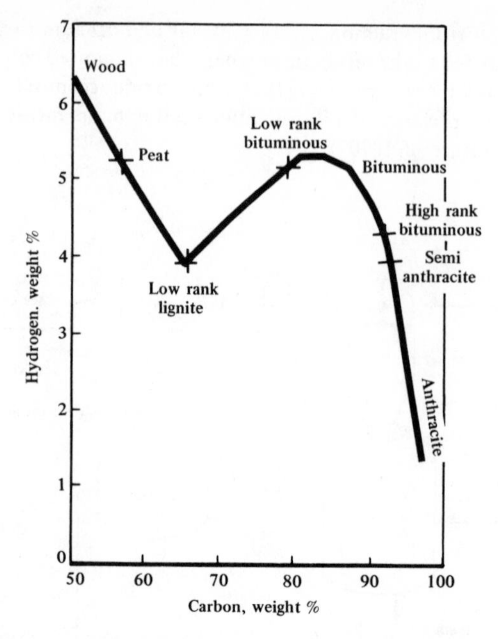


Carbon and hydrogen content of typical American coals, wood, and peat. (*Courtesy Mining Enforcement and Safety Administration*)

**coal augering.** A surface mining method used when coal lies in high walls that were prepared for this operation or were left by stripping, or when the coal outcrops to the surface. The mining machines consist of large single and double augers which drill horizontally into the seams to extract the coal.

**coal bank.** An exposed seam of coal.

**coal basin.** Depressions in older rock formations in which coal-bearing strata have been deposited.

**coalbed.** A bed or stratum of coal. Coal seam is more commonly used in the United States and Canada.

**coal blasting.** A method of breaking coal with explosives by either blasting cut coal and blasting off the solid (the method most commonly used), or grunching.

**coalbreaker.** A building containing the machinery for breaking coal with toothed rolls, sizing it with sieves, and cleaning it for market; also a machine for breaking coal.

**coal breccia.** Coal broken into angular fragments by natural processes occurring within the coal bed.

**coal briquettes.** Coal made more suitable for burning by a process which forms it into regular square- or oval-shaped pieces.

**coal classification systems.** In all countries the basis for classification is content of volatile matter. Anthracite is 10 percent volatile; lean coal, semi-anthracite, or dry-steam coal is 10–13 percent volatile; variously designed coal is 14–20 percent volatile; coking coal is 20–30 percent volatile.

**coal cleaning plant.** A plant where raw or run-of-mine coal is washed, graded, and treated to remove impurities and to reduce ash content.

**Coalcon process.** Low-temperature, intermediate-pressure process for hydrocarbonization of finely divided low-rank coal or high-boiling tars in a fluidized bed to produce chars, tars, and gases. It was originally designed for a subbituminous coal having high tar and potentially high phenolic yields during carbonization, but it is currently being developed for high-sulfur, high-volatile bituminous coals.

**coal conversion.** Converting coal to liquid and gaseous fuels through the process of increasing the hydrogen-to-carbon ratio of the coal by breaking down the complex coal molecule to simpler molecules and adding hydrogen to the mixture.

**coal costs.** Since coal is purchased on the basis of its heat content, its cost is measured by computing the "cents per million Btu" of the fuel consumed. It is the total cost of fuel consumed divided by its total Btu content, and the answer is multiplied by one million.

**coal cutter.** The longwall chain coal cutter, now almost universal, is a power-operated machine which draws itself by rope haulage along the face, usually cutting out a thin strip of coal from the bottom of the seam, in preparation for shot firing and loading or a cutter loader

**coal deposits.** Also called beds, seams, and veins. They range from a fraction of an inch to several hundred feet in thickness. The differences in coals are due to age, pressure (folding and/or depth of burial), and heat, which may have been supplied by transecting dikes or by movement in the rocks.

**coal drill.** Usually an electric rotary drill of a light, compact design; may also be a light percussive drill operated by compressed air, or a hand-operated drill.

**coal dryer.** A plant or vessel in which water or moisture is removed from fine coal by dewatering classifiers, or by vacuum filtration.

**coal equivalent or other fuels burned.** The Btu content of other fuels divided by the representative heat value per ton of coal burned.

**coal fuel ratio.** The content of fixed carbon divided by the content of volatile matter is called the fuel ratio. According to their fuel ratios, coals have been classed: anthracite, not less than 10; semianthracite, 6 to 10; semibituminous, 3 to 6; and bituminous, 3 or less.

**coal gas.** Flammable gas derived from coal either naturally in place or by induced methods of industrial plants and underground gasification.

**coal gasification.** The conversion of coal (a solid) to a gas which is suitable for use as a fuel. The basic process involves crushing coal to a powder, then heating this material in the presence of steam and oxygen. The gas produced is then refined to reduce the content of sulfur and other impurities and to increase the methane content. Also under investigation are methods of burning coal in situ; that is, burning the coal while it remains underground to produce gas. This eliminates the need for physically recovering coal from areas difficult or impossible to mine.

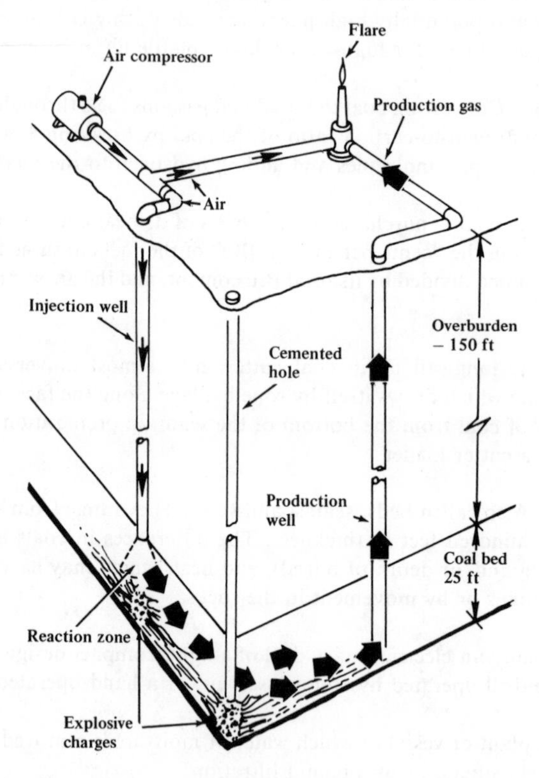


Drawing showing the main features of a coal gasification concept. (*Courtesy Lawrence Livermore Laboratory*)

**coal horizon.** The stratigraphic position where a coal should occur.

**coalification.** Those processes involved in the genetic and metamorphic history of coalbeds. The plant materials that form coal may be present in vitrinized or fusinized form. Materials contributing to coal differ in their response to diagenetic and metamorphic agencies. The three essential processes of coalification are called incorporation, vitrinization, and fusinization.

**coalite; semicoke.** A trade name for a smokeless fuel produced by carbonizing coal at a temperature of about 600° C. It has a calorific value per pound of about 13,000 British thermal units, and is used for domestic purposes.

**coal liquefaction.** The conversion of coal into liquid hydrocarbons and related compounds by hydrogenation at elevated temperatures and pressures. This involves putting pulverized bituminous coal into an oily paste, which is treated with hydrogen gas under appropriate conditions of temperature and pressure, to form the liquid molecules of carbon and hydrogen which constitute oil.

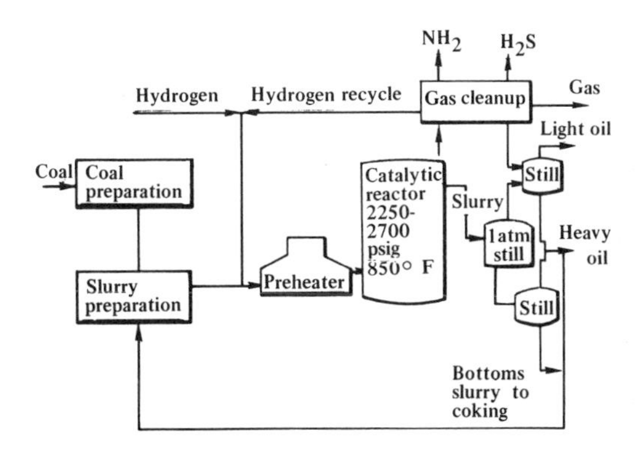


Hydrogen-coal liquefaction process. (*Courtesy U.S. Council on Environmental Quality*)

**Coal Measure unit.** The Coal Measure strata disclose a rough repetition or cycle of different kinds of rock in the same regular manner. Broadly, the cycle of strata upwards is coal, shale, sandstone, and coal—a sequence sometimes referred to as a unit.

**coal mine.** Any and all parts of the property of a mining plant, on the surface or underground, which contribute, directly or indirectly, to the mining or handling of coal under one management. (See Fig., p. 86)

**coal mining methods.** Over the years, a very large number of methods of mining coal have been developed to suit the seam and local conditions. They may be divided, broadly, into the longwall and pillar methods of working.

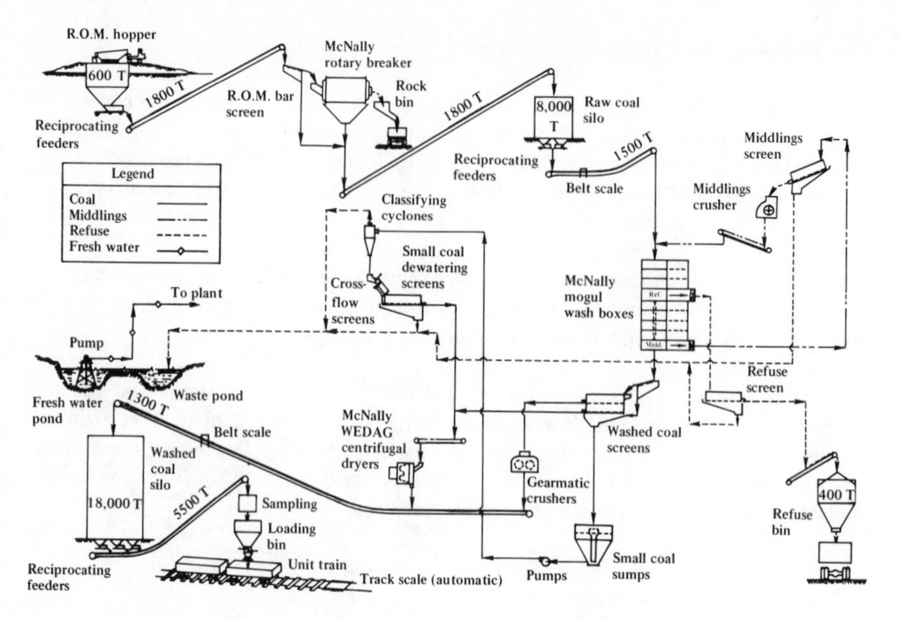


Coal preparation plant at the Lehy Mine of Amax Coal Company. (*Courtesy Van Nostrand's Scientific Encyclopedia*)

**crude oil.** The crude oil obtained by the destructive distillation of bituminous coal; that distillate obtained from such a crude oil which is kerosine.

**coal preparation.** A collective term for the physical and mechanical processes applied to coal to make it suitable for a particular use.

**coal rank.** Classification according to the degree of metamorphism or progressive alteration in the natural series from lignite to anthracite; higher rank is classified according to fixed carbon on a dry basis; lower rank is according to British thermal units on a moist basis. (See Fig., p. 87)

**coal rash.** Very impure coal containing much argillaceous material, fusain, etc.

**coal saw.** A coal cutter employing a very thin chain and bits, or saw, which cuts a kerf 2 inches wide, in comparison with a normal chain and bit kerfs which are 5 to 7 inches wide. The coal saw is for use where hydraulic devices could be employed to break down the coal and thus eliminate the use of explosives ordinarily required.

**coal seam.** A bed or stratum of coal.

**coal slurry.** Finely crushed coal mixed with sufficient water to form a fluid. To use coal slurry pumped through a pipeline as fuel, expensive drying and dewatering pretreatment is necessary, it can, however, be fired in a cyclone furnace as it is received from a pipeline in the form of a coal and water mixture.

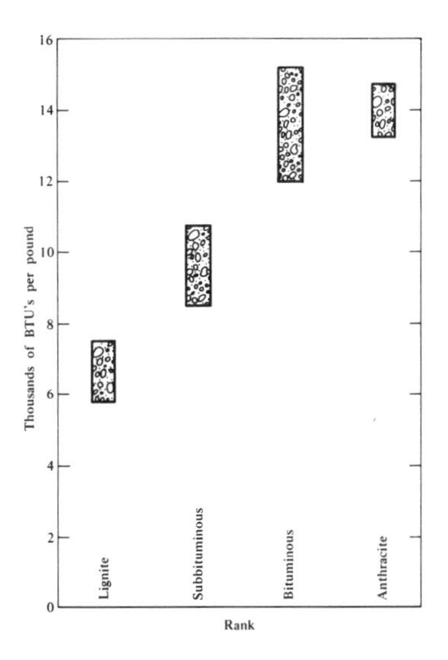


Coal rank in thousands of Btu's per pound. (*Courtesy U.S. Council on Environmental Quality*)

**coal slurry pipeline.** A pipeline which transports coal in pulverized form suspended in water.

**coal tar.** A gummy, black substance produced as a byproduct when bituminous coal is distilled. About 9 gallons of coal tar result from one ton of coal. Among the numerous constituents derived from coal tar are benzene, cumene, pyridine, quinoline naphthalene, phenol, cresols, anthracene, and creosote. (See Fig., p. 88)

**coal type.** A variety of coal, such as common banded coal, cannel coal, algal coal, and splint coal. The distinguishing characteristics of each type of coal arise from the differences in the kind of plant material that produced it.

**coal washing.** The process of removing impurities from small sizes of coal or ore by washing with water.

**cobalt bomb.** A nuclear weapon encased in cobalt.

**coccidioidomycosis.** A potentially significant safety problem which could be encountered during Solar Thermal Electric plant construction in southwestern desert regions, coccidioidomycosis is an infectious fungal disease which is caused by a soil fungus (Coccidiodes immitis) and is contracted exogenously through contact with contaminated soil. The disease is especially prevalent during the dry summer months and among outdoor laborers engaged in earth disturbing operations such as construction. The careful application of fugitive dust control methods during Solar Thermal Electric plant construction should prove adequate to control the spread of this disease. In addition, proper protective clothing worn by the construction personnel should significantly reduce the risk of direct contact with contaminated soil.

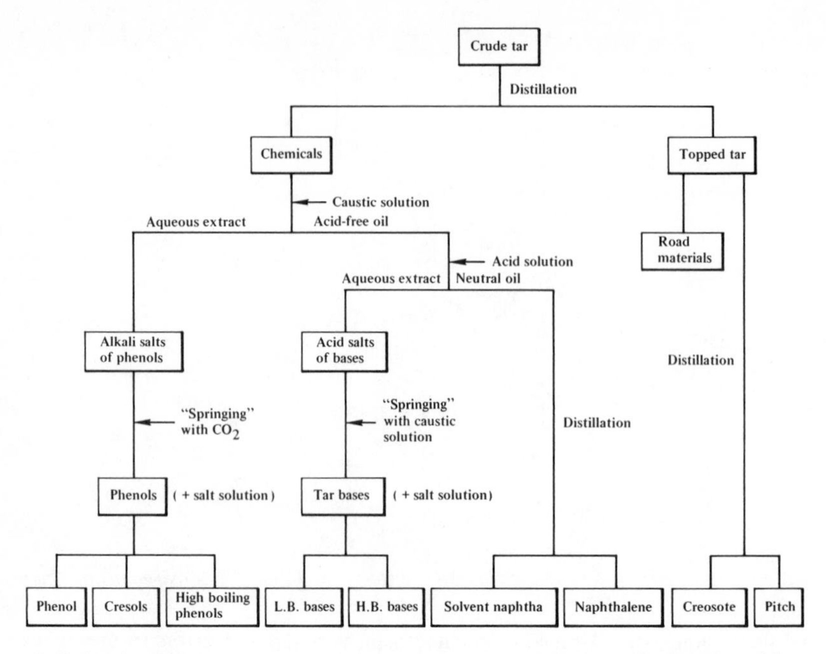


Bulk fractions from crude coal tar. (*Courtesy Van Nostrand's Scientific Encyclopedia*)

**Cockcroft-Walton accelerator.** A device for accelerating charged particles by the action of a high direct-current voltage on a stream of gas ions in a straight insulated tube; the voltage is generated by a voltage multiplier system consisting essentially of a number of condenser pairs connected through switching devices (vacuum tubes). The particles (which are nuclei of an ionized gas, such as protons from hydrogen) gain energies of up to several million electron volts from the single acceleration so produced.

**COED process.** A char-oil energy development process (COED). Pulverized coal is heated using successively higher temperatures in a series of four fluidized bed pyrolytic reactors (carbonizers). In the first stage, coal is heated by hot fluidizing gases. Devolatilized char from the first reactor flows toward hotter reactors while steam and oxygen introduced in the last reactor flow countercurrent to the main stream. Vapors from the second stage are separated into a liquid product and a gas product in a product recovery section. Process pressures are from 5 to 10 pounds per square inch gauge. Process temperatures range from 500°F in the first stage to 1600°F in the fourth stage.

**coefficient of expansion.** The factor which expresses the change per unit length of any material for each degree of temperature.

**coefficient of performance.** Ration of effect produced to the energy supplied; expressed in the same thermal units.

**coefficient permeability.** The rate of flow of water under laminar flow conditions through a unit cross section of a porous medium under a unit hydraulic gradient and a standard temperature, usually 20°C.

**coefficient of thermal expansion.** The fractional change in the length of a material per degree of temperature change.

**coffin.** A heavily shielded shipping cask for spent (used) fuel elements. Some coffins weigh as much as 75 tons.

**COGAS process.** A fluidized bed char gasifier produces a medium-Btu gas which can be cleaned up and upgraded to pipeline quality gas. A portion of the feed char is burned in a combustor with air. Char fines or inert pellets are thereby heated and then fed to the gasifier to provide endothermic heat for a steam-carbon reaction which produces the medium-Btu gas. A coal pyrolysis process, such as COED, would produce liquid fuels and char for the production of high-Btu gas without the use of an oxygen plant. Reaction pressures are 0 to 30 pounds per square inch gauge and reaction temperatures are 871° to 927°C.

**cohesion.** That property by which molecules of the same kind or of the same body are held together in opposition to forces tending to separate them.

**coincidence counting.** A method for detecting or identifying radioactive materials and for calibrating their disintegration rates by counting two or more characteristic radiation events (such as gamma ray emissions) which occur together or in a specific time relationship to each other.

**coincidence factor.** The ratio of the maximum demand of a group, class, or system to the sum of the individual maximum demands of the several components of the group, class, or system.

**coke.** The solid, combustible residue left after the destructive distillation of coal or crude petroleum.

**coke breeze.** Coke particles of a size passing a $\frac{1}{2}$-inch or $\frac{3}{4}$-inch screen opening.

**coking.** Distillation to dryness of a product containing complex hydrocarbons, such as tar or crude petroleum, which break down in structure during distillation to form tar or crude petroleum. The residue of the process is coke.

**coking coal.** The most important of the bituminous coals, it burns with a long yellow flame, giving off more or less smoke, and creates an intense heat when properly attended. It is usually quite soft and does not bear handling well. In the fire, it swells, fuses, and finally runs together in large masses which are rendered porous by the evolution of the contained gaseous hydrocarbons.

**cold testing.** In the nuclear field, a method of testing employing inactive materials or materials containing radioactive tracers.

**collapsible cladding.** A fuel element cladding which is designed to achieve direct contact with the fuel.

**collector.** That component of a solar system which collects solar radiation. It may consist of an insulated box (metal, wood, plastic, or fiber glass) with or without glazing containing an absorber plate and flow passages which carry the transfer medium (either liquid or gaseous) to be heated. The collector may also be integrated with storage, for example, south glass with water drums or water tubes.

**collector/concentrator subsystem.** The collector/concentrator subsystem has as its basic function the interception, redirection, and concentration of direct solar radiation to a receiver/heat transfer subsystem. This subsystem consists of a field of heliostats and a tracking control system to maintain continuous focus on the central receiver. Unlike distributed collectors, the heliostats of the central receiver system must track the sun in two dimensions.

**collector efficiency.** The fraction of incoming solar radiation captured by the collector. For example, if the system captures half of the incoming radiation, the system is said to be 50 percent efficient. Efficiency is the capability of a collector to capture heat under various climatic conditions. There is no way a collector can be 100 percent efficient, that is, capture all the heat that falls on the collector; 55 percent is good under desirable weather conditions.

**collector efficiency (instantaneous).** The ratio of the amount of energy removed by the heat transfer fluid per unit of aperture over a time period of five minutes or less to the total incident solar radiation on the collector for the same time period under steady state conditions (test method described in ASHRAE 93-77).

**collector-heat exchanger correction factor.** An index ranging in value from 0 to 1 indicating the penalty in useful energy collection resulting from heat exchange between the collector and the storage tank in liquid solar heating systems.

**collector heat removal efficiency factor.** The ratio of the actual useful energy gain of a flat-plate solar collector to the energy gain if the entire collector plate were at the temperature of the inlet fluid.

**collector overall energy loss coefficient.** A parameter characterizing the energy losses of the collector to the surroundings.

**collector subsystem.** The assembly used for absorbing solar radiation, converting it into useful thermal energy, and transferring the thermal energy to a heat transfer fluid.

**collector plates.** In a solar collector system, the primary function of the solar collector plates is to absorb as much of the radiation reaching it as possible, and to lose as little heat as possible when transferring the retained heat to the transport fluid.

**collector tilt.** The angle at which a solar heat collector is tilted to face the sun for better performance.

**collectron.** A neutron detector.

**collimator.** A device for focusing or confining a beam of particles or radiation within an assigned angle.

**Collins miner.** A type of remote-controlled, continuous miner for thin seam extraction. The coal seam is extracted in a series of parallel stalls. The extraction is controlled entirely from the roadway at the entrance of each stall.

**collision.** A close approach of two or more particles, photons, atoms, or nuclei, during which such quantities as energy, momentum, and charge may be exchanged.

**colloid.** A substance composed of extremely small particles, ranging from 0.2 micron to 0.005 micron which, when mixed with a liquid, does not settle, but remains permanently suspended. The colloidal suspension thus formed has properties that are quite different from the simple, solid-liquid mixture or a solution.

**colloidal fuel.** A mixture of finely pulverized coal and fuel oil, which remains homogeneous in storage, has a high calorific value, and is used in oil-fired boilers as substitute for fuel oil alone.

**combination gas.** Natural gas rich in oil vapors; wet gas.

**combined cycle electric generating plant.** A plant that utilizes waste heat from large gas turbines (driven by gases from the combustion of fuels) to generate steam for conventional steam turbines, thus extracting the maximum amount of useful work from fuel combustion. It may produce fuel (char) as well as electric power.

**Combined Cycle Gasification process.** An air-blown, two-stage, entrained flow gasifier is used to produce low-Btu gas. Coal is fed into the top stage of the reactor where it is entrained and partially gasified by hot gases from the lower stage. Char is separated from the raw product gases by cyclones and fed to the lower stage where complete gasification occurs with air and steam. Product gases are cooled and cleaned and can then be used to fuel a gas turbine. Waste gases are then cooled in a waste heat boiler producing steam for a steam turbine resulting in a combined power system. The process operates at a pressure of 500 pounds per square inch gauge and temperatures of 1800° F in the top stage and 2800° F in the lower stage. (See Fig., p. 92)

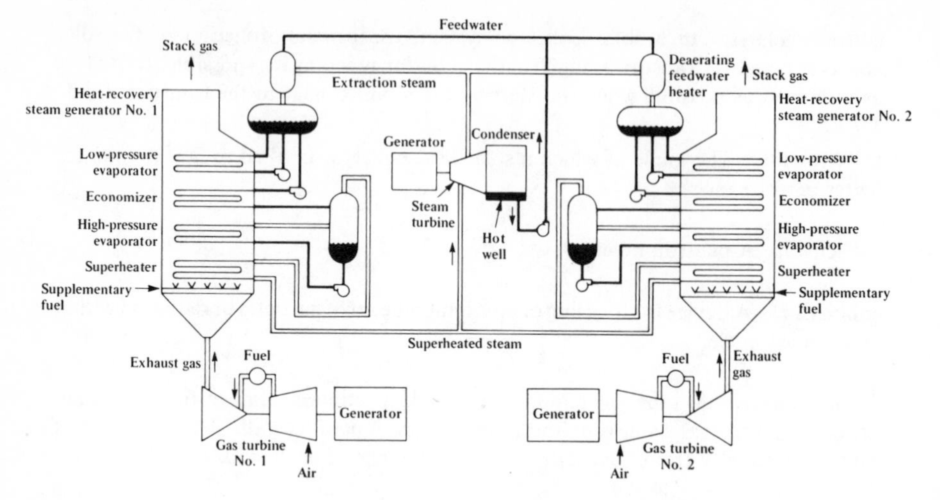


Combined cycle power plant. (*Courtesy Van Nostrand's Scientific Encyclopedia*)

**combustibility.** An assessment of the speed of combustion of a coal under specified conditions.

**Combustible constituents.** The components of a fuel that will burn.

**Combustible schist.** Another name for carbonaceous or bituminous shale.

**combustion.** Burning; any very rapid chemical reaction in which heat and light are produced. Most familiar combustions are oxidations—unions with oxygen.

**combustion control.** Equipment which proportions combustion air to fuel over the entire operating range of the burner.

**combustion rate.** The weight of fuel burnt per square foot of grate area per hour.

**combustor.** A vessel in which combustion of gaseous products from a fuel takes place.

**comfort energy.** Used to describe any form of energy whose end use is the heating and cooling of buildings and homes.

**comminution.** The breaking, crushing, or grinding of coal, ore, or rock; also the mechanical shredding or pulverizing of waste.

**common banded coal.** The common variety of bituminous and subbituminous coal. It consists of a sequence of irregularly alternating layers of (a) homogeneous black material having a brilliant vitreous luster; (b) grayish-black, striated material of silky luster; and (c) thinner bands of soft, powdery, and fibrous particles of mineral charcoal.

**compaction.** Decrease in volume of sediments as a result of compressive stress; also reduction in bulk of solid waste by rolling and tamping.

**components.** An individually distinguishable product that forms part of a more complex product (i.e., subsystem or system).

**compound parabolic concentrators.** A type of concentrating collector using parabolic reflectors which does not form an image of the sun on the receiving surface.

**compound engine.** A steam engine in which high-pressure steam expands and does work in a succession of cylinders or turbine chambers.

**compound nucleus.** A highly excited nucleus of short lifetime.

**compressed air.** Air compressed in volume and transmitted through pipes for use as motive power for underground machines.

**compressed air storage.** A method of storing reserve energy by compressing air in large underground caverns. A small water reservoir is necessary to maintain constant air pressure, and the air system is used only in conjunction with gas turbine generators. Off-peak electricity is used to run a large air compressor which compresses air to about 40 atmospheres about 1500 feet underground. This air is later released to run the gas turbine.

**compressibility.** The property of a material pertaining to its susceptibility to decrease in volume when subjected to an increase in pressure.

**compression.** The action on a material which decreases its volume as the pressure to which it is subjected increases.

**compression gasoline.** Natural gasoline made by compressing natural gas.

**compression ignition engine.** An internal combustion engine in which ignition of the liquid fuel injected into the cylinder is performed by the heat of compression of the air charge.

**compression ratio.** The ratio of absolute pressure after and before compression.

**compression refrigerating system.** A refrigerating system in which the cooling effect results from expansion of a refrigerant after mechanical compression.

**compressor.** A mechanical device for raising the pressure of a gas; also a machine which compresses air.

**compressor station.** Any permanent combination of facilities of production, transmission, and distribution companies which supplies the energy to move gas in transmission lines or into storage by increasing the pressure.

**Compton effect.** The elastic scattering of a photon by an electron. In each such process the electron gains energy and recoils, and the photon loses energy. This is one of three ways photons lose energy upon interacting with matter.

**concentrating collectors.** Solar collectors designed to focus large amounts of solar radiation upon a relatively small collection area to produce higher temperatures than those attainable by flat-plate collectors.

**concentration.** The percentage of a specified constituent in a mixture to the total quantity of the mixture.

**concrete solar collector.** Concrete, at roughly one cent a pound, is the least expensive material generally available as a solar collector cast panel. Payne Inc. has produced a trickle solar collector design composed of expanded shale aggregate and U.S. Steel "Fibrecon" fibers for reinforcement. Incorrectly formulated concrete is susceptible to both thermal shock cracking and freeze-thaw damage.

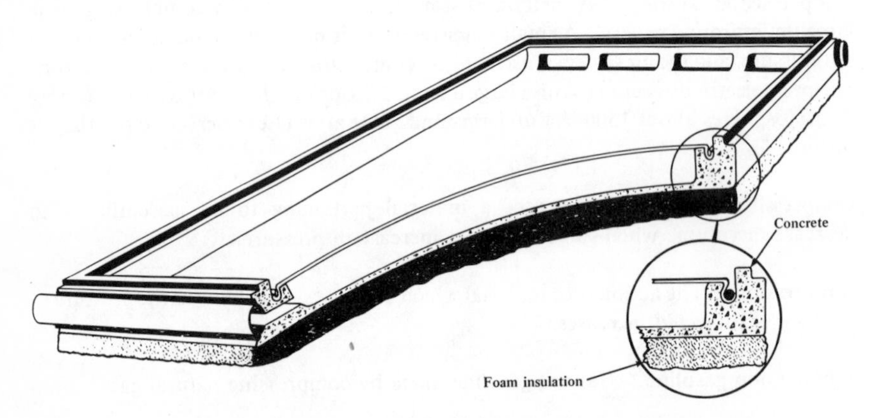


Cutaway drawing of a 2 feet by 8 feet Payne concrete solar collector panel. The aggregate is expanded shale, and U.S. steel "Fibrecon" fibers provide reinforcement to give a material having roughly the same density as wood. (*Courtesy Payne, Inc., Annapolis, MD*)

**condensate.** Liquid hydrocarbon obtained by the combustion of a vapor or gas produced from oil or gas wells and ordinarily separated at a field separator and run as crude oil.

**condensation point.** The temperature at which the vapor changes into its liquid state (i.e., steam into water).

**condenser.** An apparatus for removing heat from a gas (steam) so as to cause the gas to revert to the liquid state (water).

**conduction.** The transfer of heat through matter by the transfer of kinetic energy from particle to particle rather than by a flow of heated material; the transmission of energy directly from molecule to molecule. It is the way in which electricity travels through a wire or heat moves from a warm body to a cool one when the two bodies are placed in contact.

**conductivity.** The ease with which heat will flow through a material as determined by the material's physical characteristics.

**configuration control.** Adjusting the configuration of the fuel, coolant, or moderator in reactor control.

**connected load.** The sum of the capacities or ratings of the gas or electric power-consuming apparatus connected to a supplying system, or any part of the system under consideration.

**conservation.** Conserving, perserving, guarding, or protecting; keeping in a safe or entire state; using in an effective manner or holding for necessary uses, (i.e., mineral resources).

**conservation of energy.** The total energy of an isolated system remains constant irrespective of whatever internal changes may take place. Energy disappearing in one form reappears in another.

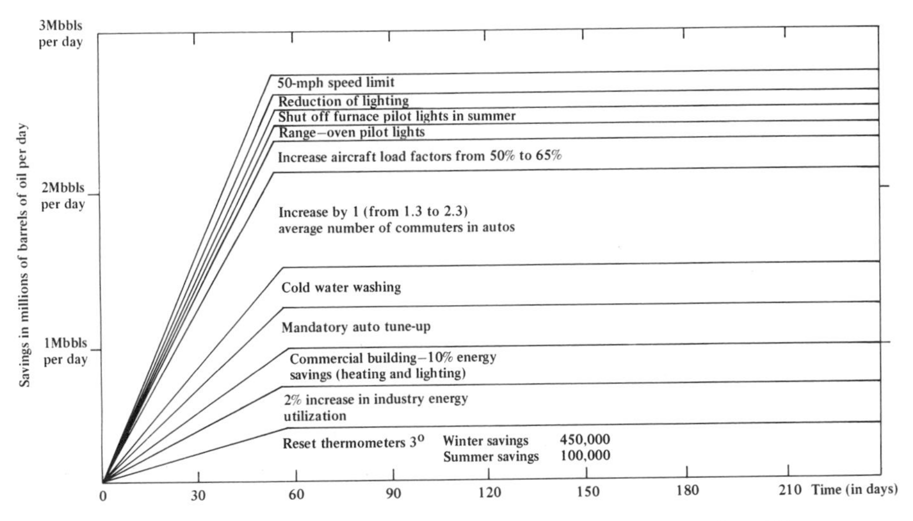


Potential energy saving of emergency conservation programs, assuming 100 percent cooperation. (*Courtesy National Science Teachers Association*)

**conservation strategy.** The actions taken by the Federal Government to reduce demand for energy, particularly petroleum, and possibly including the setting of minimum mileage standards for new automobiles, and the provision of incentives and standards to increase residential insulation and energy-use efficiency.

**Consol Stirred Bed process.** Fluidized-bed carbonization of ground coal in a vessel equipped with stirrer blades.

**construction spread.** A group organized to handle all phases of pipeline construction.

**consumer conservation products.** The DOE, working with industry, has initiated programs to improve product energy efficiency, and in so doing, incur substantial savings in both energy and money. Such products include additional insulation for water heaters; dual purpose appliances in which waste heat from one will operate the other; more efficient electric and gas heaters and motor compressor units for refrigerators, freezers, and room air conditioners; long lasting fluorescent light bulbs; and new types of windows that, in winter, will allow the sun's rays to pass through and keep the heat from bouncing back outside, and, in summer, will automatically change to reflect heat and light.

**consumption charge.** That portion of a utility charge based on energy actually consumed, as distinguished from the demand charge.

**containment.** The provision of a gastight shell or other enclosure around a reactor to confine fission products that otherwise might be released to the atmosphere in the event of an accident. The reactor building itself may be sealed off as a secondary containment system.

**containment structure.** A massive reinforced and prestressed concrete structure, usually of a right circular cylindrical shape or lightbulb shape depending upon type of reactor, in which the reactor is housed. Structures are designed to withstand high pressure and high temperature and, in an emergency, to contain the radioactivity released.

**containment vessel.** A gas tight shell or other enclosure around a reactor.

**contaminants (hazardous).** Materials (solids or liquids or gases) which when added unintentionally (or intentionally) to the potable water supply cause it to be unfit for human or animal consumption.

**continental shelf.** The submerged shelf of land sloping gradually from the exposed edge of a continent. It is usually defined as those areas where the water is less than 200 meters (600 feet) deep.

**continuously variable transmission.** A transmission system which continuously varies the power/speed relationship of the engine to maximum efficiency. A 26 percent improvement in fuel economy, compared to a conventional 3-speed automatic transmission, can be achieved.

**continuous miner.** A mining machine designed to remove coal from the face and to load that coal into cars or conveyors without the use of cutting machines, drills, or explosives.

**continuous mining.** A system of mining in which a mining machine (continuous miner) is used to cut or rip coal from the face and load it into cars or conveyors.

**contour mining.** Mining coal alongside steep slopes in mountainous areas.

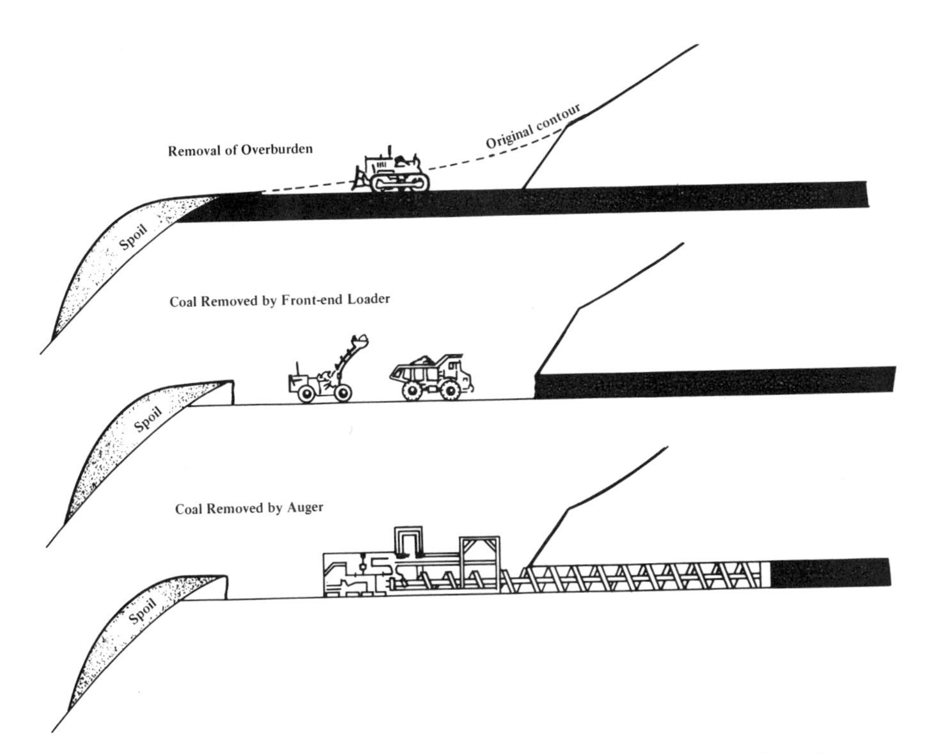


Contour mining with bulldozer and auger. (*Courtesy U.S. Energy Research and Development Administration, Division of Coal Conversion and Utilization*)

**contour stripping.** Removing the overburden and mining from a seam of coal that lies close to the surface at approximately the same elevation in steep or mountainous terrian.

**contracted reserves.** Natural gas reserves dedicated to the fulfillment of gas purchase contracts.

**control.** A mechanism by which the gas, air, water and/or electrical supply to a gas-consuming device is regulated.

**control drive.** A mechanism by means of which a control member in a nuclear reactor can be moved.

**controlled area.** In the nuclear industry, a specified area under the supervision of a person who is responsible for protecting personnel from exposure to radiation.

**controlled crude oil.** Domestically produced crude petroleum that is subject to the ceiling price for crude oil. For a particular property which is not a stripper well lease, the volume of controlled oil equals the base production control level minus an amount of released oil equal to the new oil production from that property.

**controlled thermonuclear reaction.** Controlled fusion, that is, fusion produced under research conditions or for production of useful power; a reaction in which two light nuclei combine to form a heavier atom, thus releasing a large amount of energy; also referred to as controlled fusion experiment.

**control member.** That part of a nuclear reactor which is used for reactor control.

**control rod.** A control member in the form of a rod; a rod, plate, or tube containing a neutron absorbing material (hafnium, boron, etc.) used to control the power of a nuclear reactor. By absorbing neutrons, a control rod prevents the neutrons from causing further fission.

**control system.** A system composed of a number of elements of any kind to control any operation or equipment.

**convection.** The transfer of heat by means of the upward motion of the particles of a liquid or a gas which is heated from beneath; transmission of heat by moving masses of air.

**convector.** In heat transfer, a surface designed to transfer its heat to a surrounding fluid largely or wholly by convection. Such a surface may or may not be enclosed or concealed.

**conventional gas.** Natural gas as contrasted to synthetic gas.

**conventional mining.** The cycle of operations which includes cutting the coal, drilling the shot holes, charging and shooting the holes, loading the broken coal, and installing the roof support. Also known as cyclic mining.

**conventional oil.** Crude oil and condensate as contrasted with synthetic oil from shale or coal.

**conversion.** The process of nuclear transformation of a fertile substance into a fissile substance; also the chemical processing of uranium concentrates into uranium hexafluoride gas.

**conversion burner.** Fuel-burning devices (usually oil or gas) intended for installation in a wide variety of boilers or furnaces. The firing door type is a conversion burner designed specifically for boiler or furnace firing door installation. The inshot type is a conversion burner normally designed for boiler or furnace ash pit installation and fired in a horizontal position. The upshot type is a conversion burner normally designed for boiler or furnace ash pit installation and fired in a vertical position at approximately grate level.

**conversion efficiency.** The actual net output provided by a conversion device divided by the gross input required to produce the output.

**conversion fuel factor.** A number facilitating statement of units of one system in corresponding values in another system. The following energy equivalents are among those commonly used:

*Coal:* Anthracite = 26.0 million Btu/ton. Bituminous = 24.8 million Btu/ton. Subbituminous = 19.0 million Btu/ton. Lignite = 13.4 million Btu/ton.
*Petroleum:* Crude petroleum = 5.60 million Btu/bbl (42 gas). Residual Fuel Oil = 6.29 million Btu/bbl. Distillate Fuel Oil = 5.83 million Btu/bbl. Gasoline (including aviation) = 5.25 million Btu/bbl. Jet Fuel (kerosine type) = 5.67 million Btu/bbl. Jet Fuel (naphtha-type) = 5.36 million Btu/bbl. Kerosine = 5.67 million Btu/bbl. Asphalt and Road Oil = 6.64 million Btu/bbl.
*Natural Gas:* Dry = 1031 Btu/cu ft at STP. Wet = 1103 Btu/cu ft at STP. Liquids (avg.) = 4.1 million Btu/bbl.
*Fissionable Material:* = 74 million Btu/gm U-235 fissioned.

**conversion of nuclear fuels.** The process by which nonfissionable fertile materials (Th-232, U-238) are converted into fissionable material (U-233, Pu-239) in a reactor.

**conversion process.** A process by which energy is converted from one form to another, such as radiant energy to heat or electric energy. (See Fig., p. 100)

**conversion system.** A device or process that converts a raw energy form into another, more useful form of energy. Examples: conversion of wood into methanol or sunlight into electricity.

**conversion ratio.** The ratio of the number of atoms of new fissionable material produced in a converter reactor to the original number of atoms of fissionable fuel consumed.

**conversion unit.** A unit used to convert heating equipment from one fuel to another.

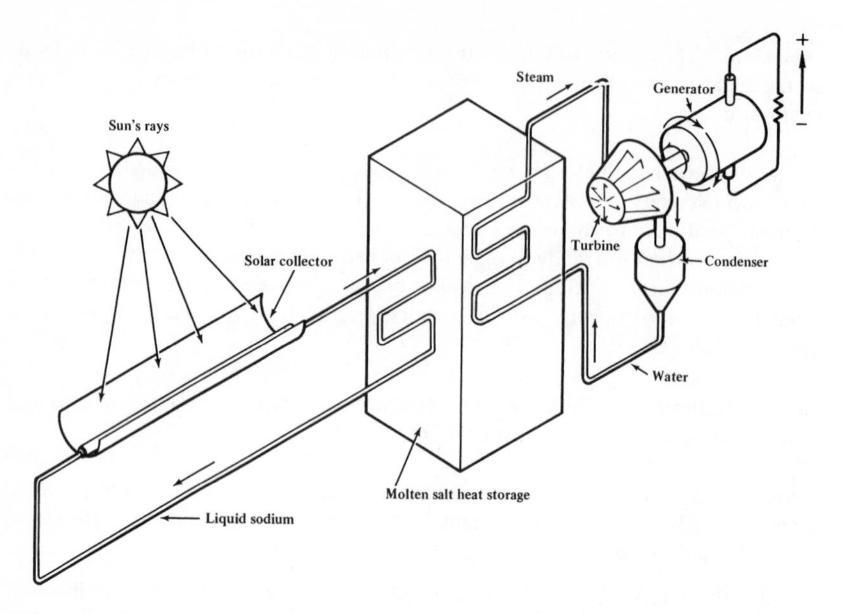


Solar thermal-conversion power system. (*Courtesy U.S. Council on Environmental Quality*)

**converter.** An apparatus for transforming the quality or quantity of electrical energy.

**converter diffuser.** The assembly containing separative elements of one stage of a gaseous diffusion cascade.

**converter plate.** A device placed in a flux of slow neutrons to produce fission neutrons.

**converter reactor.** A reactor in which significant conversion takes place; a reactor that produces some fissionable material, but less than it consumes. In some usages, a reactor that produces a fissionable material different from the fuel burned, regardless of the ratio.

**coolant.** A substance (such as water, air, carbon dioxide, liquid sodium, and sodium-potassium alloy) circulated through a nuclear reactor to remove or transfer heat; also any medium (such as air, water, gas, oil, mud, etc.) used as a circulation medium in a drilling operation.

**cooling capacity.** The quantity of heat that a room air conditioner is capable of removing from a room in one hour's time.

**cooling coil.** A coil of pipe or tubing used as a heat exchanger to cool material inside or outside the coil.

**cooling degree day.** A measure of the need for air conditioning (cooling based on temperature and humidity).

**cooling pond.** An artificial pond used to receive and dissipate waste heat, usually from a steam-electric power plant. Approximately one acre of pond surface is needed per megawatt of electric output for a modern steam-electric power plant.

**cooling tower.** A device to remove excess heat from water; a unit or structure, usually built of wood, for the cooling of water by evaporation; a unit or structure for cooling water by conduction and convection into the air.

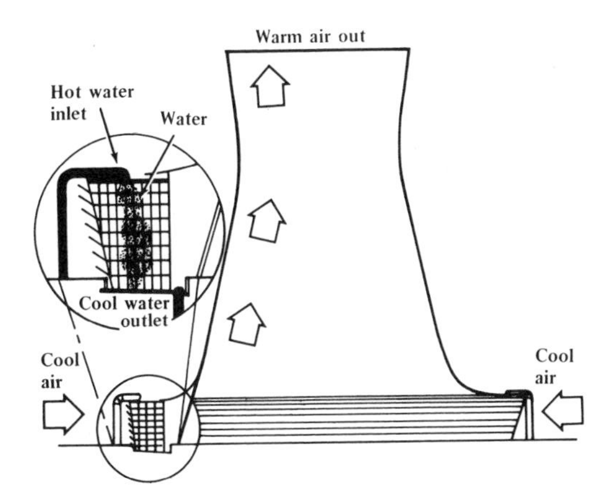


Hyperbolic natural draft cooling tower. (*Courtesy ERDA-69*)

**Copenhagen water.** The standard used in determining the chlorinity of water. It is prepared by the Hydrographical Laboratories in Copenhagen, Denmark.

**copperas water.** Water containing ferrous sulfate by suspension or in solution.

**cord of wood.** A stack of wood 4 feet by 4 feet by 8 feet. Burned, it produces approximately 5 million calories of energy.

**core.** The central portion of a nuclear reactor containing the fuel elements and usually the moderator, but not the reflector; in mining, a cylindrical sample of rock obtained in core drilling. (See Fig. p. 102)

**core analysis.** Laboratory examination of geological samples taken from the well bore. As used by the petroleum industry, a study of a core sample to determine its water and oil content, porosity, permeability, etc.

**core barrel.** A hollow cylinder attached to a drill pipe and supported by a bit to receive and retain the rock or other strata penetrated in core drilling.

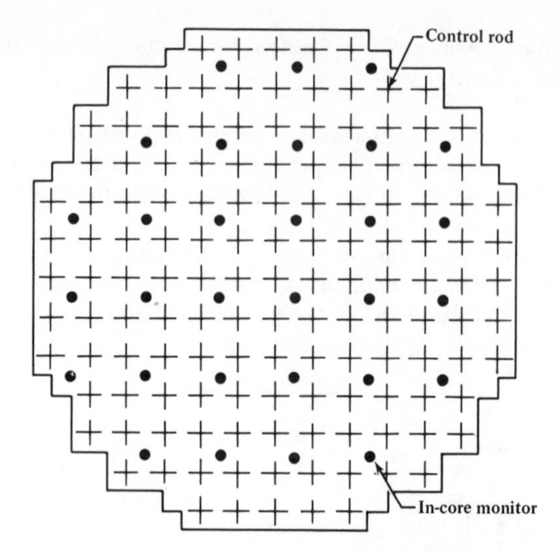


Core arrangement in a boiling water reactor. (*Courtesy Van Nostrand's Scientific Encycolpedia*)

**core drilling.** The process of obtaining cylindrical samples of rock and other strata through the use of a hollow drilling bit and core barrel.

**core oil.** An oil used as a binder in making sand cores in foundry work.

**Coriolis force.** An apparent force caused by the earth's rotation which deflects a moving body on the surface of the earth to the right in the northern hemisphere and to the left in the southern hemisphere. Named for the French engineer and mathematician, G. G. Coriolis. Because the first complete explanation of this apparent force was by the American meteorologist, W. Ferrel, it is also called Ferrel's law.

**corrosion.** The gradual destruction of a material by a chemical process such as oxidation or solution. The deterioration of a substance or its properties because of a chemical or electrochemical reaction with its environment.

**cosmic rays.** Radiation of many sorts but mostly atomic nuclei (protons) with very high energies, originating outside the earth's atmosphere. Cosmic radiation is part of the natural background radiation. Some cosmic rays are more energetic than any man-made forms of radiation.

**Cosorb process.** A process for the separation of carbon monoxide from gaseous mixtures by selective adsorption in a unique solvent.

**Co-Steam process.** A process in which the pulverized lignite in some of the product oil is pumped with process gas (carbon monoxide and steam) into a stirred reactor. The process utilizes the natural alkalinity of lignite to catalyze the carbon monoxide + water reaction to produce hydrogen, which liquifies lignite. The product goes to a receiver where oil, gas (carbon monoxide and hydrogen), and unreacted solids are separated. Low sulfur fuel oil yield is 69.4 to 77 percent by weight. Product gases provide the hydrogen for the hydrogenation process. Reactor operating conditions are 427°C and 4000 pounds per square inch gauge.

**Cottrell precipitator.** An electrostatic device whereby negatively charged dusts or mists are removed from gases.

**coulomb.** The System International (SI) practical meter-kilogram-second unit of electrical charge equal to the quantity of electricity transferred by a current of 1 ampere in 1 second.

**Coulomb's law.** In electrostatics, the force between two charges in vacuo is proportional to the product of their magnitude divided by the square of the distance between them; in electromagnetics, the force between two poles is proportional to the product of their pole strengths divided by the square of the distance between them.

**counter.** A general designation applied to radiation detection instruments or survey meters that detect and measure radiation in terms of individual ionizations, displaying them either as the accumulated total or their rate of occurrence.

**country rock.** The rocks surrounding and penetrated by mineral veins or invaded by and surrounding an igneous intrusion. Also used to refer to the common rock of a region.

**coupling.** A connecting device, such as a threaded sleeve used to connect two pipes; also an arrangement for transferring electrical energy from one circuit to another.

**coupon.** A piece of metal which is used to measure the rate of corrosion of the metal in a gaseous or liquid environment.

**course stacking.** The method of shovel operation in which no ground is hauled away. The shovel simply stacks the ground on the side opposite from the working cut or turns around and dumps the spoil on a bank behind.

**cracking.** A process carried out in a refinery reactor in which the large molecules in the charge stock are broken up into smaller, lower-boiling, stable hydrocarbon molecules, which leave the vessel as overhead (unfinished cracked gasoline, kerosines, and gas oils). At the same time, certain of the unstable or reactive molecules in the charge stock combine to form tar or coke bottoms. The cracking reaction may be carried out with heat and pressure (thermal cracking) or in the presence of a catalyst (catalytic cracking).

**cracking plant.** The combined equipment—furnace, reaction chamber, fractionator—for the thermal conversion of heavier charging stock to gasoline.

**crankshaft.** The engine shaft that converts the reciprocating motion and force of pistons and connecting rods to rotary motion and torque.

**creep.** The slow and imperceptible movement of rock debris or soil from higher to lower levels because of gravity.

**creosote.** A colorless to yellowish oily liquid compound consisting of a mixture of phenols distilled from wood. It has a smoky odor and a burning taste.

**Cresap process.** A process wherein crushed coal is mixed with a recycled solvent, heated, and fed to the extractor. The coal is dissolved in solvent and the product from the extractor is separated. The solid product is carbonized to char, which may be used to produce hydrogen for the process. The liquid is subjected to hydrogenation and fractionation to produce low sulfur oil and solvent for recycle. No catalyst is required for solvent extraction. Reactor operating conditions are 407° C and 150 pounds per square inch gauge.

**crest.** The summit of any eminence, such as the top of a dam or spillway to which water must rise before passing over the structure.

**crevice corrosion.** An intense form of localized corrosion which frequently occurs within crevices, under deposits or gaskets. It is usually associated with small volumes of stagnant solution, e.g., under-gasket corrosion of stainless steel in seawater service.

**critical.** The state of a nuclear reactor when it is sustaining a nuclear chain reaction. A nuclear reactor is critical when the rate of neutron production is equal to the rate of neutron loss.

**critical area.** In prospecting work, an area found to be favorable from geological age and structural considerations.

**critical area of extraction.** The area of coal required to be worked to cause a surface point to suffer all the subsidence possible from the extraction of a given seam.

**critical assembly.** An assembly of sufficient fissionable material and moderator to sustain a fission chain reaction at a very low power level. This permits study of the behavior of the components of the assembly for various fissionable materials in different geometrical arrangements.

**critical coefficient.** The ratio of the critical temperature to the critical pressure.

**critical density.** The density of a substance at its critical temperature and under its critical pressure.

**critical depth.** A given quantity of water in an open conduit may flow at two depths having the same energy head. When these depths coincide, the energy head is a minimum, and the corresponding depth is Belanger's critical depth.

**critical experiment.** An experiment conducted to verify or supplement calculations of the critical size and other physical data affecting the reactor design. The power is kept very low so that a system for removing heat is not required.

**critical facility.** A facility where critical experiments are conducted.

**critical flow.** A condition of flow for which the mean velocity is at one of the critical values.

**criticality.** The condition of being critical; the state of a nuclear reactor when it is sustaining a chain reaction.

**critical mass.** The smallest mass of fissionable material, such as uranium-235 or plutonium-239, that will support a self-sustaining chain reaction under stated conditions.

**critical organ.** That organ in which the dose equivalent would be most significant because of organ's radiosensitivity.

**critical point.** The point at which the properties of a liquid and its vapor become indistinguishable. It is generally synonymous with critical temperature.

**critical potential.** A potential which produces a sudden change in magnitude of the current.

**critical pressure.** The pressure at which a gas may just be liquefied at its critical temperature. Above this temperature, no matter what pressure is applied, the gas cannot be liquefied.

**critical size.** The physical dimensions found to be favorable for a reactor core or an assembly to be made critical.

**critical slope.** The maximum angle with the horizontal at which a sloped bank of soil or given height of soil will stand unsupported.

**critical state.** An unstable condition of a substance when on the point of changing from a liquid to a vapor (or vice versa), defined by its critical temperature and its critical pressure.

**critical temperature.** That temperature above which a substance can exist only in the gaseous state, no matter what pressure is exerted.

**critical velocity.** Reynold's critical velocity is that velocity at which the flow changes from laminar to turbulent, and where friction ceases to be proportional to the first power of the velocity and becomes proportional to a higher power—practically the square. Belanger's critical velocity is that condition in open channels for which the velocity head equals one-half the mean depth. Kennedy's critical velocity is that velocity in open channels which will neither deposit nor pick up silt.

**critical volume.** The specific volume of a substance in its critical state.

**crop coal.** Coal of inferior quality near the surface.

**cropping.** Coal cutting beyond the normal cutting plane.

**crosscut.** In mining, a passageway driven at right or other angles to the main entry.

**cross section.** A profile portraying an interpretation of a vertical section of the earth explored by geophysical and/or geological methods; also, a measure of the probability that a nuclear reaction will occur; usually measured in barns, it is the apparent (or effective) area presented by a target nucleus (or particle) to an on-coming particle or other nuclear radiation, such as a photon or gamma radiation.

**crosswind.** A wind blowing in a direction not parallel to a course.

**crude.** A substance in its natural unprocessed state, such as crude ore or crude oil; in a natural state, not cooked or prepared by fire or heat, not altered or prepared for use by any process, not refined.

**crude gas.** Gas containing a wide range of impurities.

**crude naphtha.** Light distillate made in the fractionation of crude oil.

**crude oil.** Raw petroleum as it comes from the earth, in its natural unprocessed, unrefined state. It is composed principally of hydrocarbons with traces of sulfur, nitrogen, or oxygen compounds. It is liquid at atmospheric pressure after passing through surface separating processes and does not include natural gas products. It includes the initial liquid hydrocarbons produced from tar sands, gilsonite, and oil shale.

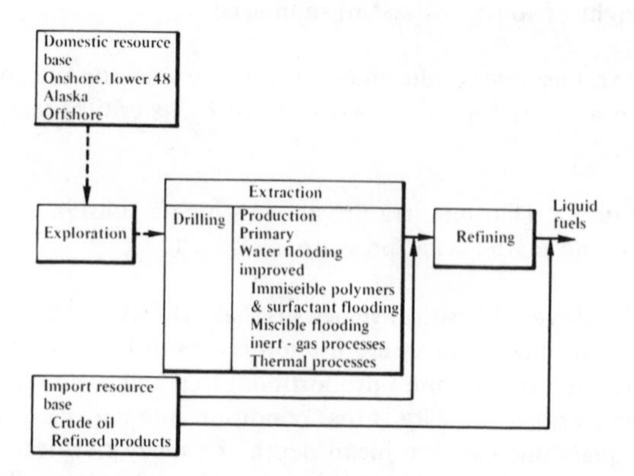


Crude oil resource development. (*Courtesy U.S. Council on Environmental Quality*)

**crude oil domestic production.** The volume of crude oil flowing out of the ground. Domestic production is measured at the wellhead and includes lease condensate, which is a natural gas liquid recovered from lease separators or field facilities.

**crude oil imports.** The monthly volume of crude oil imported which is reported by receiving refineries and including crude oil entering the United States through pipelines from Canada.

**crude oil stocks.** Stocks held at refineries and at pipeline terminals. Does not include stocks held on leases (storage facilities adjacent to the wells), which historically total approximately 13 million barrels.

**crude ore.** The unconcentrated ore as it leaves the mine.

**crude shale oil.** The oil obtained as a distillate by the destructive distillation of oil shale.

**cryogenic cables.** Cable supercooled to temperatures typically around −196°C to reduce the conductor's resistance to electricity and permit its current carrying capacity to be multiplied above the normal carrying capacity of conventional underground cables.

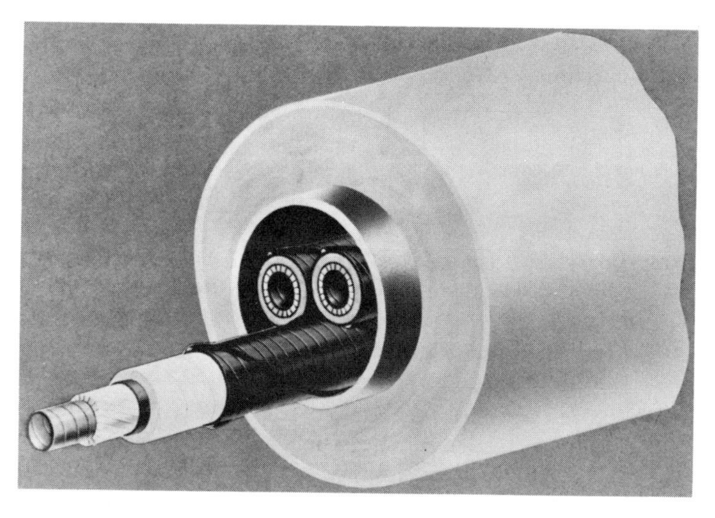

General Electric's concept of a highly efficient conductor of electricity follows a design wherein the transmission cables are cooled to the temperature of liquid nitrogen (−320° F) and are kept cold by an outer layer of urethane foam thermal insulation. (*Courtesy G. E. Research and Development Center, Schenectady, NY*)

**cryogenic techniques.** Techniques involving the use of extremely low temperatures to keep certain fuels, such as liquefied hydrogen, methane, and propane, in a liquid form.

**crystallization.** Oil refining separation process through which crystalline phases separate out of the liquid.

**Csiro process.** Commonwealth Scientific and Industrial Research Organization process for fluidized-bed hydrocarbonization of noncaking brown coal to produce methane, tar, and residual char.

**cubic foot.** The most common unit of measurement of gas volume. It is the amount of gas required to fill a volume of one cubic foot under stated conditions of temperature, pressure, and water vapor. One cubic foot of natural gas equals 1000 British thermal units under standard conditions of atmosphere and temperature.

**cubic meter.** A measure of volume in the metric system. One cubic meter equals 8.6 barrels (U.S. liquids); 35.3 cubic feet; 1.3 cubic yards; 264.1 gallons (U.S.); and 999.9 liters.

**cubic yard.** A measure of volume. One cubic yard equals 27 cubic feet and 0.76 cubic meters.

**culm.** Carbonaceous shale; varieties of anthracite smalls; the waste or slack of the Pennsylvania anthracite mines.

**cultural eutrophication.** Acceleration by man of the natural aging process of bodies of water.

**culvert.** A covered channel or a pipe of large diameter taking a watercourse below ground level.

**cumulative fission yield.** The fraction of fissions which have resulted in the production of a given nuclide either directly or indirectly up to a specified time.

**curb cock.** A shutoff valve in a gas service line.

**curie.** The basic unit to describe the intensity of radioactivity in a sample of material. The curie is equal to 37 billion disintegrations per second, which is approximately the rate of decay of 1 gram of radium. A curie is also a quantity of any nuclide having 1 curie of radioactivity. Named for Marie and Pierre Curie, who discovered radium in 1898.

**Curie point; Curie temperature.** The temperature at which there is a transition in a substance from one phase to another of markedly different magnetic properties. Specifically, the temperature at which there is a transition between the ferromagnetic and paramagnetic phases.

**current.** The flow of electric charge; also the part of a fluid body (as water or air) moving continuously in a certain direction.

**current density.** The amount of current per unit area of electrode.

**current efficiency.** The proportion of current used in a given process to accomplish a desired result.

**current meter.** An instrument, as a galvanometer, for measuring the strength of an electric current; also an instrument for measuring the velocity of moving water or fluid.

**current reservoir capacity.** The total volume of gas which a storage reservoir can contain.

**curtailment.** Reductions, from contract quantities or normal historical quantities, of deliveries to natural gas end-use customers.

**cut.** To intersect a vein or working; to excavate coal; in longitudinal excavation made by a strip-mining machine, to remove overburden in a single progressive line from one side or end of the property.

**cutie pie.** A common radiation survey meter used to determine exposure levels or to locate possible radiation hazard.

**cut-in speed.** The wind speed at which a wind machine is activated.

**cuttings.** Solid material removed from a drilled hole.

**cycle.** In the coal industry, the complete sequence of face operations required to get coal; in the oil industry, a series of events (or processes) which repeat themselves; in the electrical industry, the current goes from zero potential or voltage to a maximum in one direction, back to zero, then to a maximum in the other direction and then back again to zero, in one cycle of alternating electric current.

**cycle stock.** Unfinished product taken from a stage of a refinery process and recharged to the process at an earlier period in the operation.

**cyclone.** A cleaning device which uses a circular flow to separate the heavier particulates from stack gases.

**cyclone collector.** An apparatus for the separation by centrifugal means of fine particles suspended in air or gas.

**cyclone separator.** A funnel-shaped device for removing material from an airstream by centrifugal acceleration.

**cyclotron.** A particle accelerator in which charged particles receive repeated synchronized accelerations by electrical fields as the particles spiral outward from their source. The particles are kept in the spiral by a powerful magnetic field.

**cylinder.** In hydraulic systems, a hollow metal cylinder containing a piston, piston rod, and end seals, and fitted with a part or parts to allow entrance and exit of fluid.

**cylinder oil.** Mixture of mineral oil with 5 to 15 percent of animal or vegetable oils.

# D

**Dalton's law.** In a mixture of gases, the total pressure is equal to the sum of the pressures that the gases would exert separately.

**damp.** Any mine gas or mixture of gases, particularly those deficient in oxygen.

**damper.** A device used to regulate the flow of air or other gases.

**damping.** In seismology, a resistance contrary to friction; a force opposing vibration, damping acts to decrease the amplitudes of successive free vibrations.

**Darcy's law.** In fluid dynamics, the velocity of flow of a liquid through a porous medium because of a difference in pressure, is proportional to the pressure gradient in the direction of flow.

**data base.** A set of numbers, variables and information used to provide the operational criteria for processing and decision making.

**daughter element.** The element formed when another element undergoes radioactive decay.

**dead-weight tons.** The total lifting capacity of a ship expressed in long tons (2,240 pounds). For example, the oil tanker Universe Ireland is listed as 312,000 deadweight tons, which means it can carry 312,000 tons of oil or about 1.9 million barrels.

**deasphalting.** Process for removing asphalt from petroleum fractions such as reduced crude. A common deasphalting process introduces liquid propane, in which the nonasphaltic compounds are soluble while the asphalt settles out.

**decay chain.** A radioactive series.

**decay cooling.** The storage of irradiated fuel elements to allow for the radioactive decay of short-lived radioisotopes prior to the initiation of fuel reprocessing.

**decay curve.** A graphic presentation of the manner in which a quantity decays with time; usually the decay of electrical signals.

**decay heat.** The heat produced by the decay of radioactive nuclides.

**decay probability.** The number of disintegrations per second per nucleus of a radioactive substance. In equation form, decay probability $= -(1/N)(dN/dt) = 0.6931/T$, where T is the radioactive half-life.

**decay product.** A nuclide resulting from the radioactive disintegration of a radio nuclide, formed either directly or as the result of successive transformations in a radioactive series. A decay product may be either radioactive or stable.

**declination.** The angular position of the sun at solar noon with respect to the plane of the equator, i.e., the angular position of the sun north or south of the equator; a function of the time of year.

**declining block rates.** A rate structure in which the charge for energy decreases as the amount of energy consumed increases.

**decomposition.** Reduction of the net energy level and change in chemical composition of organic matter through actions of aerobic or anaerobic microorganisms.

**decomposition value.** Minimum voltage at which continuous electric current flows through an electrolytic solution of normal strength.

**decontamination.** The removal of radioactive contaminants from surfaces or equipment, as by cleaning and washing with chemicals.

**decontamination factor.** The ratio of the initial concentration of contaminating radioactive material to the final concentration resulting from a decontamination process.

**deep chiseling.** A surface treatment that loosens compacted soils.

**deep mining.** The exploitation of coal or mineral deposits of depths exceeding about 3000 feet. It would appear that the deepest coal mine in the world is the Rieu de Coeur colliery at Quaregnan, Belgium (4462 feet) with a rock temperature of 52°C. (See Fig., p. 113)

**definite proportions law.** A chemical compound always contains the same elements in the same proportions by weight.

**deflation.** Removal of loose material by the wind, leaving the rocks bare to the continuous attack of the weather.

**deflocculant.** Any organic or inorganic material which is used as an electrolyte to disperse nonmetallic or metallic particles in a liquid; basic materials such as calgonate, sodium silicate, and soda ash are used as deflocculants in clay slips.

**degasification.** Progressive loss of gases in a substance leading to the formation of a more condensed product. Applied primarily to the formation of solid bitumens from liquid bitumens, but also used in connection with coal formation.

**degradation.** The breakage of coal during cutting, loading, and transportation.

Deep mining attracts young people due to high wages in relation to depressed local economies. (*Courtesy Federal Energy Administration*)

**degree day.** A unit measuring the extent to which the outdoor mean daily dry-bulb temperature falls below or rises above an assumed base, normally taken as 18°C for heating and for cooling unless otherwise designated.

**dehumidify.** To reduce, by any process, the quantity of water vapor contained in a solid or gas.

**deionization.** Removal of ions from solution by chemical means.

**delayed coking.** A process in which coal is subjected to a long period of carbonization at moderate temperatures to form coke.

**delayed neutron fraction.** The ratio of the mean number of delayed neutrons per fission to the mean total number of prompt plus delayed neutrons per fission.

**delayed neutron precursor.** A nuclide whose nuclei undergo decay followed by neutron emission.

**delayed neutrons.** Neutrons emitted by radioactive fission products in a reactor over a period of seconds or minutes after a fission takes place. Fewer than 1 percent of the neutrons are delayed, the majority being prompt neutrons. Delayed neutrons are important considerations in reactor design and control.

**delay tank.** A tank or reservoir for the temporary holdup of radioactive fluids to permit their activity to decay.

**deliverability.** The volume of gas that a well, field, pipeline, or distribution system can supply in a given period of time; also the practical output from a storage reservoir.

**demand.** The rate at which electric energy is delivered to or by a system expressed in kilowatts, kilovoltamperes, or other suitable unit, at a given instant or averaged over any designated period of time. The amount of energy required to satisfy the needs of a stated sector of the economy.

**demand charge.** That part of a utility service charged on the basis of the possible demand as distinguished from the actual energy consumed.

**demand day.** That 24-hour period specified by a supplier-user contract for purposes of determining the purchaser's daily quantity of gas used. This term is primarily used in pipeline-distribution company agreements. It is similar to, and usually coincides with, the distribution company "sendout day."

**demand diversity.** The overall variation in the time at which individual demands occur.

**demand factor.** The ratio of the maximum demand to the connected load.

**demand interval.** The period of time during which the flow is averaged in determining demand.

**demand meter.** System utilizing a mechanical mechanism to register the highest level of power consumed between meter readings. This maximum-kilowatt figure determines the electric rate the consumer pays for the billing period. (See Fig. p. 115)

**demand metering.** Concept of higher electric rates during peak demand periods and lower electric rates during low demand periods.

**demand rates.** Applies to any method of charge for gas service which is based upon, or is a function of, the rate of use or size, of the customer's installation or maximum demand during a given period of time. The term "flat demand rate" applies to a charge for gas service based upon the customer's installation of gas-consuming devices. The term "Hopkinson demand rate" applies to that method of charge which consists of a demand charge based upon demand (either estimated or measured) or connected load, plus a commodity charge based upon the quantity of gas used. Any of the fore-

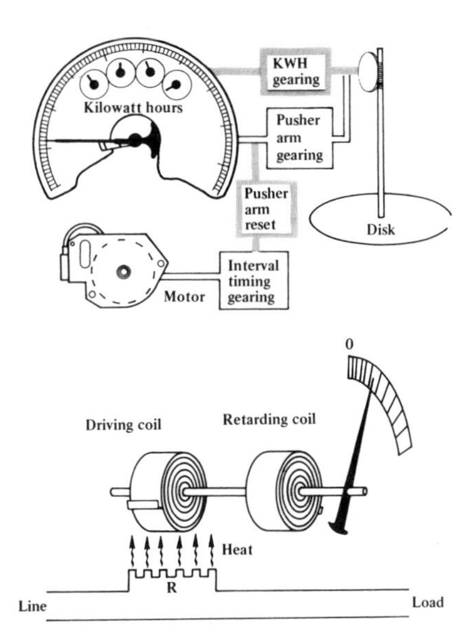


How demand meters work. Watt-hour-meter disks turn at a speed proportionate to the power consumption. (*Courtesy Popular Science, September 1975, p. 52*)

going types of rates may be modified by the addition of a customer charge. When such a charge is introduced in the Hopkinson demand rate, it becomes a "three part rate" or "three charge rate," which consists of a charge per customer or per meter plus demand and commodity charges. The term "Wright demand rate" applies to that method of charge which was the first to recognize load factor conditions. Under this rate the consumer pays a different unit charge for each successive block of consumption. The consumer's maximum daily or hourly demand is a factor in the determination of the block size.

**demineralization.** Removal of mineral matter, such as ash from coal, by solvent extraction.

**demonstrated coal reserve base.** Measured and indicated in-place quantities of bituminous coal and anthracite located in beds with thicknesses of 28 inches or more, and subbituminous coal in beds with thicknesses of 60 inches or more which are located in depths up to 1000 feet. The demonstrated coal reserve base also includes small quantities of coal located in beds thinner and/or deeper than coal presently mined, for which there is evidence that mining is commercially feasible at this time. The data for lignite includes beds 60 inches or thicker that can be surface mined. These are generally located at depths no greater than 120 feet. In general, the amount of coal that can be recovered from a deposit ranges from 40 to 90 percent of the reserve base.

**demonstration reactor.** A reactor designed to demonstrate the technical feasibility and explore the economic potential of a given type of reactor.

**Department of Energy.** In October 1977, the Department of Energy (DOE) was created to consolidate the multitude of energy-oriented government programs and agencies. The Federal Energy Administration and the Energy Research and Development Administration are now part of DOE. The Department will carry out the National Energy Plan policy through a unified organization that will coordinate and manage energy conservation, supply development, information collection and analysis, regulation, research, development, and demonstration. An Energy Information Administration within the Department will organize and analyze information so that it can be used by governments, industry, and the public.

**density.** The weight of a unit volume, usually expressed as pounds per cubic foot.

**depleted uranium.** Uranium having a smaller percentage of uranium-235 than the 0.7 percent found in natural uranium. It is obtained either from the spent (used) fuel elements, or as byproduct tails or residues of uranium isotope separation.

**depletion.** Reduction of the concentration of one or more specified isotopes in a material.

**depletion allowance.** A federal tax exemption for a portion of the net income received from producing a natural resource. The tax allowance is extended to the owner of exhaustible resources based on an estimate of the permanent reduction in value caused by the removal of the resource.

**deposit.** Anything laid down. Formerly applied only to matter left by the agency of water, but now includes mineral matter in any form that is precipitated by chemical or other agencies, as the ores in veins. Mineral deposit or ore deposit is used to designate a natural occurrence of a useful mineral or ore in sufficient extent and degree of concentration to invite exploitation.

**depth of stratum.** The vertical distance from the surface of the earth to a stratum.

**derived fuel.** A fuel, such as coke, charcoal, benzol, and petrol, that is obtained from a raw fuel by some process of preparation.

**desalting.** Any process for making potable water from sea water or other salt waters. Reuse of vapors through compressive distillation or multiple-effect evaporation is practiced in order to limit heat consumption. Distillation is the oldest method. Electrodialysis is another. Other methods are: freezing by direct contact of refrigerant with sea water, foam separation, liquid-liquid extraction, various nonelectric membrane processes, and ion exchange. In the oil industry, the term applies to the removal of mineral salts (mostly chlorides) from crude oils.

**desiccant.** A substance having an affinity for water. Used for drying purposes.

**desiccant-type spacers.** Small objects containing a water-absorbing material that are used to hold materials apart. In solar energy usage, they serve to free the solar collector panel from the formation of condensation, which would keep out some of the sun's energy.

**designated solar surface.** A south facing wall within 45° of true south, or a south tilted surface within 60° of true south, or a horizontal surface all of which receive adequate direct solar radiation to provide for the intended use. The designated surface need not be a wall or roof but can include the hypothetical plane under the colllector as it rests on mounting brackets.

**design day.** A 24-hour period of the greatest theoretical energy demand, used as a basis for designing purchase contracts, and/or production facilities, and/or delivery capacity.

**design day availability.** The amount of each type of energy arranged to be available on the design day and the maximum combination of such supplies.

**day temperature.** The mean temperature assumed for the design day.

**design heating load.** The maximum probable space heating needs of a building.

**design horsepower.** The specified horsepower for a chain drive multiplied by a service factor. It is the value used to select the chain size for the drive.

**design load.** The load generally taken as the worst combination of forces and loads which a structure is calculated to sustain. The term is similarly applied to projects such as air conditioning.

**design pressure.** The maximum operating pressure permitted by various codes, as determined by the design procedures applicable to the material and location involved.

**design temperature.** The temperature which an apparatus or system is design to (a) maintain or (b) operate against, under the most extreme conditions. The former is the inside design temperature; the latter, the outside design temperature.

**design temperature difference.** The maximum probable difference between the indoor and the ambient temperatures.

**design voltage.** The nominal voltage for which a line or piece of equipment is designed; a reference level of voltage for identification and not necessarily the level at which it operates.

**desorption.** The reverse of adsorption whereby adsorbed matter is removed from the adsorbent. The term is also used as the reverse process of absorption.

**destressed area.** In strata control, a term used to describe an area where the force is much less than would be expected after a consideration of the depth and type of strata.

**destructive distillation.** The distillation of solids accompanied by their decomposition. The destructive distillation of coal results in the production of coke, tar products, ammonia, gas, etc.

**desulfurization.** The removal of sulfur or sulfur-bearing compounds from a hydrocarbon by any one of a number of processes, such as hydrotreating.

**desulfurize.** To free from sulfur; to remove the sulfur from an ore or mineral by some suitable process, such as roasting.

**detector.** Material or device that is sensitive to radiation and can produce a response signal suitable for measurement or analysis; radiation detection instrument.

**detergent oil.** A lubricating oil having special sludge-dispersing properties for use in internal-combustion engines. These properties are usually conferred on the oil by the incorporation of special additives which give it the ability to hold sludge particles in suspension as well as to promote engine cleanliness.

**detonation.** A violent explosion. The term is also used to describe the knock-producing type of combustion in spark-ignition, internal-combustion engines.

**detrital.** Descriptive of minerals occurring in sedimentary rocks that were derived from preexisting igneous, sedimentary, or metamorphic rocks.

**deuterium.** An isotope of hydrogen whose nucleus contains one neutron and one proton and is therefore about twice as heavy as the nucleus of normal hydrogen, which is only a single proton. Deuterium is often referred to as heavy hydrogen. It occurs in nature as 1 atom to 6500 atoms of normal hydrogen. It is nonradioactive and expected to be the primary fuel for fusion power plants.

**deuteron.** The nucleus of a deuterium atom. It consists of one proton and one neutron and has a mass of 2 and a positive charge of 1. Because of their higher mass, deuterons are used to bombard other nuclei to produce radioactive isotopes (i.e., in a cyclotron).

**development well.** A well drilled in order to obtain production of a gas or oil known to exist.

**devolatilization.** Progressive loss of the volatiles by the substance undergoing coalification process.

**dewatering coal.** The removal of moisture from coal after it has passed through the washer. The product may be further dewatered in centrifuges.

**dewpoint.** The temperature, at constant pressure and constant water vapor content, to which air must be cooled in order for saturation or condensation to occur. Also called saturation point.

**dewpoint temperature.** The temperature at which a vapor begins to condense and deposit as liquid.

**dialysis.** The separation of substances in solution by means of semipermeable membranes (such as parchment, cellophane, or living cells) through which the smaller molecules and ions diffuse readily while the larger molecules and colloidal particles diffuse very slowly or not at all. Such separations are important in nature (i.e., in living organisms and soil).

**diaphragm meter.** A meter which uses a flexible diaphragm in a bellows-type arrangement to measure the volume of gas.

**diastrophism.** The process of deformation that produces the continents and ocean basins, plateaus and mountains, folds of strata, and faults in the earth's crust.

**diatomic.** Consisting of two atoms; having two atoms in the molecule; having two replaceable atoms or radicals.

**dielectric.** An insulator; a material which offers relatively high resistance to the passage of an electric current but through which magnetic or electrostatic lines of force may pass. Most insulating materials (i.e., air, porcelain, mica, and glass) are dielectrics. A perfect vacuum would constitute a perfect dielectric.

**diesel engine.** A type of internal-combustion engine that burns heavy oil and needs no spark to ignite the fuel. It works in the following way: Air is introduced into the cylinders and compressed until it reaches a temperature above the ignition point of

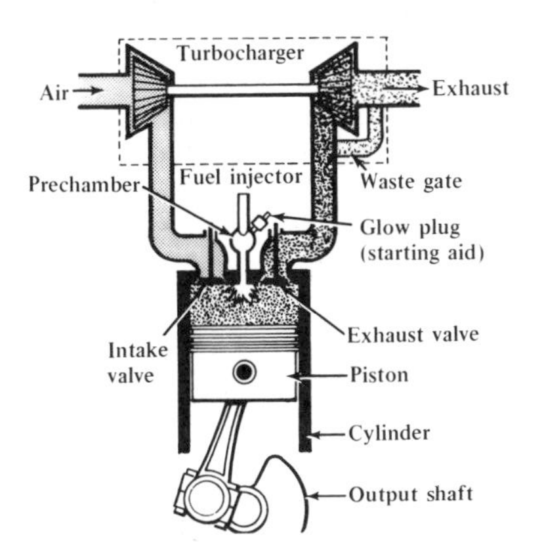


Diesel engine uses heat of compression for ignition. (*Courtesy Popular Science, April 1976, p. 83*)

the fuel; fuel oil is sprayed into the air; it is ignited by the high temperature and burned; the pressure of the hot expanding gases pushes the piston downward; the gases produced by combustion are evacuated through the exhaust valves.

**dieseling.** In a compressor, explosions of mixtures of air and lubricating oil in the compression chambers or in other parts of the air system.

**diesel oil.** Fuel for diesel engines obtained from the distillation of petroleum. It is composed chiefly of aliphatic hydrocarbons. Its volatility is similar to that of gas oil. Its efficiency is measured by cetane number.

**difference of potential.** The difference in electrical pressure existing between any two points in an electrical system or between any point of such a system and the earth. Determined by a voltmeter.

**differential cross section.** The cross section for an interaction involving one or more outgoing particles with specified direction of energy per unit interval of solid angle or energy.

**differential particle flux density.** That part of the particle flux density resulting from particles having a specified direction, energy, or both, per unit interval of solid angle, energy, or both.

**differential pressure.** The pressure difference between two points in a system; for example, the difference in pressure between the upstream and downstream taps of an orifice plate is used to measure volume passing through the orifice.

**diffused sky radiation.** Indirect sunlight.

**diffuser.** A device or structure that diffuses a windstream.

**diffuse radiation.** Solar radiation which is scattered by air molecules, dust or water droplets before reaching the ground and which is not capable of being focused.

**diffusion.** The permeation of one substance through another, such as gas through gas, liquid, or solid; solute through solvent; liquid through liquid or solid; and solid through solid.

**diffusion barrier.** A porous structure, the small pore size of which restricts ordinary gas flow but permit diffuse flow.

**diffusion coefficient for neutron flux density.** The ratio of the neutron current density at a particular energy to the negative gradient of the neutron flux density at the same energy in the direction of that current.

**diffusion cooling.** The decrease of the average energy of neutrons in a finite assembly due to the preferential leakage of neutrons with higher energies.

**diffusion heating.** The increase in the average energy of neutrons in an assembly due to either spectral hardening or the preferential diffusion or higher energy neutrons from an external neutron source.

**diffusion length.** The square root of the diffusion area.

**diffusion theory.** An approximate theory for the diffusion of particles, especially neutrons, based on the assumption that in a homogenous medium the current density is proportional to the gradient of the particle flux density.

**diluent.** A neutral fluid added to another fluid to reduce the concentration of the second fluid in a mixture.

**diluvium.** Sand, gravel, clay, etc., in surficial deposits.

**dimensionless variable.** A quantity which does not have dimensional units and is therefore has the same value in any system of units.

**dip.** The angle at which coal beds, stratified rocks, or other planar features, such as faults, are inclined from the horizontal.

**dipole.** Coordinate valence link between two atoms; electrical symmetry of a molecule. When a molecule is formed by the sharing of two electrons between a donor atom and an acceptor, it is more positive at the donor end and more negative at the acceptor end and has a dipole moment of the order of $10^{-18}$ electrostatic units.

**direct current.** An electric current, such as that produced by a battery, in which the electrical potential does not change its sign, and the voltage is often invariant with time. In a direct current, therefore, energy is carried by a continuous, unidirectional flow of electrons through a conductor.

**direct-cycle reactor system.** A nuclear power plant system in which the coolant or heat transfer fluid circulates through the reactor and then passes directly to the turbine in a continuous cycle.

**direct energy conversion.** The generation of electricity from an energy source which does not involve transference of energy to a working fluid. The conversion methods have no moving parts and usually produce direct current. Some methods include thermoelectric conversion, thermionic conversion, and magnetohydrodynamic conversion.

**direct-fired.** A heating unit in which the combustion products are mixed with the air or liquid being heated.

**directional drilling.** Rotary drilling technique in which the course of a borehole is controlled by deflection wedges or other means. The technique is used to deflect a deviated borehole back on to course, to bypass an obstruction in the hole and to reach side areas.

**directly ionizing particles.** Charged particles (such as electrons, protons, or alpha particles) having sufficient kinetic energy to produce ionization by collision.

**direct solar conversion (photovoltaics).** The process of directly converting sunlight to electricity through solar cells.

**dirty bomb.** A fission bomb or any other weapon which would distribute relatively large amounts of radioactivity upon explosion, as distinguished from a fusion weapon.

**disabling injury.** One which renders the employee unable to return to a regularly assigned job in one or more days after the injury.

**discriminator.** An electronic circuit which selects signal pulses according to their pulse height or voltage. It is used to delete extraneous radiation counts or background radiation, or as the basis for energy spectrum analysis.

**discounted cash flow.** The present worth of a future payment.

**disintegration.** The breaking up and crumbling away of a rock, caused by the action of moisture, heat, frost, air, and the internal chemical reaction of the component parts of rocks when acted upon by these surface influences.

**disintegration energy.** For a given nuclear disintegration, the amount of energy released.

**disordering.** Any process by which atoms are displaced from or rearranged among their positions in a crystal lattice, e.g., by ionizing radiation.

**dispatch.** The commitment of a utility's generating units to meet demand for electricity in a fashion determined to be most efficient by the system controllers.

**dispersant.** A substance used to break up concentrations of organic material.

**dispersion fuel.** A fuel mixture in which the fuel bearing phase is dispersed in a matrix of nonfissionable material.

**displacement.** A general term for the change in position of any point on one side of a fault plane relative to any corresponding point on the opposite side of the fault plane.

**Disposable Catalyst Hydrogenation process.** A process in which a low cost disposable catalyst is tested in a Synthoil style fixed bed reactor and in a stirred reactor.

Low sulfur fuel oil is the primary product. Reactor operating conditions (460° C and 250 to 700 atmospheres) are significantly less than those of the Bergius process.

**dissection.** The effect of erosion in destroying the continuity of a relatively even surface by cutting ravines or valleys into it.

**dissolution.** The taking up of a substance by a liquid with the formation of a homogenous solution.

**dissolved natural gas.** Natural gas in solution in crude oil in the reservoir.

**distillate.** That portion of a liquid which is removed as a vapor and condensed during a distillation process. As fuel, distillates are generally within the 204° to 353°C boiling range and include Nos. 1 and 2 fuel, diesel, and kerosine.

**distillate fuel oil.** The lighter fuel oils distilled off during the refining process. Included are products known as ASTM grades Nos. 1 and 2 heating oils, diesel fuels, and No. 4 fuel oil. The major uses of distillate fuel oils include heating, fuel for on- and off-highway diesel engines, and railroad diesel fuel. Minor quantities of distillate fuel oils produced and/or held as stocks at natural gas processing plants are not included.

**distillation.** A process of evaporation and recondensation whereby liquids are separated into various fractions according to their boiling points or boiling ranges.

**distillation of petroleum.** Process of separating hydrocarbons in crude oil into narrow boiling range fractions which can be further processed into useful products. In a modern distillation unit, the heated crude oil—a mixture of vapors and liquids—is discharged into a fractionating tower. The vapors rise upward through distillation trays at various levels. At the same time, part of the liquid condensate from the top of the tower (called reflux) is returned to the system and flows down over the trays, establishing the proper conditions for fractional distillation. As the operation continues, different fractions of the crude oil condense on different trays, depending upon the temperatures at which they change from vapor to liquid. Heavier fractions with higher boiling points condense on the lower (and hotter) trays; lighter fractions with lower boiling points condense on the higher (and cooler) trays. The unvaporized residue is drawn off at the bottom for further processing.

**distributed collectors.** One basic configuration of the distributed collector system is a series of modular cylindrical (parabolic) or segmented mirror focusing collectors with an interconnected absorber pipe network to carry the solar-heating working fluid to a heat exchanger unit. There the working fluid generates steam in order to power the turbogenerator. Since distributed collectors focus direct solar radiation on the absorber pipe, they must incorporate at least a one-dimensional tracking mechanism for purposes of tracking the sun. Distributed collector systems of the parabolic or segmented mirror configuration are generally designed to operate with working fluid temperatures in the vicinity of 300°C. Another distributed collector system is the paraboloidal dish. This system consists of a series of circular dish-shaped collectors in the form of a parabola, with the absorber located at the focus. The absorber units are interconnected with insulated piping. Paraboloidal dish systems are capable of generating temperatures of approximately 800°C. These systems generally have overall conversion efficiencies in the range of approximately 20 to 25 percent.

Highly reflective curved metal plates on this collector cause the sun rays to converge on the glass tube in the center. Water or other suitable liquid is heated by the rays and circulated to utilize the solar energy. (*Courtesy Sandia Laboratories*)

**distribution factor.** A term used to express the modification of the effect of radiation in a biological system attributable to the nonuniform distribution of an internally deposited isotope, such as radium's being concentrated in bones.

**distribution system.** Equipment which carry or control the supply of fuel.

**diurnal variation.** The daily variation in the earth's magnetic field; in tides, having a period or cycle of approximately 1 lunar day (24.84 solar hours).

**diversity.** That characteristic of variety of electric loads whereby individual maximum demands usually occur at different times. Diversity among customer's loads results in diversity among the loads of distribution transformers, feeders, and substations, as well as between entire systems.

**diversity factor.** The ratio of the sum of the individual maximum loads during a period to the simultaneous maximum loads of all the same units during the same period.

**DOE.** Department of Energy.

**doldrums.** A part of the ocean near the equator abounding in the light shifting winds.

**dollar.** A unit of reactivity. One dollar is the maximum amount of reactivity in a reactor due to delayed neutrons alone.

**domestic coke.** Normally a byproduct of coal-gas plants and commercial byproduct plants. The general characteristics of the coal, therefore, are fixed by the requirements for gas and coking coals. Domestic coke varies greatly in quality.

**domestic hot water.** The hot water used for conventional purposes such as bathing and washing.

**domestic water heating system.** A solar heating system which supplies a portion of the energy needed to heat water used for domestic purposes, such as bathing and washing.

**domestic noncontrolled crude oil.** That portion of domestic crude oil production including new, released, and stripper oil, which may be sold at a price exceeding the ceiling price.

**donor.** An atom or ion which supplies an electron pair to another atom, thus creating a covalent bond.

**dope.** Slang expression for additive in the oil industry; absorbent material used in certain manufacturing processes.

**Doppler-averaged cross section.** A cross section averaged over energy, employing appropriate weighting factors, to take into account the effect of thermal motion of the target particles. The product of the average cross section so obtained and the flux density in the laboratory system then gives the correct reaction rate.

**Doppler broadening.** In spectroscopy, the observed broadening of a spectral line resulting from the thermal motion of the molecules, atoms, or nuclei. In reactor technology, it is the observed broadening of the energy width of a cross section resonance resulting from the thermal motion of the target particles.

**Doppler coefficient.** That part of the temperature coefficient of reactivity which arises from Doppler broadening.

**Doppler effect.** A shift in the measured frequency of a wave pattern caused by movement of the receiving device or the wave source. The moving receiver will intercept more or fewer waves per unit time, depending on whether it is moving toward or away from the source of the waves. Also, in the nuclear field, it is the shift with temperature of the interaction rate between neutrons and reactor materials, such as fuel rods, structural materials, and fertile materials.

**dose.** The amount of ionizing radiation energy absorbed per unit mass of irradiated material at a specific location, such as a part of the human body.

**dose equivalent.** A term used to express the amount of effective radiation when modifying factors have been considered. It is the product of absorbed dose multiplied by a quality factor multiplied by a distribution factor expressed numerically in rems.

**dose equivalent residual.** The accumulated dose corrected for such physiological recovery as has occurred at a specific time. It is based on the ability of the body to recover to some degree from radiation injury following exposure. It is used only to predict immediate effects.

**dose rate.** The radiation dose delivered per unit time and measured in rems per hour.

**dosimeter.** Any instrument which measures radiation dose; also called dosimeter.

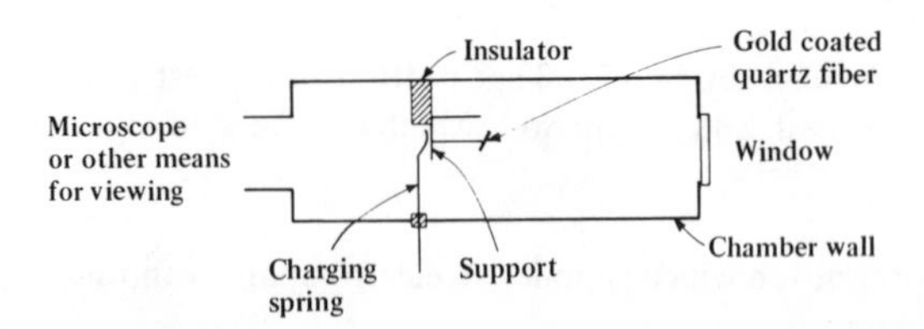


Structure of the typical electrostatic dosimeter. (*Courtesy Basic Nuclear Engineering, 3rd Edition, Arthur R. Foster and Robert L. Wright, Jr., Allyn and Bacon, Inc., Boston, Mass.*)

**double block and bleed system.** A valving system wherein a full-flow vent valve is located on the piping between two shutoff valves in series for the purpose of bleeding to atmosphere excess pressure between the valves.

**double glazing.** Windows having two sheets of glass with an airspace between.

**doubling dose.** Radiation dose eventually causing a doubling of gene mutations.

**doubling time.** The time it takes for a quantity to double; the time required for a breeder reactor to produce as much fissionable material as the amount usually contained in its core plus the amount tied up in its fuel cycle (as in fabrication and reprocessing.). Doubling time depends on the breeding gain and the specific power at which the reactor operates. It is estimated as 10 to 20 years in typical reactors.

**downdraft.** A flow of air down the chimney or flue because of adverse draft conditions; a downward current of air or other gas as in a mine shaft, kiln, or carburetor.

**downhole well-logging instruments.** Instruments which measure characteristics of formations such as electrical resistivity, radioactivity, and density. The information is used to evaluate the formations for petroleum content.

**downstream.** Any point in the direction of flow of a fluid or gas from the reference point.

**downwind.** On the opposite side from the direction from which the wind is blowing.

**draft.** A difference of pressure which tends to cause a flow of air and/or flue gases through the boiler, flue connector, breeching, flue, or chimney.

**draft hood.** A device built into an appliance, or made a part of the flue or vent connector from an appliance, which is designed to (a) assure the ready escape of the products of combustion from the combustion chamber in the event of no draft, back draft, or stoppage beyond the draft hood; (b) prevent a back draft from entering the combustion chamber of the appliance; and (c) neutralize the effect of stack action of the chimney or gas vent upon the operation of the appliance.

**draft regulator.** A device which functions to maintain a desired draft in the appliance by automatically reducing the draft to the desired value.

**drag.** The frictional resistance of a current of air in a mine.

**dragline.** A type of excavating equipment which casts a rope-hung bucket a considerable distance, collects the dug material by pulling the bucket toward itself on the ground with a second rope, elevates the bucket, and dumps the material on a spoil bank, in a hopper, or on a pile.

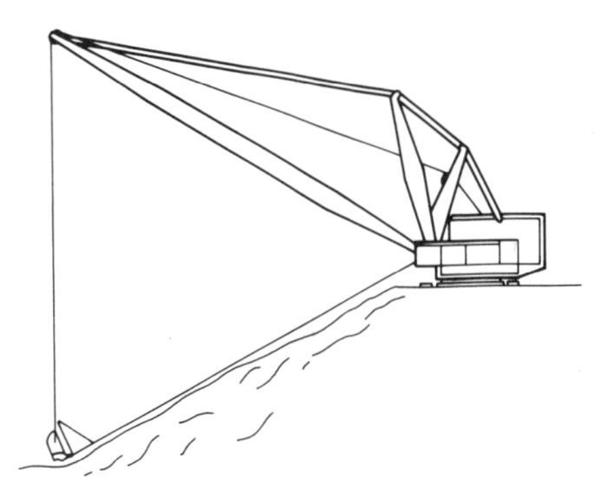

Dragline. (*Courtesy U.S. Council on Environmental Quality*)

**drag-type devices.** Devices that are actuated by aerodynamic drag in a windstream.

**drainage basin.** The area from which water is carried off by a drainage system, a watershed, or a catchment area.

**draw.** Strictly speaking, the distance on the surface to which the subsidence or creep extends beyond the workings; also, the effect of creep upon the pillars of a mine.

**drawdown.** The difference between the static and the flowing bottom hole pressure.

**draw slate.** A soft slate, shale, or rock approximately 2 inches to 2 feet in thickness, above the coal, and which falls with the coal or soon after the coal is removed.

**dredging.** The removal of soil from under water to deepen streams, swamps, or coastal waters. The resulting mud is usually deposited in marshes in a process called filling. Dredging and filling can disturb natural ecological cycles.

**drift.** Water lost from an evaporative cooling tower as liquid droplets are entrained in the exhaust air; also a deep mine entry driven directly into a horizontal or near horizontal mineral seam or vein when it outcrops or is exposed at the ground surface.

**drift mine.** A mine that opens into a level or nearly level seam of coal.

**drift mining.** The working of relatively shallow coal seams from the surface by drifts, which are generally inclined and may be driven in rock or in a seam. Drift mining may be viewed as intermediate between opencast coal mining and shaft or deep mining.

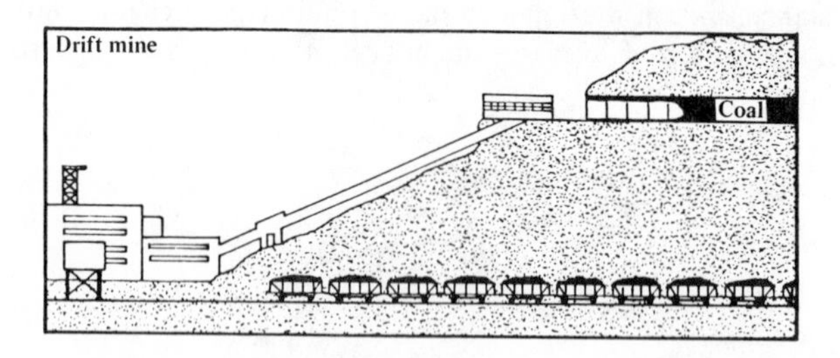


Drift mines are underground mines found in hilly areas.

**drilling bit.** A drilling tool that cuts the hole in the earth in drilling a well. Bits are designed on two basic and different principles: the cable tool bit, which moves up and down to pulverize; and the rotary bit, which rotates to cut or grind.

**drilling fluid.** The thick fluid kept circulating in a borehole to clear the chippings and cool the chisel. (See Fig., p. 129)

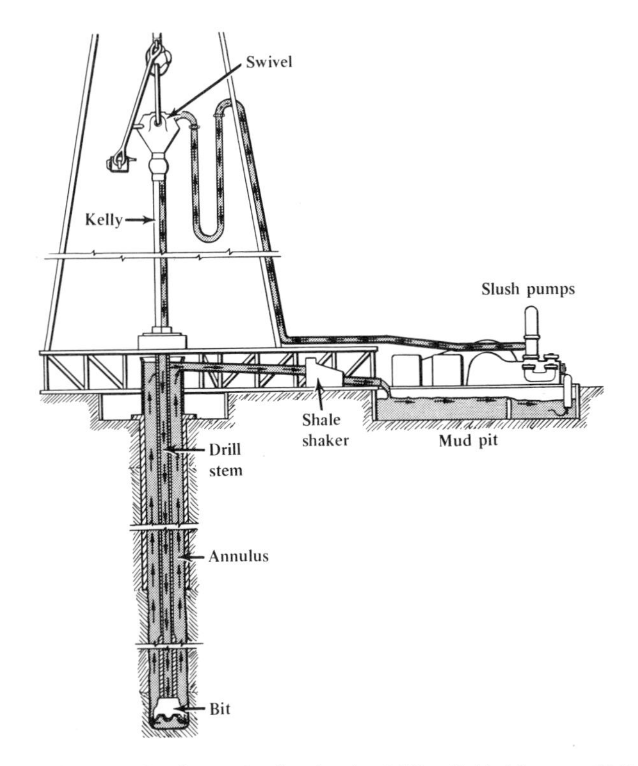


Drilling and mud system showing path taken by the drilling fluid. (*Courtesy U.S. Council on Environmental Quality*)

**drilling rig.** A drill machine complete with all tools and accessory equipment needed to drill boreholes or service an oil well. Also called drill rig.

**drill pipe.** In rotary well drilling, the rigid pipe connection between the collar of the drill at the working level and the rotary table on the derrick platform. In addition to transmitting the driving power to the drill bit, the open drill pipe is used to force mud to and through the perforated drill bit for the purpose of cooling and lubricating the bit and picking up the cuttings so that they can be washed to the surface and removed.

**drill string.** A column of pipe that connects to a bit used to bore (drill) holes for wells.

**drip.** A container, or segment of piping, placed at a low point in a system to collect condensate, dust, or foreign material, enabling their removal.

**drive.** To excavate horizontally or at an inclination, as in a drift, adit, or entry.

**driver fuel.** The nuclear fuel in the driver zone.

**driver zone.** That zone which principally sustains the chain reaction in a multizone reactor core.

**dross.** Small coal which is inferior or worthless, and often mixed with dirt.

**dry bulb temperature.** The temperature registered by the dry bulb thermometer of a psychrometer. It is identical with the temperature of the air.

**dry cell.** A primary cell which does away with the liquid electrolyte so that it may be used in any position.

**drycleaned coal.** Coal from which impurities have been removed mechanically without the use of liquid media.

**dry cooling.** Cooling in which waste heat is dissipated directly to the atmosphere.

**dry criticality.** Reactor criticality achieved without a coolant.

**dry gas.** A natural gas consisting principally of methane and ethane and devoid of the heavier hydrocarbons; a gas which does not contain fractions that may easily condense under normal atmospheric conditions; also applied to gas that has been produced and from which liquid components have been removed.

**dry hole.** A drilled well which does not yield gas and/or oil quantities or condition to support commercial production; also a drill hole in which no water is used for drilling, as a hole driven upward.

**dry limestone process.** The process of controlling air pollution caused by sulfur oxides by exposing the polluted gases to limestone, which combines with oxides of sulfur to form manageable residues.

**dry steam.** Steam containing no moisture—either saturated or superheated steam; an energy source obtained when hot water boils in an underground reservoir, steam rises, some of it condensing on surrounding rock and dry steam approaches the surface to be tapped and used in a turbine. At present, only three areas in this work have been revealed by drilling and the capacity of these areas is limited.

**dry steam energy system.** A geothermal energy source; a vapor-dominated hydrothermal convective system generally having a temperature greater than 300°F; a natural steam field producing superheated steam. The major geothermal power plants in the United States are located at a natural steam field in Northern California known as "the Geysers."

**dual-cycle reactor system.** A reactor-turbine system in which part of the steam fed to the turbine is generated directly in the reactor and part in a separate heat exchanger; a combination of direct-cycle and indirect-cycle reactor systems.

**dual-purpose reactor.** A reactor designed to achieve two purposes; for example, to produce both electricity and new fissionable material.

**duct.** A passageway made of sheet metal or other suitable material, not necessarily leak-tight, used for conveying air or other gas at low pressures.

**dummy assembly.** An assembly without nuclear fuel intended to replace or to represent a fuel assembly.

**dump.** The place of deposit of solid waste in a manner that does not protect the environment.

**dust.** Fine-grain particulate matter.

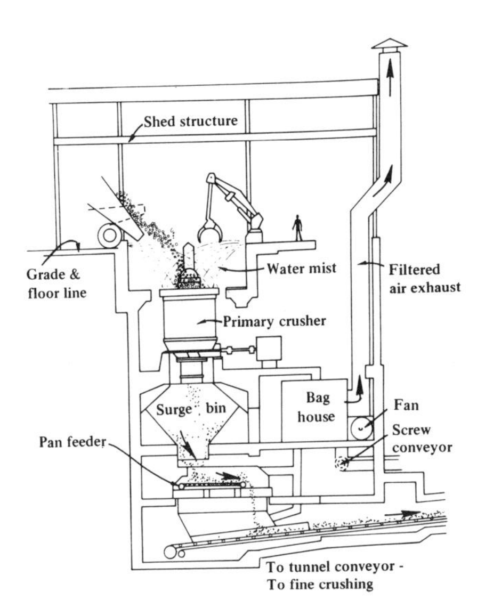


Primary crusher dust control. (*Courtesy U.S. Council on Environmental Quality*)

**dutchman.** A filler-piece used to close a gap between two pieces of pipe.

**dynamite.** A general term relating to explosives in which the principal explosive ingredient is nitroglycerin contained within an absorbent substance.

**dynamo.** A machine for converting mechanical energy into electrical energy by magnetoelectric induction. A dynamo may also be used as a motor.

**dyna-ship.** A clipper type, wind-powered ship, designed by Mr. Wilhelm Prölss, which has, in effect, a continuous airfoil from the top of the mast to the bottom. The angle of the sails is changed by turning the whole mast hydraulically.

**dyne.** Unit of force; the force that will give a mass of 1 gram an acceleration of 1 centimeter per square second.

**dystrophic lakes.** Shallow lakes having high humus content, high organic matter content, low nutrient availability, and high biochemical oxygen demand (BOD).

# E

**earth coal.** An earthy brown coal; a name sometimes given to lignite; a mineral coal as distinguished from charcoal.

**earth current.** A light electric current which appears to traverse the earth's surface but which in reality exists in a wire grounded at two ends, due to small potential differences between the two points at which the wire is grounded.

**earthquake.** A local trembling, shaking, undulating, or sudden shock of the surface of the earth, sometimes accompanied by fissuring or by permanent change of level.

**earth tremor.** A slight earthquake.

**ebb current.** The movement of the tidal current away from shore or down a tidal stream.

**EBR-I.** The experimental breeder reactor EBR-I was the first nuclear reactor to produce electricity. This occurred on December 20, 1951.

**ebullition.** The act, process, or state of boiling or bubbling up.

**eccentric.** A device for converting continuous circular motion into reciprocating rectilinear motion, consisting of a disk mounted out of center on a driving shaft, and surrounded by a collar or strap connected with a rod. Rotation of the driving shaft gives the rod a back-and-forth motion.

**echo ranging.** Locating underwater objects by sending sound pulses into water. Target range is derived by measuring transit time of sound pulse.

**ecology.** The study of interrelationships of animals and plants to one another and to their environment.

**economic coal reserves.** The reserves in coal seams which are believed to be workable with regard to thickness and depth. In most cases, a maximum depth of about 400 feet and a minimum thickness of about 2 feet is taken as workable. The minimum economic thickness varies according to quality and workability.

**economizer.** A heat exchanger for recovering heat from flue gases and using it to heat feedwater or combustion air.

**ecosystem.** The interacting system of a biological community and its nonliving environment.

**eddy.** A circular movement of water or air. Eddies may be formed where currents pass obstructions or between two adjacent currents flowing counter to each other.

**eddy loss.** Energy lost by eddies as distinct from that lost by friction.

**edge retaining system.** The metal channel which holds in place the edges of the various layers of a solar collector panel.

**edge water.** Subsurface water that surrounds gas and oil in reservoir structures.

**edge well.** A well so located as to be at the edge of oil or gas accumulation or at the edge of a leased reservoir; a well at or near the contact of oil and/or gas and water.

**Edison, Thomas.** Thomas Edison (1847–1931) American inventor. Edison patented a total of 1093 inventions—more than any other man in American history. Among the most important were the incandescent electric light bulb, the phonograph, the movie projector, and an amplifier for the telephone.

**Edwards balance.** An instrument for determining the specific gravity of gases.

**effective capacity.** The maximum load which a system is capable of carrying under existing service conditions.

**effective cutoff energy.** For a specific absorbing cover surrounding a given detector, that energy value which satisfies the condition that if the cover were replaced by a hypothetical cover black to neutrons with energy below this value and transparent to neutrons with energy above this value, the observed detector response would be unchanged.

**effective delayed neutron fraction.** The ratio of the mean number of fissions caused by delayed neutrons to the mean total number of fissions caused by delayed plus prompt neutrons.

**effective half-life.** The time required for a radionuclide contained in a biological system, such as a man or an animal, to reduce its activity by half, as a combined result of radioactive decay and biological elimination.

**effectiveness.** The ratio of actual heat transfer in a heat exchanger to the maximum possible heat transfer.

**effective horsepower.** The amount of useful energy that can be delivered by an engine.

**effective relaxation length.** The distance within a material in which the radiation intensity is reduced to $1/e$ of its initial value. This term includes geometric effects.

**effective stack height.** The effective stack height equals the sum of the actual height, the rise attributed to the density difference between the stack gases and the atmosphere, and the rise attributed to the velocity of the stack gases.

**effective temperature.** An arbitrary index which combines into a single value the effect of temperature, humidity, and air movement on the sensation of warmth or cold felt by the human body. The numerical value is that of the temperature of still, saturated air which would induce an identical sensation.

**effective thermal cross section.** A fictitious cross section for a specified interaction which, when multiplied by the conventional flux density, gives the correct reaction rate.

**efficiency.** That percentage of the total energy content of a power plant's fuel which is converted into electricity, the remaining energy being lost to the environment as heat. The efficiency of an energy conversion is the ratio of the useful work or energy output to the work or energy input. (See Fig., p. 136)

$$E = \frac{\text{work or energy out}}{\text{work or energy in}} \times 100 \text{ percent}$$

**efflorescence.** In geology, the formation of crystals by the evaporation of water from solutions brought to the surface by capillarity.

**effluent.** A liquid, solid, or gaseous product, frequently waste, discharged from a process; generally used in regard to discharges into waters.

**effusion.** That property of gases which allows them to pass through porous bodies, that is, the flow of gases through larger holes than those to which diffusion is strictly applicable.

**egress.** The provision of two or more exits from a confined space containing machinery in order to minimize the risk of a person being trapped in the event of an outbreak of fire, or escape of steam or noxious gases. Also applies to mine workings.

**Ekman spiral.** An idealized mathematical description of the wind distribution in the planetary boundary layer, within which the earth's surface has an appreciable effect on the air motion. The model is simplified by assuming that within this layer the eddy viscosity and density are constant, the motion is horizontal and steady, the isobars are straight and parallel, and the geostrophic wind is constant with height.

**elasticity.** The property or quality of being elastic, as when an elastic body returns to its original form or condition after a displacing force is removed. A measure of the percentage impact of a change in one economic variable on another.

**elastomer.** A stretchable substance, resembling rubber, that will return to approximately its original length.

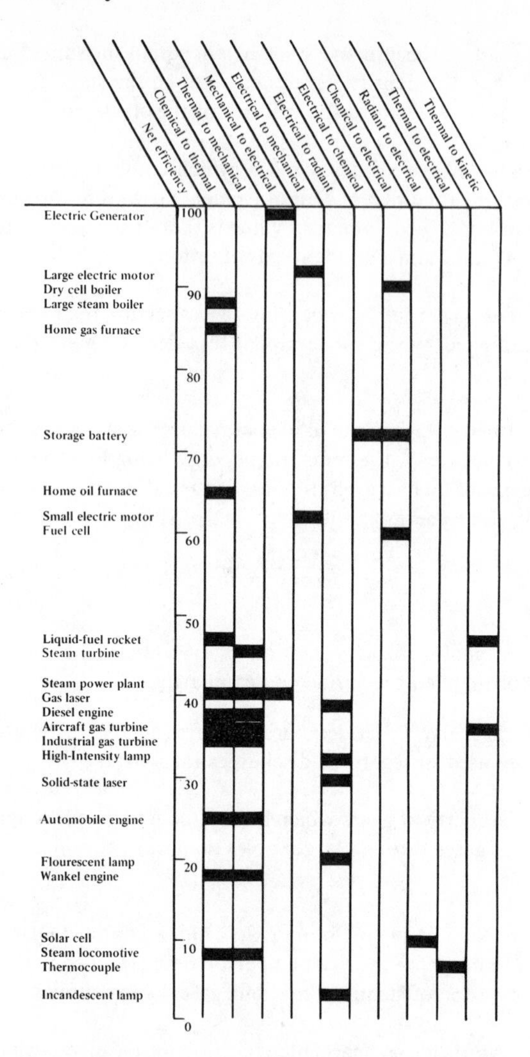


Efficiency of energy converters. (*Courtesy "Energy and the Future," Hammond, A. L., Matz, W. D., and Maugh II, T. N., American Association for the Advancement of Science, 1973, Fig. 43)*

**electrical conductivity.** The transfer of electricity within a substance from points of higher potential to points of lower potential. Conductivity is usually measured by its inverse—resistivity. The unit of conductivity is the mho.

**electrical energy.** The energy of moving electrons.

**electrical heat.** When a current flows in a circuit which contains resistance, heat is produced and the resistance and conductors of the circuit are raised in temperature.

**electrical system.** A system in which all the conductors and apparatus are electrically connected to a common source of electromotive force.

**electric charge.** A property of matter resulting from an imbalance between the number of protons and the number of electrons in a given piece of matter. The electron has a negative charge; the proton, a positive charge. Like charges repel each other; unlike charges attract.

**electric energy.** The energy associated with electric charges and their movements. Measured in watt-hours or kilowatt-hours. One watt-hour equals 860 calories. Available heat in electricity; one kilowatt hour equals 3412.97 British thermal units.

| Source | Percent of total |
|---|---|
| Coal | 45.5 |
| Natural gas | 19.8 |
| Petroleum | 17.3 |
| Hydropower | 14.7 |
| Nuclear | 4.3 |

Sources of electric power. (*Courtesy National Science Teachers Association*)

**electric generator.** A source of electricity, especially one that transforms mechanical or heat energy into electric energy.

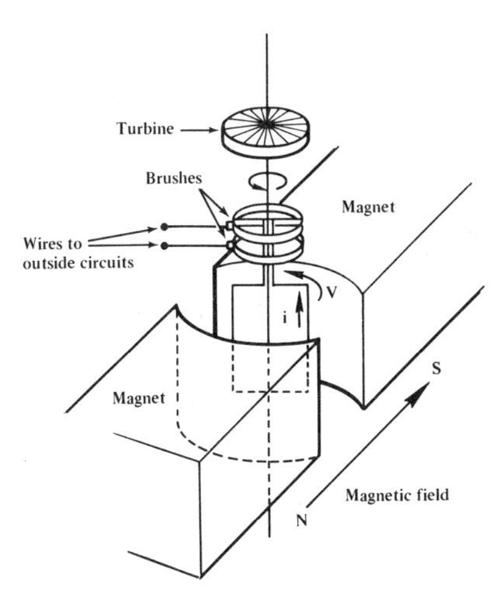


Diagram of an electric generator. (*Courtesy National Science Teachers Association*)

**electricity.** The general name for all phenomena that arise out of the flow or accumulation of electrons in matter. The unit of quantity is the coulomb; that of pressure, the volt; and that of flow, the ampere. (See Fig., p. 138)

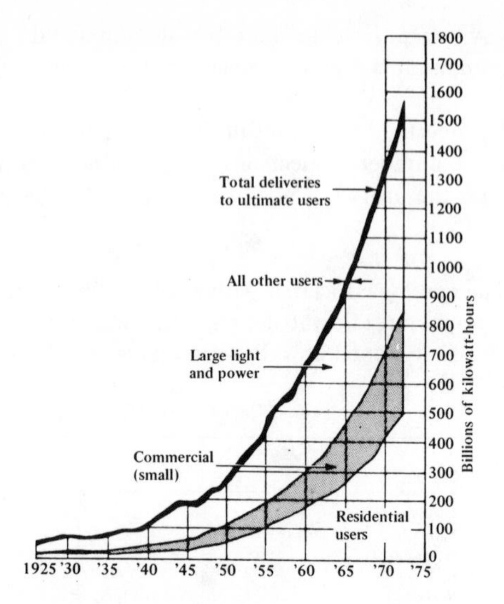


Major traditional classifications of electric power users. (*Courtesy Van Nostrand's Scientific Encyclopedia*)

**electric power transmission.** The transmission of energy through conductors is a flow of electrons under pressure. The flow is measured in amperes, and the pressure or electrical potential is measured in volts. The amount of energy transmitted is proportional to the product of the flow and the pressure.

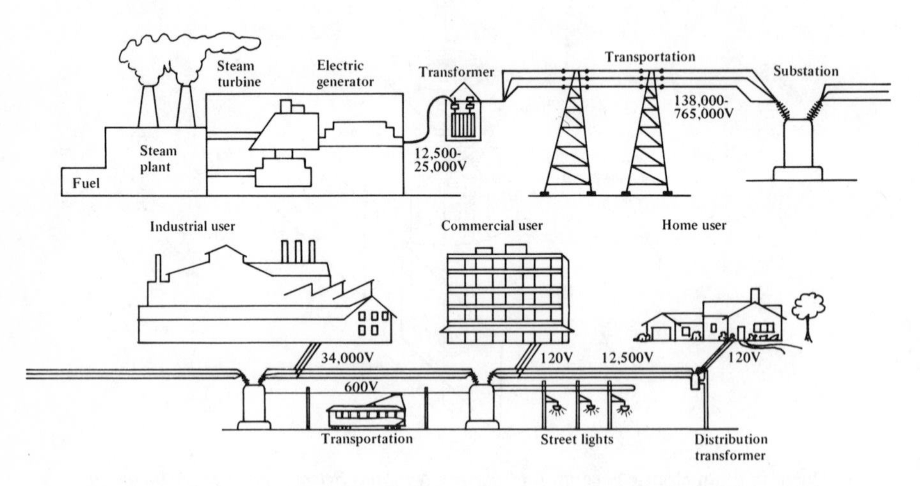


The transmission of electricity. (*Courtesy National Science Teachers Association*)

**electric utility company fuel costs.** The total cost of fuel consumed divided by its total Btu content, multiplied by one million.

**electric vehicles.** Electric-powered vehicles. Advanced battery systems which have a three- to four-fold increase in range over existing electric vehicles. These vehicles will be able to travel 180 miles on one charge.

Four passenger electric test car designed for stop and go urban driving. A flywheel is featured for better acceleration. (*Courtesy AiResearch Manufacturing Company, Torrance, CA*)

**electrochemical.** Chemical action employing a current of electricity to cause or to sustain the action.

**electrochemical cell (battery).** A battery consisting of a combination of two or more cells, each cell containing two conducting electrodes, one positive and the other negative, and made of unlike materials (metals or their salts). These electrodes are immersed in chemical solutions called electrolytes, whose function is mainly to carry positive ions from the negative pole to the positive. To neutralize the charge that results, a current of electrons will then flow through external conducting wires from one pole to the other. This process can be made to give up energy if sent through a proper energy-converting device such as an electric motor.

**electrode.** Reactive materials, such as metals and metal oxides, attached to grids that conduct electricity.

**electrodeless fluorescent bulb.** An energy-efficient fluorescent light developed by Lighting Technology Corporation in a research and development project supported by DOE. The LITEK bulb contains no electrodes or filaments and uses one third the energy needed for the more conventional incandescent lighting. Visible light is emitted through the excitation of atoms by a magnetic field. This magnetic field is set up by a radio frequency signal produced from an electronics package located in the base of the bulb.

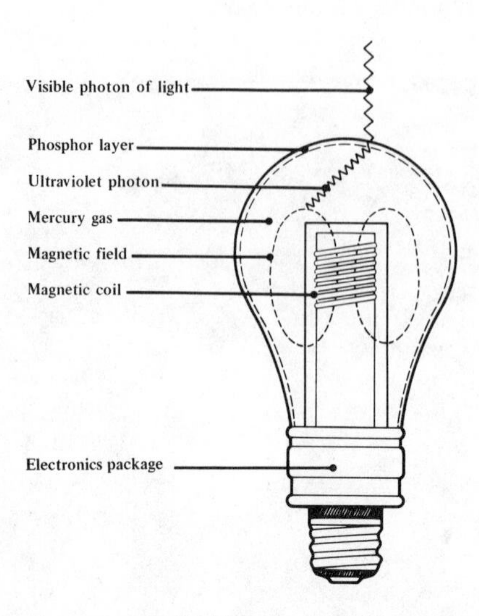


The LITEK electrodeless fluorescent bulb. (*Courtest U.S. Energy Research and Development Administration*)

**electrodialysis.** A process employing an electrical current and an arrangement of permeable membranes for separating soluble minerals from water.

**electrolysis.** The chemical decomposition of a substance when electricity is passed through it in solution or in the molten states. In a pipeline, the term applies to the decomposition or destruction of the pipe wall by stray electrical currents.

**electrolyte.** Materials which when placed in solution make the solution conductive to electrical currents. These include aqueous solutions of salts, acids, or bases, and some molten salts.

**electrolytic cell.** An assembly consisting of a vessel, electrodes, and an electrolyte in which electrolysis can be carried out.

**electromagnet.** A core of magnetic metal (as soft iron) that is surrounded wholly or in part by a coil or wire that is magnetized when an electric current is passed through the wire, and that retains its power of attraction only while the current is flowing.

**Electrolytic Process.** A process using electric current for the decomposition of chemical compounds, for example, the transformation of alumina (aluminum oxide) into aluminum.

**electromagnetic pump.** A pump in which electromagnetic fields are used directly to pump a liquid metal.

**electromagnetic radiation.** The wide range of radiant energy in the universe that makes up the total electromagnetic spectrum.

**electromagnetic separation process.** A process for separating isotopes which depends on their behavior in an electromagnetic field.

**electromagnetic spectrum.** The entire range of electrical energy extending from the extremely long rays of radio and electricity at one end to the extremely short x-rays at the other.

**electromagnetism.** Every electric current generates a magnetic field which is in a plane perpendicular to the current. The strength of the field is proportional to the current and, in the case of a long, straight wire is inversely proportional to the distance from the wire. This principle is important in magnetic prospecting insofar as it forms the basis for certain types of geomagnetic instruments.

**electromotive force.** A measure of the intensity of electrical energy needed to produce a current in a circuit. The practical unit is the volt.

**electron.** An elementary particle with a rest mass of $9.1 \times 10^{-28}$ grams, bearing a negative electric charge. Electrons orbit the atomic nucleus; their transfer or rearrangement between atoms underlies all chemical reactions. Either negative or positive electrons (sometimes called positrons) may be emitted from atomic nuclei during nuclear reactions; they are then called beta particles.

**electron capture.** A mode of radioactive decay of a nuclide in which an orbital electron is captured by and merges with the nucleus, thus forming a new nuclide in which the mass number remains unchanged but the atomic number is decreased by 1.

**electronegative.** Descriptive of an element or a group which ionizes negatively, or acquires electrons, becoming thereby a negatively charged anion.

**electron volt.** The amount of kinetic energy gained by an electron when it is accelerated through an electric potential difference of 1 volt. It is equivalent to $1.603 \times 10^{-12}$ erg. It is a unit of energy or work, not of voltage.

**electropositive.** Positively charged; having more protons than electrons.

**electrorefining.** The process of anodically dissolving a metal from an impure anode and depositing it in a more pure state at the cathode.

**electrostatic precipitator.** A device that removes particulate matter by imparting an electrical charge to particles in a gas stream for mechanical collection on an electrode. The accumulated particles can then be washed away.

**electrostatics.** Science of electric charges captured by bodies which then acquire special characteristics because of their retention of such charges.

**element.** A substance consisting entirely of atoms of the same atomic number.

**elementary particles.** The simplest particles of matter and radiation. Most are short-lived and do not exist under normal conditions (exceptions are electrons, neutrons, protons, and neutrinos). Originally this term was applied to any particle that could not be subdivided, or to constituents of atoms; now it is applied to nucleons (protons and neutrons), electrons, mesons, muons, baryons, strange particles, the antiparticles of each of these, and to photons, but not to alpha particles or deuterons. Also called fundamental particles.

**elkerite.** A variety of bitumen formed through a slow oxidation of petroleum.

**elutriation.** A process of purification or sizing by washing and pouring off the lighter or finer matter suspended in water while leaving the heavier or coarser portions behind.

**eluvium.** Atmospheric accumulations in situ, or shifted only by wind, in distinction to alluvium, which requires the action of water.

**emanation.** The escape of radioactive gases from the materials in which they are formed; for example, radon from radium and krypton and xenon from a substance undergoing fission.

**emergency core cooling system.** A safety system in a nuclear reactor. It consists of a reserve system of pipes, valves, and water supplies designed to flood water into the core to prevent the fuel in the reactor from melting if a sudden loss of coolant occurs.

**emergency shutdown.** The act of shutting down a reactor suddenly to prevent or minimize a dangerous condition.

**emergency spillway.** A spillway designed to convey water in excess of that impounded for flood control.

**emission factor.** The average amount of a pollutant emitted from each type of polluting source in relation to a specific amount of material processed.

**emission inventory.** A list of air pollutants emitted into a community's atmosphere, in amounts (usually tons) per day, by type of source.

**emission rate.** The number of particles of a given type and energy leaving a given radiation source per unit time.

**emission standard.** The maximum amount of a pollutant legally permitted to be discharged from a single source.

**emissivity.** The ratio of radiant energy emitted by a body to that emitted by a perfect blackbody. A perfect blackbody has an emissivity of 1; a perfect reflector, an emissivity of 0.

**emittance.** The ratio of the radiant energy (heat) emitted from a surface at a given temperature to the energy emitted by a perfect black body at the same temperature.

**emulsification.** The phenomenon of holding finely divided particles of a liquid in suspension with the body of another liquid.

**encroachment.** To work coal beyond the boundary which divides one mine area from another; also the advancement of water, replacing withdrawn oil or gas in a reservoir.

**endogenetic.** Pertaining to rocks resulting from physical and chemical reactions, their origin being due to forces within the material.

**endothermic.** Heat absorbing. An endothermic reaction is one in which heat must be supplied to further the reaction.

**end point.** The temperature at which the last portion of oil has been vaporized in ASTM or Engler distillation. Also called the final boiling point.

**end-use demand.** The amount of energy used by final consumers. (See Fig. p. 144)

**energy.** The capacity to do work. It takes such forms as potential, kinetic, heat, chemical, electrical, nuclear, and radiant energy. Potential energy arises by virtue of the position or configuration of matter. Kinetic energy is energy of motion. Heat energy is the kinetic energy of molecules. Chemical energy arises out of the capacity of atoms to evolve heat as they combine or separate. Electrical energy arises out of the capacity of moving electrons to evolve heat, electromagnetic radiation, and magnetic fields. Nuclear energy arises out of the elimination of all or part of the mass of atomic particles. Radiant energy is energy in transit through space; it is emitted by electrons as they change orbit and by atomic nuclei during fission and fusion; on striking matter, such energy appears ultimately as heat. Only radiant heat can exist alone; all other forms require the presence of matter. Some forms of energy can be converted into other forms, and all forms are ultimately converted into heat. Energy is measured in ergs.

**energy absorption.** Generally, it refers to the energy adsorbed by any material subjected to loading. Specifically, it is a measure of toughness or impact strength of a material—the energy needed to fracture a specimen in an impact test. It is the difference in kinetic energy of the striker before and after impact, expressed as total energy (foot-pound or inch-pound) for metals and ceramics and as energy per inch for plastics and electrical insulating materials.

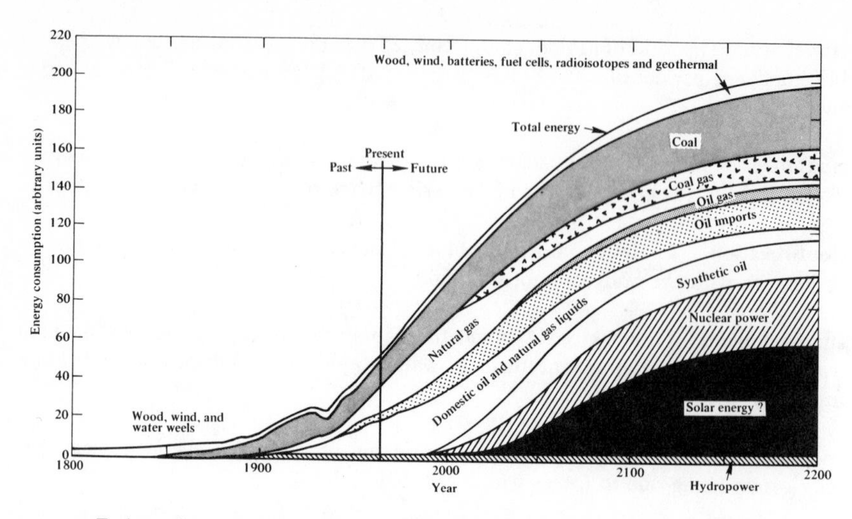


End-use demand requirements starting about year 2030. (*Courtesy DOE*)

**energy band.** Energy spectrum of valence electrons in a polyatomic material. Conduction is not significant if the energy band is filled.

**energy conservation plan.** Energy consumption reduction measures proposed by President Carter; including gasoline taxes, insulation rebates to homeowners and businesses, tax credit on solar energy equipment for residential and business use, natural gas increase in price, domestic oil increase in price, and the requirement of the burning of coal rather than natural gas or oil for utilities and industry.

**energy consumption.** The amount of energy consumed in the form in which it is acquired by the user. The term excludes electrical generation and distribution losses. Also called net energy consumption. (See Table, p. 145)

**energy conversion.** Transformations of energy from one form to another. (See Fig. top of p. 146)

**energy density.** Energy per unit of weight contained in a fuel.

**energy dissipation.** The transformation of mechanical energy into heat energy.

**energy efficiency ratio (EER).** Value representing the relative electrical efficiency of room air conditioners. The higher the energy efficiency ratio number, the less electricity the air conditioning unit will use to cool the same amount of air. Typical efficiency ratios are from 4 to 12; a unit with an EER of 4 will cost about 3 times as much to operate as one with an EER of 12. The Government is developing a pilot program to encourage the use of energy conservation and alternative energy technologies at

| Appliance | Average wattage | Average hours per year | Kilowatt hours per year | Average cost (1) per year |
|---|---|---|---|---|
| Air cleaner | 50 | 4320 | 216 | $ 6.48 |
| Air conditioner | 566 | 887 | 1389 | 41.67 |
| Blanket | 177 | 831 | 147 | 4.41 |
| Blender | 386 | 39 | 15 | .45 |
| Broiler | 1436 | 69 | 100 | 3.00 |
| Clock | 2 | 8500 | 17 | .51 |
| Clothes dryer | 4856 | 204 | 993 | 29.79 |
| Coffeemaker | 894 | 119 | 106 | 3.18 |
| Dehumidifier | 257 | 1467 | 377 | 11.31 |
| Dishwasher (2) | 1201 | 302 | 363 | 10.89 |
| Fan, attic | 370 | 786 | 291 | 8.73 |
| Fan, circulating | 88 | 489 | 43 | 1.29 |
| Fan, window | 200 | 850 | 170 | 5.10 |
| Freezer, 15 cu. ft. | 341 | 3504 | 1195 | 35.85 |
| Freezer, frostless, 15 cu. ft. | 440 | 4002 | 1761 | 52.83 |
| Frying pan | 1196 | 157 | 188 | 5.64 |
| Hair dryer | 381 | 37 | 14 | .52 |
| Heater, portable | 1322 | 133 | 176 | 5.28 |
| Heating pad | 65 | 154 | 10 | .30 |
| Hot plate | 1257 | 72 | 90 | 2.70 |
| Humidifier | 177 | 921 | 163 | 4.89 |
| Iron, hand | 1008 | 143 | 144 | 4.32 |
| Mixer | 127 | 102 | 13 | .39 |
| Oven, microwave | 1500 | 200 | 300 | 9.00 |
| Oven, self-cleaning | 4800 | 239 | 1146 | 34.38 |
| Radio | 71 | 1211 | 86 | 2.58 |
| Radio/record player | 109 | 1000 | 109 | 3.27 |
| Range | 8200 | 128 | 1175 | 35.25 |
| Refrigerator, 12 cu. ft. | 241 | 3021 | 728 | 21.84 |
| Refrigerator, frostless, 12 cu. ft. | 321 | 3791 | 1217 | 36.51 |
| Refrigerator/freezer, 14 cu. ft. | 326 | 3488 | 1137 | 34.11 |
| Refrigerator/freezer, frostless, 14 cu. ft. | 615 | 2974 | 1829 | 54.87 |
| Roaster | 1333 | 154 | 205 | 6.15 |
| Sewing machine | 75 | 147 | 11 | .33 |
| Shaver | 14 | 129 | 2 | .06 |
| Sun lamp | 279 | 57 | 16 | .48 |
| Toaster | 1146 | 34 | 39 | 1.17 |
| Toothbrush | 7 | 7 | 1 | .03 |
| Trash compactor | 400 | 125 | 50 | 1.50 |
| TV, black and white | 237 | 1527 | 362 | 10.86 |
| TV, color | 332 | 1512 | 502 | 15.06 |
| Vacuum cleaner | 630 | 73 | 46 | 1.38 |
| Waffle iron | 1116 | 20 | 22 | .66 |
| Washing machine, automatic (2) | 521 | 198 | 103 | 3.09 |
| Washing machine, nonautomatic (2) | 286 | 266 | 76 | 2.29 |
| Waste disposer | 445 | 67 | 30 | .90 |
| Water heater | 2475 | 1705 | 4219 | 126.57 |
| Water heater, quick recovery | 4474 | 1075 | 4811 | 144.33 |

(1) At 3 cents per kilowatt hour.
(2) Does not include hot-water heating.

Listing of major appliances and their energy consumption. (*Courtesy Help: The Useful Almanac 1976–1977, Consumer News Inc., Washington, DC 20045*)

the state and local levels. The program focuses primarily on homeowners, small businesses, public institutions, and state and local governments. The Service will be similar to the USDA extension service. (See Fig. p. 146)

**energy farm.** A concept involving the farming of selected plants for the purpose of providing biomass that can be used as a fuel or converted into other fuels or energy products.

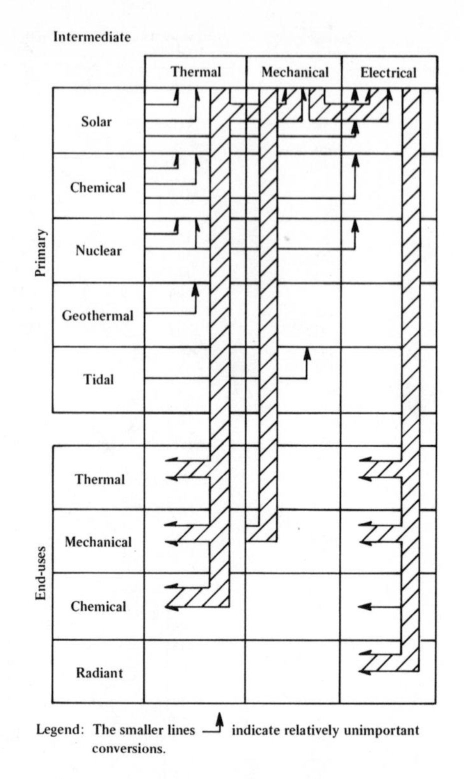


Energy conversion. (*Courtesy National Science Teachers Association*)

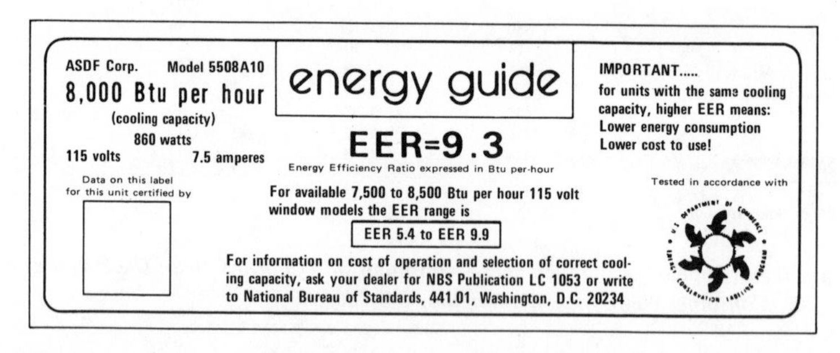


Department of Commerce energy guide label. Diagram shows the contents and design of labels on air conditioners. (*Courtesy Electric Power Research Institute and the Applied Nucleonics Co., Inc.*)

**energy flux density.** At a given point, the sum of energies, exclusive of rest energy, of all particles or photons incident per unit time on a small sphere centered at that point, divided by the cross-sectional area of that sphere. It is identical with the product of the particle flux density and the average energy of the particles.

**energy gap.** Forbidden part of the energy spectrum of valence electrons. If the lower energy band is filled, electrons must be activated across this gap before electronic conductance is realized.

**energy imparted to matter.** The difference between the sum of the energies of all the ionizing particles or photons which have entered a volume and the sum of the energies of all those which have left it, minus the energy equivalent of any increase in rest mass that took place in nuclear or elementary particle reactions within the volume.

**energy intensiveness.** In transportation, the relative amount of energy required to move one unit (one passenger or one ton of cargo) a distance of one mile. In industry, the ratio of total energy consumed for each dollar of production goods shipped out.

**energy level.** The distance from an atomic nucleus at which electrons can have orbits. It may be thought of as a shell surrounding the nucleus.

**energy load.** The amount of energy it takes to accomplish a task.

**Energy Research and Development Administration (ERDA).** The independent executive agency of the Federal Government with responsibility for management of research and development in all energy matters. Now absorbed into the Department of Energy.

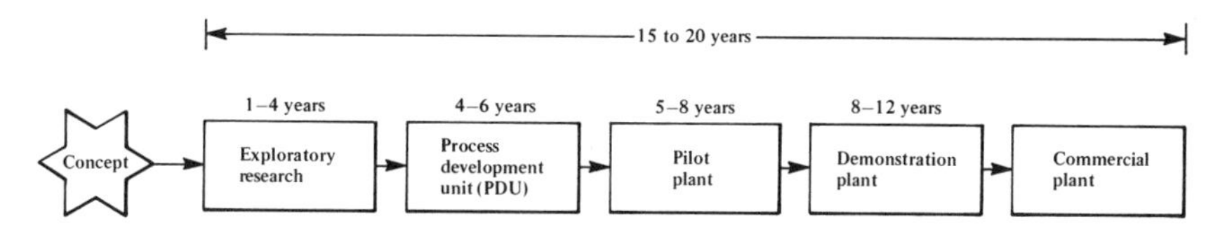


Typical process development sequence. (*Courtesy U.S. Energy Research and Development Administration*)

**energy reserves.** Refers to the bank of natural resources such as natural gas, natural gas liquids, petroleum, coal, lignite, and energy available from water power and solar and geothermal energy. (See Fig. top of p. 148)

**energy storage.** The ability to convert energy into other forms, such as heat or a chemical reaction, so that it can be retrieved for later use; also the development, design, construction, and operation of advanced devices for storing energy until needed. The technology includes devices such as batteries, pumped storage for hydroelectric generation, flywheels, and compressed gas. (See Fig. bottom of p. 148)

**energy supply.** The total amount of primary energy resources used.

**energy technology.** Covers the application of scientific and engineering knowledge to the development of various forms of energy.

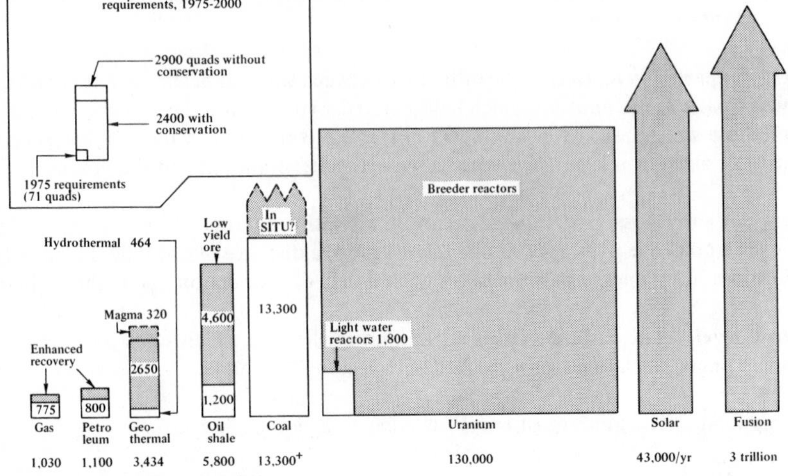


Potentially recoverable domestic energy resources. (*Courtesy U.S. Energy Research and Development Administration*)

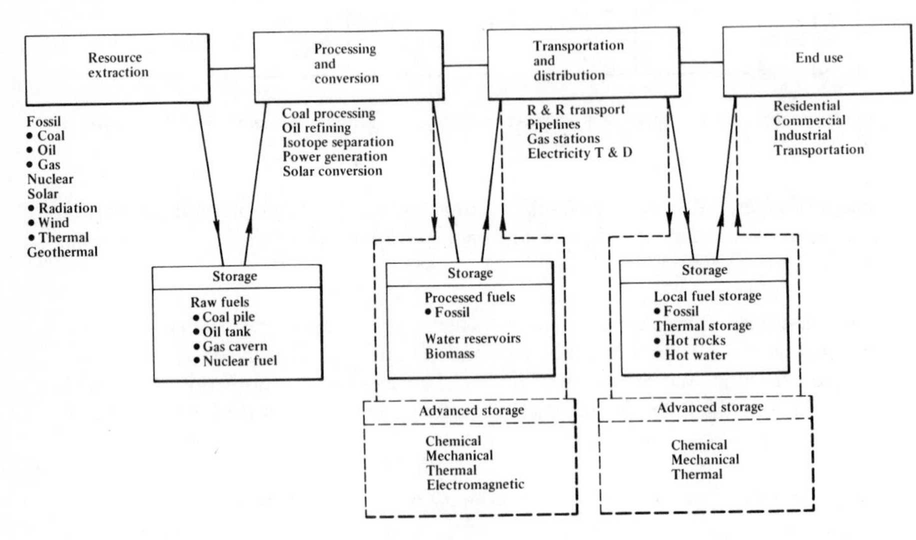


Energy flow and storage. (*Courtesy Annual Review of Energy, Volume 1, Jack M. Hollander (ed.), Annual Reviews Inc., Palo Alto, CA 94306*)

**energy threshold.** The limiting kinetic energy of an incident particle below which a specified process cannot take place.

**energy transfer coefficient.** Of a substance for a parallel beam of specified indirectly ionizing particles incident normally on a thin layer of this substance, the sum of the kinetic energies of all charged particles liberated in this layer divided by the layer thickness and by the sum of the kinetic energies of the incident particles. The energy transfer coefficient is a function of the energy of the radiation.

**engine.** A device for turning energy of different kinds into kinetic energy, usually with the intention of putting it to work. An electric engine or motor uses the power of electric current to create a magnetic field. An internal-combustion engine works by converting chemical energy in gasoline or diesel-fuel molecules into heat which is then used to expand gases in a cylinder. The expanded gases drive a piston (as in a steam engine); the piston, in turn, drives a crankshaft.

**engineered safeguards.** Safety features built as an integral part of any reactor facility. These include: ceramic fuel, fuel cladding, primary coolant system, safeguard systems, and a containment system surrounding the entire reactor.

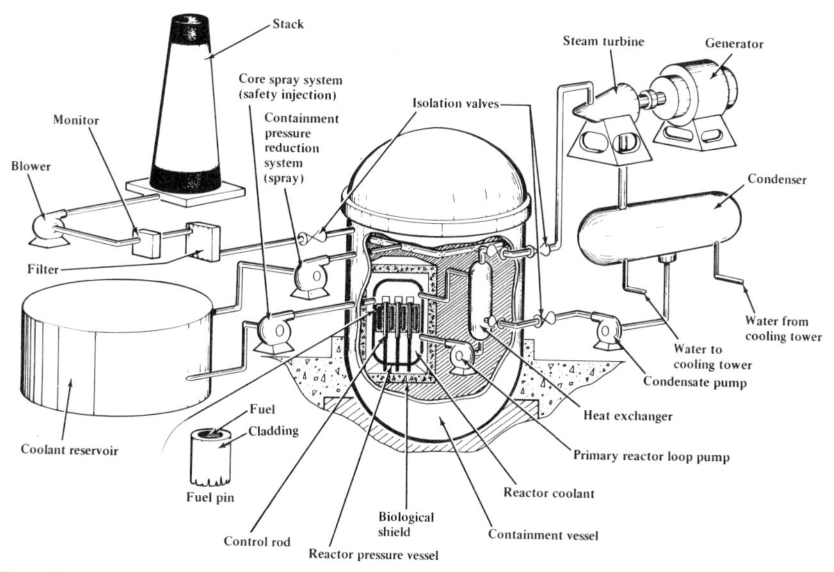


Drawing of a typical water reactor power plant with engineered safeguards. (*Courtesy National Reactor Testing Station*)

**enhanced oil recovery.** Increased recovery of crude oil (and natural gas in the case of enhanced gas recovery) from a reservoir, which is achieved by the external application of physical or chemical processes that supplement naturally occurring or simple fluid injection processes.

**enriched fuel.** Nuclear fuel containing uranium which has been enriched in one or more of its fissile isotopes or to which chemically different fissile nuclides have been added.

**enriched material.** Material in which the percentage of a given isotope present has been artificially increased so that it is higher than the percentage of that isotope naturally found in the material.

**enriched uranium.** Uranium in which the amount of uranium-235 present has been artificially increased above the 0.71 percent found in nature. Uranium enriched between 3 and 6 percent is a common fuel for civil nuclear power stations. Uranium enriched to 90 percent or more is used for nuclear propulsion of warships and submarines and in atomic bombs.

**enriching.** Increasing the heat content of a gas by mixing with it a gas of higher Btu content; also a nuclear energy term applied to isotopic enrichment.

**enrichment.** A process by which the proportion of the fissionable uranium isotope, uranium-235, is increased above the 0.7 percent contained in natural uranium. The principal method of enrichment is gaseous diffusion, but gaseous centrifugation is also receiving much attention, particularly in Europe. Also the addition of nitrogen, phosphorous, and carbon compounds or other nutrients into a lake or other waterway that greatly increases the growth potential for algae and other aquatic plants. Most frequently, enrichment results from the inflow of sewage effluent or from agricultural runoff.

**enthalpy.** Total heat content of air; the sum of the enthalpies of dry air and water vapor, per unit weight of dry air; measured in British thermal units per pound.

**entrained bed.** A coal combustion or gasification process in which pulverized coal is carried along in a gas stream; also a bed in which solid particles are suspended in a moving fluid and progressively carried over in the effluent stream.

**entropy.** A measure of the unavailable energy in a system; energy that cannot be converted into another form of energy. The change in entropy, $\Delta S$, of any system is equal to the increment of heat added divided by its absolute temperature, so that:

$$dS = \frac{dQ}{T} \frac{\text{joules}}{\text{kelvin}}.$$

The total entropy of the universe is continually increasing.

**entry driver.** A mining machine, designed to work in entries and other narrow places, to load coal as it is broken down.

**environment.** The aggregate of all the surrounding conditions, influences, or forces affecting the life, development, and survival of an organism.

| Energy source | Effects on land | Effects on water | Effects on air | Biological effects | Supply |
|---|---|---|---|---|---|
| Coal | Disturbed land<br>Large amounts of solid waste<br>Mine tailings | Chemical mine drainage<br>Increased water tempreature | Sulfer oxides<br>Nitrogen oxides<br>Particulates<br>Some radioactive gases | Respiratory problems from air pollutants | Large reserves |
| Oil | Wastes in the form of brine<br>Pipeline construction | Increased water tempreture<br>Oil spills | Nitrogen oxides<br>Some sulfur oxides | Respiratory problems from air pollutants | Limited domestic reserves |
| Gas | Pipeline construction | Increased water tempreature | Some oxides of nitrogen | None detectable | Extremely limited domestic reserves |
| Uranium | Disposal of radioactive wastes<br>Mine tailings | Increased water tempreature<br>Some radioactive liquids | Some release of radioactive gases | None detectable in normal operation | Large reserves if breeders are developed |

Environmental effects of electric power generation. (*Courtesy U.S. ERDA-69*)

**environmental carcinogens.** Potential cancer causing agents in the environment. They include among others: industrial chemical compounds found in food additives, pesticides and fertilizers, drugs, toys, household cleaners, toiletries, and paints; also naturally occurring ultraviolent solar radiation.

**environmental development plans (EDP)** An environmental development plan (EDP) is the basic DOE management document for the planning, budgeting, managing, and reviewing of the broad environmental implications of each energy technology alternative for each major DOE research, development, and demonstration and commercialization program. The EDP is designed to identify environmental issues, problems, and concerns as early as possible during the program's development; to analyze the available data and assess the current state of knowledge related to each issue, problem, and concern; to set forth strategies to resolve these; to set forth the processes by which the public is involved in identification and resolution of these issues, problems, and concerns; and to designate significant milestones for resolution of these issues, problems, and concerns. The timing of the EDP's milestones reflects the sequencing of the technology development. EDP's, once completed, are made available to the public.

**environmental impact assessments (EIA).** An environmental impact assessment (EIA) is a written report, prepared by an Assistant Administrator or an DOE program office, which evaluates the environmental impacts of proposed DOE actions to assure that environmental values are considered at the earliest meaningful point in the decision-making process and which, based upon the evaluation, determines whether or not an environmental impact statement should be prepared. The EIA is intended to be a brief, factual, and objective document describing the proposed action, the environment which may be impacted, the potential environmental impacts during construction, operation, and site restoration, potential conflicts with Federal, State, regional, or local plans, and the environmental implications of alternatives.

**environmental impact statements (EIS).** An environmental impact statement (EIS) is a document prepared at the earliest meaningful point in the decision-making process which analyzes the anticipated environmental impacts of proposed DOE actions and of reasonably available alternatives, and which reflects responsible public and governmental views and concerns. An EIS is prepared in response to plans in the program's EDP or after the review of an EIA which identifies potentially significant impacts. The EIS goes through a specific preparation process involving agency and public review. The EIS goes through four steps during its preparation. The preliminary draft is reviewed within DOE, the draft is distributed to the public for review and comment, the preliminary final incorporating comments submitted to DOE in response to the draft is reviewed within DOE, and the final EIS is issued reflecting the agency's final review and deliberations. This final EIS is then officially filed with the Council on Environmental Quality and distributed to the public. Except in special cases, no DOE action subject to EIS preparation can be taken sooner than 30 days after the final EIS has been issued. Required by the National Environmental Policy Act (NEPA), section (102(2) (c).

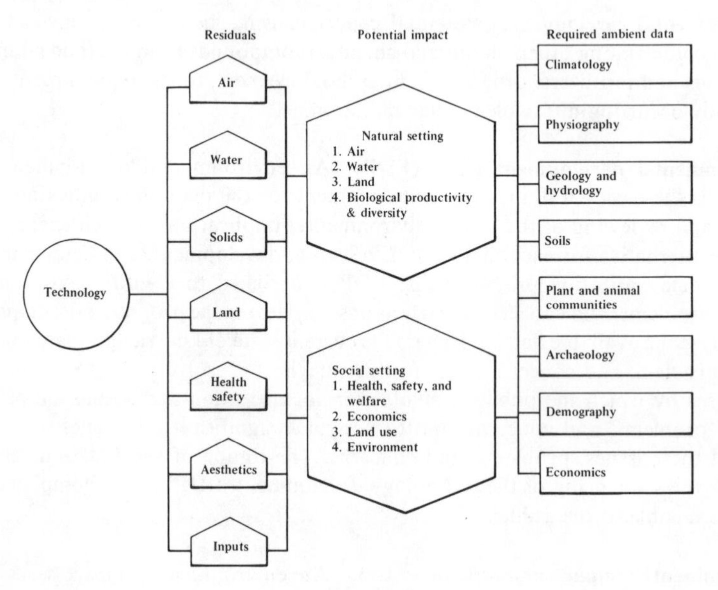


The impact of technology on the environment. (*Courtesy U.S. Council on Environmental Quality*)

**Environmental Protection Agency (EPA).** A Federal agency created in 1970 to permit coordinated governmental action for protection of the environment by the systematic abatement and control of pollution through integration of research, monitoring, standard setting, and enforcement activities.

**eolation.** The process by which wind modifies land surfaces, both directly by transportation of dust and sand and by the work of sandblasts, and indirectly by wave action on shores.

**eolian.** Of, relating to, formed by, or deposited from the wind or currents of air. Eolian was formerly spelled aeolian.

**epidermal deformation.** Deformations affecting the outer cover of the earth's crust, such as slumping, volcanotectonic collapse, sliding, and compressive settling.

**epigene.** Formed, originating, or taking place on or not far below the surface of the earth.

**epithermal neutron.** An intermediate neutron.

**epithermal reactor.** A reactor in which fission is induced predominantly by epithermal neutrons; an intermediate reactor.

**epoch.** A geologic time unit corresponding to a series; a subdivision of a period; formerly used for other smaller divisions of geologic time.

**equation of motion.** The Newtonian law of motion states that the product of mass and acceleration equals the vector sum of the forces. In meteorological and oceanographic use, both sides of the equation of motion are divided by mass to give force per unit mass. The forces considered in ocean currents are gravity, Coriolis force, pressure gradient force, and frictional forces.

**equilibrium.** When two or more forces act upon a body in such a manner that the body remains at rest, the forces are said to be in equilibrium or perfect balance.

**equilibrium cycle.** For the purpose of calculating the economics of a nuclear power plant, an assumed fuel cycle in which the feed and waste materials have constant compositions.

**equinox.** Either of the two times each year when the sun crosses the equator and night and day are everywhere of equal length. The spring equinox occurs about March 21 and the fall equinox about September 23.

**equivalent direct radiation.** Heat expressed in terms of a square foot of steam radiator surface emitting 240 British thermal units per hour.

**equivalent length of pipe.** The resistance of pipe valves, controls, and fittings to flow expressed as equivalent length of pipe or pipes of other sizes, for convenience in calculating pipe diameters.

**equivalent weight.** It equals the molecular weight of the substance or a submultiple of it, chosen according to some convention.

**erosion.** The physical and chemical processes by which the earth or rock material is loosened or dissolved and removed from any part of the earth's surface. It includes the processes of weathering.

**escalation rate.** A number which defines the annual increase in monetary value of a specified quantity.

**erosion corrosion.** Form of corrosion characterized by the acceleration of deterioration or attack of a metal due to the relative movement of the corrosive fluid and the metal surface. The fluid movement is generally rapid and mechanical wear or abrasion is involved. All types of equipment exposed to moving fluids as in solar energy systems could be subject to erosion/corrosion, piping, particularly in turbulent flow areas such as bends, elbows, and tees; valves; pumps; heat exchangers and measuring devices.

**ester.** A compound formed by the reaction of an alcohol and an acid usually with elimination of water.

**estuary.** Areas, such as bays, mouths of rivers, salt marshes, and lagoons, where the fresh water meets salt water.

**ethane.** A colorless hydrocarbon gas of slight odor having a heating value of 1604 British thermal units per cubic foot and a specific gravity of 1.0488. It is a normal constituent of natural gas. Chemical formula $C_2H_6$.

**ethanol.** Ethyl alcohol or grain alcohol, $C_2H_5OH$. It is the alcohol contained in intoxicating beverages. Ethanol can be produced from biomass by the conversion process called fermentation.

**ethylene.** Colorless, flammable, unsaturated gas with characteristic sweet odor and taste. It is contained in illuminating gas.

**eudiometer.** An instrument for the volumetric measurement and analysis of gases.

**eutectic material.** A chemical which has the property of changing from a solid to a liquid at a relatively low temperature while maintaining a constant temperature. The heat energy which caused the transformation is stored in the eutectic liquid until the liquid returns to solid form and gives up heat. In application, eutectic salts, which are stored in plastic tubes, are used as thermal reservoirs to conserve and then to release solar heat.

**eutrophication.** The slow aging process by which a lake becomes so rich in nutritive compounds, especially nitrogen and phosphorus, that algae and other plant life become superabundant, thereby eventually causing the lake to dry up and disappear.

**evaporation.** The change by which any substance is converted from a liquid state and carried off in vapor.

**evaporative centrifuge.** A batch-separating device in which a mixture to be separated is introduced into the centrifuge as a liquid. The vapors are removed at a point near the axis of the centrifuge after having been separated by diffusion through the centrifugal field.

**evaporative cooling.** The adiabatic exchange of heat between air and a water spray or wetted surface. The water approaches the wet bulb temperature of the air, which remains constant during its traverse of the exchanger.

**evaporative cooling system with collectors.** A space conditioning system which accomplishes space cooling by evaporating water as it is circulated through the collectors during those hours which have the appropriate conditions for cooling (usually night and early morning).

**evaporator.** A component of a heating and cooling system in which a working fluid undergoes a change from liquid to gas, extracting heat and producing a cooling effect.

**evolution.** The act of giving off or emitting a gas or vapor as the result of a chemical reaction.

**excavation.** The act of removing overburden material; also the blasting, breaking, and loading of coal, ore, or rock in mines.

**excavator.** A power-operated digging and loading machine used in opencast mining and in quarrying.

**excess air.** Air which passes through a combustion zone in excess of the quantity theoretically required for complete combustion. In practice, complete combustion cannot be obtained without slightly more air than is theoretically necessary. The amount of this excess air varies with the design and mechanical condition of the appliance, but ranges from 15 percent upwards.

**excess reactivity.** More reactivity than that needed to achieve criticality. Excess reactivity is built into a reactor (by using extra fuel) in order to compensate for fuel burnup and the accumulation of fission-product poisons during operation.

**excess resonance integral.** The resonance integral in which the cross section excludes that part which varies inversely with neutron speed.

**exchange gas.** Gas that is received from (or delivered to) another party in exchange for gas delivered to (or received from) such other party.

**excitation.** The addition of energy to a system, thereby transferring it from its ground state to an excited state.

**excited state.** The state of a molecule, atom, electron, or nucleus when it possesses more than its normal energy. Excess nuclear energy is often released as a gamma ray. Excess molecular energy may appear as fluorescence or heat.

**exciter.** An auxiliary generator that supplies energy for the field excitation of another electric machine.

**exclusive area.** An area immediately surrounding a nuclear reactor where human habitation is prohibited to assure safety in the event of an accident.

**excursion.** A sudden, very rapid rise in the power level of a reactor caused by supercriticality. Excursions are usually quickly suppressed by the negative temperature coefficient of the reactor and/or by automatic control rods.

**exhaust fan.** A fan which sucks used air from an area and thereby causes fresh air to enter.

**exhaustion.** In mining, the complete removal of ore reserves.

**exhaust port.** In engines, the opening through which a fluid discharges out of a cylinder. In gas meters, the openings through which gas leaves the metering chamber.

**exhaust steam.** Water vapor which has not had most of the usable energy removed.

**exine.** The outer of the two layers forming the wall of certain spores.

**exinoid.** A coal constituent similar to material derived from plant exines.

**exit dose.** The absorbed dose at the surface of a body opposite to that upon which a beam is incident.

**exit temperature.** The flue gas temperature taken at the point where the gas leaves the combustion chamber.

**exosphere.** Space beyond the earth's atmosphere. It begins at a height of about 1000 kilometers.

**exothermic.** That characteristic of a chemical reaction, such as fuel combustion, in which heat is liberated.

**exothermic reaction.** A reaction which releases more energy than is required to start it. The combustion reaction (burning) is an example as are fission and fusion reactions.

**expansion loop.** Either a bend like the letter "U" or a coil in a line of pipe to provide for expansion and contraction.

**expansion ratio.** The ratio of the gas volume after expansion to the gas volume before expansion.

**expansion valve.** A special valve used in refrigerating systems through which the liquid refrigerant (under high pressure) is allowed to escape into a lower pressure and thus expand into a gas.

**experimental reactor.** A reactor operated primarily to obtain reactor physics or engineering data for the design or development of a reactor or type of reactor. Reactors in this class include: zero-power reactor (may also be a research reactor), reactor experiment, and prototype reactor.

**exploitation.** The process of winning or producing from the earth the oil, gas, minerals, or rocks which have been found as the result of exploration.

**exploration.** Generally, the act of searching for potential subsurface reservoirs of gas or oil. Methods include the use of magnetometers, gravity meters, seismic exploration, surface mapping, among others.

**exploration drilling.** Drilling boreholes by the rotary, diamond, percussive, or any other method, for geologic information or in search of a mineral deposit.

**exploratory wells.** A well drilled in search of a new, and as yet undiscovered, field of oil or gas, or with the expectation of extending the limits of a field already partly developed.

**explosion head.** A term applied to a protective device that is arranged to blow out a disk. It is usually employed if an air-gas mixture explodes in a piping system. Thereafter, the gas will escape until a shutoff valve is closed.

**explosive limits.** The lowest and highest concentrations of a specific gas or vapor in a mixture with air that can be ignited at the ordinary temperature and pressure of the mixture.

**explosive strength.** A measure of the amount of energy released by an explosive or detonation, and its capacity to do useful work.

**explosive stripping.** A method, encouraged by the introduction of low-cost ANFO explosives, in which, by using an excess of explosives in the strip mine bench, up to about 40 percent of the overburden can be removed from the coal seam by the energy of the explosive, thereby requiring no excavation.

**exponential assembly.** A subcritical assembly used for an exponential experiment.

**exponential experiment.** An experiment, performed with a subcritical assembly of reactor materials and an independent neutron source, used to determine the neutron characteristics of a configuration of these materials.

**exponential growth.** Growth for which the percentage increase for a given time interval is constant.

**exposure.** The incidence of radiation on living or inanimate material by accident or intent. For x-radiation or gamma-radiation in air, the sum of the electrical charges of all of the ions of one sign produced in air when all electrons liberated by photons in a suitably small element of volume of air are completely stopped in air, divided by the mass of the air in the volume element. It is commonly expressed in roentgens.

**exposure rate.** The exposure per unit time.

**extended bench method.** The use of a large capacity walking dragline in deep overburden operating from a machine-supporting bench formed by filling the V between the bank and the spoil. This V-section is formed from material that falls from the bank or is placed by the dragline and must be rehandled.

**extensometer.** Instrument used for measuring small deformations, deflections, or displacements.

**external combustion engines.** An engine in which the fuel is burned outside the cylinders.

**external radiation.** Ionizing radiation reaching the body from sources outside the body.

**extraction.** In the oil industry, the process of separating a material, by means of a solvent, into a fraction soluble in the solvent (extract) and an insoluble residue; in coal mining, the process of mining and removal of coal or ore from a mine.

**extraction cycle.** A series of steps involving solvent extraction, stripping, and, in some cases, scrubbing.

**extraction-hydrogenation.** Extraction carried out in the presence of hydrogen either as a gas or as derived from hydrogen donor solvents.

**extraction plant.** A plant in which a product such as propane, butane, oil, or natural gasoline, which is initially a component of the gas stream, is extracted or removed for sale.

**extra high voltage.** Voltage levels of transmission lines which are higher than the voltage levels commonly used.

**extraneous gas.** Volume of gas which is not indigenous to the storage area.

**extrapolated boundary.** A hypothetical surface outside an assembly on which the neutron flux density would be zero if extrapolated from the flux distribution neglecting the distribution within a few mean free paths of the physical surface.

**extrapolated range.** The distance from a radiation source given by extrapolation to zero flux density of a tangent to the flux density-versus-distance curve (taken at that point where the flux density has decreased to one-half of its initial value.

**extrapolation ionization chamber.** An ionization chamber of adjustable volume which permits the estimation of the limiting value of the ionization current per unit volume as the volume becomes vanishingly small.

**extravastion.** The eruption of molten or liquid material from the earth, as lava from a vent, water from a geyser, etc.

**extraterrestrial radiation.** The solar radiation which would be received on a horizontal surface if there were no atmosphere.

# F

**fabric.** The special arrangement and orientation of rock components, whether crystals or sedimentary particles, as determined by their sizes and shapes.

**fabric filters.** A device for removing dust and particulate matter from industrial emissions.

**face.** The solid surface of the unbroken portion of the coalbed at the advancing end of the working place.

**facies.** The aspect belonging to a geologic unit of sedimentation, including mineral composition, type of bedding, fossil content, etc. Sedimentary facies are really segregated parts of differing nature belonging to any genetically related body of sedimentary deposit.

**facing.** The main vertical joints often seen in coal seams. They may be confined to the coal or continued into the adjoining rocks.

**Fahrenheit scale.** A temperature scale in which the boiling point of water is 212° and its freezing point 32°. To convert °F to °C, subtract 32, multiply by 5, and divide the product by 9.

**failed fuel element.** A fuel element with a defect which allows fission products to escape.

**fail safe.** Refers to a principle of design by which, in the event of any failure in a system, the system takes up a safe condition.

**fall.** A mass of roof or side which has fallen in any subterranean working or gallery.

**fallout.** Air-borne particles containing radioactive material which fall to the ground following a nuclear explosion. "Local fallout" from nuclear detonations falls to the earth's surface within 24 hours after the detonation. "Tropospheric fallout" consists of material injected into the troposphere but not into the higher altitudes of the stratosphere. It does not fall out locally, but usually is deposited in relatively narrow bands around the earth at about the latitude of injection. "Stratospheric fallout" or "worldwide fallout" is that which is injected into the stratosphere and then falls out relatively slowly over much of the earth's surface.

**fall wind.** A strong, cold, downslope wind.

**Faraday, Michael.** Michael Faraday (1791–1867) English physicist and chemist. In 1831 he announced his discovery of electromagnetic induction; shortly afterward he showed that the five known kinds of electricity (frictional, galvanic, voltaic, magnetic, and thermal) were fundamentally the same. He also discovered the laws of electrolysis.

**farm tractor fuel.** Any petroleum product, exclusive of gasoline diesel fuel and liquefied petroleum gas, which is used for the generation of power for the operation of farm implements.

**fast breeder nuclear reaction.** A "fast neutron" hits plutonium, simultaneously releasing energy to generate electricity and a number of free neutrons. Some neutrons split additional plutonium atoms, sustaining the chain reaction; other neutrons are absorbed by "fertile" uranium-238 and thus are converted into new plutonium-239.

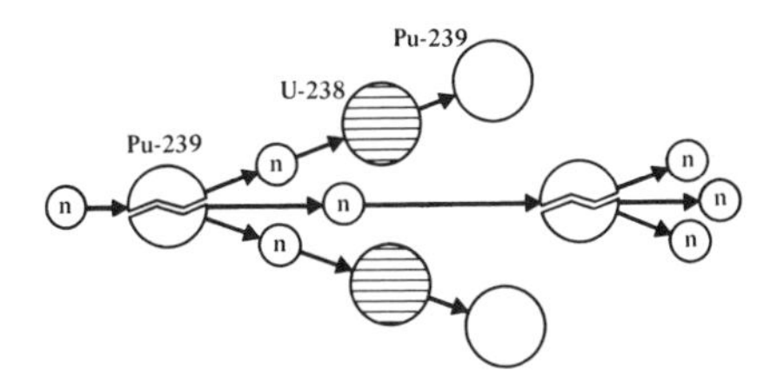


Diagram illustrating the fast breeder nuclear reaction. (*Courtesy Breeder Reactor Corporation*)

**fast breeder reactor (FBR).** A reactor that operates with fast neutrons and produces more fissionable material than it consumes.

**fast fission factor.** In an infinite medium, the ratio of the mean number of neutrons produced by fissions due to neutrons of all energies to the mean number of neutrons produced by thermal fissions only.

**Fast Flux Test Facility (FFTF).** The Fast Flux Test Facility at Hanford, Washington, will be used for the testing of fuels and components for liquid metal fast breeder reactors (LMFBR).

**fast neutron.** High energy neutron. Fast neutrons are utilized in the fast breeder actor both to produce nuclear fissions and to transform fertile material (e.g., $^{238}U$) into fissionable nuclear fuel.

**fast reactor.** A reactor in which the fission chain reaction is sustained primarily by fast neutrons rather than by thermal or intermediate neutrons. Fast reactors contain little or no moderator to slow down the neutrons from the speeds at which they are ejected from fissioning nuclei. Also called fast neutron reactor.

**fathom.** A measure used for sea depths; a unit of linear measurement that equals 6 feet or 1.828 meters.

**fathometer.** An instrument used in measuring the depth of water by the time required for a sound wave to travel from the surface to the bottom and for its echo to be returned; may also be used in measuring the rise and fall of the tides in offshore localities.

**fault.** A fracture or a fracture zone along which there has been displacement of the two sides relative to one another parallel to the fracture. The displacement may be a few inches or many miles.

**favored nations clause.** A special form of adjustment clause in gas supply contracts providing for automatic increases in purchase price if the price other parties pay or receive is increased.

**feathering.** The pitch of the blades of a wind machine is altered so that the net lift on the blades is zero to avoid unnecessary stress from severe gusts.

**Federal Energy Regulatory Commission Priority Scheme.** The order in which natural gas customers must be curtailed by interstate pipelines and gas utilities if there are insufficient gas supplies to meet all requirements.

**Federal Power Commission.** An agency of the government of the United States which has jurisdiction over energy producers that sell or transport fuels for resale in interstate commerce. With respect to the gas industry, the general regulatory principles of the FPC are defined in the Natural Gas Act, as amended.

**feeder.** Very small fissures or cracks through which methane escapes from coal; an electric line for supplying electric energy within an electric service area or subarea; also a gas main that delivers gas from a city gate station or other source of supply to the distribution system.

**feed materials.** Refined uranium or thorium metal or their pure compounds in a form suitable for use in nuclear reactor fuel elements or as feed for uranium enrichment processes.

**feed points.** Connections between gas feeder lines and distribution networks.

**feed rate.** Rate at which a drilling bit is advanced into or penetrates the rock formation being drilled. It is expressed in inches per minute, inch per bit revolution, number of bit revolutions per inch of advance, or feet per hour. Also called cutting rate.

**feedstock.** Fossil fuels used for their chemical properties, rather than their value as fuel, e.g., oil used to produce plastics and synthetic fabrics. A raw material that can be converted to one or more end-products (methanol or synthetic natural gas, for example). Biomass is an energy feedstock.

**fen.** Low peaty land area covered wholly or partly by water.

**fermentation.** The process of decomposition of carbohydrates with the evolution of carbon dioxide or the formation of acid, or both.

**Fermi age theory.** A theory of neutron slowing down in which the essential assumptions are that the slowing down process is continuous and that the spatial transport of neutrons can be treated by diffusion theory.

**Ferrel's law.** A statement of the fact that currents of air or water are deflected by the rotation of the earth to the right in the northern hemisphere and to the left in the southern hemisphere.

**ferrous iron.** A reduced or low-valence form of iron ($Fe^{+2}$). It imparts a blue-gray appearance to water and some wet subsoils on long standing.

**fertile material.** A material, not itself fissionable by thermal neutrons, which can be converted into a fissile material by irradiation in a reactor. There are two basic fertile materials, uranium-238 and thorium-232. When these fertile materials capture neutrons, they are partially converted into fissile plutonium-239 and uranium-233, respectively.

**field gas.** A district or area from which natural gas is produced.

**field pressure.** The pressure of natural gas as it is found in the underground formations from which it is produced.

**field price.** The price paid for natural gas at the wellhead or outlet of a central gathering point in a field.

**fiery mine.** A mine in which the seam or seams of coal being worked give off a large amount of methane; a gassy mine.

**film badge.** A lighttight package of photographic film worn like a badge by workers in nuclear industry or research, used to measure possible exposure to ionizing radiation. The absorbed dose can be calculated by the degree of film darkening caused by the irradiation.

**film coefficient.** The heat transferred by convection per unit area per degree temperature difference between the surface and the fluid.

**filter.** A device for separating solids or suspended particles from liquids or fine dust.

**filter air.** A low-density, inert, fibrous, or fine granular material used to increase the rate and improve the quality of filtration.

**filter cake.** The compacted solid or semisolid material separated from a liquid and remaining on a filter after filtration.

**filtrate.** The liquid matter from the filtration process.

**filtration.** A process for separating solids from liquids by passing the mixture through a suitable medium, such as finely woven cloth, paper, and sand.

**FIMA (fissions per initial metal atom).** A measure of specific burnup. It is equal to the total number of fissions that have occurred in a mass of fuel, divided by the number of fissionable atoms initially present in that mass.

**fin.** In heat exchange equipment, an extended surface to increase the heat transfer area, as metal sheets attached to tubes.

**final cut.** The pit remaining after stripping equipment has made its final pass through the deposit—usually, at the point where stripping depth to the coal has reached the economic limit. The resulting excavation has a spoil bank to one side, a highwall on the other, and a flat space between where the coal has been mined.

**fine control member.** A control member used for small and precise adjustment of the reactivity of a reactor.

**fines.** Very small material produced in breaking up large lumps, as of ore or coal.

**fin-fan cooler.** A dry cooler that passes cooling air over finned tubes through which some hot fluid is being passed during the cooling processes. A variable speed fan is used to regulate movement of air over the finned tubes.

**finished products.** Petroleum oils, or a mixture of combination of such oils, or any component or components of such oils, which are to be used without further processing.

**finned tube.** Heat transfer tube or pipe with extended surface in the form of fins, discs, or ribs.

**fire.** The manifestation of rapid combustion, or combination of materials with oxygen; also to blast with gunpowder or other explosives.

**fireball.** The luminous ball of hot gases that forms a few millionths of a second after a nuclear explosion.

**fire brick.** Heat resistant ceramic material formed into bricks and used to line the fire boxes of boilers, furnaces, or other combustion chambers.

**fire classification.** The National Fire Protection Association classifications are as follows: Class A fires are defined as those in ordinary solid, combustible materials such as coal, wood, rubber, textiles, paper, and rubbish. Class B fires are defined as those in flammable liquids such as fuel or lubricating oils, grease, paint, varnish, and lacquer. Class C fires are defined as those in (live) electric equipment such as oil-filled transformers, generators, motors, switch panels, circuit breakers, insulated electrical conductors, and other electrical devices.

**fire clay.** A clay that is high in silica or alumina, and low in iron and alkalies. Fire clay forms the seat earth of many coalbeds. Fire clay will not melt or fuse at high temperatures.

**firedamp.** A combustible gas that is formed in mines by decomposition of coal or other carbonaceous matter and that consists chiefly of methane; also the explosive mixture formed by this gas with air; also called marsh gas.

**fire flooding.** An experimental means of recovery of oil of low gravity and high viscosity which is unrecoverable by other methods. The method consists of heating the oil in the horizon to increase its mobility by decreasing its viscosity. Heat is applied by igniting the oil sand and the fire is kept alive by the injection of air. The heat breaks the oil down into coke and lighter oils, and the coke catches fire. As the combustion front advances, the lighter oils move into the bore of a producing well.

**fire point.** Minimum temperature at which a substance will continue to burn after being ignited.

**firing.** The process of initiating the action of an explosive charge or the operation of a mechanism which results in a blasting action.

**firing rate.** The rate at which fuel is fed to a burner, expressed as volume, heat units, or weight per unit time.

**firing valve.** A lubricated plug-type variable position valve which is usually operated with an attached handle or, in the large sizes, by a loose fitting key or extended handle wrench.

**firm natural gas service.** High priority gas service in which the pipeline company has the contractual option to temporarily terminate deliveries to customers by reason of claim of firm service customers or high priority users. Large commercial facilities, industrial users, and electric utilities usually fall into this category.

**firm service.** Service offered to customers under schedules or contracts which anticipate no interruptions. The period of service may be for only a specified part of the year as in Off-Peak Service. Certain firm service contracts may contain clauses which permit unexpected interruption in case the supply to residential customers is threatened during an emergency.

**first collision probability.** The probability that a neutron starting at a given point makes its first collision in some specified region.

**first law of thermodynamics.** States that energy can neither be created nor destroyed. Also called the law of conservation of energy.

**fiscal year (FY).** Government's 12-month financial year, from October through September of the following calendar year; for example, FY-1978 extends from October 1977 through September 1978.

**Fischer assay.** A standardized laboratory procedure which is used as a basis for comparing oil shale processing alternatives and shale feedstocks.

**Fischer-Tropsch gasoline.** A gasoline generally produced from coal by combining carbon monoxide and hydrogen over a cobalt-thorium oxide catalyst at 200 to 250° C.

**Fischer-Tropsch process.** A process in which hydrogenation of carbon monoxide forms hydrocarbons from coal or natural gases.

**Fischer-Tropsch Synthesis at SASOL process.** Synthesis gas (carbon monoxide and hydrogen) from Lurgi gasifiers is converted to liquid hydrocarbons via an iron catalyst in two basic reactor types. The two reactor types are the Arge Reactor System (fixed-bed synthesis) and Synthoil reactor system. Operating conditions are 300 to 360 pounds per square inch gauge and 430 to 660° F, depending on the reactor used.

**fissile.** Fissionable.

**fissile material.** Any material fissionable by neutrons of all energies, including (and especially) thermal (slow) neutrons as well as fast neutrons; for example, uranium-235 and plutonium-239.

**fission.** The splitting of a heavy nucleus into two approximately equal parts (which are nuclei of lighter elements), accompanied by the release of a relatively large amount of energy and generally one or more neutrons. Fission can occur spontaneously, but usually is caused by nuclear absorption of gamma rays, neutrons, or other particles.

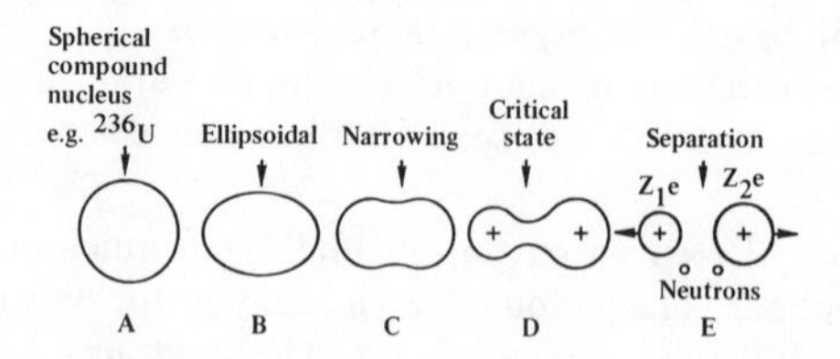


Fission mechanism according to the liquid-drop model of the nucleus. (*Courtesy Van Nostrand's Scientific Encyclopedia*)

**fissionable material.** Material that can be fissioned by fast neutrons only, such as uranium-238. Used in reactor operations to mean fuel.

**fission fragments.** The two nuclei which are formed by the fission of a nucleus. Also referred to as primary fission products. They are of medium atomic weight and are radioactive.

**fission gas.** A fission product in gaseous form.

**fission neutrons.** Neutrons originating in the fission process which have retained their original energy.

**fission poisons.** Fission products (for example, xenon-135) which have appreciable neutron absorption cross sections.

**fission-product poisoning.** The absorption or capture of neutrons by fission products in a reactor, decreasing its reactivity.

**fission products.** The nuclei (fission fragments) formed by the fission of heavy elements, plus the nuclides formed by the fission fragments' radioactive decay.

**fission weapon.** An atomic bomb.

**fission width.** The partial level width for decay through fission.

**fission yield.** The amount of energy released by fission in a thermonuclear (fusion) explosion as distinct from that released by fusion. Also the amount (percentage) of a given nuclide produced by fission.

**fissium.** Fissile material artifically mixed with fission product elements to simulate the material resulting from fission.

**fissure.** An extensive crack, break, or fracture in the rocks. A mere joint or crack persisting only for a few inches or a few feet is not usually termed a fissure by geologists or miners, although in a strict, physical sense, it is one.

**fitting.** A device, usually metal, for joining lengths of pipe into various piping systems; includes couplings, ells, tees, crosses, reducers, unions, caps, and plugs.

**fixation.** The process by which a fluid or a gas becomes or is rendered stable in consistency, and evaporation or volatilization is prevented; in nuclear field, the incorporation of radioactive elements, usually fission products, into solid materials in such a way as to insure no significant release over long periods of exposure to the natural environment.

**fixed bed.** Solid particles in intimate contact with fluid passing through them too slowly to cause fluidization.

**fixed carbon.** In the case of coal, coke, and other bituminous materials, the solid residue other than ash, obtained by destructive distillation; also a calculated figure obtained by subtracting from 100, the sum of the percentages of moisture, volatile matter, and ash.

**fixed mirror/distributed focus (FMDF) solar-to-electrical conversion.** A number of hemispherical reflectors concentrate sunlight on receivers, one of which is suspended in each bowl. The receiver is actually a steam boiler, and the steam is used to drive a conventional turbogenerator to yield electricity. The bowl-like concentrators are fixed, but the receivers track the sun, swinging through a daily arc and automatically adjusting to the sun's varying position with the seasons. This dual adjustment is made possible by a two-axis tracking mount. In this concept, an energy-storage means is used, and the electrical power output is so regulated that it can interface with an existing power transmission grid.

**fixed needle.** A tapered projection, in a fixed position, coaxial with an orifice which can be moved to regulate the flow of gas.

**fixed poison.** Nuclear poison, in the form of a solid material, attached to a process vessel or a reactor component to reduce reactivity or to prevent criticality.

**flame.** An ordinarily visible condition resulting from the rapid oxidation of a fuel and producing self-evident heat, light, or both.

**flame front.** The plane along which combustion starts; the base of the flame.

**flame geometry.** The measure of flame shape and dimension. Such shape can be produced by single or multiple burners.

**flame retaining nozzle.** Any burner nozzle with built-in features to hold the flame close to the burner at high mixture pressure or high velocities.

**flame test.** The use of the characteristic coloration imparted to a flame to detect the presence of certain elements.

**flame velocity.** The speed at which flame progresses through a fuel-air mixture.

**flare gas.** Waste gas; unutilized natural gas burned in flares at an oil field.

**flaring.** Burning of gas for the purpose of safe disposal.

**flash back.** The burning of gas in the mixing chamber of a burner, or in a piping system, usually because of an excess of primary air or too low a velocity of the combustible mixture through the burner port.

**flash burn.** A skin burn due to a flash of thermal radiation. It can be distinguished from a flame burn by the fact it occurs on unshielded parts of the body that are in a direct line with the origin of the thermal radiation.

**flash carbonization.** A carbonization process characterized by very short residence times of coal in the reactor, to optimize tar yields.

**flashing.** A thin impervious material placed in mortar joints and through air spaces in masonry to prevent water penetration or to provide water drainage.

**flashpoint.** The lowest temperature at which the vapors arising from a liquid surface can be ignited by an open flame.

**flash separation.** Distillation to separate liquids of different volatility accomplished by rapidly reducing the pressure on the liquid.

**flat plate collector.** A device for gathering the sun's heat, consisting of a shallow metal box with a glass or plastic transparent lid, where either air or liquid is circulated through the cavity of the box. The basic heat collection device used in solar heating systems; consists of a "black" plate, insulated on the bottom and edges, and covered by one or more transparent covers.

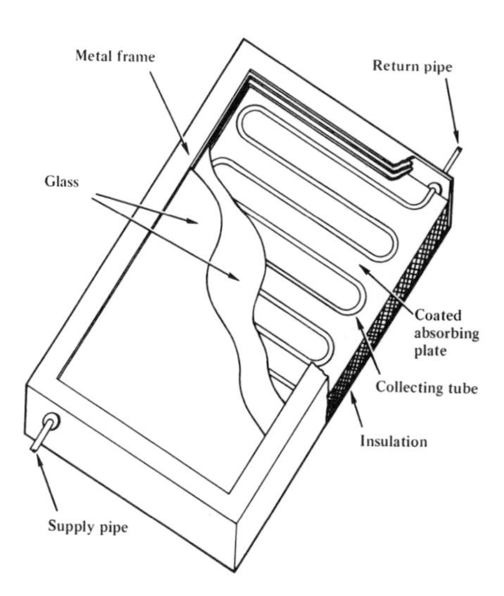


Water-type flat plate solar collector. (*Courtesy Design for a Limited Planet, 1st Edition, Norma Skurka and Jon Naar, Random House, Inc., New York 10022*)

**flat-rating.** Limit placed on the maximum output of a power source for economic or technical reasons.

**flexible connector.** A flexible tubing connecting a rigid pipe gas supply line to gas utilizing equipment.

**float coal.** Small, irregularly shaped, isolated desposits of coal imbedded in sandstone or in siltstone that appear to have been removed from the original bed by washing during the peat stage, carried a short distance, and redeposited.

**floc.** A loose, open-structure mass formed in a suspension by the aggregation of minute (colloidal) particles; a clump of solids formed in sewage by biological or chemical action.

**flocculation.** The gathering of suspended particles into aggregations; in waste water treatment, the process of separating suspended solids by chemical creation of clumps or flocs.

**flooding.** The drowning out of a well by water that sometimes results from drilling too deeply into the sand.

**flood plain.** The flat ground along a stream that is covered by water at the flood stage.

**flood tide.** The flow, or rising toward the shore, is called the flood tide, and the falling away, ebb tide; a rising tide.

**flow formula.** In the gas industry, a formula for determining the flow of gas between any two points in a pipeline under various conditions.

**flowmeter.** A device which registers rate of flow and perhaps quantity of gases, fluids, and fluid pulps; in waste water treatment, a meter that indicates the rate at which waste water flows through the plant.

**flow prover.** Apparatus used to determine the accuracy of displacement meters. Types of provers include critical flow prover and low pressure flange tap prover.

**flow rate.** Weight of dry air flowing per unit time, measured in pounds per hour.

**flue.** The exhaust channel through which gas and fumes are released from a building; also a furnace, such as a large coal fire, at or near the bottom of an upcast shaft for producing a current of air for ventilating the mine.

**flue collar.** That portion of an appliance designed for the attachment of the draft hood or vent connector.

**flue exhauster.** A device installed in and made a part of the vent to provide a positive induced draft.

**flue gas.** The gases from the fire (before the draft hood or draft regulator), or the products of combustion and excess air, consisting principally of carbon dioxide, carbon monoxide, oxygen, and nitrogen. (See Fig. p. 171)

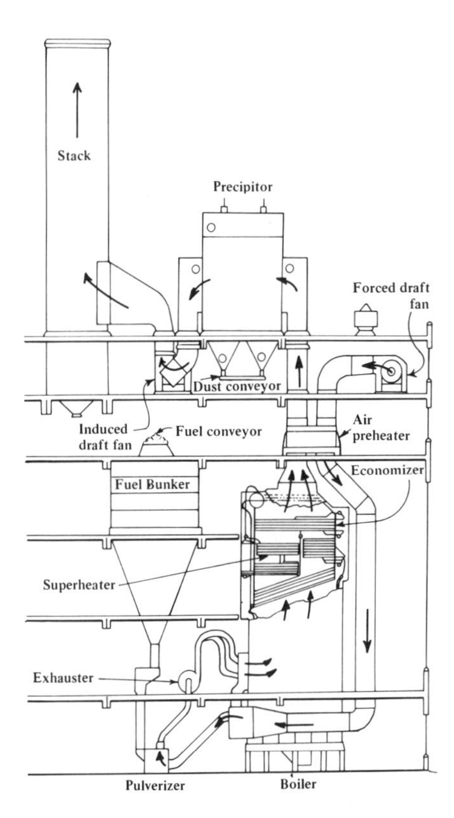


Boiler air and flue gas circulation patterns. (*Courtesy U.S. Council on Environmental Quality*)

**flue gas expander.** A turbine used to recover energy where combustion gases are discharged under pressure to the atmosphere. The pressure reduction drives the impeller of the turbine.

**flue gas vent.** A conduit or passageway for conveying flue gases to the outer air.

**fluid catalytic cracking.** A cracking process which converts a heavy oil fraction into a high-grade motor spirit by a process of thermal decomposition with the aid of a catalyst. (See Fig. p. 172)

**fluid fuel reactor.** A type of reactor (for example, a fused-salt reactor) whose fuel is in fluid form.

**fluid injection.** Injection of fluid into a reservoir to maintain the pressure in the formation or to produce an artificial fluid drive in the formation.

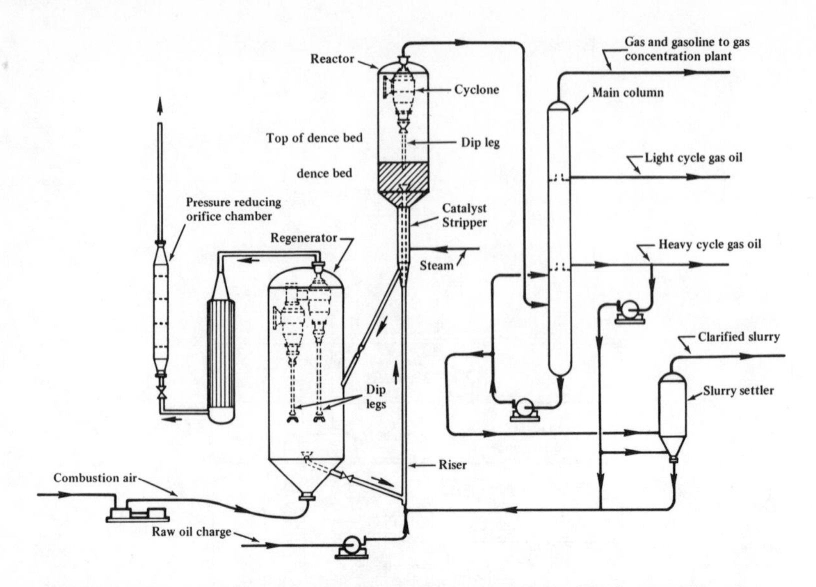


Fluid catalytic cracking unit. (*Courtesy Van Nostand's Scientific Encyclopedia*)

**fluidization.** A roasting process in which finely divided solid materials are kept in suspension by a rising current of air or other gas. This produces a fluidized bed which provides an ideal condition for gas solid reaction because each solid particle is in constant motion and in contact with the moving gas stream on all sides.

**fluidized bed.** A fluidized bed results when a fluid, usually a gas, flows upward through a bed of suitably sized solid particles at a velocity high enough to buoy the particles, to overcome the influence of gravity, and to impart to them an appearance of great turbulence. Fluidized beds are used in the petroleum industry. The Office of Coal Research is developing a coalfired fluidized bed boiler which would permit use of Western low sulfur coals without slagging and use of high sulfur coals without causing unacceptable environmental effects. (See Fig. p. 173)

**fluidized bed boiler.** A new type of boiler designed to reduce combustion product pollutants and reduce boiler size.

**fluidized-bed combustion.** Process for reducing sulfur dioxide emissions from coal combustion.

**fluidized bed reactor.** A reactor design in which the fuel ranges in size from small particles to pellets. Although the fuel particles are solid, their entire mass behaves like a fluid because a stream of liquid or gas coolant keeps them moving.

**fluid poison control.** Reactor control by adjustment of the position or quantity of a fluid nuclear poison in such a way as to change the reactivity.

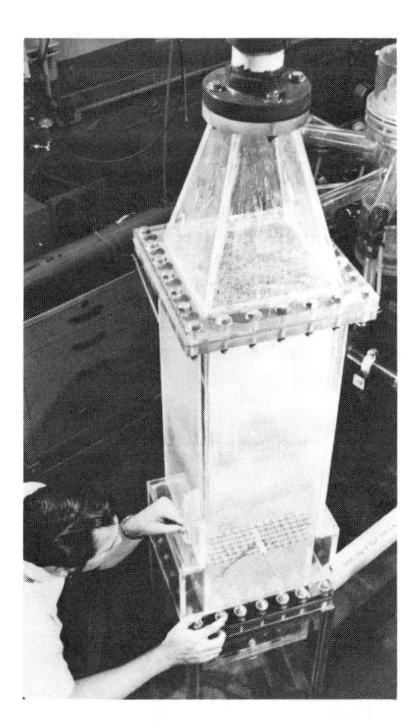

Small-scale model of a fluidized bed system. (*Courtesy U.S. Energy Research and Development Administration, Photo by Frank Hoffman*)

**flume.** An inclined channel, usually of wood and often supported on a trestle, for conveying water from a distance to be utilized for power and transportation.

**fluorescence.** Many substances can absorb energy (as from x-rays, ultra-violet light, or radioactive particles) and immediately emit this energy as an electromagnetic photon, often of visible light. This emission is fluorescence. The emitting substances are said to be fluorescent.

**fluorides.** A compound of fluorine with one other element or radical.

**fluorocarbon gases.** Used as propellants in aerosol products and as refrigerants; an agent believed to be causing depletion of the earth's ozone shield.

**flushing.** A drilling method in which water or some other thicker liquid (for instance, a mixture of water and clay) is driven into the borehole through the rod and bit. The water rises along the rod on its outer side, between the walls of the borehole and the rod, with such a velocity that the broken rock fragments are carried up by this water current (direct flushing); or, water enters the borehole around the rod and issues upwards through the rod (indirect flushing).

**flushout.** An accumulation of water removed suddenly from surface-mined lands by heavy precipitation which contains contaminants that result in pollution of the receiving stream.

**flush production.** The yield of an oil well during the early period of production.

**fluvial.** Of or pertaining to streams and rivers; produced by stream or river action, as a fluvial plane.

**flux.** The rate of flow or transfer of electricity, magnetism, water, heat, energy, etc.; the term is used to denote the quantity that crosses a unit area of a given surface in a unit of time.

**flux flattening.** The achievement of an approximately uniform neutron flux density in a reactor core.

**fly ash.** Fine solid particles introduced into the air by burning pulverized coal, which can be recovered within the stack by electrostatic precipitation. The collected ash has value for use in oil-well casings and also as a means of combating oil spills on sea water.

**flywheel.** A heavy wheel used in a rotating system to reduce surges of power input on demand by storing and releasing kinetic energy as it changes its rate of rotation.

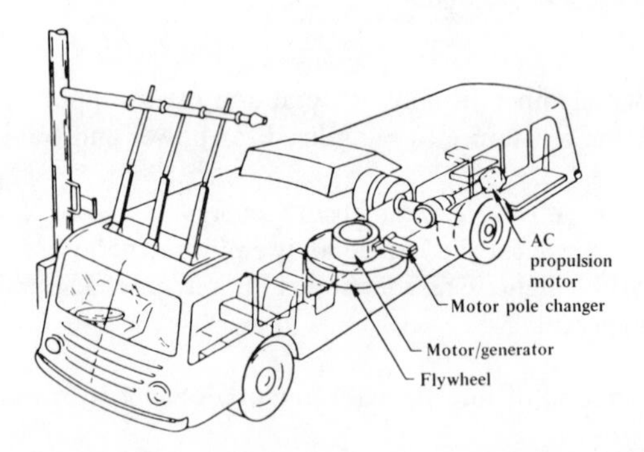


Flywheel-powered passenger bus. Basically, electrical energy drawn from an overhead pole speeds up the internal flywheel which drives a generator that powers the road wheel motors. (*Courtesy U.S. Energy Research and Development Administration*)

**foaming.** In a boiler, the carryover of slugs of water into the piping due to dirty water.

**focus.** In seismology, the source of a given set of elastic waves; the true center of an earthquake, within which the strain energy is first converted to elastic wave energy.

**focusing collectors.** Solar collector device employing semicircular reflectors made of aluminum to focus sunlight into water-bearing copper piping. A gear motor pivots the reflectors to change their angle according to the sun's azimuth.

**foehn.** A warm dry wind on the lee side of a mountain range. The warmth and dryness of the air is due to adiabatic compression upon descending the slopes. It is always associated with a strong wind blowing over a mountain ridge or chain.

**fog.** Dispersion of liquid as minute droplets in a gas.

**follower.** An extension of a nuclear control member that is intended to take the place of that member when it is withdrawn. It may contain nuclear fuel.

**food chain.** The pathways by which any material (such as radioactive material from fallout) passes from the first absorbing organism, through plants and animals, to man.

**footage** The payment of miners by the running foot of work; also the sum given; also the number of feet of borehole drilled per unit of time, or required to complete a specific project.

**footing.** That portion of the foundation of a structure that transmits loads directly to the soil; also the characteristics of the material directly beneath the base of a drill tripod, a derrick, or mast uprights.

**force.** The push or pull that alters the motion of a moving body or moves a stationary body. The unit of force is the dyne or the poundal:

$$\text{force} = \frac{\text{mass} \times \text{velocity}}{\text{time}}$$

**forced air systems.** A solar space conditioning system which utilizes mechanical equipment (pumps or blowers) to transfer solar heat from collectors to living space or to storage and/or living space.

**forced circulation water heater systems.** A solar water heating system that utilizes mechanical equipment (pumps) to transfer solar heated water from the collector to storage and, depending upon system design, to the space requiring additional heat.

**forced vibration.** Vibration of a structure, generally caused by engines or machines, sometimes by wind.

**foreshaft sinking.** The first 150 feet or so of shaft sinking from the surface, during which time the plant and services for the main shaft sinking are installed. Sometimes the main sinking does not commence until the foreshaft has been completed.

**formation.** The ordinary unit of geologic mapping consisting of a large and persistent stratum of one kind of rock; also used loosely for any local and more or less related group of rocks.

**formation drilling.** Boreholes drilled primarily to determine the structural, petrologic, and geologic characteristics of the overburden and rock strata penetrated; also called formation testing.

**form energy.** The potentiality of minerals to develop crystal form within a solid medium, such as rock.

**fossil energy programs.** Government programs involving the advancement of techniques for increasing the use of domestic fuels (coal, natural gas, petroleum, and oil shale) to supplement existing supplies. Such programs include enhancing oil and gas recovery methods, in-situ coal research, underground coal gasification research, magnetohydrodynamics (MHD), and advanced power systems.

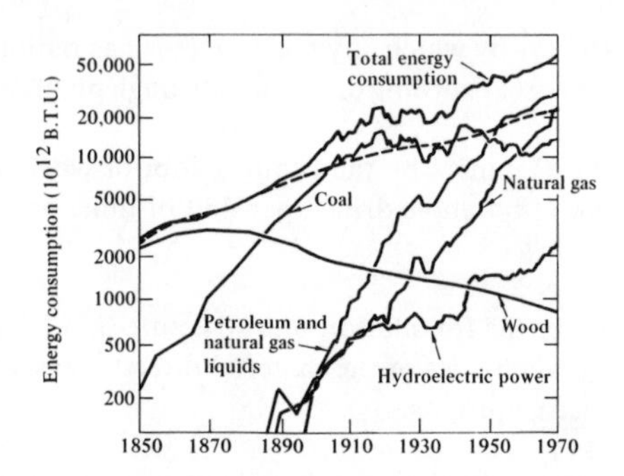


Energy from fossil fuels. (*Courtesy "Energy and the Future," Figure p. 5.*)

**fossil fuel.** Any naturally occurring fuel of an organic nature such as coal, crude oil, and natural gas. (See Fig. p. 177)

**foul.** Impure; a condition of the atmosphere of a mine so contaminated by gases as to be unfit for respiration; in a coal seam, place where the seam was washed out during deposition, leaving barren area.

**fouling.** In reactor technology, the formation of solid deposits on fuel element surfaces or on heat transfer surfaces.

**foundation wall.** That portion of a load-bearing wall below the level of the adjacent grade or below the first floor beams or joists.

**fractional distillation.** A distillation process for the separation of the various components of liquid mixtures by the use of fractionating columns attached to the still. (See Fig. p. 177)

**fractionating column.** A vertical tube or column attached to a still and usually filled with rings or intersected with bubble plates, which is used to separate various fractions of petroleum by a single distillation. The column may be tapped at different points

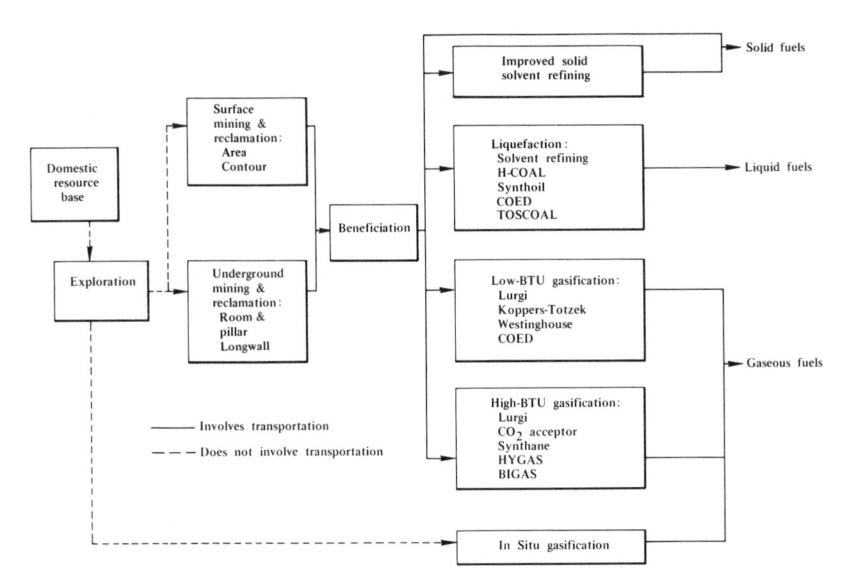


Coal resource development. (*Courtesy U.S. Council on Environmental Quality*)

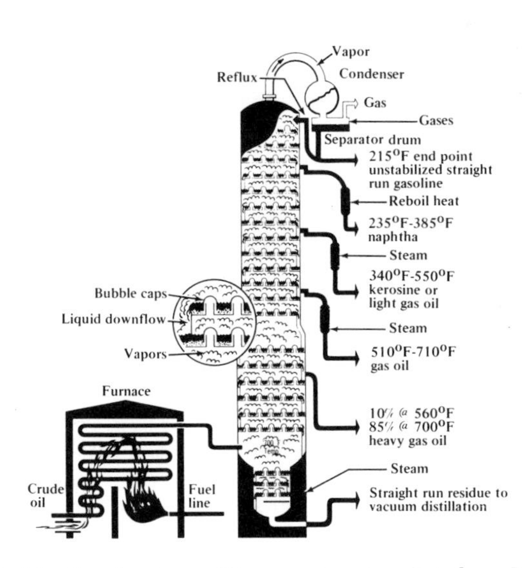


Typical fractionating tower. (*Courtesy American Petroleum Institute*)

along its length to separate various fractions in the order of their condensing temperatures (boiling points).

**fractionation.** The separation of a substance from a mixture; for example, the separation of one isotope from another of the same element.

**fractions.** Refiner's term for the portions of oils containing a number of hydrocarbon compounds within certain boiling ranges, separated from other portions in fractional distillation. They are distinguished from pure compounds which have specified boiling temperatures, rather than a range.

**fracturing.** The process of breaking a fluid-bearing strata by injecting a fluid under such pressure as to cause partings in the strata rock; also the process of increasing the permeability of strata near a well by pumping in water and sand under high pressure.

**fragmentation.** The breaking of coal, ore, or rock by blasting so that the bulk of the material is small enough to load, handle, and transport, without requiring a second blasting.

**Francis turbine.** A water turbine operating on a low and medium head, often installed in large hydroelectric schemes. Water enters the turbine radially and leaves axially.

**Frasch process.** A desulfurizing process which consists of distilling oil over lead oxide, followed by refining with sulfuric acid; also a process for mining sulfur in which superheated water is forced into the sulfur deposit to melt the sulfur which is then pumped to the surface.

**free-air ionization chamber.** An ionization chamber so designed that the observed ionization current is wholly due to ions and electrons arising from the absorption of radiation in air.

**free atmosphere.** That portion of the earth's atmosphere in which the effect of the earth's surface friction on the air motion is negligible, and in which the air is usually treated (dynamically) as an ideal gas.

**freeboard.** The space in a fluidized-bed reaction between the top of the bed and the top of the reactor; also the vertical distance between normal water level and the crest of a dam.

**free energy.** The energy which can be converted completely into work.

**free moisture.** Moisture in coal that can be removed by ordinary air drying.

**free swelling index.** A standard test that indicates the caking characteristics of coal when burned as a fuel.

**Freidel Crafts catalyst.** The catalyst employed in the synthesis of benzene hydrocarbons by the action of alkyl halides on aromatics in the presence of anhydrous aluminum chloride.

**French drain.** A covered ditch containing a layer of fitted or losse stone or other pervious material.

**freon.** A highly stable gas that does not combine readily with other chemical elements and, for that purpose, is used in heat pumps and refrigeration systems.

**fresh air.** Air free from the presence of deleterious gases.

**Fresnel lens.** Used in solar concentrator devices to focus sunlight on solar cells in converting solar energy into electricity.

A Fresnel lens is used to concentrate the sun's rays on a solar cell as part of Sandia Laboratories' research on solar cells operating at high illumination levels. (*Courtesy Sandia Laboratories*)

**friction.** A force that slows down movement and causes heat.

**friction head.** The head or energy lost as the result of the disturbances set up by the contact between a moving stream of water and its containing conduit. Friction head in pipes is the additional pressure that the pump must develop to overcome the frictional resistance offered by the pipe, by bends or turns in the pipeline, by changes in the pipe diameter, by valves, and by couplings.

**friction layer.** The layer of air, below the "free atmosphere," where the friction with the earth's surface affects its flow.

**frost.** A designation for the mouth or collar of a borehole; also, the working attachment of a shovel, as dragline, hoe, or dipper stick.

**froth floatation.** A separation process that uses the surface wetting behavior of chemicals to precipitate some materials and float others in an aerated pond; in the coal industry, a process for cleaning fine coal in which the coal, with the aid of a reagent, becomes attached to air bubbles in a liquid medium and floats as a froth.

**fuel.** Any substance that can be burned to produce heat. Sometimes includes materials that can be fissioned in a chain reaction to produce heat.

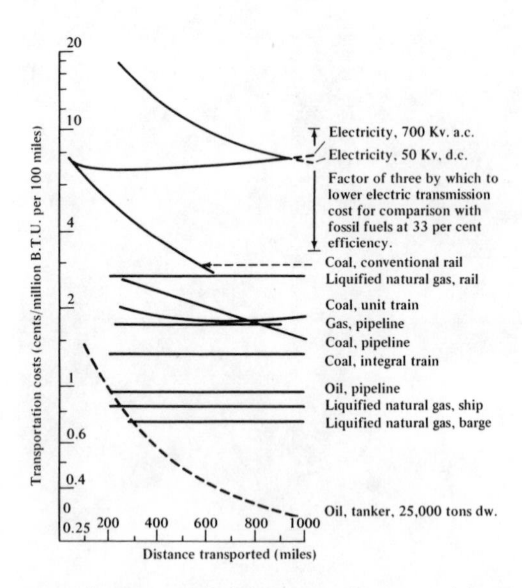


Transportation costs of energy and fuel. (*Courtesy "Energy and the Future," Fig. 30*)

**fuel assembly.** A grouping of fuel elements which is not taken apart during the charging and discharging of a reactor core.

**fuel burnout.** Severe local damage of a fuel element, due to failure of the coolant to dissipate all the heat produced in the element.

**fuel cell.** Developed initially for on-board power for the Gemini and Apollo spacecraft, fuel cells are attracting attention from utilities as small units or backup power sources. In fuel cells, hydrogen, which can be produced from just about any type of fossil fuel or the decomposition of water, is chemically reacted with oxygen from the air to produce electricity. Done electrochemically, without having to go through the inefficient combustion steps required by most other fossil-fueled electrical generating approaches, this process allows conversion efficiencies as high as 60 to 70 percent. Fuel cells emit almost no air pollutants, require no cooling water, and operate quietly.

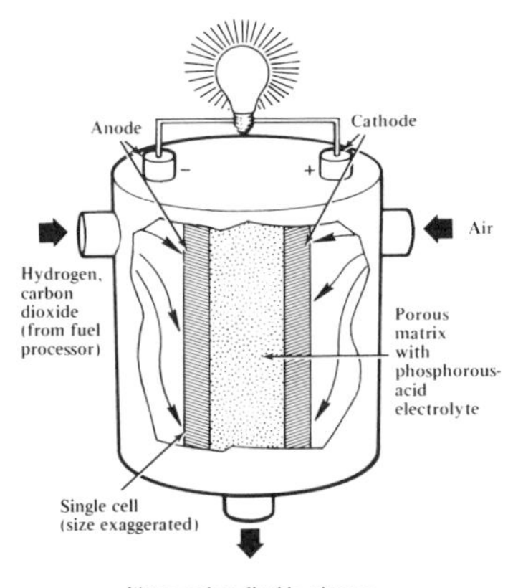


Structure of a fuel cell. (*Courtesy Popular Science, March 1977, p. 86*)

**fuel cell battery.** A chemical battery, with sodium, chlorine, and mercury among the possible ingredients, which will have self-generating tendencies.

**fuel cell power plants.** An advanced system for converting fuel directly into energy electrochemically. The basic elements of the system are fuel, oxidant, electrodes, and electrolyte chemicals. No moving parts are required, and since the reaction is chemical, fuel cells operate quietly.

**fuel channel.** A passage extending through the reactor core which is designed to contain one or more fuel assemblies and through which the primary coolant circulates.

**fuel charging machine.** A remotely operated mechanism used in reactor installations for loading (unloading) fuel assemblies.

**fuel cluster.** A group of fuel elements in the form of rods or pins, usually mounted parallel to one another.

**fuel consumption.** Automobiles consume fuel as a function of the weight and speed of the vehicle.

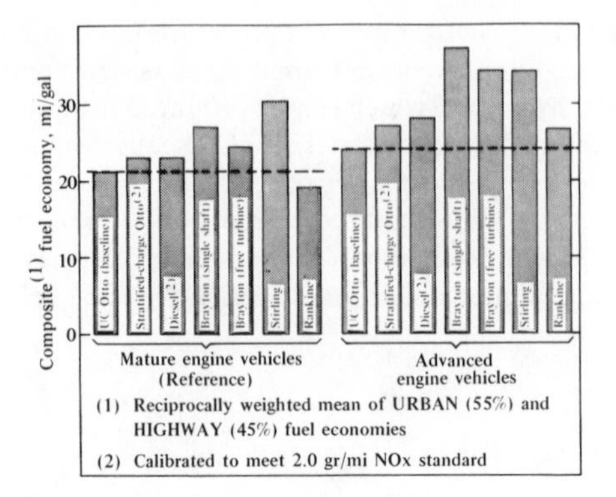


Fuel economy gains of Stirling and Brayton engines over Otto engine baseline projected for mature and advanced compact-class vehicles. (*Courtesy Popular Science, April 1976, p. 86.*)

**fuel consumption charge.** The fee for burnup and processing losses of nuclear fuel and for decreases in its value due to change in isotopic composition (including plutonium credit).

**fuel control.** Reactor control by adjustment of the properties, position, or quantity of nuclear fuel in such a way as to change the reactivity.

**fuel conversion factor.** Often used to compare converter reactors in which new fuel is produced from fertile material. A conversion factor of 1 indicates that one atom of new fuel is produced for every atom of fuel fissioned. Leakage and absorption reduce the conversion ratio in most thermal reactors. When the conversion factor is greater than 1, signifying that more fuel is produced than is consumed, the reactor is a breeder type and this ratio is the breeding factor.

**fuel cooling installation.** A large container or cell, usually fitted with coolant, in which spent nuclear fuel is set aside until its activity has decreased to a desired level.

**fuel cycle.** The series of steps involved in supplying fuel for nuclear power reactors. It includes mining, refining of uranium, the orignal fabrication of fuel elements, their use in a reactor, chemical processing to recover the fissionable material remaining in the spent fuel, reenrichment of the fuel material, and refabrication into new fuel elements. (See Fig. p. 183)

**fuel depot.** A bulk storage installation composed of storage tanks and related facilities such as docks, loading racks, and pumping units.

**fuel efficiency.** The ratio of the heat produced by a fuel for doing work to the available heat of the fuel. This efficiency is determined by the nonheat-forming materials in the fuel and the nonwork-producing heat which is developed by the fuel.

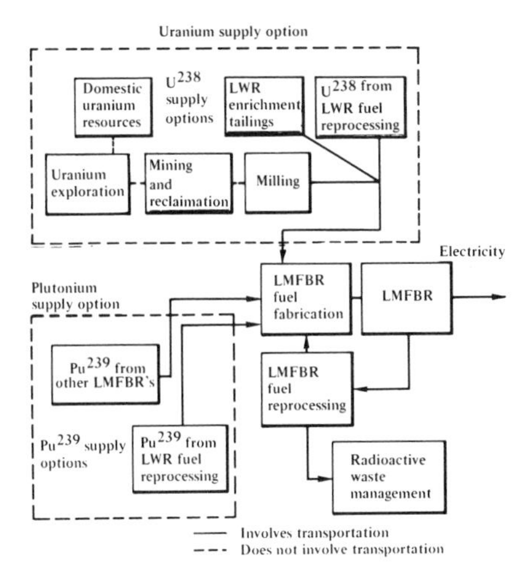


Fuel cycle of the Liquid Metal Fast Breeder Reactor. (*Courtesy U.S. Council on Environmental Quality*)

**fuel element.** A rod, tube, plate, or other mechanical shape or form into which nuclear fuel is fabricated for use in a reactor.

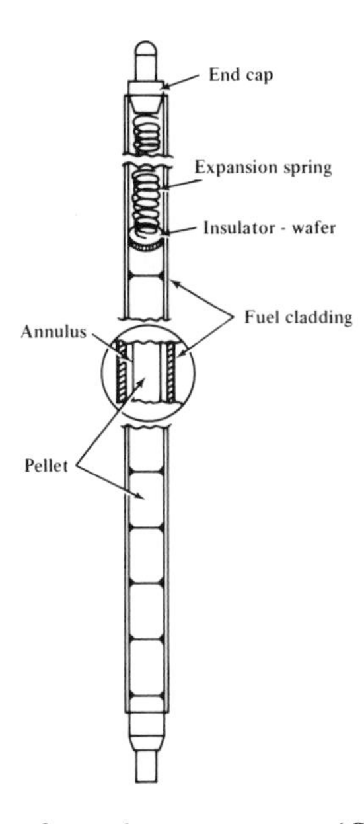


Cutaway of fuel element for nuclear reactor core. (*Courtesy U.S. ERDA-69*)

**fuel fabrication.** The manufacturing and assembly of reactor fuel elements containing fissionable and fertile nuclear material.

**fuel for electric generation.** Includes all types of fuel—solid, liquid, gaseous, and nuclear—used for the production of electric energy. Fuel for other purposes, such as building heating or steam sales, is excluded.

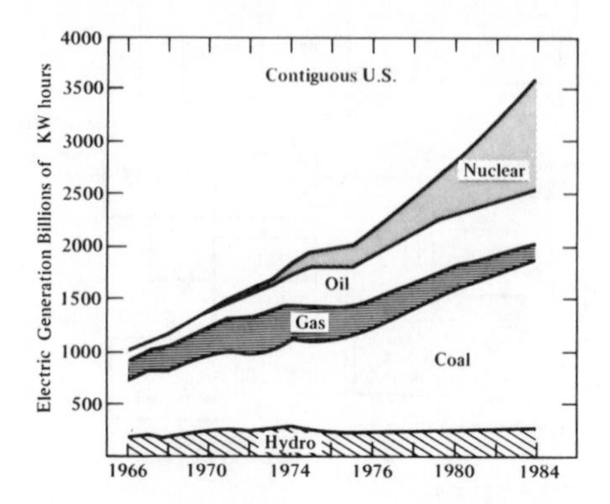


Electric generation sources in contiguous U.S. (*Courtesy U.S. Energy Research and Development Administration, Division of Coal Conversion and Utilization*)

**fuel gas.** Synthetic gas used for heating or cooling. It has less energy content than pipeline quality gas. Our government is developing a process to produce clean, low-Btu fuel gas from coal. The product could be burned in nearby power plants or used as a feed material for production of other synthetic fuels such as high-Btu pipeline gas.

**fuel mix.** The percentages of various fuels that make up total fuel consumption.

**fuel oil.** Any liquid or liquefiable petroleum product burned for the generation of heat in a furnace or firebox or for the generation of power in an engine, exclusive of oils with a flash point below 100° F, tag closed-cup tester, and oils burned in cotton or wool-wick burners.

**fuel pellet.** Small pellet of frozen deuterium and tritium for laser-induced fusion power plants.

**fuel pin.** A long (approximately 12 to 15 feet), thin tube that is approximately one-half inch in diameter. The tube is filled with nuclear fuel pellets.

**fuel rate.** The amount of fuel needed to generate 1 kilowatt-hour of electricity. In the United States electricity industry the rates are: 0.88 pounds of coal, 0.076 gallons of oil, and 10.4 cubic feet of natural gas.

**fuel reprocessing.** The processing of nuclear fuel after its use in a reactor to remove fission products and recover fissile, fertile, and other valuable material.

**fuel rod.** A fuel element in the form of a rod.

**fuel saver.** A solar device used solely to save fuel at conventional fossil fuel-burning facilities. The conventional systems provide the needed system reliability.

**fuel slug.** A fuel element in the form of a short rod.

**full-seam mining.** A mining system, brought on by the advent of mechanical loading and mechanical coal cleaning, in which the entire section is dislodged together, and the coal is separated from the rock outside of the mine by the cleaning plant.

**fume.** Tiny solid particles commonly formed by the condensation of vapors of solid matter. Sulfur trioxide and elemental sulfurs, driven off from furnaces and condensed, are classed as fume.

**fundamental particle.** Ultimate units occurring in the atom or formed by impact of high-energy particles from an outside source. In chemistry, the most important fundamental particles are the proton, the neutron, the electron, and the photon. Besides these, there are mesons of various designations, such as neutrinos and positrons, which are of more direct concern to the physicist than to the chemist.

**fungi.** Small, often microscopic plants without chlorophyll. Some fungi infect and cause disease in plants or animals; other fungi are useful in stabilizing sewage or in breaking down wastes for compost. Fungi are important elements of bioconversion of waste molecules for new energy sources.

**furling.** Stopping the rotors of a wind generator so that they cannot turn.

**furnace.** When used in a central heating system, a self-contained appliance for heating air by transfer of heat of combustion through metal to the air.

**furnace oil.** A distillate fuel primarily intended for use in domestic heating equipment.

**fusain.** This substance is recognized macroscopically by its black or gray-black color, its silky luster, its fibrous structure and its extreme friability. It is the only constituent of coal which marks and blackens objects with which it comes in contact. Fusain may include a high proportion of mineral material, which strengthens it and reduces its friability but not its silky luster. In macroscopic description of seams, only those bands having a thickness of several millimeters are recorded. Microscopic examination shows that fusain consists mainly of fusite. Fusain occurs as wide bands and lenses in almost all humic coals. It is widely distributed but not abundant.

**fused-salt reactor.** A type of reactor that uses molten salts of uranium for both fuel and coolant.

**Fushun process.** Oil shale retorting process involving direct heating by a mixture of combustion gases and reheated recycled gases.

**fusion.** A nuclear reaction in which two light atomic nuclei unite (or fuse) to form a single nucleus of a heavier atom. This process takes place in the sun and in other active stars. It has been duplicated by physicists in the hydrogen bomb, though it has not yet been controlled for useful purposes, as has the fission reaction. The term also applies to the conversion of a solid substance to its liquid state, also known as melting.

**fusion fuels.** The atoms of several light elements can be used to fuel fusion power reactors. Those of greatest interest are deuterium (D) and tritium (T), both isotopes, or varieties, of hydrogen. With existing technology, a virtually unlimited amount of deuterium can be obtained from water at a relatively small cost. Tritium must be bred from lithium, a relatively abundant element found in the earth and the sea. First generation fusion reactors probably will use a combination of deuterium and tritium (DT) while later systems may burn deuterium alone.

**fusion point.** The temperature at which melting takes place. Most refractory materials have no definite melting points, but soften gradually over a range of temperatures.

**fusion power.** In fusion, two light nuclei are united to form a heavier nucleus, thereby releasing energy. The fusion reaction uses the heavy isotopes of hydrogen known as deuterium and tritium. Deuterium can be economically separated from sea water, and tritium can be obtained in a nuclear reaction involving lithium. To make controlled fusion work, one must heat an electrified gas called a plasma to temperatures on the order of 100 to 300 million degrees Centrigade (Celcius)—hotter than the interior of the sun. This gas must then be contained in some way so that it does not touch the walls of the vessel, and held in this condition until an adequate number of fusion reactions take place. One approach is to confine a fusion plasma with the use of specially-shaped magnetic fields which control the motions of the plasma. Another potentially feasible approach is to use high-powered lasers to initiate and confine fusion reactions. (See Fig. p. 187)

**fusion reactions.** Fusion occurs in magnetic confinement when a gaseous fuel is heated sufficiently to strip electrons from their atoms and drive the remaining nuclei into each other with such force that they fuse, releasing energy in the process. The temperatures needed are in the range of 100,000,000 degrees Centigrade. Partially ionized gas at these high temperatures is called plasma. Magnetic fields are used to contain the fusion plasma and to prevent it from coming in contact with reactor walls. If it touched the walls, the plasma would lose its heat and the fusion reaction would be quenched. In order to reach the high temperatures needed for fusion reactions, it is necessary to invest energy in the plasma. This may be done by one method or by a combination of methods, including electric current (resistance heating), high frequency radio waves, magnetic compression and the injection of a beam of high energy deuterium or tritium neutral atoms (neutral beam heating). Practical production of fusion energy requires that the plasma be maintained at a minimum temperature and density for a sufficiently long time to release much more energy than was invested to heat and maintain the plasma. The required minimum densities were achieved first in 1953, minimum temperatures in 1962, and adequate confinement

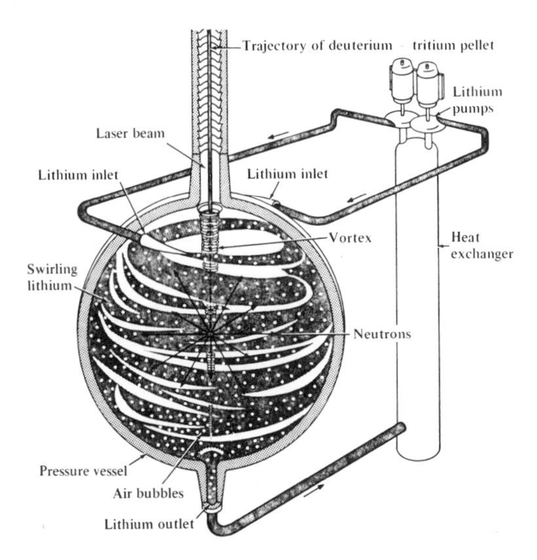


Design of a laser-fusion reactor with a lithium vortex and no interior walls. The fusion energy would be absorbed in a pool of liquid lithium contained in the pressure vessel. (*Courtesy "Energy and the Future," Fig. 29*)

in 1969. Only combinations of two of these conditions can be achieved simultaneously in today's relatively small magnetic confinement experiments.

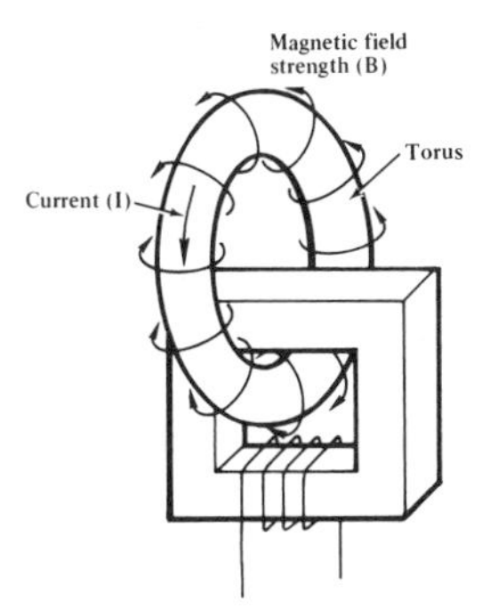


A simple fusion reactor. The torus acts as the secondary of a transformer and contains the plasma. The current (I) induces the magnetic field of strength (B) which pinches the plasma inward away from the walls. (*Courtesy Basic Nuclear Engineering, 3rd Edition*)

**fusion weapon.** An atomic weapon using the energy of nuclear fusion, such as a hydrogen bomb.

# G

**gabion.** A mesh container used in engineering; when filled with rocks or stones, it may form the foundation of a jetty.

**gallatin.** The heavy oil of coal tar used in the Bethell process for the preservation of timber. Also called deal oil.

**gallery.** A horizontal or nearly horizontal underground passage, either natural or artificial.

**gallium arsenide.** A compound used in making photovoltaic cells.

**gallon.** A unit of measure. A U.S. gallon contains 231 cubic inches, 0.133 cubic feet, or 3.785 liters. It is 0.83 times the imperial gallon.

**galvanic cell.** A cell in which chemical change is the source of electrical energy. It usually consists of either two dissimilar conductors in contact with each other and with an electrolyte, or two similar conductors in contact with each other and with dissimilar electrolytes.

**galvanic coupling.** Contact between dissimilar metals immersed in an electrically conductive solution (such as water-based heat transfer media) results in a potential (voltage) difference, causing the less noble material metal to corrode. The galvanic series which tabulates the potential differences between metals under corrosive conditions in a given environment or electrolyte form the basis for predicting corrosion tendencies.

**gamma rays.** High-energy, short-wavelength electromagnetic radiation. Gamma radiation frequently accompanies alpha and beta emissions and always accompanies fission. Gamma rays are very penetrating and are best stopped or shielded against by dense materials, such as lead or depleted uranium. Gamma rays are essentially similar to x-rays, but are usually more energetic, and are nuclear in origin.

**gangue.** Undesired minerals, mostly nonmetallic, associated with ores.

**gangway.** A main haulage road underground; frequently called entry.

**gantry.** An overhead structure that supports machines or operating parts, such as a bridge or platform carrying a traveling crane or winch and supported by a pair of towers, trestles, or side frames running on parallel tracks.

**Garrett Coal Pyrolysis process.** Pulverized coal is transported to the entrained-flow pyrolysis reactor where it is mixed with the stream of hot char coming from the char heater. Hot char provides heat for the flash pyrolysis process. Effluent from the reactor is passed through a cyclone to separate the char, part of which goes to the char heater, while the remainder goes out as product. Effluent gases are cooled. Tar is separated and hydrotreated to produce syncrude and medium-Btu gases which are purified to get product gas. The process temperature is 593°C.

**gas.** That state of matter which has neither independent shape nor volume. It expands to fill the entire container in which it is held. It is one of the three forms of matter: solid, liquid, and gas.

**gas absorption.** The extraction of a gaseous substance from an atmosphere by liquid or solid material.

**Gasahol.** A program established by the State of Nebraska to aid in the development of an alternative automotive fuel containing a blend of 10 percent agriculturally derived ethyl alcohol and 90 percent unleaded gasoline. Also a trademark registered by the Agricultural Products Industrial Utilization Committee (APIUC), a Nebraska state agency established to administer the Gasahol program.

**gas bag.** A gas-proof, inflatable bag which can be inserted in a gas pipe and inflated to seal off the flow of the gas.

**gas balance.** An instrument used for determining the specific gravity of gases.

**gas burner.** A device for the final release of air/gas, or oxygen/gas mixtures, or air and gas separately into the combustion zone. Gas burners may be classed as atmospheric burners or blast (pressure) burners.

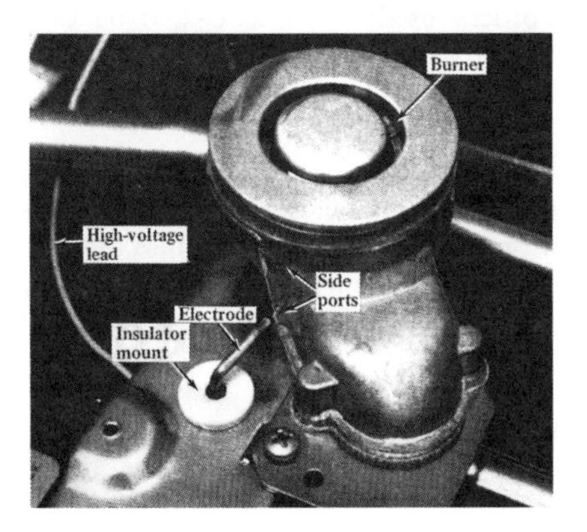


Photograph shows a direct igniter which ignites the burner. (*Courtesy Popular Science, January 1976, p. 20*)

**gas cap.** A layer of gas on top of oil in an underground structure or reservoir.

**gas centrifuge process.** A method of isotopic separation in which heavy gaseous atoms or molecules are separated from light ones by centrifugal force.

**gas conditioning.** The removal of objectionable constituents and addition of desirable constituents.

**gas consumer.** The ultimate user of gas, as contrasted to a "customer" who may purchase gas for resale.

**Gas-Cooled Fast-Breeder Reactor (GCBR).** A fast-breeder reactor which is cooled by a gas, usually helium, under pressure.

**gas-cooled reactors.** A nuclear reactor in which a gas is the coolant.

**gas cycling.** A petroleum recovery process which takes gas produced with oil and injects it back into the oil sand to aid in producing more oil.

**gas detector.** A device to show the presence of firedamp in a mine.

**gas distribution company.** Company which obtains the major portion of its gas operating revenues from the operation of a retail gas distribution system, and which operates no transmission system other than incidental connections within its own system or to the system of another company. For purposes of American Gas Association statistics, a distribution company obtains at least 95 percent of its gas operating revenues from sales to ultimate customers, and classifies at least 95 percent of mains (other than service pipe) as distribution.

**gaseous diffusion.** A process used to separate U-235 from U-238 and thus provide an enriched U-235 fuel. The process, requiring extensive equipment and huge structures when done on a large scale, separates the materials on the basis of their differences in diffusivity.

**gaseous diffusion (plant).** A method of isotopic separation based on the fact that gas atoms or molecules with different masses will diffuse through a porous barrier (or membrane) at different rates. The method is used to separate uranium-235 from uranium-238; it requires large gaseous-diffusion plants and enormous amounts of electric power.

**gaseous fuel.** Includes natural gas and the prepared varieties such as coal gas, oil gas, iron blast furnace gas, as well as producer gas.

**gas flow processes.** Oil shale retorting processes in which heat transfer is effected by a superheated steam mixed with air.

**gas "guzzler".** Motor vehicles consuming excessive fuel in accordance with the guidelines established by President Carter's energy program.

**gas holder.** A gas-tight receptacle or container in which gas is stored for future use. There are two general ways of storing gas; at approximately constant pressure (low-pressure containers) in which case the volume of the container changes; and in containers of constant volume (usually high-pressure containers) in which case the quantity of gas stored varies with the pressure.

**gas house riser.** The principal vertical pipe which conducts the gas from the meter to the different floors of the building.

**gasification.** In the most commonly used sense, gasification refers to the conversion of coal to a high-Btu synthetic natural gas under conditions of high temperatures and pressures; in a more general sense, conversion of coal into a usable gas.

**gasification of coal.** The conversion of solid coal into a gaseous fuel by chemical processes.

**gasifier.** A vessel in which gasification takes place.

**gas impurities.** Undesirable matter in gas, such as dust, excessive water vapor, hydrogen sulphide, tar, and ammonia.

**gas injection.** The process of injecting gas into a reservoir to maintain the pressure in the producing formation.

**gasket.** A piece of solid material, usually metal, rubber, plastic, or asbestos, placed between two pieces of pipe or between automotive engine cylinders and heads to make gastight joints.

**gas lift.** The effect of either natural or artificially induced gas pressure in an oil well that causes the oil to flow from the well.

**gas meter.** An instrument for either measuring and indicating or recording the volume of gas that has passed through it.

**gas oil.** A petroleum distillate obtained after kerosine; it has a flash point 76° C and is used for carbureting water gas in gas plants and for driving diesel engines.

**gas-oil ratio.** The number of cubic feet of gas produced with each barrel of oil.

**gasoline.** A volatile flammable liquid obtained from petroleum which has a boiling range of approximately 29° to 216° C, and is used as fuel for spark ignition internal-combustion engines.

**gasoline plant.** A plant in which hydrocarbon components common to the gasoline fractions are removed from "wet" natural gas, leaving a "drier" gas.

**gasometers.** Tall metal chambers (gas holders) fitted with a roof which rises or falls with the entry or removal of gas. The roof is weighted so that the gas is compressed, thus enabling it to pass along gas mains and pipes at a suitable pressure for efficient combustion.

**gasoscope.** An apparatus for detecting the presence of dangerous gas escaping into a coal mine or a dwelling house.

**gas range.** Cooking stove.

**Gas Recycle Hydrogenation process.** Gasification of distillate feedstock produced from crude oil to manufacture synthetic natural gas (SNG).

**gas sendout.** Total gas produced, purchased (including exchange gas receipts), or net gas withdrawn, from underground storage within a specified time interval, measured at the point(s) of production and/or purchase and/or withdrawal, adjusted for changes in local storage quantity. It comprises gas sales, exchange, deliveries, gas used by company, and unaccounted for gas.

**gassy.** A coal mine is rated gassy by the Government if an ignition occurs or if a methane content exceeding 0.25 percent can be detected. Work must be halted if the methane exceeds 1.5 percent in a return airway.

**gas synthesis.** A mixture of carbon monoxide and hydrogen containing small amounts of nitrogen, some carbon dioxide, and various trace impurities prepared for petrochemical synthesizing processes.

**gas transmission company.** A company which obtains at least 95 percent of its gas operating revenues from sales for resale and/or transportation of gas for others and/or main line sales to industrial customers, and classifies at least 95 percent of its mains (other than service pipe) as field and gathering, storage, and/or transmission.

**gas transported for others.** That volume of gas owned by another company received into and transported through any part of the transmission system under a transportation tariff.

**gas turbine.** A prime mover in which gas, under pressure or formed by combustion, is directed against a series of turbine blades; the energy in the expanding gas is converted into mechanical energy supplying power at the shaft. (See Fig. top of p. 193)

**gas turbine combustion engine.** A new generation engine in which a continuous flow of heated, compressed air is directed at high velocity against a bladed turbine which in turn drives the transmission and vehicle wheels. Any type of burnable liquid can be used to provide the heat. (See Fig. p. 193)

**gas turbine generating station.** An electric generating station in which the prime mover is a gas turbine engine.

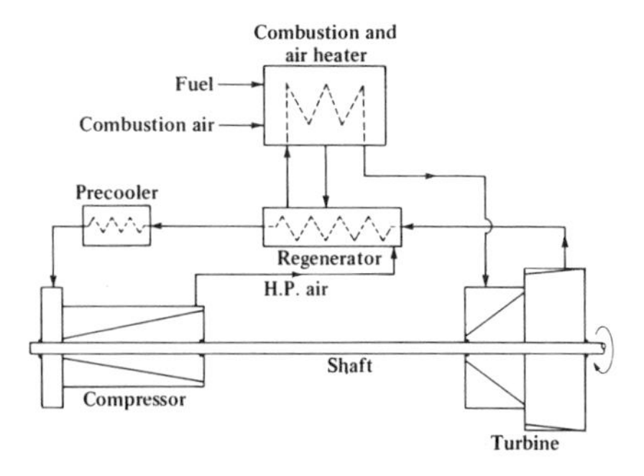


A closed-cycle gas turbine. (*Courtesy Van Nostrand's Scientific Encyclopedia*)

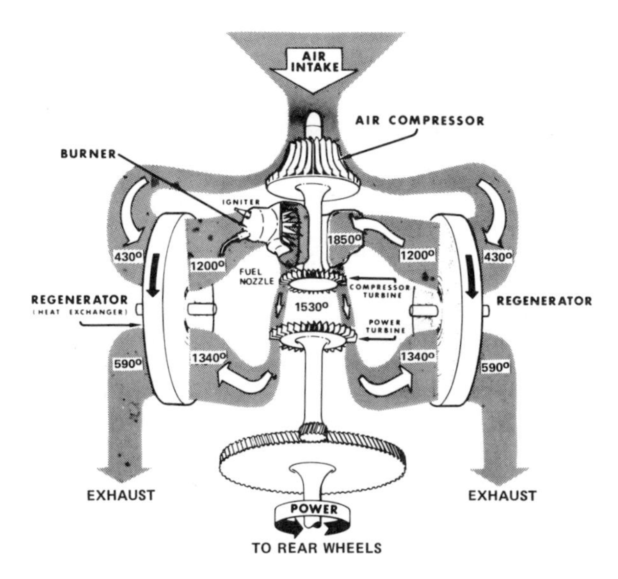


Diagram of Chrysler Corporation's sixth generation gas turbine engine. (*Courtesy U.S. Energy Research and Development Administration*)

**gas utility.** A company that is primarily a distributor of natural gas to ultimate customers in a given geographic area.

**gas well.** A well that produces chiefly natural gas; a well that produces the contents of a gas reservoir at surface conditions. Legal definitions vary among the states.

**gas works.** A plant for manufacturing gas.

**Gasynthan process.** Production of synthetic natural gas with calorific value up to 1000 British thermal units per standard cubic foot, at pressure between 300 and 500 pounds per square inch gauge, from natural gas condensates, propane-butane, refinery gases, light and full range naphtha.

**gas zone.** A porous, permeable formation containing natural gas under pressure.

**gate valve.** A full-opening valve controlled by a verticle movement of a single or pair of solid discs perpendicular to the direction of flow. Used to regulate the main supply of water.

**gathering station.** A compressor station at which gas is gathered from wells by means of suction because the pressure is not sufficient to produce the desired rate of flow into a transmission or distribution system.

**gauging.** The measurement of the thickness, density, or quantity of a material by the amount of radiation it absorbs. This is the most common use of radioactive isotopes in industry. Also spelled gaging.

**gauss.** The unit of magnetic field intensity equal to 1 dyne per unit pole. The preferred term for this unit is oersted. One oersted equals $10^5$ gammas. Gauss was used before the official adoption of the oersted in 1932.

**gauze.** The wire mesh used to prevent the passage of flame from a flame safety lamp to the external atmosphere.

**Gay-Lussac's law.** When gases react, they do so in volumes which bear a simple ratio to one another, and to the volumes of their products if these are gaseous, temperature and pressure remaining constant. Also called law of gaseous volumes.

**Gegas process.** An integrated coal gasification, gas cleaning process for the production of clean, low-Btu gas.

**Geiger-Muller counter.** A radiation detection and measuring instrument. It consists of a gas-filled tube containing electrodes, between which there is an electrical voltage but no current flowing. When ionizing radiation passes through the tube, a short, intense pulse of current passes from the negative electrode to the positive electrode and is measured or counted. The number of pulses per second measures the intensity of radiation. Known widely as Geiger counter, it was named for Hans Geiger and W. Muller who invented it in the 1920's. (See Fig. p. 195)

**Geissler tube.** A sealed and partly evacuated glass tube containing electrodes. Used for the study of electric discharges through gases.

**generation time.** The mean time for the neutrons produced by one fission to produce fissions again in a chain reaction.

**generator.** A machine that converts mechanical energy into electrical energy.

Geiger counter being used to differentiate the area of varying ore grade materials. (*Courtesy U.S. Energy Research and Development Administration, photo by Wistcott*)

**generic system.** A system that is designated a "typical" system, when there are numerous variations of the system being studied.

**genetically significant dose.** A population-averaged dose which estimates the potential genetic effects of radiation on future generations. It takes into consideration the number of people in various age groups, the average dose to the reproductive organs to which people in these groups are exposed, and their expected number of future children.

**genetic effects of radiation.** Radiation effects that can be transferred from parent to offspring. Any radiation-caused changes in the genetic material of sex cells.

**Genter filter.** A filter utilized in coal-washing plants for the recovery of fine coal particles.

**geochemical exploration.** Exploration or prospecting methods that depend on chemical analysis of the rocks or soil, or of soil gas, or of plants.

**geofault.** A large fault directly affecting the relief of the earth's surface, on land or beneath the sea.

**geognosy.** A branch of geology that deals with the materials of the earth and its general interior and exterior constitution.

**Geological Survey.** A bureau of the Department of the Interior established in 1879. The objectives of the Survey are to "perform surveys, investigations, and research covering topography, geology, and the mineral and water resources of the United States; classify land as to mineral character and water and power resources; enforce departmental regulations applicable to oil, gas, and other mining leases, permits, licenses, development contracts, and gas storage contracts; and publish and disseminate data relative to the foregoing activities."

**geometrically safe.** A system containing fissile material which is incapable of supporting a self-sustaining nuclear chain reaction by virtue of the geometric arrangement or shape of the components.

**geometric attenuation.** The reduction of a radiation quantity because of the effect of the distance between the point of interest and the source, and excluding the effect of any matter present.

**geometry.** The spatial configuration, pattern, or relationship of components in an experiment or apparatus. In reactor technology, the term refers to the shape and size of fuel elements, moderator and reflector, and their location with respect to each other. In nuclear physics, it refers to the arrangement of source and detecting equipment. In counting and scanning, the term commonly indicates the percentage of the radiation leaving a sample which reaches the sensitive volume of a counter.

**geophone.** An instrument for detecting vibrations passing through rocks, soil or ice.

**geophysical survey.** Searching and mapping of the subsurface structure of the earth's crust by use of geophysical methods to locate probable reservoir structures capable of containing gas or oil.

**geophysics.** A study of subsurface geological conditions of structure or material through the interpretation of measurement variations in density, magnetics, elasticity, electrical conductivity, temperature, and/or radioactivity.

**geopressured water.** Geologic formations—called geopressured aquifers—beneath the Texas and Louisiana Gulf Coasts contain large volumes of hot brine under high pressure. It is considered likely that these deposits also contain large amounts of entrapped natural gas. Release of these trapped geopressured waters could provide heat energy, hydraulic energy, and energy from burning the natural gas.

**Georgia V-ditch.** Preparing a ditch by grading to create draining swales midpoint between and parallel to the highwall and lowwall in order to convey water runoff to drains established to carry the water away from the spoil area.

**geostrophic motion.** Motion which is unaccelerated and frictionless.

**geostrophic wind.** That horizontal wind velocity for which the Coriolis acceleration exactly balances the horizontal pressure force.

**geostrophic wind level.** The lowest level, observed to be between 1.2 and 1.6 kilometers, at which the wind becomes geostrophic. The geostrophic wind level may be considered to be the base of the free atmosphere.

**geothermal energy.** The heat energy available in the rocks, hot water, and steam in the earth's subsurface.

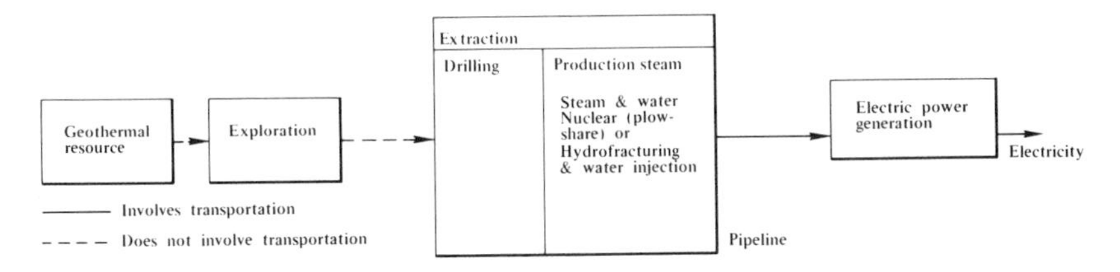


Geothermal resource development. (*Courtesy U.S. Council on Environmental Quality*)

**geothermal energy resources.** Thermal energy resources found in the earth; these include hydrothermal convective systems, pressurized water reserves, hot dry rocks, manual gradients, and magma.

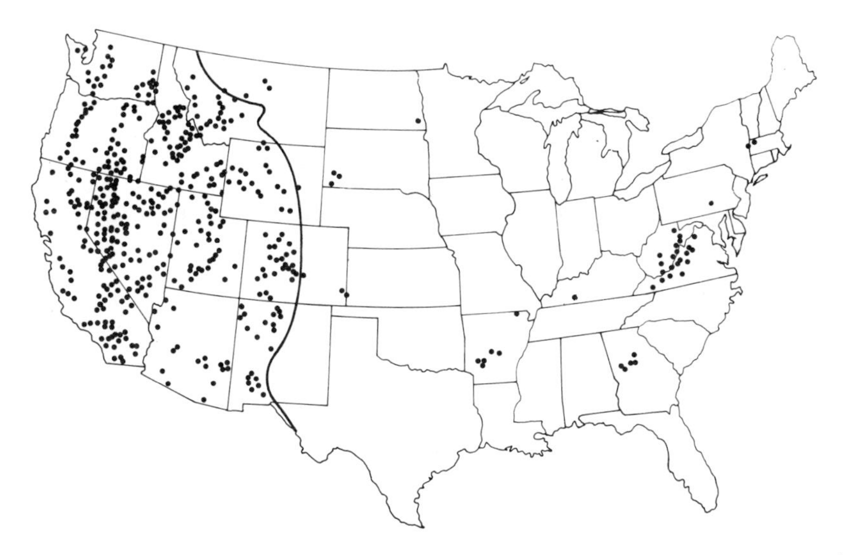

Location of geothermal resources in the U.S. (*Courtesy National Science Teachers Association*)

**geothermal generating station.** An electric generating station in which the prime mover is a steam turbine, and the source of the steam is generated in the earth by heat from the earth's magma. (See Fig. p. 198)

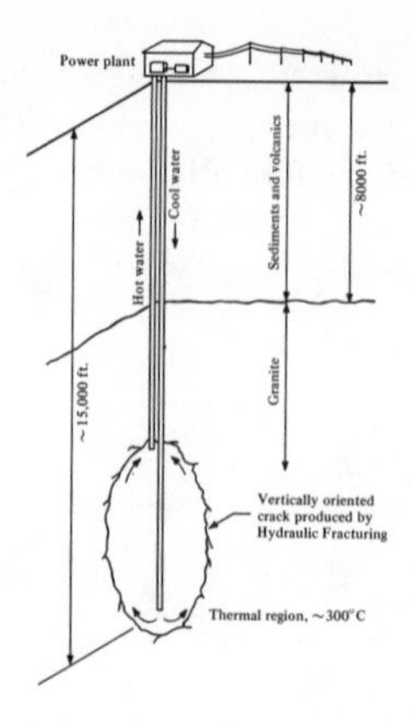


Extracting energy from a dry geothermal reservoir. (*Courtesy "Energy and the Future," Fig. 19*)

**geothermal gradient.** The change in temperature of the earth with depth, expressed either in degrees per unit depth, or in units of depth per degree.

**geothermal heat.** The natural migration of heat outward from the earth's molten core. In unbroken portions of the earth's crust, heat is transferred by conduction, a slow and diffuse process revealed by gentle temperature gradients that decrease toward the earth's surface. But where that surface is weakened, as by the fracture and faults at crustal plate boundaries, magma intrudes a hot mass transported from below. Geothermal heat is thus concentrated in geologically active regions. Hydrothermal convection cells transfer the heat up, where it vents naturally in hot springs, fumaroles, or geysers—or to a level that is accessible for extraction through wells.

**geothermal power.** Power plants using hot water or steam that is stored in the earth from volcanic activity. Sources of geothermal energy are currently under development in this country and New Zealand. Geysers in California presently produce electricity in the United States. In a few places natural steam is available. In many places, hot water can be tapped as a usable energy source. Also, there are areas of intensely hot underground rock that can be used by fracturing the rock and forcing cold water down to it. The heated water can be returned to the surface to produce steam power. (See Fig. p. 199)

**geothermal steam.** Steam drawn from deep within the earth. There are about 90 known places in the continental United States where geothermal steam could be harnessed for power. These are in California, Idaho, Nevada, and Oregon. (See Fig. p. 200)

**geothermometer.** A thermometer designed to measure temperatures in deep-sea deposits or in boreholes deep below the surface of the earth; a geologic thermometer.

**geyser.** A volcano in miniature from which hot water and steam (instead of lava and ashes) are erupted periodically during the waning phase of volcanic activity. (See Fig. p. 200)

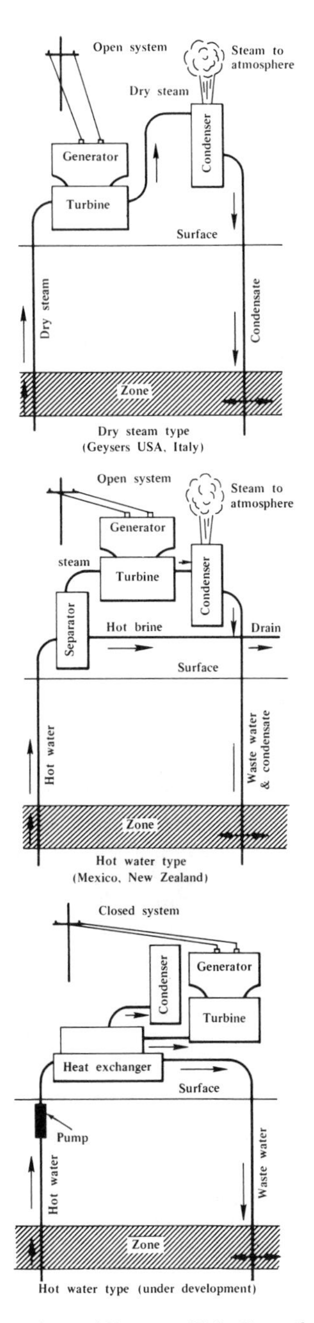


Types of geothermal power plants. (*Courtesy U.S. Council on Environmental Quality*)

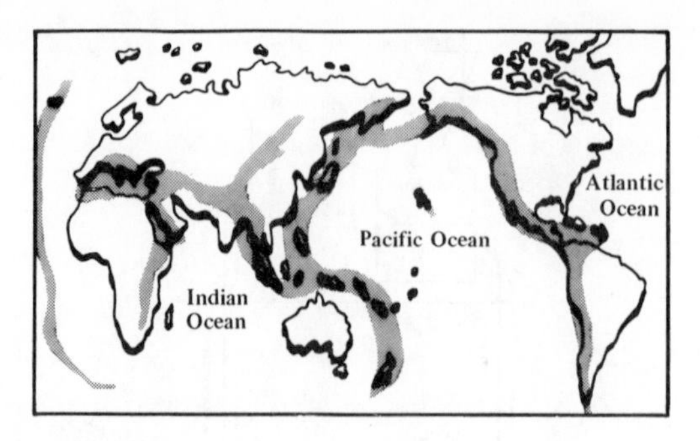


Regions of intense geothermal manifestations. (*Courtesy Annual Review of Energy, Volume 1*)

The geysers area in California is about 2 by 8 miles. Over 100 wells have been drilled, the deepest of which is more than 8,000 feet. Temperatures of the reservoirs are about 255° C (or 480° F). The basic heat source appears to be a mass of heated rocks at a 3 to 5 mile depth, covering an area of about 100 to 500 square miles. (*Courtesy U.S. Energy Research and Development Administration*)

**gib.** A temporary support, either by hand or by machine, at the face to prevent coal from falling before the cut is complete.

**Gibbs-Helmholtz equation.** A thermodynamic relationship useful in calculating changes in the energy (heat content) of a system.

**gilsonite.** An asphaltite or solidified hydrocarbon found only in the United States, Utah, and Colorado; one of the purest (99.9 percent) natural bitumens. Term also applies to very rich tar deposits and a tar sand with a very high hydrocarbon content and low mineral content.

**Girbotol process.** A wet scrubbing process for removing hydrogen sulfide from fuel gases in which aqueous solutions of aliphatic amines dissolve hydrogen sulfide and carbon dioxide from gases. The dissolved gases may be recovered by boiling.

**Glauber's salts.** A hydrous sodium sulfate, sealed in cans to store heat, which can be used to warm air circulated through an area.

**glazing materials.** Glass, plastic films, and sheets used on solar collectors. They admit as much solar radiation as possible and reduce the upward loss of heat to the lowest attainable value.

**globe valve.** A normally spherical valve equipped with an orifice and a stem attached to a plug and matching circular seat. Shutoff is obtained by direct contact of the plug and the seat.

**glove box.** A sealed box in which workers, using gloves attached to and passing through openings in the box, can handle radioactive materials safely from the outside.

**glow.** The incandescence of a heated substance, or the light from such a substance; white or red heat.

**goaf.** That part of a mine from which the coal has been worked away and the space filled up; also the refuse or waste left in a mine.

**gob.** Waste coal, rock pyrites, slate, or other unmerchantable material of relatively large size which is separated from coal and other mined material in the cleaning process; also the space left in a coal mine after the coal is removed and in which caving occurs.

**gouging.** A surface configuration wherein a channel is formed to trap precipitation, increase infiltration, and reduce erosion.

**governor.** A device for regulating the speed of an engine or motor under varying conditions of load and pressure; also a device that automatically controls the speed of the rotors in a wind generator by changing their angle. In a high wind, the governor turns the blades into the wind (called feathering) to prevent damage to the machine.

**grade of coal.** A term to indicate the nature of coal as determined mainly by the amount and nature of the ash and the sulfur content. Grade is sometimes used as a synonym for rank.

**gradient wind.** Any horizontal wind velocity that is tangent to the contour line of a constant-pressure surface.

**grading.** The degree of mixing of size classes in sedimentary material; well graded implies comparatively uniform distribution from coarse to fine; poorly graded implies uniformity in size or lack of continuous distribution.

**graduator.** An apparatus for evaporating a liquid by causing it to flow over large surfaces while exposed to a current of air.

**grain loading.** The rate of emission of particulate matter from a polluting source, measured in grains of particular matter per cubic foot of gas emitted.

**grain size.** Physical size of particle, usually determined by either sieve or hydrometer analysis. For metals, a measure of the areas or volumes of grains in a polycrystalline material. Grain sizes are reported in terms of number of grains per unit area or volume, average diameter, or as a grain-size number derived from area measurement.

**gram-roentgen.** An obsolete unit of integral absorbed dose equal to about 84 ergs.

**Granby cars.** A popular type of automatically dumped car for hand or power-shovel loading.

**graphite.** A very pure form of carbon used as a moderator in nuclear reactors.

**gravimetric survey.** An exploration method which involves interpreting the probable density of minerals in the earth by measured gravity variations.

**gravitational energy.** The energy of attraction between two material bodies. It is directly proportional to the product of the masses of the two bodies and inversely proportional to the square of the distance between their centers.

**gravitational separation.** The separation of oil, gas, and water in a reservoir rock in accordance with their relative gravities.

**gravity.** The force by which substances are attracted to each other, or fall to earth.

**gravity API.** The gravity scale developed by the American Petroleum Institute to express the density of liquid petroleum products. In this scale, water has a gravity of 10° API; and liquids lighter than water (such as petroleum oils) have API gravities numerically greater than 10.

**gravity survey.** A method which uses a gravity instrument to detect variations in the gravitational pull of rocks in the subsurface. Variations or anomalies are contoured on a map and give evidence of geologic structures.

**gray.** A body or medium absorbing a significant part, but not all, of the radiation of some specified energy incident on it.

**Great Coal Age.** Another name for the Pennsylvania (or Coal Measures) Period. So called because the greatest coal deposits of the world are found in formations of this age.

**green belts.** Those areas, restricted from use for buildings and houses, which serve as separating buffers between pollution sources and concentrations of population.

**greenhouse effect.** The heating effect of the atmosphere upon the earth. Light waves from the sun pass through the air and are absorbed by the earth. The earth then reradiates this energy as heat waves that are absorbed by the carbon dioxide in the air which behaves like glass in a greenhouse, allowing the passage of light but not of heat. Many scientists theorize that an increase in the atmospheric concentration of carbon dioxide can eventually cause an increase in the earth's surface temperature.

**grey.** A unit of absorbed dose. (One Gy = J/kg = 100 rads.)

**grid.** A wire bottomed mining sieve; a screen; electrodes which are placed in the arc stream and to which a control voltage may be applied; the layout of a gas distribution system of a city or town in which pipes are laid in both directions in the streets and connected at intersections.

**grille.** A covering over an inlet or outlet with openings through which a fluid passes.

**gross calorific value.** In the case of solid and liquid fuels of low volatility, the heat produced by combustion of unit quantity, at constant volume, in any oxygen-bomb calorimeter under specified conditions.

**gross collector area.** The maximum projected area that encompasses the collector and its associated manifolds, and other elements normally part of the complete collector module.

**Gross National Product (GNP).** The total market value of the goods and services produced by the nation before the deduction of depreciation charges and other allowances for capital consumption; a widely used measure of economic activity.

**gross national product implicit price deflator.** A measure of the increase in the aggregate price of gross national product over the price in a base year.

**ground state.** The state of a nucleus, atom, or molecule at its lowest (normal) energy level.

**ground temperature.** In the gas industry, the temperature of the earth at pipe depth.

**groundwater.** Water below the ground surface, such as the supply of fresh water under the earth's surface, in an aquifer or soil that forms the natural reservoir for man's use.

**groundwater runoff.** Groundwater discharged into a stream channel as spring or seepage water.

**ground zero.** The point on the surface of land or water vertically below or above the center of a burst of a nuclear explosion.

**group removal cross section.** The weighted average cross section, characteristic of a neutron energy group, that accounts for the removal of neutrons from that group by all processes.

**grout.** A pumpable slurry of neat cement or a mixture of neat cement and fine sand, commonly forced into a borehole to seal crevices in a rock to prevent ground water from seeping or flowing into an excavation, or to seal crevices in a dam foundation.

**grouting.** The act or process of injecting a grout into a rock formation through a borehole to mend cracks and make the surface water-resistant.

**growing season.** The season which is warm enough for the growth of plants. On the whole, the growing season is confined to that period of the year when the daily means are above 6°C.

**Gulf CCL process.** Coal is slurried in recycle oil, mixed with hydrogen, and fed to the fixed-bed reactor. Coal molecules are depolymerized in the presence of hydrogen and a catalyst. Gas, liquid, and solid products are then separated. A heavy fuel oil with a heating value of 17,900 British thermal units per pound and a light fuel oil with a heating value of 18,800 British thermal units per pound are produced. The yield of heavy fuel oil is 2.3 barrels per ton, and the yield of light fuel oil is 0.9 barrels per ton. Products contain approximately 0.04 percent sulfur.

**gully erosion.** Removal of soil by running water with the subsequent formation of deep channels.

**gum.** Small coal, slack, screenings; also a resinous material formed in regulators, meters, and orifices from the polymerization of certain gas components present in manufactured gas.

**gun perforator.** A device for firing bullets or explosive powder through tubings, casings, and cement into the producing formation of a well to provide channels for flow of gas and/or oil into the well.

**gusher.** An oil well with a strong natural outflow; a geyser.

**gust.** A sudden brief increase in the speed of the wind followed by a lull or slackening in wind speed.

**G-value.** The number of specified chemical changes in an irradiated substance produced per 100 electron volts of energy absorbed from ionizing radiation. Examples of such chemical changes are cross-linking, production of particular molecules, and production of free radicals.

# H

**habitat.** The place where an animal or plant naturally or normally lives and grows.

**hade.** The angle of inclination of a vein measured from the vertical; dip is measured from the horizontal.

**half-life.** The time in which half the atoms of a particular radioactive substance disintegrate to another nuclear form. Measured half-lives vary from millionths of a second to billions of years; for example, the half-life of radium is 1620 years.

**half mask.** The part of a mine rescue or oxygen-breathing apparatus which covers the nose and mouth only and through which the wearer breathes the oxygen furnished by the apparatus.

**half-thickness.** The thickness of any given absorber that will reduce the intensity of a beam of radiation to one-half its initial value.

**Hall effect.** When a conductor or semiconductor carrying a current is placed in a magnetic field that has a component normal to the direction of the current density, the distribution of the current density over the cross section of the conductor or semiconductor is changed, and a potential gradient is set up in a direction normal to the directions of both the current density and the field.

**halogens.** Members of the family of very active chemical elements consisting of bromine, chlorine, fluorine, and iodine, chemically resembling each other closely: all are monovalent, nonmetallic, and capable of forming negative ions; also elements that react with metals to form salts.

**hammermill shredder.** A cylindrical machine which is lined with spike-shaped projections which are utilized to tear and break up organic waste material.

**hammerpick**. A compressed-air-operated hand machine consisting mainly of a pick and a hammer and used by miners to break up the harder rocks in a mine.

**hand and foot counter.** A monitoring device arranged to give a rapid radiation survey of the hands and feet of persons working with radioactive materials to detect radioactive contamination.

**hand cleaning.** The removal by hand of impurities from coal.

**hard coal.** All coal of higher rank than lignite. In the United States the term is restricted to anthracite.

**hardness scale.** Quantitative units by means of which the relative hardness of minerals and metals can be determined.

**hard radiation.** Ionizing radiation of short wavelength and high penetration.

**hard water.** Water containing dissolved minerals such as calcium, iron, and magnesium.

**Haring cell.** A four-electrode cell for measurement of electrolyte resistance and electrode polarization during electrolysis.

**haulage.** The drawing or conveying in cars or otherwise, or the movement of men, supplies, ore, and waste both underground and on the surface.

**haul road.** A road built to carry heavily loaded trucks at a good speed; a road from pit to loading dock, tipple, ramp, or preparation plant used for transporting mined material by truck.

**haunt.** Coal sold at the pithead.

**hazardous air pollutant.** A pollutant which may cause or contribute to an increase in mortality or serious illness. Asbestos, beryllium, and mercury have been declared hazardous air pollutants.

**H-bomb.** A hydrogen bomb.

**H-Coal process.** Process in which pulverized coal is slurried with coal-derived recycled oil mixed with hydrogen, and fed into an ebullated bed with cobalt-molydbate catalyst. Liquids and gases are produced. Hydrogen consumption requirements are from 12,200 to 18,600 standard cubic feet per ton where, respectively, low sulfur fuel oil or syncrude are to be produced. Synthetic crude yield may be 4.06 to 4.38 barrels per ton depending upon the type of coal used. Product oil contains less than 0.1 percent sulfur (by weight). Reactor operating conditions are 454° C and 3000 pounds per square inch gage.

**head.** Development openings in a coal seam; an advance main roadway driven in solid coal; the top portion of a seam in the coal face. The term also applies to the differential of pressure causing flow in a fluid system, usually expressed in terms of the height of a liquid column that the pressure will support.

**headbox.** A device for distributing a suspension of solids in water to a machine, or for retarding the rate of flow such as to a top-feed filter, or for eliminating by overflow some of the finest particles.

**header.** A large pipe into which one set of boilers is connected by suitable nozzles or tees, or similar large pipes from which a number of smaller ones lead to consuming points. Headers are essentially branch pipes with many outlets, which are usually parallel. In a solar heating system, a header is a section of pipe which carries the main liquid flow at the top and the bottom of a solar collector panel, and from which smaller pipes or channels extend to the panel. The bottom header carries cool water to the panel(s). The top header carries sun-heated water away from it. A header is usually at least one and one-half inches in diameter and may be 2-inch pipe.

**header pipe.** A pipe or fitting that interconnects a number of branch pipes.

**"head of hollow" method.** A method of reclamation whereby solid residuals are deposited in a naturally-occurring deep canyon.

**head up.** To tighten the bolts on a hatch cover or manhole plate so that no leakage will occur from or into the vessel when operating.

**health physics.** That branch of science dealing with the protection of personnel from the harmful effects of ionizing radiation.

**heat.** A form of kinetic energy that flows from one body to another because of a temperature difference between them. It may be generated or transferred by combustion, chemical reaction, mechanical means, or passage of electricity. Heat is usually measured in calories or British thermal units (Btu).

**heat balance.** The accounting of the energy output from a system to equal the energy input.

**heat budget.** The total amount of the sun's heat received on the earth during any one year must equal exactly the total amount which is lost from the earth by reflection and radiation into space.

**heat capacity.** Quantity of heat required to raise the temperature of a unit quantity of a substance one degree.

**heat conduction.** The oceans receive heat by conduction through the sea bottom. Since the amount is very small (50 to 80 gram calories per square centimeter per year), it is neglected when considering the heat budget. This amount of heat coming from the bottom does not affect sound propagation.

**heat content.** The rate of change in heat content or enthalpy of air per unit time, measured in British thermal units per hour.

**heat energy.** Energy in the form of heat.

**Heat engine.** A device that converts heat energy into mechanical energy.

**heat exchanger.** A device that transfers heat from one fluid (liquid or gas) to another, or to the environment. (See Fig. p. 209)

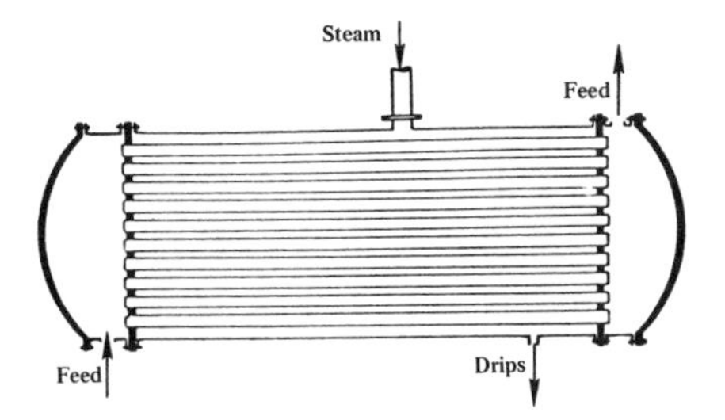


Cross section of a heat exchanger. (*Courtesy Van Nostrand's Scientific Encyclopedia*)

**heat fusion joint.** A joint made in thermoplastic piping by heating the parts sufficiently to permit fusion of the materials when the parts are pressed together.

**heat gain.** The increase of heat in a space resulting from direct solar radiation and from heat given off by some other source such as lights, equipment, machinery, and people.

**heating capacity.** The quantity of heat necessary to raise the temperature of a system or substance by 1 degree of temperature, expressed in calories per degree centigrade; also the quantity of heat in British thermal units that a room air conditioner is capable of adding to a room in one hour's time.

**heating degree day.** A measure of the coldness of the weather experienced, based on the extent to which the daily mean temperature falls below a reference temperature, usually 18° C. For example, on a day when the mean outdoor dry-bulb temperature is 2°C, there would be −1 degree days experienced. A daily mean temperature usually represents the sum of the high and low readings divided by two.

**heating oils.** A trade term for the group of distillate fuel oils used in heating homes and buildings as distinguished from residual fuel oils used in heating and power installations. Both are burner-fuel oils.

**heating tendency.** The ability of coal to fire spontaneously. This phenomenon can occur whenever the heat generated from oxidation reactions in coal exceeds the heat dissipated. This characteristic varies for different types of coals and even for coals of the same classification but of different origin.

**heating value.** The amount of heat produced by the complete combustion of a unit quantity of fuel. The gross or higher heating value is that which is obtained when all of the products of combustion are cooled to the temperature existing before combustion, the water vapor formed during combustion is condensed, and all the necessary corrections have been made. The net or lower heating value, is obtained by subtracting the latent heat of vaporization of the water vapor formed by the combustion of the hydrogen in the fuel from the gross, or higher, heating value.

**heat island effect.** An air circulation problem, relative to cities where tall buildings, heat from pavements, and concentrations of pollutants create a haze dome that prevents rising hot air from being cooled at its normal rate. In the absence of relatively strong winds, the heat island can trap high concentrations of pollutants and present a serious health problem.

**heat liberation rate.** The amount of heat which is liberated per unit time per cubic foot of combustion space.

**heat loss.** The amount of heat lost in a space through the walls, windows, and roof.

**heat map.** Map showing values of heat flow from the earth's interior, and locations of active volcanoes and earthquake epicenters. Issued by the National Oceanic and Atmospheric Administration (NOAA).

**heat of combustion.** The heat released when a substance is completely burned in oxygen.

**heat of fusion.** The heat lost or gained by a substance in passing from a liquid to a solid state or a solid to a liquid state, at a constant temperature.

**heat of reaction.** The quantity of heat consumed or liberated in a chemical reaction, as in heat of combustion, heat of neutralization, or heat of formation.

**heat of vaporization.** The latent heat required to change a liquid to a gas.

**heat pipe.** Self-contained, closed system capable of transporting thermal energy from a source to a sink.

**heat pump.** A reversible heating and cooling mechanism that can produce additional usable heat from the amount stored, such as a mechanical refrigerating system which is used for air cooling in the summer and which, when the evaporator and condenser effects are reversed, can absorb heat from the outside air or water in the winter and raise it to a higher potential so that it can also be used for winter heating. (See Fig. p. 211)

**heat register.** The grilled opening into a room by which the amount of warm air from a furnace can be directed or controlled; it includes a damper assembly.

**heat sink.** Anything that absorbs heat such as metal; also may be part of the environment such as air, river, or outer space. Concrete, adobe, other masonry walls, and large barrels or tanks of water are also effective heat sinks.

**heat tax.** Applies to the heat energy that becomes unavailable for further use whenever energy is converted from one form to another.

**heat transfer.** The methods (radiation, convection, and conduction) by which heat may be propagated or conveyed from one place to another. This term is sometimes used to mean rate of heat transfer.

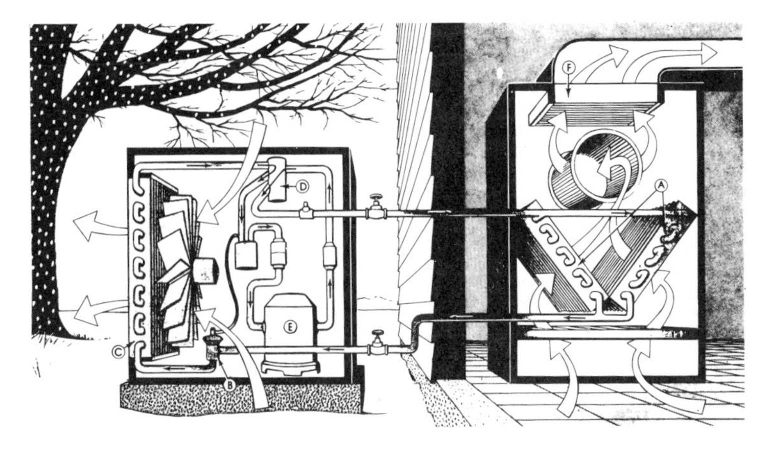

Heating cycle for air air-to-air heat pump system for residential installations. (*Courtesy Popular Science, September 1976, p. 92*)

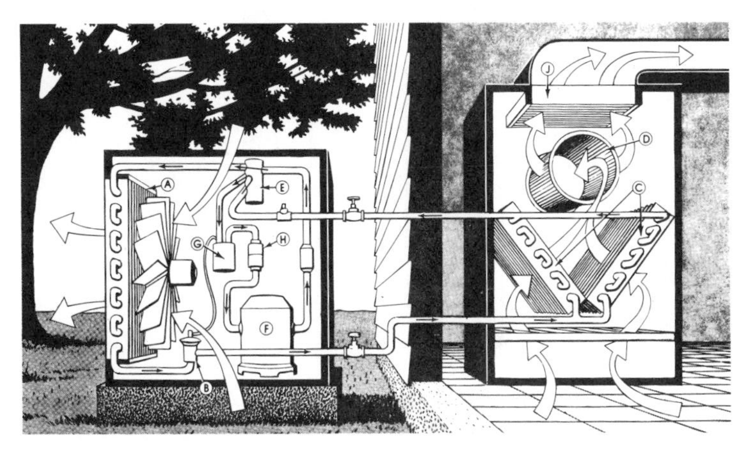

Cooling cycle for air air-to-air heat pump system for residential installations. (*Courtesy Popular Science, September 1976, p. 93*)

**heat transfer coefficient.** The quantity of heat transferred through a unit area of a material in a unit time, per unit of temperature difference between the two sides of the material.

**heat transfer medium.** A medium, liquid or air or solid, used to transport thermal energy.

**heat transport.** Meteorological phenomena account for most heat transport; ocean currents, however, are considered of major importance. The transport of heat by a unit volume of ocean water is a function of the specific heat, density, temperature, and north-south component of the velocity if the transport of heat from the equator toward the poles is being considered.

**heat treatment.** Heating and cooling a solid metal or alloy in such a way as to obtain desired conditions or properties of the metal to improve its functions.

**heat unit.** A unit of quantity of heat, such as the British thermal unit or the calorie.

**heat value.** The amount of heat obtainable from a fuel and expressed, for example, in British thermal units per pound.

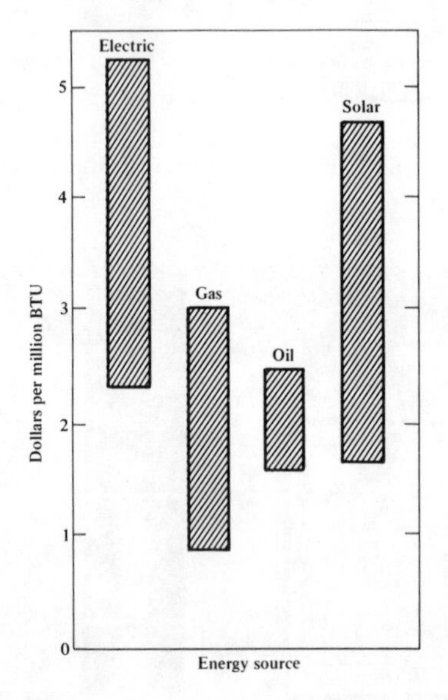


Costs of space heating with different energy sources. (*Courtesy "Energy and the Future," Fig. 20*)

**heavy ends.** The highest-boiling portion of a gasoline or other petroleum oil.

**heavy metals.** Metallic elements with high molecular weights. They are generally toxic in low concentrations to plant and animal life. Examples include mercury, chromium, cadmium, arsenic, and lead.

**heavy oil.** Oils distilled from coal tar between 230° C and 330° C; heavy, thick, and viscous oils; usually refinery residuals commonly specified as grades 5, 6, and Bunker C.

**heavy water.** Deuterium oxide, $D_2O$. Water containing significantly more than the natural proportion (one in 6500) of heavy hydrogen (deuterium) atoms to ordinary hydrogen atoms. Heavy water is used as a moderator in some reactors because it slows down neutrons effectively and has a low cross section for the absorption of neutrons.

**heavy-water-moderated reactor.** A reactor that uses heavy water as its moderator. Heavy water is an excellent moderator and thus permits the use of inexpensive natural (unenriched) uranium as a fuel.

**helical screw expander.** A spiral shaped machine for driving a generator through which hot water and steam expand.

**heliochemical process.** The process by which solar energy is utilized through photosynthesis.

**helioelectrical process.** The process by which energy is utilized by photovoltaic converters.

**heliostat.** A device that contains a mirror moved by a control mechanism to reflect the light of the sun in a particular direction.

Portion of heliostat field at ERDA's Solar Thermal Test Facility. Each heliostat has 25 four-foot square mirror facets. (*Courtesy Sandia Laboratories*)

**heliothermal process.** The process by which solar energy can be utilized to provide thermal energy.

**heliothermometer.** An instrument that measures the sun's heat.

**heliotropic wind.** A diurnal component of the wind velocity leading to diurnal shift of the wind or turning of the wind with the sun, produced by the east-to-west progression of daytime surface heating.

**helium.** A colorless, odorless, inert element having a specific gravity of 0.1368 and found in some natural gas. Chemical symbol He.

**Helmholtz coil.** A pair of similar coaxial coils with their distance apart equal to their radium, permitting an accurate calculation of the magnetic field between the coils. Used in the calibration of magnetometers.

**hermetic.** Made impervious to air and other fluids by fusion. Originally applied to the closing of glass vessels by fusing the ends and, by extension, to any mode of airtight closure.

**hertz.** A unit of frequency used in electrical and electronic measurements and equal to one cycle or one wavelength of electrical energy per second.

**heterogeneous reactor.** A reactor in which the fuel is separate from the moderator and is arranged in discrete bodies, such as fuel elements. Most reactors are heterogeneous.

**hexane.** Any of five isomeric, volatile, liquid paraffin hydrocarbons found in petroleum. Chemical formula $C_6H_{14}$.

**high Btu gas.** Manufactured gas with value of 1000 British thermal units or more per cubic foot.

**high Btu oil-gas process.** A manufactured gas process in which oil is converted into a fuel gas having a higher heating value than that of coal gas or carbureted water gas.

**high explosive.** An explosive with a nitroglycerin base. A detonator is required to initiate the explosion, which is violent and practically instantaneous.

**high fire.** An expression for the design maximum rate of fuel input to a burner.

**high-flux reactor.** A reactor capable of providing neutron flux densities greater than approximately $10^{14}$ neutrons per square centimeter per second.

**high heating value.** The heat produced during a combustion process in which the product water vapor is condensed to a liquid.

**high potential.** The Mine Safety Code states that voltages in excess of 650 volts are considered high potential.

**high pressure.** A liquid or aeriform gas pressurized to more than 150 pounds per square inch.

**high pressure air burner system.** A system using the momentum of a jet of high pressure air (in excess of 5 pounds per square inch gage) to entrain gas or a combination of air and gas, to produce a combustible mixture.

**high-rank coals.** Coals containing less than 4 percent of moisture in the air-dried coal or more than 84 percent of carbon (dry ash-free coal). All other coals are considered as low-rank coals.

**high sulfur coal.** Coal with a sulfur content greater than 1.68 pounds of sulfur per million Btu.

**High Temperature Gas-Cooled Reactor (HTGCR).** An approach to commercial nuclear power which would permit more efficient use of uranium and some use of thorium in its fuel cycle. Also offers greater thermal efficiency than light water reactors.

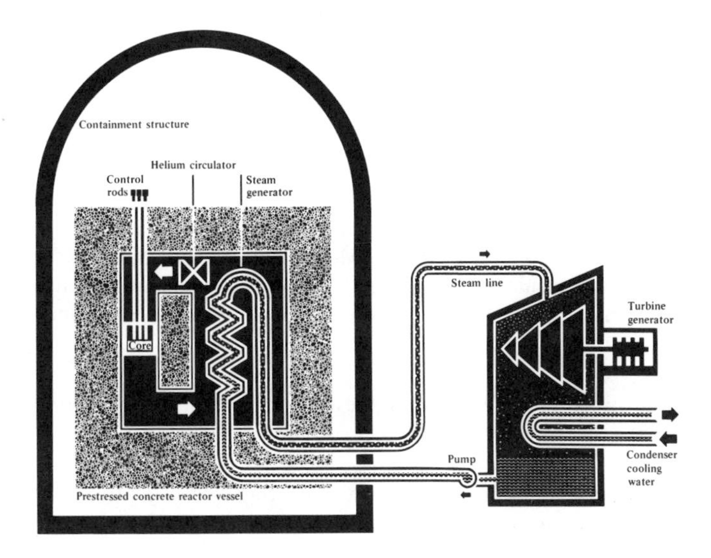


High temperature gas-cooled reactor. (*Courtesy U.S. Council on Environmental Quality*)

**high temperature gas reactor (HTGR).** A nuclear reactor which differs from light water reactors in having graphite as a moderator and helium as a coolant. Although not primarily a breeder reactor, thorium-232 can be added to the core and converted into uranium-233, which is a reactor fuel.

**high-volatile A bituminous coal.** Nonagglomerating bituminous coal having less than 69 percent of fixed carbon, more than 31 percent of volatile matter, and 14,000 or more British thermal units.

**high-volatile B bituminous coal.** Nonagglomerating bituminous coal having 13,000 or more, and less than 14,000 British thermal units.

**high-volatile C bituminous coal.** Either agglomerating or nonweathering bituminous coal having between 11,000 and 13,000 British thermal units.

**high-volatile coals.** Coals containing over 32 percent of volatile matter with a coal rank code No. 400 to 900.

**highwall.** The face or bank on the uphill side of a contour strip mine.

**Hinshaw Amendment.** An amendment to the Natural Gas Act which exempts from Federal Power Commission regulation the transportation and sale for resale of natural gas received within the boundaries of a state, provided (a) all such gas is ultimately consumed within the state, and (b) the facilities and rates are regulated by the state.

**hitch.** A fault; fractures and dislocations of strata common in coal measures, accompanied by some displacement.

**HITEC.** A eutectic combination of potassium nitrate, sodium nitrite, and sodium nitrate, used as a working fluid in solar thermal electric systems.

**hi-volume sampler.** A device used in the measurement and analysis of suspended particulates.

**hole.** A drill hole, borehole, or well. Mouse hole and rat hole are shallow bores under the derrick in which the kelly joint and joints of drill pipe are temporarily suspended while making a connection. Rat hole also refers to a hole of reduced size in the bottom of the regular well bore.

**holiday.** A discontinuity or break in the anticorrosion protection on pipe or tubing that leaves the bare metal exposed to corrosive processes.

**holiday detector.** An electronic device for locating discontinuities or breaks in the corrosion protection of pipe or tubing.

**homogeneous reactor.** A reactor in which the fuel is mixed with the moderator or coolant.

**homogenizing.** Holding at high temperature to eliminate or decrease chemical segregation by diffusion.

**homopolar generator.** A solid rotor machine without commutators or permanent magnets.

**hood.** A protective device, usually providing special ventilation to carry away gases, in which dangerous chemical, biological, or radioactive materials can be safely handled.

**hoop stress.** The stress imposed in the wall of a cylindrical tube in the circumferential direction by internal pressure.

**hopper.** A vessel into which materials are fed. It is usually constructed in the form of an inverted pyramid or cone terminating in an opening through which the materials are discharged (not primarily intended for storage).

**horizon.** Identifiable rock stratum regionally known to contain or be associated with rock containing valuable minerals.

**horizon mining.** A system of mine development which is suitable for inclined, and perhaps faulted, coal seams. Main stone headings are driven, at predetermined levels, from the winding shaft to intersect and gain access to the seams to be developed. The stone headings, or horizons, are from 100 to 200 yards apart vertically, depending on the seams available and their inclination. The life of each horizon varies from 10 to 30 years. Also called horizontal mining.

**horizontal axis wind turbine.** The most promising near-term technology for harnessing wind power to produce electricity appears to be the large horizontal axis wind turbine with propeller-type blades. The Smith-Putnam machine was of this type. Large horizontal-axis machines would be used by utilities in wind areas to supplement their generating power. An experimental 100-kilowatt (KW) wind turbine generator was built for ERDA in 1975 by the National Aeronautics and Space Administration's (NASA) Lewis Research Center at its Plum Brook test site near Sandusky, Ohio. Technical improvements resulting from the testing of this machine will be incorporated in large wind turbines to be built in 1977–78. The 100 KW wind turbine, called "Mod-O", has two slender rotor blades that look like airplane wings, mounted on an open truss tower 100 feet (30 meters) high. Each aluminum blade is 62.5 feet (18.8m) in length, and the blades together span 125 feet (37.5m). The machine starts to generate power in an eight mph (12.8 kmph) and operates at its rated power in an 18 mph (28.8 kmph) wind. It could provide enough electricity for about 30 average-sized homes. (See Fig. p. 218)

**horsepower.** A unit of power, equivalent to 33,000 foot-pounds per minute, or 550 foot-pounds per second (mechanical horsepower), or 746 watts.

**horsepower hour.** One horsepower expended for one hour, or the horsepower multiplied by the number of hours. One horsepower hour equals 1,980,000 foot-pounds, 0.745 kilowatt-hours, 2545 British thermal units (mean).

**hot.** An expression commonly used to mean highly radioactive; also applied to a mine or part of a mine that generates methane in large quantities.

**hot atom.** An atom in an excited energy state or having kinetic energy above the thermal level of the surroundings.

**hot cell.** A heavily shielded enclosure in which radioactive materials can be handled by persons using remote manipulators and viewing the materials through shielded windows or periscopes.

Artist's concept of a plant powered by a horizontal axis wind turbine to produce nitrogen fertilizer from air and water. (*Courtesy Lockheed—California Company, Divison of Lockheed Aircraft Corporation*)

**hot channel.** The coolant channel in a nuclear reactor core with the highest temperature.

**hot channel safety factor.** The safety factor corresponding to the ratio of the maximum change of enthalpy in case of an incident to the nominal variation of enthalpy in the channel in which the temperature is the highest (hot channel).

**hot dry rock.** Evidence suggests that the largest potential geothermal energy resource may be hot rock that has not come into contact with sufficient underground water to be categorized as hydrothermal. Before this resource can be developed, a technology must be perfected to extract heat from deposits of dry rock. One promising method under study employs a hydro-fracturing technique comparable to that used in petroleum recovery. Through this method, high pressure water is used to create a large crack in hot rock, such as granite, which overlays an area of high heat flow from the earth's interior. Pressurized water is then circulated through the crack to extract the heat. Another method of extracting heat from hot dry rock may be to use explosives instead of hydrofracturing to form cavities with increased surface area for water contact. Both methods—explosion and hydrofracture—require extensive research programs before being put into use.

**hot laboratory.** A laboratory designed for the safe handling of radioactive materials and usually containing one or more hot cells.

**hot rock reservoir.** A potential source of geothermal power. The "hot rock" system requires drilling deep enough to reach heated rock then fracturing it to create a reservoir into which water can be pumped.

**hot spot.** A surface area of higher-than-average radioactivity.

**hot-spot safety factor.** The safety factor by which the nominal temperature difference between the can and the coolant at a given point is multiplied to obtain the value of that difference resulting from the heat-transfer parameters, taking their maximum permissible values in case of an incident. The point chosen is that at which the highest temperature increase would occur (hot spot).

**hot tap.** The connection of branch piping to an operating line and the tapping of the operating line after it is under gas pressure.

**hot testing.** Testing of method, process, apparatus, or instrumentation under normal working conditions and at expected activity levels.

**hot water heater.** Used to heat water for home use. More than 400 kilowatt hours per year can be saved by lowering the thermostat setting on the electric hot water heater from 150 degrees to 130 degrees. Insulated hot water tanks and bare pipes further reduce heat loss.

**hot water wells.** Geothermal wells in the Geysers region and elsewhere which produce dry steam free of water, and more commonly a mixture of steam and hot brines; a wet steam, liquid-dominated, hydrothermal convection system which has a temperature range from below 88°C to greater than 349°C and can produce a mixture of vapor and liquid, or liquid alone. The thermal energy from these hot water wells is used to generate electricity and, more notably, to provide heating.

**hour angle.** 15° times the number of hours from solar noon.

**humacite.** A group name for bitumens which vary from gelatinous to hard resinous or elastic, and are insoluble in organic solvents. Believed to represent an emulsion of high acidic (humic acids) hydrocarbons with a varying amount of water (as high as 90 percent).

**humic coals.** A group of coals, including the ordinary bituminous varieties, which have been formed from accumulation of vegetable debris that have maintained their morphological organization with little decay. The majority of them are banded and have a tendency to develop jointing or cleat. Chemically, humic coals are characterized by hydrogen varying between 4 and 6 percent.

**humid heat.** Ratio of the increase in total heat per pound of dry air to the rise in temperature, with constant pressure and humidity ratio.

**humidifier.** A mechanical means of increasing the relative humidity by injecting water or water vapor into the air.

**humidistat.** A regulating device, actuated by changes in humidity, used for the automatic control of relative humidity.

**humidity.** The weight of water per unit weight of moisture-free gas or air.

**humins.** In coal, amorphous brown to black substances formed by natural decomposition from vegetable substances; insoluble in alkali carbonates, water, and benzol.

**humus.** Dark-colored, well-decomposed, organic soil material consisting of the residues of plant and animal materials together with synthesized cell substances of soil organisms and various inorganic elements.

**hunger.** Dirty, mottled clay, formed from the weathering of shale; also crystalline calcium carbonate found in the joints of coal seams.

**hybrid solar system.** Solar energy systems that combine both direct thermal (passive) and indirect solar design elements and equipment into one system; for example, solar collectors on a roof providing heat to the direct thermal storage component located inside the insulated shell of the structure.

**hydrafac.** An operation whereby producing formations are fractured by hydraulic pressure to increase productiveness.

**HYDRANE process.** A two-stage fluidized-bed reactor process which produces high-Btu gas from caking coals with pretreatment. Coal is fed into the top stage where it reacts with hydrogen in a free-fall zone (dilute-phase hydrogenation). Resultant char reacts further with hydrogen in the fluidized bed of the lower stage, producing methane. Hydrogen is produced in a separate reactor by steam-oxygen gasification of excess char from the gasifier. Gasifier pressures may exceed 1000 pounds per square inch gage and hydrogasification temperatures are generally 872°C.

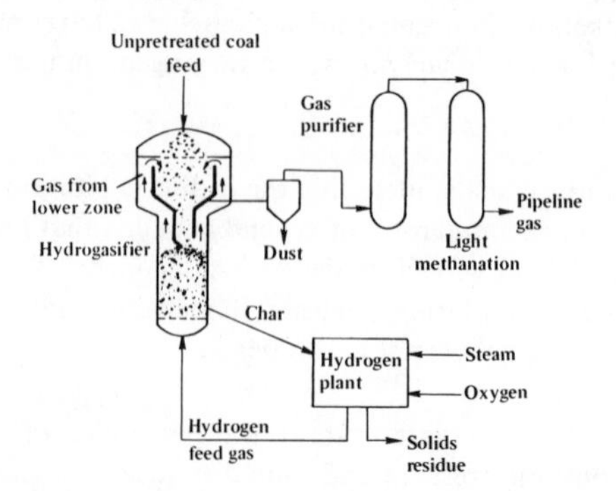


Flowsheet of HYDRANE process. (*Courtesy Van Nostrand's Scientific Encyclopedia*)

**hydrate.** A solid material resulting from the combination of a gas with water under pressure. The greater the pressure in the equipment, the higher the temperature at which the hydrate will form.

**hydration.** The chemical process of combination or union of water with another substance.

**hydraulic.** Strictly, having to do with water in motion, but the term is extended to cover all liquids which convey, store, or transfer pressure energy to reactants.

**hydraulic capacity.** The rating of a hydroelectric generating unit.

**hydraulic extraction.** A term which has been given to the processes of excavating and transporting coal or other material by water energy. Also called hydroextraction.

**hydraulic fluid.** A fluid used in hydraulic systems; may be of petroleum or nonpetroleum origin. Low viscosity, low rate of change of viscosity with temperature, and low pour point are desirable characteristics.

**haudraulic fracturing.** A general term for the fracturing of rock in an oil or gas reservoir by pumping a fluid under high pressure into the well. The purpose is to produce artificial openings in the rock in order to increase permeability. (See Fig. p. 222)

**hydraulic mining.** Mining by washing sand and dirt away with water which leaves the desired mineral; in underground hydraulic mining, the extraction of coal by high velocity water jets.

**hydraulic theory.** A theory of oil and gas migration that suggests that migration is caused by the movement of underground water which carries along oil and gas.

**hydraulic turbine.** An enclosed rotary type of prime mover which produces mechanical energy by the force of water directed against blades fastened to a vertical or horizontal shaft.

**hydro.** Meaning water or the presence of hydrogen; a prefix used to identify a type of generating station in which the prime mover is operated by water power.

**hydrocarbon.** Any of a large class of organic compounds composed solely of carbon and hydrogen. The compounds having a small number of carbon and hydrogen atoms in their molecule are usually gaseous; those with a larger number of atoms are liquid and the compounds with the largest number of atoms are solid.

**hydrocarbon fuels.** Fuels that contain an organic chemical compound of hydrogen and carbon.

**Hydrocarbonization process.** A process wherein coal is crushed, dried, and preheated to 399°C in a stream of hot flue gas. A hydrogen stream produced from char

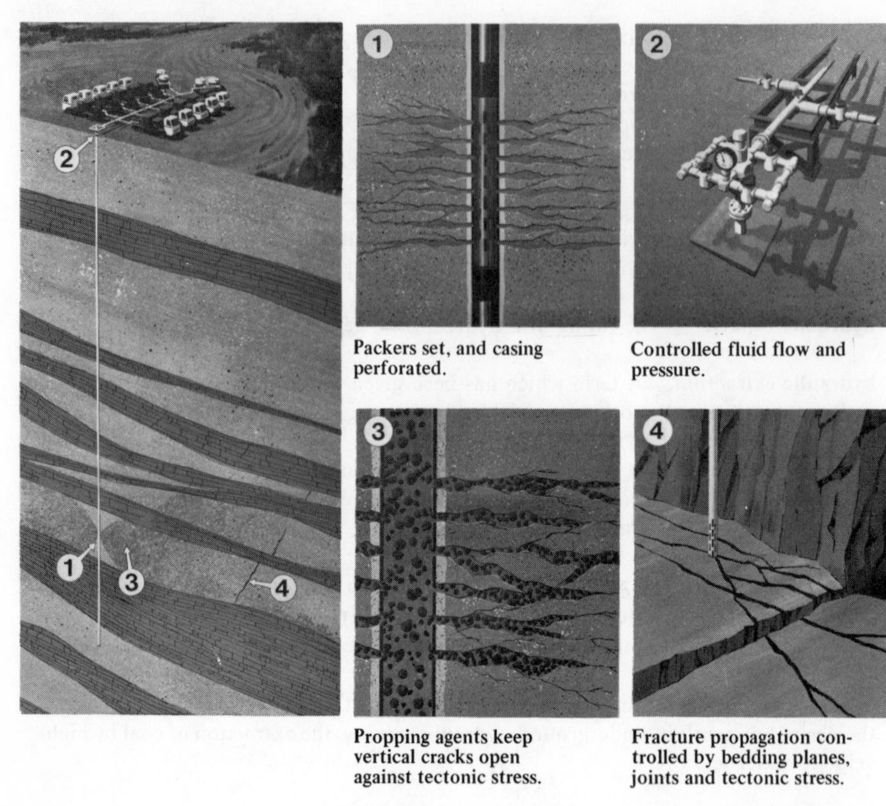

Packers set, and casing perforated.
Controlled fluid flow and pressure.
Propping agents keep vertical cracks open against tectonic stress.
Fracture propagation controlled by bedding planes, joints and tectonic stress.


Hydraulic fracturing of gas-bearing formations. (*Courtesy Lawrence Livermore Laboratory*)

carries the coal into the fluidized bed hydrocarbonization reactor where coal is converted to gas, liquid, and char. Hydrocarbonization consists of a combination of coal devolatilization and hydrogenation of various constituents of volatile matter. The product, after char removal, is sent to a fractionator. Heavy and light oils are separated. Gas, after passing through purification, is methanated to produce pipeline gas. Process conditions are 243° to 304°C and 300 to 1000 pounds per square inch gage. Hydrogen is produced by a Koppers-Totzek gasifier.

**hydrocoking.** Coking under hydrogenating conditions to form liquid products.

**hydrocracking.** A process combining cracking or pyrolysis, with hydrogenation. Feedstocks can include crude oil, residue, petroleum tars, and asphalts.

**hydrocyclone.** A cyclone separator in which a spray of water is used.

**hydrodesulfurization.** Process of removing sulfur compounds from hydrocarbon feedstock.

**hydrodynamics.** The branch of science that deals with the cause and effect of fluid migration; that branch of hydraulics which relates to the flow of liquids over weirs, or through pipes, channels, and openings.

**hydrodynamometer.** An instrument for determining the velocity of a fluid in motion by its pressure.

**hydroelectric.** System in which the potential energy of natural water is used after harnessing it by impounding in dams, by releasing it through turbogenerators.

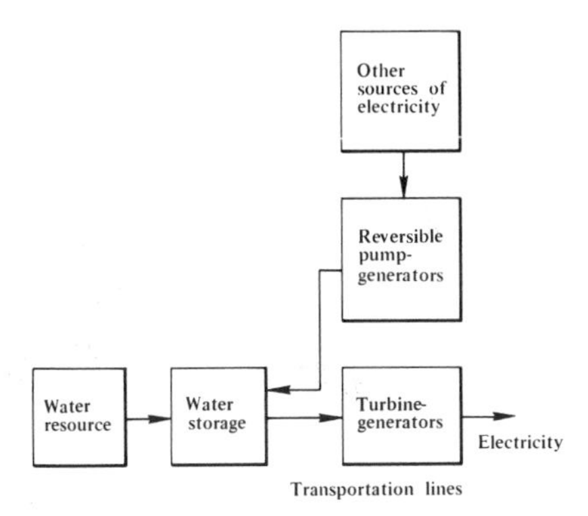


Hydroelectric resource development. (*Courtesy U.S. Council on Environmental Quality*)

**hydroelectric plant.** An electric power plant in which the energy of falling water is converted into electricity by turning a turbine generator.

**hydroelectric power.** Electrical energy obtained by using the energy of falling water. (See Fig. p. 224)

**hydroelectric scheme.** A complete project for water power development which will include the design and construction of a dam, tunnels, spillways, power station intakes, and many other constructional works over a wide area.

**hydrofining.** A fixed-bed catalytic process to desulfurize and hydrogenate a wide range of charge stocks from gases through waxes.

**hydroforming.** The process of converting low-octane hydrocarbons into a stable fuel of higher octane rating by heating them under pressure in the presence of hydrogen and catalysts. Hydroforming is an important operation in the production of gasoline.

**hydrofracturing.** The process of using water to free trapped petroleum.

**hydrogasification.** The addition of hydrogen to the products of primary gasification to optimize formation of methane.

Hydroelectric energy. Aerial view looking southwest to Hoover Dam. (*Courtesy of Bureau of Reclamation*)

**hydrogen.** The lightest element, No. 1 in the atomic series. It has two natural isotopes of atomic weights 1 and 2. The first is ordinary hydrogen, or light hydrogen; the second is deuterium, or heavy hydrogen. A third isotope, tritium, atomic weight 3, is a radioactive form produced in reactors by bombarding lithium-6 with neutrons. Chemical symbol H.

**hydrogenation.** A process whereby hydrogen atoms are added to an organic molecule to form a new compound; such reactions usually require heat and pressure in the presence of a catalyst.

**hydrogen bomb.** A nuclear weapon that derives its energy largely from fusion.

**hydrogen-chlorine battery.** An advanced battery system in early research stage, which uses gaseous hydrogen and chlorine as reactants.

**Hydrogen Donor Solvent process.** A process under development which uses a separately hydrogenated solvent which exchanges hydrogen with coal. The hydrogen donor solvent process uses coal heated in the presence of hydroaromatic material at

371° to 454°C and 200 to 1000 pounds per square inch. Liquid products can be upgraded by catalytic hydrogenation.

**hydrogen economy.** Use of hydrogen (which has been separated from water electrically at central production centers) as an energy source to be piped to local storage or usage areas. The hydrogen would be burned to produce heat, used as a transportation fuel, in industrial processes, or in fuel cells to generate electricity.

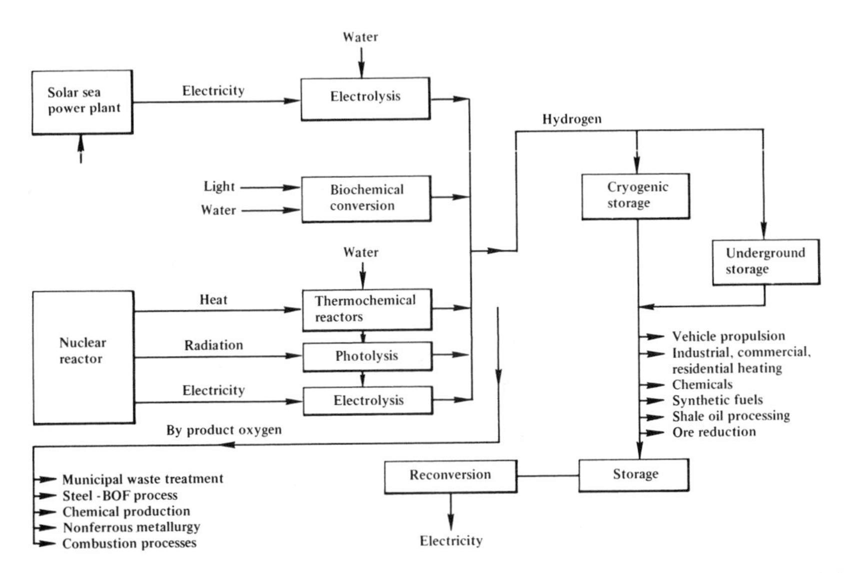


Major elements of hydrogen fuel economy. (*Courtesy Van Nostrand's Scientific Encyclopedia*)

**hydrogeneous.** Formed or produced by water; applied to rocks formed by the action of water, in contradistinction to pyrogenous rocks, or rocks formed by the action of heat.

**hydrogeneous coal.** Coal high in volatile matter, for example, gas coal or sapropelic coal; coals containing a large quantity of moisture, for example, brown coal.

**hydrogen fuel.** Use of hydrogen gas as a fuel. Hydrogen is virtually an ideal fuel since it burns in air to form nonpolluting water vapor. Hydrogen would be easily transported in existing natural gas lines and readily stored near where it is needed for power generation. (See Fig. p. 226)

**hydrogen sulfide.** A poisonous, corrosive compound consisting of two atoms of hydrogen and one of sulfur, gaseous in its natural state. It is found in manufactured gas made from coals or oils containing sulphur and must be removed. It is also found to some extent in some natural gas. It is characterized by the odor of rotten eggs. Chemical formula $H_2S$.

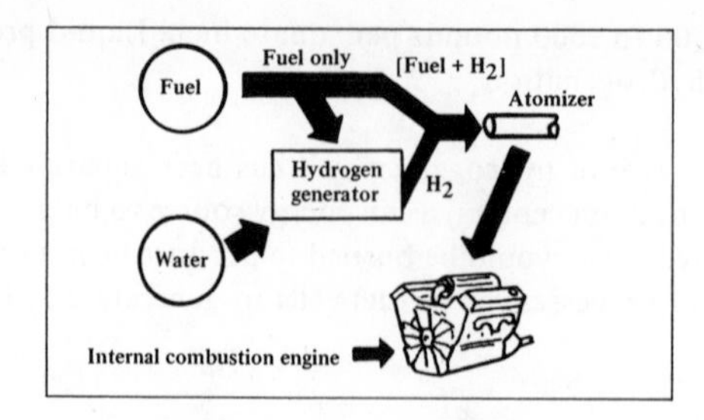


Use of hydrogen in internal combustion engines. (*Courtesy National Aeronautics and Space Administration*)

**hydrologic cycle.** The complete cycle of phenomena through which water passes, commencing as atmospheric water vapor, passing into liquid and solid form as precipitation, thence along or into the ground surface, and finally again returning to the form of atmospheric water vapor through evaporation and transpiration. Also called the water cycle.

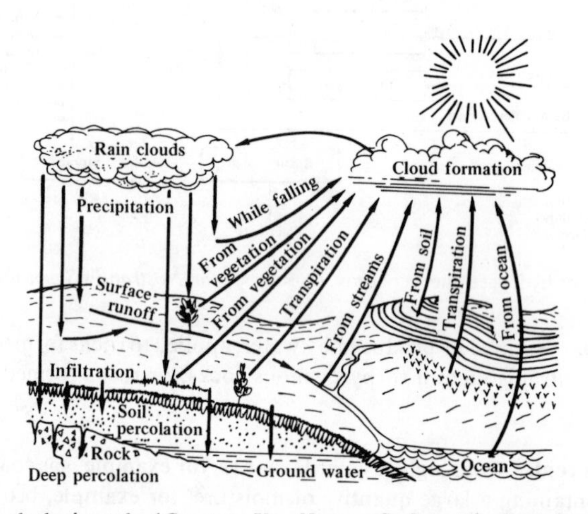


The hydrologic cycle. (*Courtesy Van Nostrand's Scientific Encyclopedia*)

**hydrolysis.** The chemical process that breaks complex organic molecules into simple molecules. For example, starch and cellulose can be hydrolyzed by acids or enzymes to produce simple sugars such as glucose, which can be fermented to produce ethanol.

**hydrology.** The science dealing with the water systems on or beneath the surface of the earth.

**hydrometer.** An instrument used for determining the density or specific gravity of fluids, such as drilling mud or oil, by the principle of buoyancy. The instrument is in the form of a glass tube, which is floated in the fluid and sinks to a greater or lesser depth depending on the density of the fluid, the amount of submergence being indicated by gradations or divisions on the stem of the instrument. These divisions may vary according to various systems.

**hydrometrograph.** An instrument for determining and recording the quantity of water discharged from a pipe, orifice, etc., in a given time.

**hydronics.** A modern term for heating and/or cooling with circulated water as the transfer medium.

**hydropower.** Power produced by falling water.

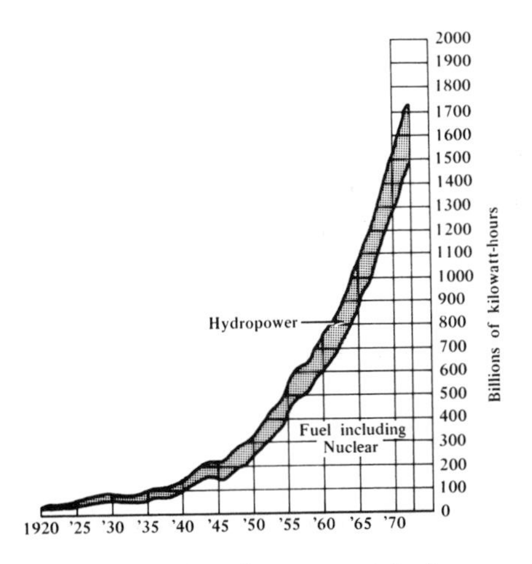


Electricity generated in the U.S. shaded portion represents hydropower; lower portion of chart includes all thermal energy (coal, fuel oil, natural gas, SNG, nuclear). (*Courtesy Van Nostrand's Scientific Encyclopedia*)

**hydrosphere.** The aqueous envelope of the earth including the ocean, all lakes, streams, and underground waters, and the water vapor in the atmosphere.

**hydrostatic design stress.** The estimated maximum tensile stress due to internal hydrostatic pressure in the wall of the pipe in the circumferential orientation that can be applied continuously with a high degree of certainty that failure of the pipe will not occur.

**hydrostatic head.** The pressure created by the weight of a height column of water.

**hydrostatic pressure.** The pressure of, or corresponding to, the weight of a column of water at rest.

**hydrostatic test.** A strength test of equipment (pipe) in which the item is filled with liquid, subjected to suitable pressure, and then shut in and the pressure monitored; a test to determine whether a container will hold a certain pressure; also a test to locate leaks or to prove that there are no leaks.

**hydrotator.** A coal washer of the classifier type whose agitator or rotator consists of hollow arms radiating from a central distributing manifold or center head. This agitator is suspended in a cylindrical tank and water is pumped through it under pressure.

**hydrothermal coal process.** A coal cleaning process developed at Battelle and designed to produce a clean solid fuel by removing up to 99 percent of inorganic and 70 percent of organic sulfur.

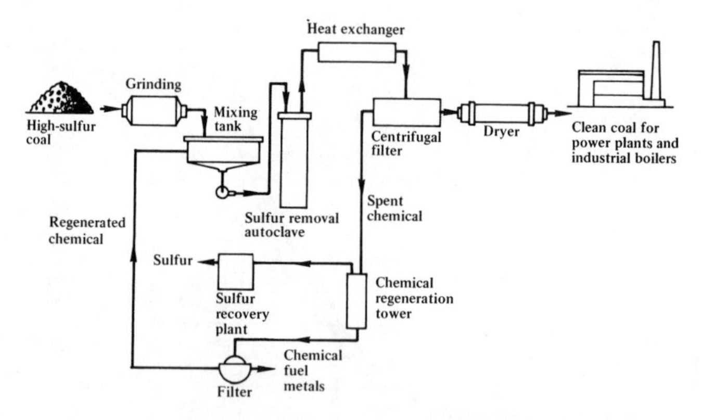


Hydrothermal coal process used in removing sulfur from high-sulfur coal. (*Courtesy Popular Science, October 1975, p. 18*)

**hydrothermal convective systems.** Source of geothermal energy; systems containing relatively high temperature fluid at shallow depths.

**hydrotorting process.** Finely crushed oil shale is retorted under high pressure in the presence of hydrogen.

**hydrotreating.** Using a catalyst, high temperature, and high pressure to change the structure of a molecule through the addition of hydrogen. An additional benefit may be the removal of sulfur as hydrogen sulfide in the process. (See Fig. p. 229)

**HYGAS process.** A two-stage, fluidized-bed gasifier process, which is used to produce a raw gas that can be upgraded to pipeline gas. Caking coals are pretreated in a fluidized bed to produce a nonagglomerating coal which is slurried in light oil and fed to a low temperature fluidized drying bed. Coal from the drying bed then passes into the first stage of the gasifier where it is devolatilized and partially methanated. Char then falls to the lower stage where it is gasified at high temperatures in the pres-

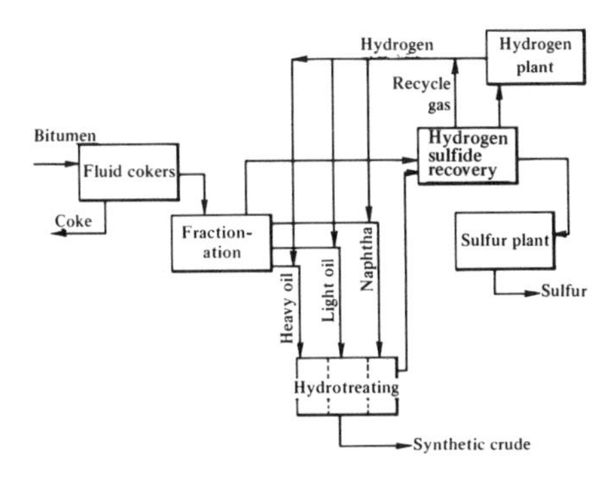


Steps involved in upgrading bitumen to synthetic crude oil. (*Courtesy U.S. Council on Environmental Quality*)

ence of hydrogen and steam. Hydrogen for the process can be generated by any one of three fluidized bed gasification methods currently under investigation. Process pressure is 1000 to 1500 pounds per square inch gage in the drying bed and in the gasifier. Gasifier temperatures are 704° to 816°C in the top stage and 927° to 987°C in the lower stage. The hydrogen producing gasifiers will also operate at high pressure and temperatures.

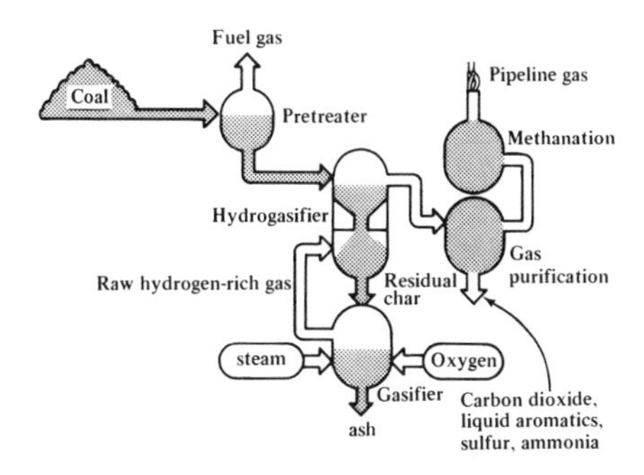


HYGAS coal gasification process. (*Courtesy National Science Teachers Association*)

**hygrometer.** An instrument for determining the relative humidity of the air.

**hygroscopic.** Having the property of readily absorbing moisture from the atmo-

**hyperon.** One of a class of short-lived elementary particles with a mass greater than that of a proton and less than that of a deuteron. All hyperons are unstable and yield a nucleon as a decay product.

**hypersonic.** Speeds greater than the speed of sound in air.

**hysteresis.** The time lag exhibited by a body in reacting to changes in the forces affecting it.

# I

**ideal cascade.** The theoretical minimum number of separative stages necessary to perform a given separation.

**ideal gas law.** The ideal gas law is the combination of the volume, temperature, and pressure relationships of Boyle's and Charles' laws. Real gases deviate by varying amounts from the ideal gas law.

**igneous rock.** Rock formed by the solidification of molten material that originated within the earth.

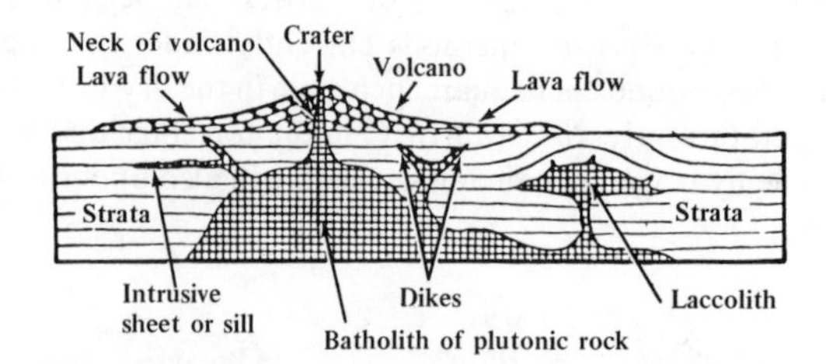


Diagrammatic structure section illustrating modes of occurrence of igneous rocks. (*Courtesy Van Nostrand's Scientific Encyclopedia*)

**ignite.** To heat a gaseous mixture to the temperature at which combustion takes place.

**ignition control.** Gasoline additives to reduce deposits in the combustion chamber.

**ignition point.** Of solids and liquids, the minimum temperature at which combustion can occur, but at which it is not necessarily continuous; of combustible gases, the flash point.

**ignition temperature.** The temperature at which a substance such as gas will take fire and continue burning with adequate air supply.

**illuminants.** The group of unsaturated or heavy hydrocarbons in a manufactured gas, such as ethylene and benzene, which burn with a luminous flame.

**illuminating gas.** Coal and carbureted water gases and their various mixtures; the different classes of oil gas; acetylene, gasoline gas, and producer gas. Producer gas is the most important for fuel and power gas.

**image furnace.** Apparatus for the production of a very high temperature in a small space by focusing the radiation from the sun (solar furnace) or from an electric arc (arc-image furnace) by means of mirrors and/or lenses.

**immersion length.** The length from the free end of a thermometer bulb or well to the point of immersion in the medium, the temperature of which is being measured.

**immiscible.** Pertaining to substances that are not capable of mixing; for example, oil and water.

**impact energy.** The amount of energy required to fracture a material, usually measured by means of an Izod test.

**impact parameter.** In elastic scattering, the minimum distance at which two interacting particles would pass each other if there were no scattering.

**impeller.** Rotating member of centrifugal pump, which receives inflowing water or ore pulp at or near its center and accelerates it radially to the periphery, where it is discharged with the kinetic energy (initial + added) needed to carry it through the pumping system. They are used to aerate and mix pulps in flotation cells, leach tanks, and mixing vats.

**impermeable.** Having a texture that does not permit water to move through it perceptibly under the head difference ordinarily found in subsurface water. Synonym for impervious.

**impermeable (hot) dry rock.** Geothermal regions where the heat is contained almost entirely in impermeable rock of very low porosity; a source of thermal energy.

**impervious.** Impassable; applied to impermeable strata, such as clays and shales, which will not permit the penetration of water, petroleum, or natural gas.

**impervious bed.** Geologically, one which caps pervious rocks, and thus prevents uprise of water, crude oil, or gas.

**implosion weapon.** A weapon in which a quantity of fissionable material, less than a critical mass at ordinary pressure, has its volume suddenly reduced by compression (a step accomplished by using chemical explosives) so that it becomes supercritical, producing a nuclear explosion.

**impounding area.** An area which may be defined through the use of berms, dikes, or the topography at the site for the purpose of containing any accidental spill of liquefied natural gas (LNG) or flammable refrigerants.

**impoundment.** A reservoir for collection of water, such as by damming a stream for irrigation purposes. Also used in connection with the storage of tailings from hydraulic mines.

**impregnation.** The treatment of porous castings with a sealing medium to stop pressure leaks.

**improved accessory drives.** The power to run auto heaters, cooling fans, wipers, air conditioners, and lights is drawn from the engine and uses energy. The effect is a loss of fuel economy. A new variable ratio belt drive for accessories is being developed and has been tested for use with current or new engines. In a conventional system, the accessory drive speeds up as the engine speeds up even though the extra power is not needed to run the accessories. In the new system, the ratio between the engine speed and the accessory drive varies so that the accessory drive operates at a fixed speed and does not needlessly consume power and therefore fuel. Preliminary road tests have shown an 8 percent improvement in gas mileage compared to the present day belt system.

**impulse turbine.** A turbine driven by high velocity jets of water or steam which impinge on some kind of vane or bucket attached to a wheel. The high velocity jets are produced by forcing the water and steam through a nozzle.

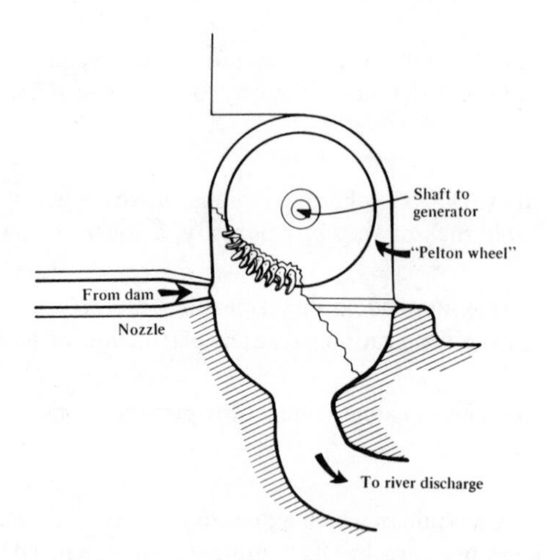


Impulse turbine. (*Courtesy U.S. Council on Environmental Quality*)

**inch of mercury.** A pressure unit representing the pressure required to support a column of mercury one inch high at a specified temperature; 2.036 inches of mercury (at 0°C and standard gravity of 32.174 feet per square second) is equal to one pound per square inch.

**inch of water.** A pressure unit representing the pressure required to support a column of water one inch high. Usually reported as inches W.C. (water column); 27.68 inches of water is equal to one pound per square inch.

**incident.** In reference to solar radiation, those light rays falling on or striking a surface.

**incident angle.** In solar calculations, the incident angle determines the intensity of the direct radiation component striking the surface and the ability of the surface to reflect, transmit, or absorb the sun's rays. The angle between the perpendicular to a surface and the direction of the solar radiation.

**incident radiation.** The amount of radiant energy incident on a surface per unit time and unit area.

**incident sunlight.** Direct sunlight.

**incineration.** Burning; the process by which combustible wastes are burned and changed into gases; reduction to ashes.

**incinerator.** A furnace for the destruction of refuse. It may be fired by electricity, gas, oil, or solid fuel.

**inclinometer.** An instrument used to determine whether or not the well bore is proceeding in a vertical orientation at any point. In most drilling operations either regulation of government bodies or contract stipulations, or both, provide a maximum deviation of the well bore from the vertical; commonly this maximum is three degrees. When deviation is in excess of the allowable, it is necessary to modify drilling procedure to bring it back in line.

**incoalation.** The process of coal formation which begins after peat formation is completed without there being any sharp boundary between the two processes. Same as coalification.

**incombustible.** Applies to substances that will not burn.

**incomplete combustion.** A term applied to combustion in which all of the fuel is not burned; for example, leaving unburned carbon in ashes.

**incorporation.** A process by which material contributing to coal formation responds to diagenetic and metamorphic agencies of coalification and becomes a part of the coal without undergoing any material modification.

**incremental valley energy costs.** The cost of producing added power increments at or near minimum power demand levels.

**independent.** As used in the oil industry, usually refers to a nonintegrated producing company. The integrated company usually operates production, transportation, refining, and marketing facilities. The independent producer has operations only in the field of petroleum production, as a rule.

**index fossil.** A fossil which, because of its wide geographic distribution and restricted time range, can be used to identify and to date the strata or the succession of strata in which it occurs.

**index of refraction.** The ratio of the velocity of light, or of other radiation, in the first of two media to its velocity in the second as it passes from one medium into the other, the first medium usually being taken to be a vacuum or air. Also called refractive index.

**indicated horsepower.** The horsepower determined from the pressure-volume indicator diagram. This is the power developed within the cylinder of the engine and is more than the power delivered at the driving shaft, by the amount of mechanical friction.

**indigenous.** Originating in a specific place; in situ.

**indirect-cycle reactor system.** A nuclear reactor system in which a heat exchanger transfers heat from the reactor coolant to a secondary coolant to produce useful power.

**indirect energy.** Any form of energy used in the manufacture of goods or provision of services which themselves may or may not use energy.

**indirect-fired.** A heater in which combustion products do not come in contact with the material to be heated. Heating of the material is accomplished by radiation or conduction from the heated surfaces.

**indirectly ionizing particles.** Uncharged particles, such as neutrons or photons, which can liberate directly ionizing particles.

**indirectly ionizing radiation.** Radiation consisting of indirectly ionizing particles.

**indirect oven.** One in which the flue gases do not flow through the oven compartment.

**indirect oven thermostat system.** A control system of two or more integrated automatic devices to maintain an oven temperature as selected by the operator. That portion of the system responsive to oven temperature causes operation of another portion of the system to turn on or shut off the gas supply to the oven burner.

**induced nuclear reaction.** Because of interaction with a particle or photon, a nucleus is changed in mass, charge, or energy state.

**induced radioactivity.** Radioactivity that is created when substances are bombarded with neutrons, as from a nuclear explosion or in a reactor, or with charged particles produced by accelerators.

**induction.** The production of magnetization or electrification in a body by the proximity of magnetized or electrified bodies, or of an electric current in a conductor by the variation of the magnetic field in its vicinity.

**induction meter.** An instrument for measuring alternating current circuits that will measure both power and energy. It is the type used for household and other consumer electricity meters.

**industrial air conditioning.** Air conditioning in industrial plants where the objective is the furtherance of a manufacturing process rather than the comfort of persons.

**industrial calorific value.** The calorific value obtained when coal is burned under a boiler.

**inert.** Not acted upon chemically by the surrounding environment. Nitrogen and carbon dioxide are examples of inert constituents of natural gases; they dilute the gas and do not burn and thus add no heating value.

**inert gas.** A gas, such as nitrogen or carbon dioxide, that is chemically inactive, especially in not burning or in not supporting combustion, under ordinary conditions; also one of the helium group of gases comprising helium, neon, argon, krypton, xenon, and sometimes radon.

**inertia.** The tendency of a body to resist force.

**inertial confinement fusion.** Concept which uses lasers, electron beams, or ion-beams to initiate the fusion process. Inertial confinement fusion is being considered as a possible commercial power source and in nuclear weapon design testing.

**inertial separator.** An air pollution control device which uses the principle of inertia to remove particulate matter from a stream of air or gas.

**infiltration.** In heat transmission, applies to the flow of air into and out of a building through the openings of doors and windows, cracks, and crevices; the term also applies to the deposition of water in the ground water system through the land surface.

**influent stream.** A stream, or the reach of a stream, is influent with respect to ground water if it contributes water to the zone of saturation.

**infracrustal rock.** Rock that originated at great depth either by consolidation from magma or by granitization.

**infrared radiant heater.** A self-contained, vented or unvented heater used to convert the flame energy to radiant energy, a substantial portion of which is in the infrared spectrum, for the purpose of direct heat transfer.

**infrared radiation.** Radiant energy of wavelengths exceeding 7,600 angstroms and not visible to the human eye. Sunlight contains some 60 percent of such rays.

**infusible.** Not transformable by heating from solid to liquid state under specified conditions of pressure, temperature, and time.

**ingate.** The point of entrance from a shaft to a level in a coal mine.

**inhibitor.** Any agent which inhibits or prevents; a substance which when present in an environment substantially decreases corrosion. Corrosion inhibitors are used widely in drilling and producing operations to prevent corrosion of metal equipment exposed to hydrogen sulfide gas and salt water. In some drilling operations a corrosion inhibitor is added to the drilling fluid to protect the drill pipe.

**inhour.** A unit of reactivity equal to the increase in reactivity of a critical reactor which produces a reactor time constant of one hour.

**inhour equation.** The equation which relates the reactivity of a reactor to its time constant.

**initial conversion ratio.** The conversion ratio in a reactor before significant burnup has taken place.

**initial nuclear radiation.** Radiation emitted from the fireball of a nuclear explosive during the first minute (an arbitrary time interval) after detonation.

**initiation.** The process of causing a high explosive to detonate. The initiation of an explosive charge requires an initiating point, which is usually a primer and an electric detonator, or a primer and a detonating cord or fuse.

**inject.** To introduce, under pressure, a liquid or plastic material into cracks, cavities, or pores in a rock formation.

**injection tube.** A tube with a Venturi throat which leads from the primary air port and gas orifice of a gas burner to mixing chamber and burner ports. As the gas passes from the gas orifice through the tube, it draws air through the primary air port into the mixing chamber, after which the mixture is burned at the burner ports.

**inoculum.** Material, such as bacteria, added to some medium to initiate biological action.

**inorganic.** All substances that do not contain carbon as a constituent, such as metals, rocks, minerals, and a variety of earths.

**in-pile.** A term used to designate experiments or equipment inside a reactor.

**in-pile test.** An irradiation test conducted within the core of a reactor.

**input rate.** The rate at which gas is supplied to an appliance. It may be expressed in British thermal units per hour, thousands of British thermal units per hour, cubic feet per hour, or thousands of cubic feet per hour.

**input rating.** The gas-burning capacity of an appliance in British thermal units per hour as specified by the manufacturer. Appliance input ratings are based on sea level operation and need not be changed for operation up to 2000 feet; input ratings should be reduced at the rate of 4 percent for each 1000 feet above sea level.

**insequent.** Developed on the present surface but not consequent on or controlled by the structure, such as a type of drainage in which young streams flowing on a nearly level plain wander irregularly.

**inserts.** Plastic, copper, etc. tubing inserted into a run of laid pipe, thereby eliminating the need for a new trench to be dug.

**in situ.** In the natural or original position or location. In situ production of oil shale, for instance, is an experimental technique in which a region of shale is drilled, fractured, and set on fire. The volatile gases burn off, the oil vaporizes, then condenses and collects at the bottom of the region from which it can be recovered by a well. There also has been experimentation with in situ conversion of coal.

**in-situ combustion.** An experimental means of recovery of oil of low gravity and high viscosity which is unrecoverable by other methods. The essence of the method is to heat the oil in the horizon to increase its mobility by decreasing its viscosity. Heat is applied by igniting the oil sand and keeping the fire alive by the injection of air. The heat breaks the oil down into coke and lighter oils, and the coke catches fire. As the combustion front advances, the lighter oils move ahead of the fire into the bore of a producing well.

**in-situ origin theory.** The theory of the origin of coal that holds that a coal was formed at the place where the plants from which it was derived grew.

**in-situ recovery.** Refers to methods to extract the fuel component of a deposit without removing the deposit from its bed.

**isolation.** The absorption of solar energy by the surface of the earth. The rate at which energy reaches the earth's surface from the sun. Usually measured in Btu/FT$^2$/Day.

**inspissation.** Drying up. An inspissated oil deposit is one from which the gases and lighter fractions have escaped, and only the heavier oils and asphalt remain.

**installed capacity.** The maximum load for which a machine, apparatus, device, plant or system is designed or constructed, not limited by existing service conditions.

**installed reserves.** Utility generating capacity that is in excess of maximum anticipated loads.

**instantaneous detonator.** A detonator in which there is no designed delay period between the passage of an electric current through the detonator and its explosion.

**instantaneous efficiency.** The instantaneous efficiency of a solar collector is the amount of energy removed by the transfer fluid per unit of gross collector area during this specified time period divided by the total solar radiation incident on the collector per unit area during the same test period, under steady or quasi-steady state.

**instantaneous peak demand.** The maximum demand at the instant of greatest load.

**instrument piping.** All piping, valves, and fittings used to connect instruments to main piping, to other instruments and apparatus, or to measuring equipment.

**insufflator.** An injector for forcing air into a furnace.

**insulate.** To separate or to shield (a conductor) from conducting bodies by means of nonconductors, so as to prevent transfer of electricity, heat, or sound.

**insulated flange.** A set of flanges made or faced with nonelectric conducting materials, for the purpose of stopping flow of electric current through the pipe.

**insulation.** If electric, separation of a conductor or charged body from earth or from other conductors by means of a nonconducting barrier; if thermal, prevention of passage to or from a body of external heat, by use of a nonconducting envelope (for example, filter material placed within the wall, floor, or ceiling of a building to reduce the loss of warm or cold air from within a building to the outside).

| | Economic optimum | |
|---|---|---|
| Insulation specification | Gas | Electric |
| Wall insulation thickness, mm (inches) | 89 (3-1/2) | 89 (3-1/2) |
| Ceiling insulation thickness, mm (inches) | 89 (3-1/2) | 152 (6) |
| Floor insulation | Yes | Yes |
| Storm windows | Yes | Yes |
| Reduction of energy use (%) | 49 | 47 |

Insulation requirements and energy savings. (*Courtesy Electric Power Research Institute and Applied Nucleonics Co., Inc.*)

**insulator.** A substance in whose atoms the outer orbital electrons are so tightly bound that small potential difference will not cause the electrons to migrate. Good electrical insulators are usally good thermal insulators too.

**intake.** The passage by which the ventilating current enters a mine.

**intangible drilling costs.** Expense items that are written off in the year incurred for tax purposes.

**integral absorbed dose.** The integral of absorbed dose over the mass of irradiated matter in the volume under consideration. It is identical with the energy imparted to matter in that volume.

**integral collector/storage (breadbox) system.** An integrated solar water heating system that combines the collection and storage of solar energy in a water tank(s). A typical breadbox system consists of an insulated box containing water tanks painted black with solar glazing and insulating lids.

**integral reactor.** A reactor in which the reactor vessel contains the heat exchanger between the primary and secondary coolant circuits.

**integrated community energy system (ICES).** An advanced system for producing heat and electricity, with the electricity connected to the local utility's distribution network and the heat used for domestic hot water and space heating and cooling for buildings in the plant's vicinity. (See Fig. p. 240)

**integrated company.** A company which obtains a significant portion of its gas operating revenues from the operations of both a retail gas distribution system and a gas transmission system.

**integrated demand.** The demand averaged over a specified period, usually determined by an integrating demand meter or by the integration of a load curve. It is the average of the instantaneous demands during a specified demand interval.

**integrated neutron flux.** Flux multiplied by time, usually expressed as nvt, when n = the number of neutrons per cubic centimeter, v = their velocity in centimeters per second, and t = time in seconds.

**integrated train.** A long string of permanently-coupled cars that shuttles back and forth between one mine and one generating plant without stopping to load or unload, since rotary couplers permit each car to be flipped over and dumped as the train moves slowly across a trestle.

**integrating ionization chamber.** An ionization chamber which indicates total ionization over a period of time.

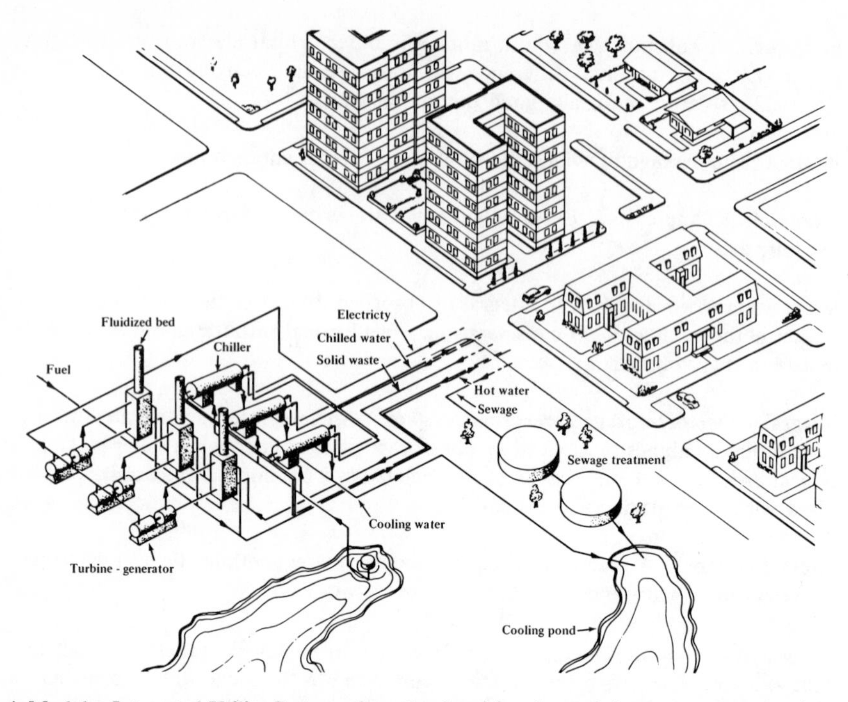


A Modular Integrated Utility System where the electricity, domestic hot water, space heating, and air conditioning for a high-density housing complex are supplied by a closed-cycle gas turbine coupled to an oil-burning fluidized bed furnace. (*Courtesy Oak Ridge National Laboratory*)

**intensity.** The energy or the number of photons or particles of any radiation incident upon a unit area or flowing through a unit of solid material per unit of time; in connection with radioactivity, the number of atoms disintegrating per unit of time.

**intensity of radiation.** At a given place, the energy per unit time entering a sphere of unit cross-sectional area centered at that place. The unit of intensity is the erg per square centimeter second or the watt per square centimeter.

**interbedded.** Occurring between beds, or lying in a bed parallel to other beds of a different material.

**intercalated.** Descriptive of a body of rock or mineral interbedded or interlaminated with another body of different rock or different mineral.

**interchangeability.** In the gas industry, this term generally refers to the combustion characteristics of gases in customers' gas-burning equipment. A perfectly interchangeable gas is defined as a gas which will produce the same effect in the customers' equipment as a reference gas. Interchangeability considerations include: production of carbon monoxide, flame stability, production of soot, and heat released by combustion.

**interfacial energy.** Tension at interfaces between the various phases of a system which may include solid, liquid, and gas interfaces, of varying combinations and qualities.

**intergranular corrosion.** This form of corrosion is a result of excessive localized attack at and adjacent to the grain boundaries of the metal, with dramatic reduction in the component strength.

**intermediate energy forms.** Energy in transit from its natural form to the form in which it is useful. For example, energy of coal or falling water (the primary energy form) is converted to electricity (the intermediate form) which in turn is converted to heat energy (the final form) in a radiator.

**intermediate neutron.** A neutron having energy greater than that of a thermal (slow) neutron but less than that of a fast neutron. The range is generally considered to be between about 0.5 and 100,000 electron volts.

**intermediate reactor.** A reactor in which fission is induced predominantly by intermediate neutrons.

**intermediate units.** Utility generating equipment used to supply power for regularly recurring loads in excess of the continuous base load.

**intermittent stream.** A stream that flows only in direct response to precipitation, such as a rainstorm or melting snow. It receives little water from springs and is dry for a large part of the year.

**internal combustion.** Pertaining to any engine in which the heat or pressure necessary to produce power is developed in the engine cylinder by the burning of a mixture of air and fuel, and converted into mechanical work by means of a piston. The automobile engine is a common example. (See Fig. p. 242)

**internal energy.** Of a gas, the total heat energy stored in a unit mass due to the motion and position of the molecules of the gas. This energy is measured by a thermometer.

**internal radiation.** Radiation originating within the body from radionuclides in body tissues.

**International Atomic Energy Agency (IAEA).** Applies nuclear material utilization safeguards to foreign nations. Under IAEA safeguards, countries are required to maintain accountability and permit IAEA to verify findings of their national nuclear material accounting systems in order to ascertain that there has been no diversion of nuclear material from peaceful uses.

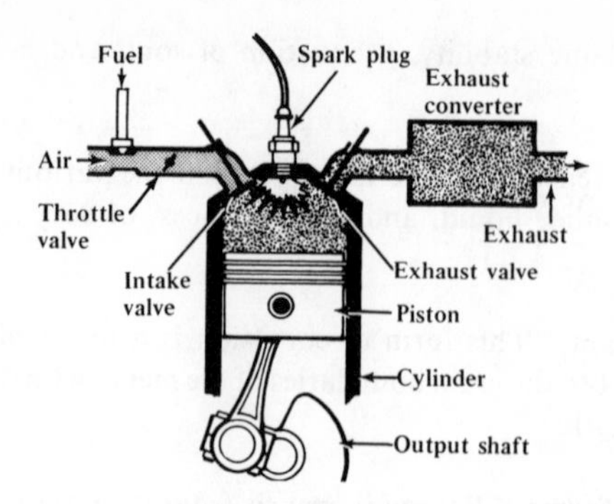


The Otto engine is an early form of internal combustion engine. (*Courtesy Popular Science, April 1976, p. 83*)

**interruptible service.** Low priority service offered to customers under schedules or contracts which anticipate and permit interruption on short notice, generally in peak-load seasons, by reason of the claim of firm service customers and higher priority users. Unlike off-peak service, gas is available at any time of the year if the supply is sufficient.

**interstate gas.** Natural gas which enters interstate commerce and hence is subject to Federal price controls. Natural gas sold to pipelines under the jurisdiction of the FERC.

**interstate waters.** Waters defined as: (a) rivers, lakes, and other waters that flow across or from a part of state or international boundaries; (b) waters of the Great Lakes; and (c) coastal waters, whose scope has been defined to include ocean waters seaward to the territorial limits and waters along the coastline (including inland streams) influenced by the tide.

**interstice.** An opening in anything or between things; especially, a narrow space between the parts of a body or things close together; a crack, a crevice, a chink, or a cranny.

**interstitial implants.** Solid or encapsulated radiation sources made in the form of seeds, wires, or other shapes to be inserted directly into tissue that is to be irradiated.

**in-the-seam mining.** The conventional system of mining in which the development headings are driven in the coal seam.

**intrastate gas.** Natural gas which is both produced and consumed within the same state. It is not subject to Federal (FERC) price controls.

**intratelluric.** Formed or occurring within the earth; descriptive of the constituents of an effusive rock formed before its appearance on the surface, or of the period of its formation.

**intrinsically safe apparatus.** Apparatus that is so constructed that, when installed and operated under the conditions specified by the certifying authority, any electrical sparking that may occur in normal working, either in the apparatus or in the circuit associated therewith, is incapable of causing an ignition of the prescribed flammable gas or vapor.

**intrinsic tracer.** An isotope, present naturally in a given sample, that may be used to trace a given element through chemical and physical processes.

**intrusion.** In geology, a mass of igneous rock which, while molten, was forced into or between other rocks; a mass of sedimentary rock occurring in a coal seam.

**intumesce.** To enlarge, to expand, to swell, or to bubble up (as from being heated).

**inundate.** To flood an entire mine or a large section of the workings with water.

**inversion.** An increase of air temperature with increased, rather than the usual decreased, elevation. Also the conversion of direct current electricity to alternating current.

**inverter.** A device, used in conjunction with the fuel cell power plant, that converts direct current to alternating current.

**ion engine.** An engine which provides thrust by expelling accelerated or high velocity ions. Ion engines using energy provided by nucelar reactors are proposed for space vehicles.

**ion exchange.** A chemical process involving the reversible exchange of ions contained in a crystal for different ions in solution without destruction of crystal structure or disturbance of electrical neutrality.

**ionization.** The removal of some or all electrons from an atom, leaving the atom with a positive charge, or the addition of one or more electrons, resulting in a negative charge.

**ionization chamber.** An instrument that detects and measures ionizing radiation by measuring the electrical current that flows when radiation ionizes gas in a chamber, making the gas a conductor of the electricity. (See Fig. p. 244)

**ionization counter.** An ionization chamber which has no internal amplification by gas multiplication and which is used for counting ionizing particles.

**ionization potential.** Energy in volts needed to remove an electron from a normal atom to leave it positively charged. To ionize is to dissociate a molecule or a compound into ions of opposite charge.

**ionized gas.** A gas that is capable of carrying an electric current.

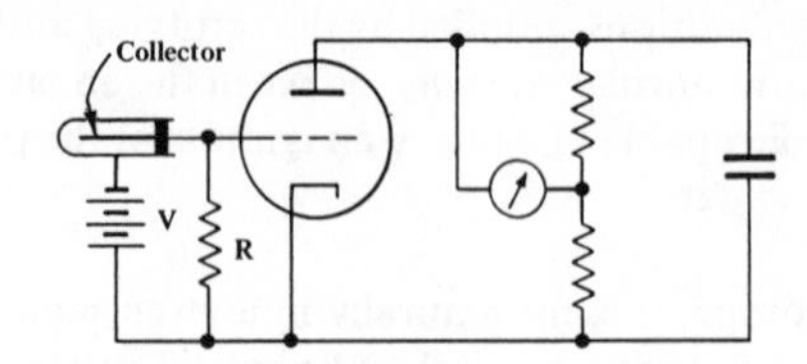


Sketch of a simple ionization detector. (*Courtesy Basic Nuclear Engineering*)

**ionizing event.** Any occurrence in which an ion or group of ions is produced; for example, by passage of a charged particle through matter.

**ionizing radiation.** Any radiation displacing electrons from atoms or molecules, thereby producing ions. Examples: alpha, beta, gamma radiation; short-wave ultra-violet light. Ionizing radiation may produce severe skin or tissue damage.

**ion pair.** A closely associated positive ion and negative ion (usually an electron) having charges of the same magnitude and formed from a neutral atom or molecule by radiation.

**ironshot.** Of a mineral, streaked or speckled with iron or an iron ore.

**irradiated fuel.** Nuclear fuel that has been used in a nuclear reactor.

**irradiation.** Incident radiation; exposure to radiation, as in a nuclear reactor.

**irradiation channel.** A hole through a reactor shield into the interior of the reactor in which irradiations are carried out.

**irradiation reactor.** A reactor used primarily as a source of nuclear radiation for irradiation of materials or for medical purposes.

**irradiation rig.** An assembly, for insertion in a reactor, which contains materials for experimental irradiation together with instruments for the measurement (and sometimes control) of the conditions under which the irradiation is carried out.

**irrespirable.** Not fit to be breathed.

**isallobaric wind.** The wind velocity whose Coriolis force exactly balances a locally accelerating geostrophic wind.

**isobar.** An imaginary line or a line on a map or chart connecting or marking places on the surface of the earth where the height of the barometer reduced to sea level is the same either at a given time or for a certain period.

**isobutane.** A hydrocarbon of the same chemical formula as butane, but of different molecular structure resulting in different physical properties, notably lower boiling points. Chemical formula $C_4H_{10}$.

**isocals.** Lines of equal calorific value in coal as drawn on a map or diagram.

**isodose.** Descriptive of every point of which the absorbed dose is the same.

**isodose curves.** Curves or lines drawn to connect points where identical amounts of radiant energy reach a certain depth in tissue.

**isointensity contours.** Imaginery lines on the surface of the ground or water, or lines drawn on a map, joining points in a radiation field which have the same radiation intensity at a given time.

**isomer.** One of two or more nuclides with the same numbers of neutrons and protons in their nuclei, but with different energies. A nuclide in the excited state and a similar nuclide in the ground state are isomers.

**isomeric state.** An excited nuclear state having a mean life long enough to be observed.

**isomerization.** A refining method which converts a straight-chain-saturated hydrocarbon to a corresponding branch chain to form high-octane products.

**isotherm.** A line on a map or chart of the earth's surface connecting points having the same temperature at a given time or the same mean temperature for a given period.

**isotone.** One of several nuclides having the same number of neutrons but a different number of protons in their nuclei.

**isotope.** Any of two or more species of atoms having the same number of protons in the nucleus (or the same atomic number) but with differing numbers of neutrons. All isotopes of an element have the same number of electrons and have identical chemical properties, but the different nuclear masses produce slightly differing physical properties. Since nuclear stability is governed by nuclear mass, one or more isotopes of an element may be radioactive or fissionable, while other isotopes of the same element may be stable. In the usual notation, isotopes of the same element are identified by the total of neutrons and protons in the nucleus; for example, uranium-235 and uranium-238.

**isotope effect.** The effect on the atomic properties of isotopes due to their differing masses.

**isotope farm.** A carbon-14 growth chamber, or greenhouse, arranged as a closed system in which plants can be grown in a carbon-14 dioxide atmosphere.

**isotope separation.** The process of separating isotopes from one another, or changing their relative abundances, as by gaseous diffusion or electromagnetic separation. Isotope separation is a step in the isotopic enrichment process.

**isotope separation cut.** The ratio of the flow of enriched material from a separative element to the flow of feed into the element.

**isotopic abundance.** The relative number of atoms of a particular isotope in a mixture of the isotopes of an element, expressed as a fraction of all the atoms of the element.

**isotopic enrichment.** A process by which the relative abundances of the isotopes of a given element are altered, thus producing a form of the element which has been enriched in one particular isotope; for example, enriching natural uranium in the uranium-235 isotope.

**isotopic power generator.** A device for generating useful power from the heat produced by the radioactive decay of a radionuclide. The heat is usually converted to electricity by thermoelectric or thermionic devices.

**isotopic tracer.** An isotope of an element, a small amount of which may be in incorporated into a sample of material in order to follow the course of that element through a chemical, biological, or physical process, and thus also follow the larger sample. The tracer may be radioactive, in which case observations are made by measuring the radioactivity. If the tracer is stable, mass spectrometers, density measurement, or neutron activation analysis may be employed to determine isotopic composition. Tracers also are called labels or tags, and materials are said to be labeled or tagged when radioactive tracers are incorporated in them.

**Izod test.** An impact test in which the specimen, usually notched, is fixed at one end and broken by a falling pendulum. The energy absorbed, as measured by the subsequent rise of the pendulum, is a measure of impact strength or notch toughness.

# J

**jacket.** The enclosure on an appliance such as a water heater, furnace, or boiler; the space surrounding a cylinder of an engine through which a cooling liquid flows; also an external layer of material applied directly to nuclear fuel or to other material by mechanical bonding to provide protection from a chemically reactive environment, to provide containment of radioactive products produced during the irradiation of the composite, or to provide structural support.

**jacket water.** In a compressor or engine, the water used for cooling the cylinder head and/or walls.

**jackknife or folding mast rig.** The type mast that can be folded for moving, as contrasted with the standard derrick, which has to be completely dismantled and re-erected.

**jackline man.** In petroleum production, one who pumps several oil wells from a central powerplant (jack plant), engaging or disengaging the rods or cables by which power is transmitted to separate wells.

**jamb.** A vein or bed of earth or stone, which prevents the miners from following a vein of ore; a large block.

**jars.** A device used in cable drilling, shaped like two elongated links, attached to the drilling tool, and used to jar the bit on the upward stroke, thus preventing the bit from sticking in the well-bore; also used to increase the impetus of a force exerted to free objects stuck in the well-bore.

**Jeppe's tables.** A series of tables especially compiled for mining work that includes tables of density, vapor pressure, and absolute humidity.

**jet.** A hydraulic device operated by pump pressure for the purpose of cleaning fluid out of the pits and tanks on a rotary drilling location; also a form of coal.

**jet engine.** An engine that produces motion as a result of discharging a fast-moving stream of heated air and exhaust gases.

**jet fuel.** Includes both naphtha-type and kerosine-type fuels meeting standards for use in aircraft turbine engines. Although most jet fuel is used in aircraft, some is used for other purposes, such as for generating electricity in gas turbines.

**jet propulsion.** The name given to forward motion produced by the discharge of a stream of liquid or gas at the rear. The four main types of jet engines used in aircraft are the turbojet, turboprop, pulse-jet, and ramjet.

**jetting.** The process of burying offshore or river crossing pipelines by hydraulically blowing sand or dirt from beneath the pipelines. The term is also applied to the injection of gas into a stratum for the purposes of pressure maintenance and secondary recovery.

**jig.** A device which separates coal from foreign matter by means of their difference in specific gravity in a water medium.

**jinny road.** Underground gravity plane.

**jobber price.** The price at which a petroleum jobber purchases refined product from a refiner or terminal operator.

**Johannsen's classification.** A mineralogical classification of igneous rocks in which a rock is characterized by a number, the Johannsen number, contisting of three or four digits, each of which has a specific mineralogical significance.

**joint.** A line of cleavage in a coal seam.

**joint compounds.** Materials to be used on pipe joints, primarily to lubricate the threads and secondarily to prevent leakage.

**joist.** One of a series of evenly spaced horizontal structural beams, usually made of lumber, steel, or concrete, for supporting floors and ceilings.

**Joliot-Curie, Frederic.** Frederic Joliot-Curie (1900–1958), French physicist. With his wife, Irene, he discovered three new radioactive elements—isotopes of nitrogen, phosphorous, and aluminum. He also observed the emission of neutrons in nuclear fission.

**Jøtul.** One of several models of cast-iron wood-burning stoves imported from Scandinavia that burns logs in an enclosed heat chamber.

**joule.** The absolute meter-kilogram-second (mks) unit of work or energy that equals $10^7$ ergs or approximately 0.7375 foot-pound or 0.2390 gram calorie.

**Joule's law.** The rate at which heat is produced by a steady current in any part of an electric circuit is jointly proportional to the resistance and to the square of the current; also, the internal energy of an ideal gas depends only on its temperature and is irrespective of volume and pressure.

**Joule-Thomson effect.** The cooling which occurs when a compressed gas is allowed to expand in such a way that no external work is done. The effect is approximately 7°F per 100 pounds per square inch for natural gas.

**Joule-Thomson expansion.** The throttling effect produced when expanding a gas or vapor from a high pressure to a lower pressure with a corresponding drop in temperature.

**Joy double-ended miner.** A cutter loader for continuous mining on a longwall face. It consists of two cutting heads fixed at each end of a caterpillar-mounted chassis. The heads are pivoted and controlled hydraulically for vertical movement. Each head comprises two bores and a frame or loop cutter that trims the bottom, face side, and top.

**Joy microdyne.** A wet-type dust collector for use at the return end of tunnels or hard headings. It may be either 6,000 or 12,000 cubic feet per minute capacity. It wets and traps dust as it passes through the appliance, and releases it in the form of a slurry which is removed by a pump.

**Joy miner.** A continuous miner mainly for use in coal headings and extraction of coal pillars. It weighs about 15 tons and comprises a turntable mounted on caterpillars, a ripper bar, and a discharge boom conveyor. The ripper bar has six cutter chains with picks running vertically to the plane of the seam. An intermediate conveyor behind the ripper bar delivers the coal into a small hopper, and a discharge conveyor takes it to the outby end of the machine. The latter conveyor can be swung 45° to right or left to facilitate conveying.

**Joy transloader.** A rubber-tired, self-propelled loading, transporting, and dumping machine

**Joy walking miner.** A continuous miner with a walking mechanism instead of caterpillar tracks. The walking mechanism was adopted to make the machine suitable for thin seams. It can work in a 2-foot 6-inch seam.

**judd.** A block of coal, holed and cut ready for breaking down; also a portion of the coal laid out and ready for extraction.

**jug.** A colloquial equivalent of detector or geophone.

**junking.** The process of cutting a passage through a pillar of coal.

**kaolin.** A clay, mainly hydrous aluminum silicate, from which porcelain may be made. Also called China clay; porcelain clay.

**kaon.** An elementary particle (contraction of K-meson); a heavy meson with a mass about 970 times that of an electron.

**kaplan turbine.** A water turbine, of propeller type, having blades of variable pitch automatically adjustable to accord with the load.

**kata.** A prefix used with metamorphic names to indicate an origin in the deepest zone of metamorphism.

**katabatic wind.** Any wind blowing down an incline. If the wind is warm, it is called a foehn; if cold, it may be a fall wind, such as the bora.

**katamorphism.** The breaking down processes of metamorphism, as contrasted with the building up processes of anamorphism.

**K-capture.** The capture by an atomic nucleus of an orbital electron from the first (innermost) orbit of shell, or K-shell, surrounding the nucleus.

**keg.** A cylindrical container made of steel or some other substance, which contains 25 pounds of blasting powder or gunpowder; also, any small cask or barrel having a capacity of 5 to 10 gallons.

**kelly.** The heavy square or hexagonal steel pipe which goes through the rotary table and turns the drill string.

**kennel coal.** A coal that can be ignited with a match to burn with a bright flame.

**kerf.** The undercut usually made in the coal to facilitate its fall.

**kerma.** The sum of the initial energies of all the charged particles released per unit mass of material from interactions of indirectly ionizing radiation (primarily neutrons and photons).

**kerma rate.** The time rate of accumulating kerma.

**kerogen.** A solid, largely insoluble organic material occurring in oil shale which yields oil when it is heated but not oxidized.

**kerosine; kerosene.** An oily liquid obtained in the distilling off after gasoline in a temperature range from 174 to 288°C; a hydrocarbon of specific gravity of 0.747 to 0.775. Used for burning in lamps, heaters, furnaces, and as a fuel or a fuel component for jet engines.

**ketone.** Any of various organic compounds containing a carbonyl group (C) and a hydrocarbon group such as $CH_3$. The carbonyl group is linked to the hydrocarbon groups in the middle of the chain resulting in at least one hydrocarbon group on each side of the carbonyl group.

**key bed.** A rock stratum that can be identified over large areas and from which measurements can be taken to determine geologic structure.

**key seat.** In drilling a well, a channel or groove cut in the side of the hole and parallel to the axis of the hole. Key seating takes place as a result of dragging action of pipe on a dog-leg.

**kicker line.** A small diameter pipeline connected to the inlet side of a sending scraper trap which contains gas pressure exceeding that in the main pipeline for the purpose of propelling a cleaner, scraper into a main gas stream.

**killing a well.** The act of bringing under control a well which is blowing out; also applied to the procedure of circulating water and mud into a completed well before starting well operations.

**kiln.** An oven, furnace, or heated enclosure used for processing a substance by burning, firing, or drying.

**kiloton energy.** The energy of a nuclear explosion which is equivalent to that of an explosion of 1000 tons of TNT. (See Fig. p. 252)

**kilovoltampere.** A unit of electric measurement equal to the product of a volt and an ampere; one thousand volt-amperes.

**kilowatt.** A unit of power equal to 1,000 watts or to energy consumption at a rate of 1,000 joules per second. It is usually used for electrical power. An electric motor rated at one horsepower uses electrical energy at a rate of about 3/4 kilowatt.

**kilowatt-hour.** The unit of energy equal to that expended in one hour at a rate of 1 kilowatt or 3,413 British thermal units.

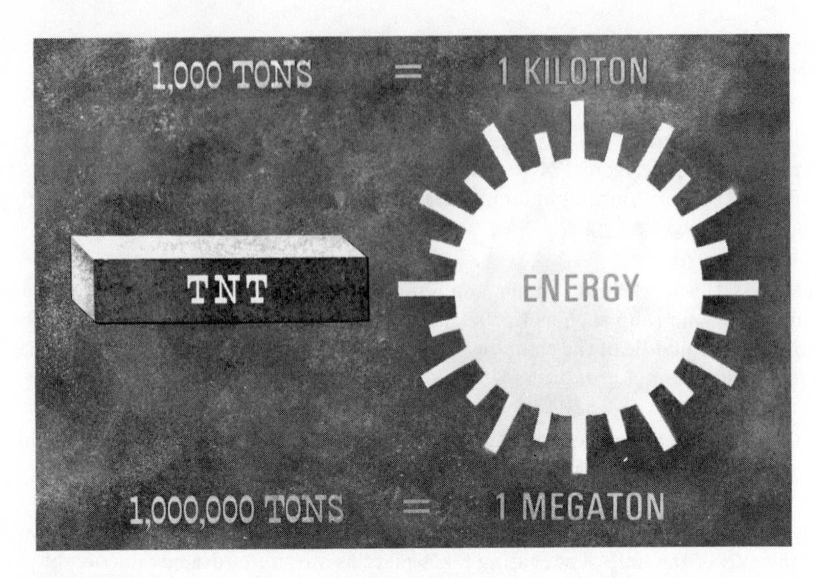


Graphic presentation of the basis of kiloton and megaton designations. (*Courtesy Lawrence Radiation Laboratory, Livermore*)

**kinematic viscosity.** The absolute viscosity divided by the density at the temperature of the viscosity measurement. The metric units of kinematic viscosity are the stoke and centistoke, which correspond to the poise and centipoise of absolute viscosity.

**kinetic energy.** The form of mechanical energy a body possesses by virtue of its velocity. It is determined by the mass and the speed of the object.

**kirwanite.** A variety of anthracite with a metallic luster.

**knapping.** The act of breaking stone; also improving the grade of ore by removing low-grade material manually with a hammer.

**knock.** The sound or "ping" associated with the autoignition in the combustion chamber of an automobile engine of a portion of the fuel-air mixture ahead of the advancing flame front.

**knockout.** Fractionating system for removal of such heavy hydrocarbons as paraffins, hexanes, pentanes, and mercaptans.

**Koppers-Totzek (K-T) process.** A process wherein an entrained flow reactor can produce a medium-Btu gas from coal. The feed coal is partially oxidized in suspension with oxygen and steam. Product gases can be upgraded to pipeline gas. Reactor temperatures may be 1927°C with pressures up to 150 pounds per square inch gage.

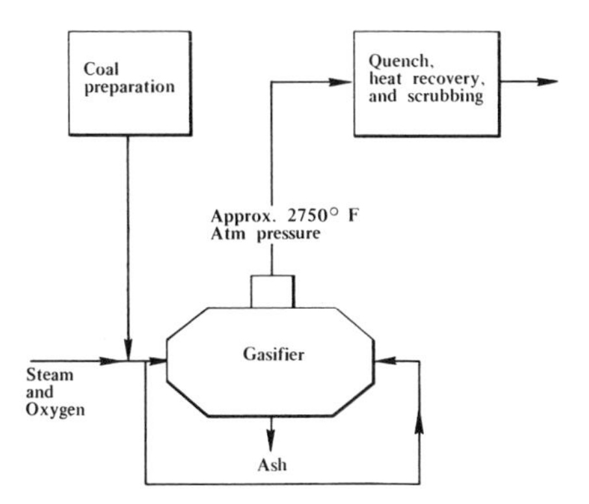


The Kopper-Totzek (K-T) coal gasification process. (*Courtesy U.S. Council on Environmental Quality*)

**krypton-85.** An inert radioactive gas which is a fission product of uranium-235 or plutonium-239.

# L

**labeled atom.** One rendered radioactive and thus traceable through a chemical process of a flow line. Also called tagged atom.

**labeling.** The incorporation of a tracer, radioactive or stable, into a molecular species or macroscopic sample for purposes of detection.

**labile.** Applied to those particles in a rock that decompose easily; unstable; easily displaced or altered.

**labor.** A shaft, cavity, or other part of a mine from which ore is being or has been extracted.

**lacing.** The timber or other material placed behind and around the main supports.

**lag.** A flattish piece of wood or other material to wedge the timber or steel supports against the ground and to secure the area between the supports.

**lagging.** Asbestos and magnesia plaster used on process equipment and piping as a thermal insulation.

**lag of ignition.** The time which elapses during the preflame period.

**lagoon.** A marsh, shallow pond, or lake; in waste water treatment, a shallow pond, usually man-made, where sunlight, bacterial action, and oxygen interact to restore waste water to a reasonable state of purity.

**lamination.** Stratification on a fine scale—each thin stratum, or lamina, being a small fraction of an inch in thickness. Typically exhibited by shales and fine-grained sandstones.

**lampblack.** A black or gray pigment made by burning low-grade heavy oils or similar carbonaceous materials with insufficient air in a closed system so that the soot can be collected in settling chambers. Used as a black pigment for cements and ceramic ware, an ingredient in liquid-air explosives, in lubricating composition, and as a reagent in the cementation of steel.

**landed cost.** The cost of imported crude oil equal to actual cost of crude at point of origin plus cost of transportation to the United States.

**land reclamation.** The process of restoring to use land which has been strip mined.

**Land Reclamation Laboratory.** Is responsible for managing a coordinated field and laboratory research program concentrated on land reclamation and land use issues both regionally and nationally.

**land subsidence.** The sinking of a land surface as the result of the withdrawal of underground material. It results from underground mining and is a hazard of the development of geothermal fields.

**land use planning.** The development of plans for the uses of land that will best serve the general welfare.

**langley.** A unit of solar radiation equivalent to 1 gram calorie per square centimeter of irradiated surface.

**lanthanide series.** The series of elements beginning with lanthanum, element No. 57, and continuing through lutetium, element No. 71, which together occupy one position in the Periodic Table of the elements. These are the "rare earths", which all have chemical properties similar to lanthanum. They also are called the "lanthanides".

**lap.** Term usually applied to an interval in the cased hole (of an oil or gas well) where the top of a liner overlaps the bottom of a string of casing.

**laser.** A device making use of a technique for obtaining exceedingly intense and coherent beams of visible radiation, by use of the fluorescent properties of ruby, emerald, or other chromium phosphors. Coherent radiation or light is produced when the atoms or molecules in a material are forced to radiate at the same time, in step with one another, rather than randomly and independently as in a normal light bulb.

**laser-induced fusion.** A concept in which the high temperature and pressure required for initiating fusion in a plasma are produced by bombarding fuel pellets with short, intense bursts of radiation from one or more lasers. Energy released in these explosions could be absorbed in a heat sink of molten lithium or lithium salts and used for production of steam and generation of electricity. (See Fig. p. 256)

**latch-on.** To attach elevators to a section of pipe; also, a slang term meaning to take hold of a variety of different objects around the drilling rig.

**latent heat.** Change in heat content of a substance when its physical state is changed without a change in temperature.

**latent heat of vaporization.** The quantity of heat required to change a unit weight of liquid to vapor with no change in temperature.

**latent period.** The period between the time of exposure to an agent and the beginning of the response.

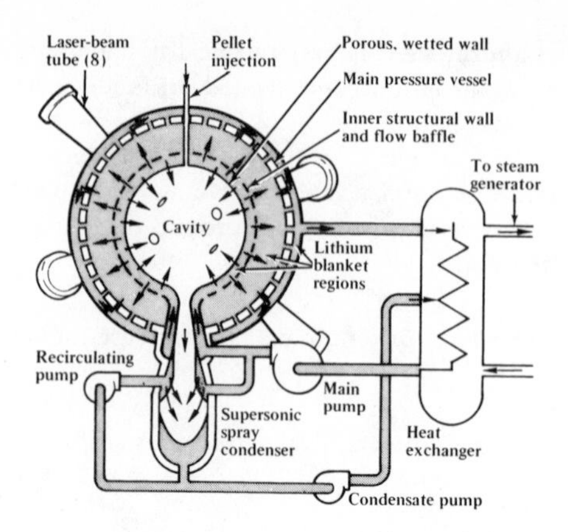


A laser-fusion reactor. (*Courtesy Popular Science, December 1976, p. 69*)

**lateral.** A pipe in a gas distribution or transmission system which branches away from the central and primary part of the system.

**lateral development.** Any system of development in coal seams or thick ore bodies in which headings are driven horizontally across the coal or ore and connected to main haulage drifts, entries, or shafts. There are many variations and modifications depending on the thickness, shape, and inclination of the deposit.

**lateral movement.** Method of removing coal by stripping and augering with no material placed on the downslope.

**lateral support.** Means whereby walls are braced either vertically or horizontally by columns, pilasters, crosswalls, or floor or roof constructions, respectively.

**laterlog.** The electrical resistivity of coal appears to decrease with ash content. The laterlog measures what is virtually the true resistivity of the coal and may ultimately provide information on seam quality. The laterlog uses a sheet of current which is focused on each formation and so measures the resistivity of that formation.

**lattice.** An orderly array or pattern of nuclear fuel elements and moderator in a reactor or critical assembly; also, the arrangement of atoms in a crystal.

**law of gravitation.** The law, discovered by Sir Isaac Newton, that every particle of matter attracts every other portion of matter, and the stress between them is proportional to the product of their masses divided by the square of their distance apart.

**law of motion.** A statement of dynamics; a body at rest remains at rest and a body in motion remains in uniform motion in a straight line unless acted upon by an external force. The acceleration of a body is directly proportional to the applied force and is in the direction of the straight line in which the force acts. For every force there is an equal and opposite force or reaction.

**law of the cube.** This formula states that at any given moment the power content of the wind is proportional to the cube of the wind velocity. When the wind speed doubles, the power output from a wind machine will increase 8 times.

**laying mains.** The complete operation of installing piping systems in towns or cities including trenching, joining sections of pipe, placing pipe in trenches, back-filling trenches, and cleaning up.

**$LC_{50}$.** Median lethal concentration, $LC_{50}$, a standard measure of toxicity, is the concentration required of a substance that will kill 50 percent of a group of experimental insects or animals.

**leachate.** The liquid that has percolated through the soil or other medium and has extracted, dissolved, or suspended materials from it.

**leaching.** The removal in solution of the more soluble minerals by percolating waters through soil.

**lead-acid battery.** Conventional automobile battery; uses lead and lead oxide for electrodes and sulfuric acid for the electrolyte; secondary or storage battery which can be recharged electrically, as opposed to a primary or dry cell battery which is used until the chemical compounds that produce the electricity have been exhausted and is then discarded.

**leaded gasoline.** Gasoline containing tetraethyl lead, an important constituent in antiknock gasoline.

**lead susceptibility.** The increase in octane number of gasoline imparted by the addition of a specified amount of tetraethyllead.

**leakage.** In nuclear engineering, the escape of neutrons from a reactor core. Leakage lowers a reactor's reactivity.

**leakage survey.** In the gas industry, a systematic search for the purpose of locating leaks in a gas piping system.

**leak clamp.** A clamp used to press and hold tight a gasket against a leaking section of pipe or pipe joint to seal the leak.

**leak limiter.** A device to limit the escape of gas from the vent opening of a regulator in the event of a diaphragm failure.

**leak vibroscope.** An instrument which detects leaks in water, oil, gas, steam, and air lines by amplifying the sound produced by the escaping fluid.

**lean gas.** Gas containing little or no liquefiable hydrocarbons; the residual gas from the absorber after the condensable gasoline has been removed from the "wet" gas;

also called dry gas. In the coal industry a term used in several European countries for coal with a low volatile matter.

**lean mixture.** A gas-air mixture of which the air content is more than adequate for complete combustion and the resultant combustion gases will contain an excess of oxygen.

**Leclanche cell.** A zinc-carbon primary cell whose exciting liquid is a solution of salammoniac.

**Leduc's law.** The volume of a gas mixture is equal to the sum of the volumes that would be occupied by each of the components of the mixture if at the temperature and pressure of the mixture.

**leg.** In mine timbering, a prop or upright member of a set or frame.

**leg wire.** One of the two wires attached to and forming a part of an electric blasting cap or squib.

**Lehmann process.** A process for treating coal by disintegration and separation of the petrographic constituents (fusain, durain, and vitrain).

**length.** The horizontal dimension of the unit in the face of a wall; a piece of pipe of the length delivered from the mill. Each piece is called a length regardless of its actual dimensions. This is sometimes called "joint", but "length" is preferred.

**lens.** A rock formation of local extent, formed by variation in sedimentation in the original formation of sedimentary beds.

**lenticle.** A rock stratum or rock bed which, from being thin at the edges, is lens-shaped. Nearly all undeformed strata are lenticles.

**lepton.** One of a class of light elementary particles (having small mass); specifically, an electron, a positron, a neutrino, and antineutrino, a muon, or an antimuon.

**lethal dose.** A dose of ionizing radiation sufficient to cause death. Median lethal dose is the dose required to kill within a specified period of time (usually 30 days) half of the individuals in a large group of organisms similarly exposed. The lethal dose for man is about 400 to 450 roentgens.

**level terrace.** A broad surface channel or embankment constructed across sloping soil on the contour.

**level width.** A quantity assigned to each mode of decay when the decay of a resonance level can proceed in several different ways.

**licensed material.** Source material, special nuclear material, or byproduct material received, possessed, used, or transferred under a general or special license.

**life-cycle cost.** The accumulation of all funds spent for the purchase, installation, operation, and maintenance of a system over its useful life. The accumulation generally includes a discounting of future costs to reflect the relative value of money over time.

**life of mine.** The time in which, through the employment of the available capital, the ore reserves—or such reasonable extension of the ore reserves as conservative geological analysis may justify—will be extracted.

**lift.** The vertical height traveled by a cage in a shaft; any of the various gangways from which coal is raised at a slope colliery; one of the movable sections of a liquid-sealed gas holder; the vertical distance a liquid is pumped; in a sanitary landfill, a compacted layer of solid waste and the top layer of cover material.

**lift-type devices.** Devices that use air-foils or other types of shapes that provide aerodynamic lift in a windstream.

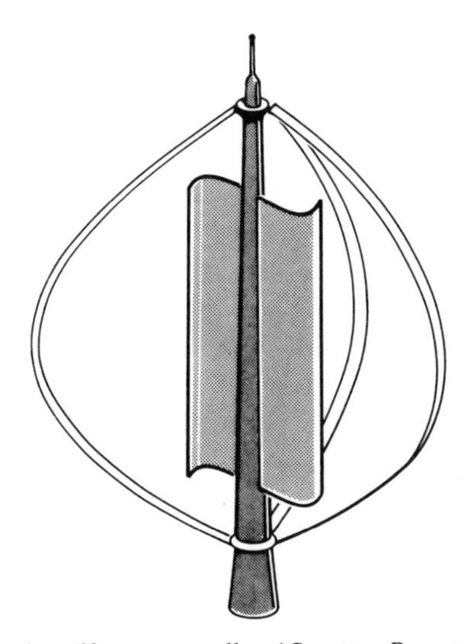

Savonious/Darrieus lift-type propeller. (*Courtesy Department of Energy*)

**light.** A form of radiant energy which gives rise to the sensation of sight. The portion of the electromagnetic spectrum considered visible light is wavelengths from approximately 430 millimicrons to 690 millimicrons. (1 millimicron = $10^{-9}$ meter.)

**light metal.** The low-density metals and alloys; especially aluminum and magnesium and their alloys.

**lightning arrester.** A protective device for limiting surge voltages on equipment by discharging or bypassing surge current. It prevents continued flow of follow current to ground and is capable of repeating these functions as specified.

**light oil.** Any of the products distilled or processed from crude oil up to, but not including, the first lubricating oil distillate.

**light water.** Ordinary water ($H_2O$), as distinguished from heavy water or deuterium oxide ($D_2O$).

**Light Water Reactor (LWR).** Nuclear reactor in which water is the primary coolant/moderator with slightly enriched uranium fuel. There are two commercial light-water reactor types—the Boiling Water Reactor (BWR) and the Pressurized Water Reactor (PWR).

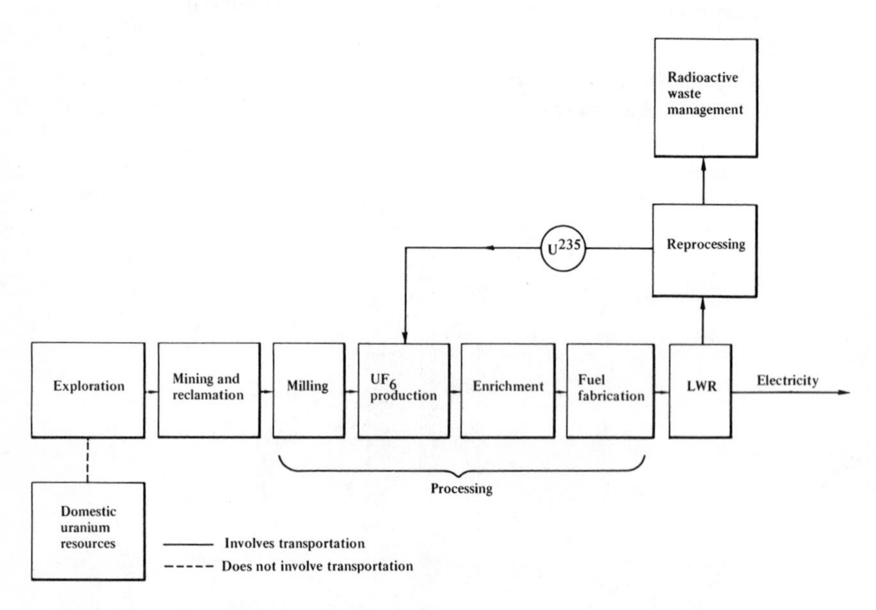


Light Water Reactor fuel cycle. (*Courtesy U.S. Council on Environmental Quality*)

**lignite.** A brownish-black coal in which the alteration of vegetal material has processed further than in peat but not so far as subbituminous coal; an early stage in the formation of coal; the lowest-rank coal, with low heat content and fixed carbon, and high percentages of volatile matter and moisture.

**lignitious coal.** Coal containing 75 percent to 84 percent of carbon (ashless, dry basis).

**limestone.** A bedded sedimentary deposit consisting chiefly of calcium carbonate which yields lime (quick lime) when burned.

**linear accelerator.** A long straight tube (or series of tubes) in which charged particles (ordinarily electrons or protons) gain in energy by the action of oscillating electromagnetic fields.

**linear energy transfer (LET).** A measure of the ability of biological material to absorb ionizing radiation; the radiation energy lost per unit length of path through a biological material. In general, the higher the LET value, the greater is the relative biological effectiveness of the radiation in that material.

**linear ionization.** The total number of ion pairs, including those created by secondary ionizing processes, produced by a directly ionizing particle.

**line drop.** Loss in voltage owing to the resistance of conductors conveying electricity from a power station to the consumer.

**line loss.** The amount of gas lost in a distribution system or pipeline.

**line pack.** Inventory of gas in a pipeline or a gas distribution system.

**line packing.** Increasing the amount of gas in a line section by increasing pressure to meet a heavy demand, usually of short duration.

**line pipe.** Steel pipe generally used to construct pipelines to transport petroleum and natural gas.

**linerider.** An employee who inspects a pipeline right-of-way for leaks or potential hazards.

**line transformer.** A transformer classified as distribution line equipment, generally having a rated primary voltage of 2,300 to 15,000 volts.

**lip screen.** A common term applied to stationary screens installed in the loading chutes over which the coal flows as it is loaded into railroad cars for market.

**liquation.** The process of separating by heat a fusible substance from a substance less fusible; also a method of recovering sulfur by liquefying under pressure and heat, drawing off the molten sulfur, and allowing it to solidify.

**liquefaction.** Conversion of a solid to a liquid by heat, or of a gas into a liquid by cold or pressure. With coal this invariably involves hydrogenation to depolymerize the coal molecules to simple molecules. (See Fig. p. 262)

**liquefied natural gas (LNG).** Natural gas cooled to −160°C so it forms a liquid at approximately atmospheric pressure. As natural gas becomes liquid, it reduces volume by a factor of almost 600, thus allowing both economical storage and economical long distance transportation in high pressure cyrogenic containers. (See Fig. p. 262)

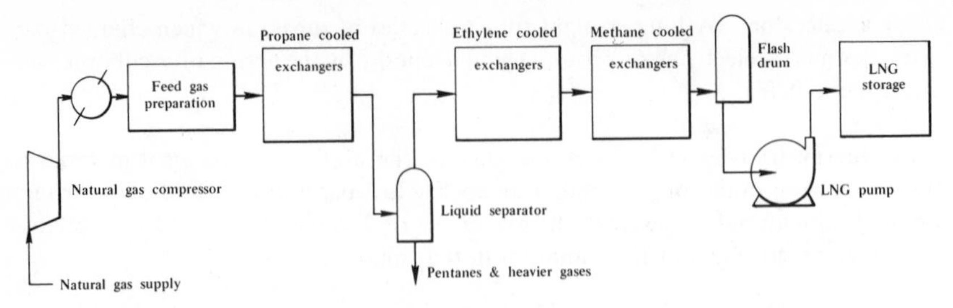


Cascade cycle liquefaction plant. (*Courtesy U.S. Council on Environmental Quality*)

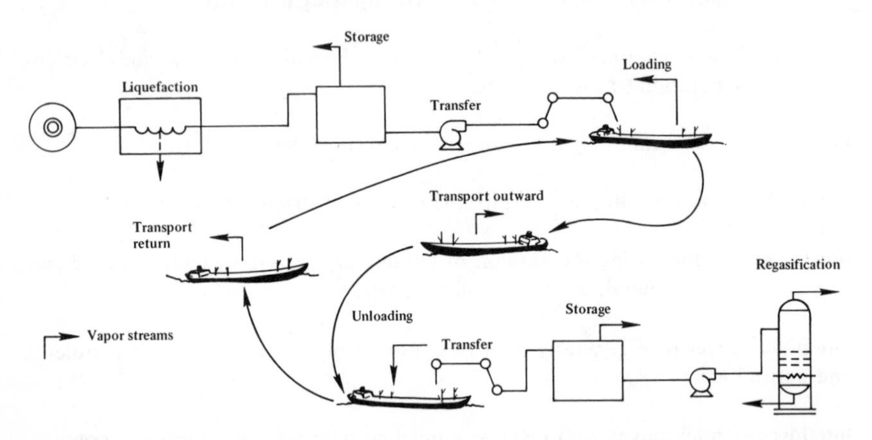


Integrated liquid natural gas operation. (*Courtesy U.S. Council on Environmental Quality*)

**liquefied petroleum gas (LPG).** Also known as bottled gas, liquefied petroleum gas consists primarily of propanes and butanes recovered from natural gas and the refining of petroleum. It has an energy content of 2000 to 3500 British thermal units per standard cubic foot. LPG is widely used as a fuel for internal combustion engines in applications where pollution must be minimized such as in buildings and mines, but its largest use is as a substitute for natural gas in areas not served by pipelines. (See Fig. p. 263)

**liquid.** Liquid storage accommodates heat in proportion to its specific heat capacity and temperature increase.

**liquid-based solar heating system.** A solar heating system in which liquid, either water or an antifreeze solution, is heated in the solar collectors.

**liquid-dominated convective hydrothermal resources.** Hydrothermal convective systems in the upper part of the earth's crust which transfer heat from a deep igneous source to a depth sufficiently shallow to be tapped by drill holes.

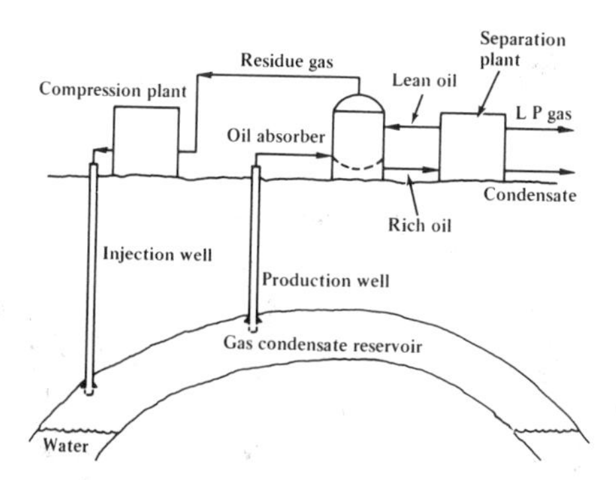


Cycling operation in the production of liquefied petroleum gas. (*Courtesy U.S. Council on Environmental Quality*)

**liquid fuel.** Any liquid used as fuel which can be poured or pumped. Petroleum or crude oil is a natural liquid fuel, while distilled oil, coal tar, and residual oil are prepared liquid fuels.

**liquid-liquid extraction.** A process in which one or more components are removed from a liquid mixture by intimate contact with a second liquid, which is itself nearly insoluble in the first liquid and dissolves the impurities but not the substance to be purified.

**Liquid Metal Fast Breeder Reactor (LMFBR).** A nuclear breeder reactor cooled by molten sodium in which fission is caused by fast neutrons. It uses uranium-235 or plutonium-239 as fuel and produces additional plutonium-239 from the uranium-238 in a "blanket" around the reactor core.

**Liquid Metal Fast Breeder Reactor (LMFBR) program.** The review of this program by DOE is being undertaken at the direction of President Carter. The President's energy priorities, reflected in the revision of ERDA's FY 1978 budget request, stress conservation and nearer-term supply technologies. These priorities suggest that past plans for expansion of the LMFBR program may no longer be viable. Furthermore, serious questions have been raised about the LMFBR technology and the structure of the current LMFBR program. The energy potential of this option must be weighed against the safety questions associated with the LMFBR and the dangers of nuclear proliferation from plutonium reprocessing needed by LMFBRs. (See Fig. p. 264)

**Liquid-Phase Methanation process.** Process being developed with the overall objective of finding practical and useful processes for converting coal-derived synthesis gases to methane as a major constituent of synthetic natural gas (SNG) using liquid fluidized beds.

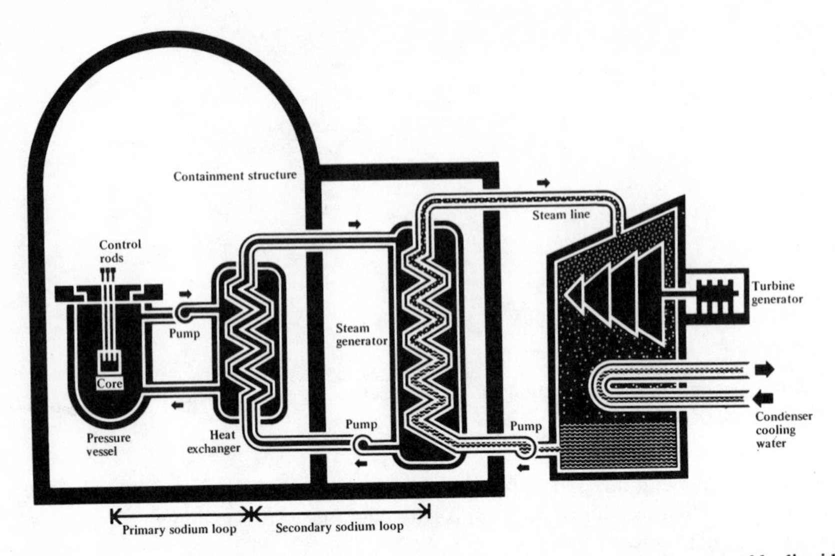


The heat generated in the fuel rods of a Liquid Metal Fast Breeder Reactor is removed by liquid sodium flowing through the core. For safety reasons, this heat is transferred from the radioactive primary sodium loop to a nonradioactive secondary loop before used to generate steam. This steam is used to spin a conventional turbine-generator to make electricity. (*Courtesy Atomic Industrial Forum*)

**liquid-to-liquid heat exchanger.** A device for heat exchange between two liquid streams.

**liter.** The primary standard of capacity in the metric system, equal to the volume of one kilogram of pure water at maximum density, at approximately 4° C, and under normal atmospheric pressure. One liter = 0.264 gallon, 1.05 quarts, or 2.11 pints.

**lithification.** The complex of processes that converts a newly deposited sediment into an indurated rock; a type of coalbed termination in which the disappearance takes place because of a lateral increase in impurities resulting in a gradual change into bituminous shale or other rock.

**lithium.** As found in nature, lithium consists of a mixture of two stable isotopes—lithium-6 (7.5%) and lithium-7 (92.4%). Lithium-6 is of interest as a possible fuel or fuel source for the generation of power from a controlled thermonuclear reaction.

**lithium anode cell.** Formed when a lithium anode is brought into contact with a cathode composed of iodine and poly-2-vinylpyridine (P2VP). A chemical reaction between these components forms the electrolyte, solid lithium iodide. Used in pacemaker batteries.

**lithium-sulfur battery.** An advanced battery, which may be able to store large amounts of energy per unit weight and size. It consists of a series of cells which contain lithium in the negative electrode and a metal sulfide in the positive electrode. The electrolyte is a molten salt. They operate at about 400 to 425° C. Present prototype lithium-sulfur cells of 200 to 250 watt-hour size have a storage capacity of about 40 to 50 watt-hour per pound; experimental versions have demonstrated 75 watt-hours per pound (in comparison, present lead-acid battery has a storage capacity of about 12 watt-hours per pound). Being considered primarily for use in electric vehicles.

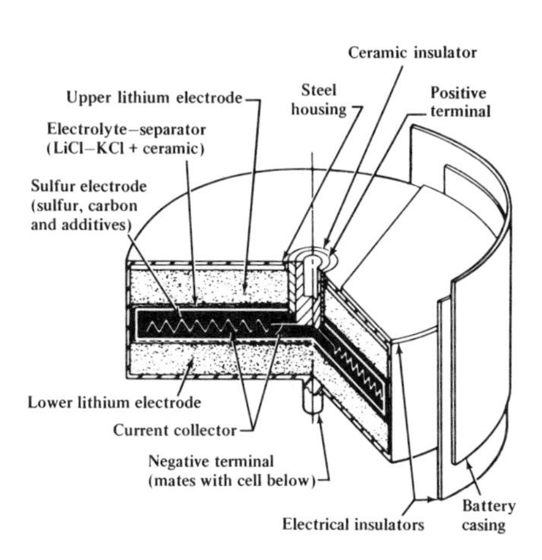


A lithium-sulfur battery cell. (*Courtesy U.S. Energy Research and Development Administration*)

**live steam.** Steam direct from the boiler and under full pressure; distinguished from exhaust steam, which has been deprived of its available energy.

**load.** The energy tapped from any power source; in water quality use, the quantity of material carried by flowing water generally expressed as pounds per day; in the gas industry, the amount of gas delivered or required at any specified point or points on a system; in the electric industry, the amount of electric power delivered or required at any specified point or points on a system. Load originates primarily at the gas- or power-consuming equipment of the customers.

**load center.** A point at which the load of a given area is assumed to be concentrated.

**load curve.** A graph in which the load of a gas system or segment of a system is plotted against intervals of time; also a curve showing power (kilowatts) supplied, plotted against time of occurrence, and illustrating the varying magnitude of the load during the period covered.

**load density.** The concentration of gas load for a given area expressed as gas volume per unit of time and per unit of area.

**load diversity.** The difference between the sum of the peaks of two or more individual loads and the peak of the combined load.

**load duration curve.** A graph made by plotting data in order of magnitude, against time intervals, for a specified period. The ordinate may be an absolute quantity or percentage.

**load factor.** The ratio of average load carried by an electric power plant or system during a specific period to its peak load during that period.

**load-frequency control.** The regulation of the power output of electric generators within a prescribed area in response to changes in system frequency or tie-line loading.

**load growth.** The growth in energy and power demands by a utility's customers.

**load leveling.** The demand for electricity varies from season to season and even hour to hour. Power companies must have generating equipment available for the times of peak electrical consumption, such as weekday daylight hours. Generally, this means operating "peaking" turbines to provide supplemental generating capacity during times of high demand. Such turbines have the advantage of being able to come on line quickly, but they are expensive to operate and are normally fueled by natural gas or highly refined oil, presently in relatively short domestic supply. Power companies throughout the nation are studying ways to reduce the large fluctuations that occur in electricity demand. One important way of "load-leveling"—reducing the peaks and valleys of electrical demand—would be to store excess electricity in a large bank of thousands of advanced batteries during low-demand periods for use during periods of higher demand.

**load power.** The load power of an energy load is the average rate of flow of energy through the terminals of that load when connected to a specified source.

**load water.** Water used to prime a well after acidizing.

**local mean time.** Hour angle of mean sun. Local apparent noon is the transit time at which the upper limb of the sun crosses the local meridian. When making sun observations, this discrepancy must be corrected by use of the equation of time from the Nautical Almanac.

**local relaxation length.** The relaxation length at a particular location when the radiation is polyenergetic, and the relaxation length is therefore not constant throughout the material.

**local storage.** The storage facilities, other than underground storage, that are an integral part of a distribution system; that is, on the distribution side of the city gate, whether for manufactured, mixed, natural, liquefied petroleum, or liquefied natural gases.

**local winds.** Winds that differ, over a small area, from those appropriate to the general pressure distribution.

**lock hopper.** A device for introducing solids, such as coal, into a pressurized system.

**lock in (unlock).** Generally, to unseal a gas meter and start gas service by opening the meter stop (valve).

**lock out (lock).** Generally, to seal and lock a gas meter and shut off the stop (valve) so that gas cannot be used.

**lockout timing.** That period of time between the initial ignition trial and the lockout by the ignition system.

**lock-up or lock-off.** The point at which a regulator or governor shuts off completely.

**Löf house.** A residential solar energy home designed by Dr. George O. G. Löf. Dr. Löf has built two of the early solar-heated houses in Colorado. Forced air heating is supplied by a 2300-square foot solar collector mounted on the roof. The house uses a 23,460-pound granite pebble heat storage bin. Dr. Löf is a pioneer in U.S. Solar Energy Development.

**log.** A record of performance, such as the record of an engine, boiler, or other test; also a record of the progress in drilling a well.

**logarithmic velocity profile.** The variation of the mean wind speed with height in the surface boundary layer.

**long screw.** A device sometimes used in place of a union. It consists of a nipple with one long straight thread, equipped with a recessed coupling and a packing nut which holds the packing against the coupling to prevent leakage.

**longwall mining.** A method of working coal seams that originated in England in the 17th century. The seam is removed in one operation by means of a long working face, or wall. The workings advance (or retreat) in a continuous line. The space from which the coal has been removed is either allowed to collapse or is completely or partially filled or stowed with stone and debris. (See Fig. p. 268)

**longwall stripping.** Longwall mining accomplished in areas of shallow cover where surface mining might normally have been conducted. The outby end, where the longwall controls, pumps, and face conveyor discharge end are located, is located in a ditch that is exposed to the surface. Roof checks are used to protect the mining area and the roof (or overburden) is allowed to settle into the mined-out section.

**long-wave radiation.** Infrared or radiant heat.

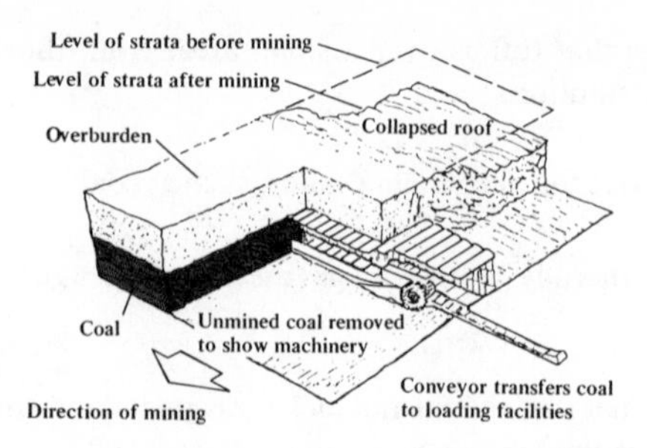


Longwall mining. (*Courtesy U.S. Energy Research and Development Administration, Division of Coal Conversion and Utilization*)

**loop.** A closed circuit of pipe in which materials and components may be placed to test them under different conditions, such as of temperature and irradiation, in a reactor, a piping system through which a fluid may flow as a part of reactor operation or for experimental purposes. If part of an experimental loop is in the core, such a loop is usually called an in-pile loop. If the loop also contains fissionable materials, it is called an active loop (hot loop).

**loss of coolant accident.** An accident in a nuclear reactor where the coolant is lost from the reactor core.

**Loss of Fluid Test (LOFT) Facility.** Facility designed to study the nuclear, thermal, hydraulic, and structural processes occurring in a loss-of-coolant accident in a pressurized water reactor. The LOFT program is funded by the Nuclear Regulatory Commission.

**loss of head.** The decrease of energy head between two points such as that caused by friction.

**louvers.** Slanted overlapping slats arranged to allow the circulation of air.

**Low-Btu/fixed-bed gasifier process.** A process in which three fixed-bed gasifier configurations are to be tested to develop a clean low-Btu gas process for combined power plant systems. The gasifiers will be developed to utilize a highly agglomerating run of mine coal. High temperature and low temperature sulfur removal methods will be used. The stirred, fixed-bed reactor under development is serving as a design basis for this gasifier.

**Low-Btu fuel gas process.** Three fluidized-bed reactors are used to produce low-Btu gas from caking and noncaking coals. Coal is devolatilized in the first stage reactor. Devolatilized coal is burned in the second stage reactor forming char, which is burned in the third stage reactor. Off-gas from the first and third stages support the primary production of fuel gas from the second stage reactor. The third stage reactor operates at a temperature of 1149° C. The second and first stages operate at lower temperatures of 1093° C and 649° C, respectively.

**low-Btu gas.** Manufactured gas with value of 100 to 200 British thermal units per cubic foot.

**lowest normal tides.** A plane of reference lower than mean sea level by half the maximum range. This does not take into account wind or barometric pressure fluctuations.

**low explosive.** An explosive in which the change into the gaseous state is effected by burning and not by detonation as with high explosives. Blasting powder (black powder or gunpowder) is the only low explosive in common use. Also called propellant.

**low-grade.** An arbitrary designation of dynamites of less strength than 40 percent. It has no bearing on the quality of the materials, as these dynamites are of as great purity and high quality as the ingredients in a so-called highgrade explosive. The term also designates coal that is high in impurities, and pertains to ores that have a relatively low content of metal compared to other, richer material from the same general area.

**low-level analysis.** A procedure to measure the radioactive content of materials with very low levels of activity using sensitive detecting instruments and good shielding to eliminate the effects of background radiation and cosmic rays.

**low population zone.** An area of low population density sometimes required around a nuclear installation to insure, with reasonable probability, that effective protection measures can be taken if a serious accident should occur.

**low pressure air burner system.** A system using the momentum of a jet of low pressure air (up to and including 5 pounds per square inch gage) to entrain gas to produce a combustible mixture.

**low pressure system.** A system in which the gas pressure in the mains and service lines is substantially the same as that delivered to the customers' appliances; ordinarily a pressure regulator is not required on individual service lines.

**low sulfur coal.** Coal with a sulfur content of less than 0.67 pounds of sulfur per million Btu.

**low sulfur crude oil.** Crude oil containing low concentrations of sulfur-bearing compounds. Crude is usually considered to be in the low-sulfur category if it contains less than 0.5 percent (weight) sulfur. Examples of low-sulfur crudes are offshore Louisiana, Libyan, and Nigerian crudes.

**low wall.** The vertical wall, on the downslope side of the mining operation, consisting of the deposit being mined.

**low-water cutoff.** A device constructed so as to automatically cut off the gas supply when the surface of the water in a boiler falls to the lowest safe water level.

**LOX.** Abbreviation for liquid oxygen explosive.

**LP gas (bottle gas).** A gas consisting of volatile hydrocarbons, from propane to pentane, mixed with hydrogen and methane under pressure. It withstands pressure and thus may be transported in steel tanks under pressure in liquefied form. May be used as a fuel to operate combustion-type engines in lieu of gasoline. Sold under various trade names but more commonly known as propane, butane, LP or bottle gas, blau or blue gas.

**LP gas-air mixtures.** Liquefied petroleum gases distributed at relatively low pressures and normal atmospheric temperatures which have been diluted with air to produce desired heating value and utilization characteristics.

**lube stock.** Refinery term for fraction of crude petroleum with suitable boiling range and viscosity to yield lubricating oils when further processed and treated.

**lubricant.** Oil, grease, graphite, and, in general, anything of the sort used to overcome friction and permit a freer action of parts.

**lubricated plug valve.** A valve of the pierced plug and barrel type provided with means for maintaining a lubricant between the bearing surfaces. It is designed so that the lapped bearing surfaces can be lubricated and the lubricant level maintained without removing the valve from service.

**lumen.** The quantity of light required to illuminate 1 square foot to an average intensity of 1 foot-candle. By multiplying the mean candlepower by 3.63, the quantity of light expressed in lumens for the solid cone with a spread of 130° is obtained.

**luminescence.** Emission of light produced by the action of biological or chemical processes, or by radiation, or by any other cause except high temperature (which produces incandescence).

**Lurgi-gasynthan process.** An SNG process developed by Lurgi.

**Lurgi process.** A coal gasification process developed in Germany and in commercial use in Europe. It produces a gas of somewhat lower heat value than natural gas. The

Lurgi process consists of roasting iron ore in a reducing atmosphere, thus forming magnetic oxide of iron which is separated first by crushing then by magnetic separation.

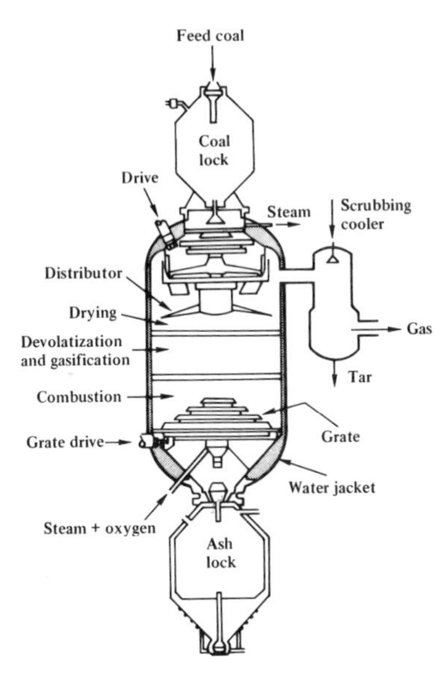


Cross section of a Lurgi gasifier. (*Courtesy "Energy and the Future," Fig. 5*)

**lysimeter.** Structure containing a mass of soil and so designed as to permit the measurement of water draining through the soil.

# M

**machine mining.** Implies the use of power machines and equipment in the excavation and extraction of coal or ore. In coal mines, the term signifies the use of coal cutters and conveyors and perhaps some type of power loader working in conjunction with face conveyors.

**machine wall.** The face at which a coalcutting machine works.

**macroporosity.** Porosity visible without the aid of a microscope, such as pipes and blowholes in ingots.

**magazine.** A storage place for explosives.

**magma.** Magma is the very hot liquid, or partly liquid, rock that is found under the entire surface of the earth, starting at depths of 6 to 9 miles. These depths are too great for economic utilization. In some places, however, the magma exists close to, or even at, the earth's surface—sometimes in large volume. No techniques are known yet for converting this energy into useful forms for commercial exploitations. Beneficial use of magma heat is expected to require a long-term research and development program.

**magnetic anomaly.** Variation of the measured magnetic pattern from a theoretical or empirically smooth magnetic field on the earth's surface.

**magnetic bottle.** A magnetic field used to confine or contain a plasma in controlled fusion (thermonuclear) experiments.

**magnetic confinement systems.** Systems in which fusion reactions take place. The most successful system to date is the tokamak, a circular device shaped like a thick

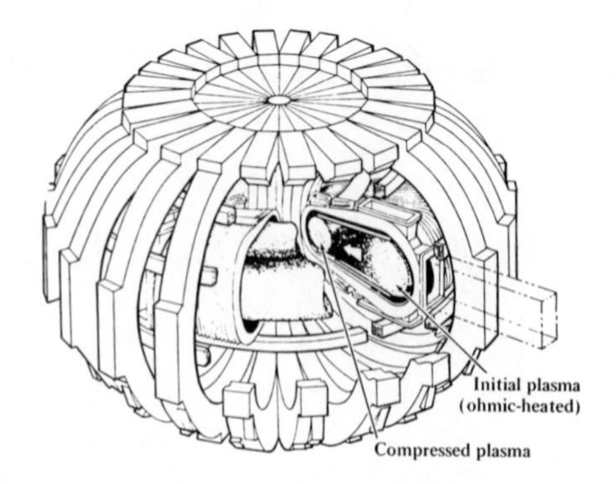


Adiabatic toriodal compressor tokamak for magnetically containing a plasma to achieve controlled fusion reaction. (*Courtesy "Energy and the Future," Fig. 24*)

doughnut. A second type is the magnetic mirror, a linear device with stronger magnetic fields at its ends to pinch the magnetic lines of force and thereby prevent the plasma from leaking. A third type is the theta pinch, which is thinner in shape than the tokamak and may be circular or linear. A program of exploratory concepts to develop and test new confinement systems that offer the possibility of better or more economical plasma containment is underway.

**magnetic field.** The field of force that exists around electrons traveling along a conductor or orbiting the atomic nucleus.

**magnetic force.** The mechanical force exerted by a magnetic field upon a magnetic pole placed in it.

**magnetic jack.** A type of linear motor. Flexible cables of control rod are moved in precise steps by switching coils in sequence, then held to walls by friction. Used in reactor control systems.

**magnetic method.** A geophysical prospecting method which maps variations in the magnetic field of the earth that are attributable to changes of structure or magnetic susceptibility in certain near-surface rocks. Sedimentary rocks generally have a very small susceptibility compared with igneous or metamorphic rocks.

**magnetic mirror.** A magnetic field used in controlled fusion experiments to reflect charged particles back into the central region of a magnetic bottle.

**magnetic shield.** A system employing high-intensity magnetic fields to prevent charged particles from entering a region that is to be protected.

**magnetohydrodynamic (MHD) power plants.** Advanced systems for generating electricity from fuels. They operate in the range from 50 to about 60 percent efficiency in contrast to conventional electric power plants which have a maximum energy conversion efficiency of about 40 percent. The magnetohydrodynamic (MHD) generator is classified as a direct energy conversion generator because it produces electrical power directly from the heat source without requiring the additional stage of steam generation, as do conventional fossil-fueled plants. In the MHD process, fuels are burned to produce gases of temperatures as high as 2760° C. Because of this very high temperature, MHD plants can produce greater fuel efficiencies than conventional plants. Through their higher temperatures and greater efficiencies MHD power plants offer the potential for large savings in electric energy costs and less pollution. (See Fig. p. 274)

**magnetohydrodynamics (MDH).** An advanced concept for improving the efficiency of generating electrical energy from fossil fuels. In this concept, hot, flowing, ionized gas is substituted for the rotating copper coils of the conventional electric generators. Gases from the high temperature combustion of fossil fuels are made electrically conductive by seeding them with suitable chemicals. This electrically conducting gas

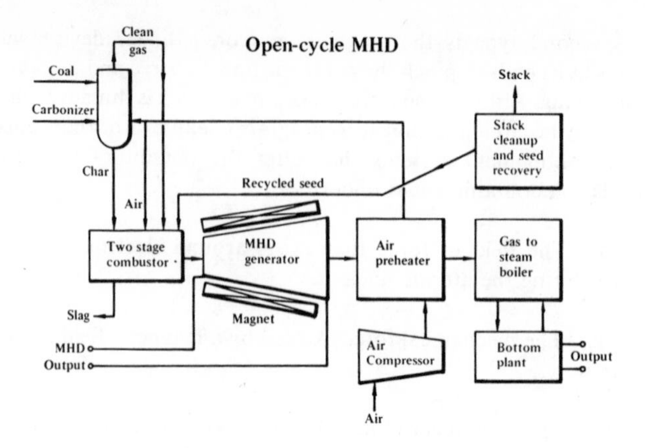


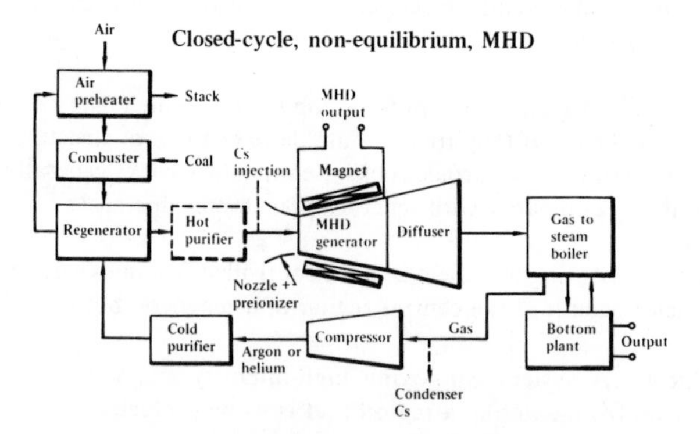


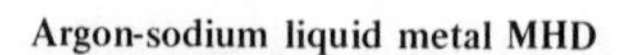


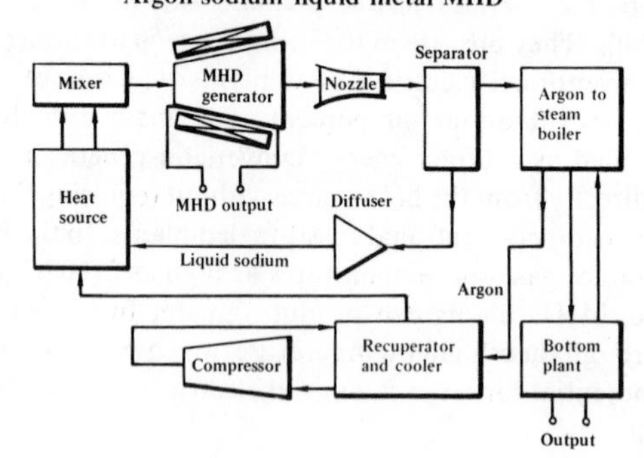


MHD power plants. (*Courtesy Annual Review of Energy, Volume 1*)

then travels at high speed through a magnetic field to produce a flow of direct current. The hot gases can then be used to fire a steam turbine generator, making the overall efficiency of the composite systems as high as 60 percent, one and one half times that of a modern fossil fuel plant. Laboratory-scale MDH generators are now operating.

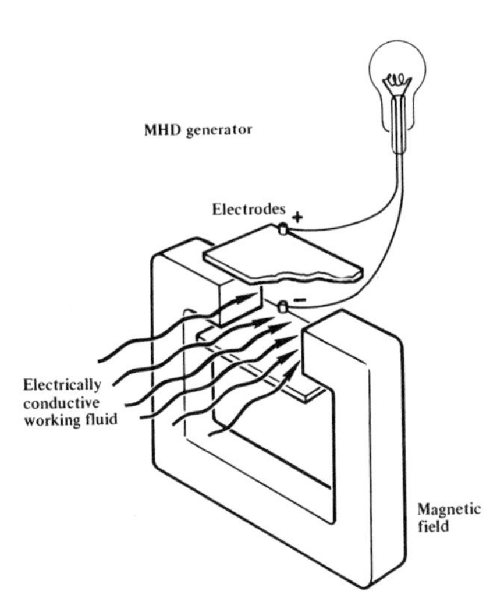


The electrical system in an MHD generator. (*Courtesy U.S. Council on Environmental Quality.*)

**magnox.** A magnesium alloy used to clad fuel elements for reduction of neutron absorption in gas-cooled reactors.

**main.** A distribution line that serves as a common source of supply for more than one service line.

**main burner control valve.** A valve which controls the gas supply to the main burner manifold.

**main extension.** The addition of pipe to an existing main to serve new customers.

**main intake.** The principal intake airway of a mine.

**major mine disaster.** Defined by the U.S. Bureau of Mines as any accident that results in the death of five or more persons.

**makeup.** To assemble, to couple or screw together; usually applied to the process of assembling the component parts of a drill string or pipe system.

**makeup water.** Water added to a tank, boiler, or other vessel to maintain a predetermined liquid level.

**man cage.** A special cage for raising and lowering men in a mine shaft.

**man car.** A certain type of car for transporting miners up and down the steeply inclined shafts of some mines.

**mandrel.** A miner's pick.

**Manhattan Project.** The program during World War II that produced the first atomic bombs. The term originated in the code-name, "Manhattan Engineer District", which was used to conceal the nature of the secret work underway. The Atomic Energy Commission, a civilian agency, succeeded the military unit on January 1, 1947.

**manhole.** An opening into a tank, boiler, furnace, vault, or other equipment through which a man can enter to service equipment; can be sealed with a removable plate or door.

**manifold.** A chamber or tube having several inlets and one outlet, or a single inlet and several outlets.

**manipulators.** Mechanical devices used for safe handling of radioactive materials, often remotely operated from behind a protective shield.

**manless coal face.** A coal face manned by remotely controlled equipment that eliminates the need for men in dangerous places.

**manlock.** An air lock through which men pass to a working chamber that is under air pressure.

**manometer.** Any instrument which measures gaseous pressure. Measures pressure or a pressure difference by balancing the applied pressure against the hydrostatic head of a column of liquid of known density.

**mantle.** A lace-like hood or envelope (sack) of some refractory material which, when placed over a flame, gives light by incandescence.

**manual input flow control valve.** A manual valve, usually with stops, which can be set to limit the gas flow to the maximum required input to the burner or burners.

**manual reset valve.** An automatic shutoff valve installed in the gas supply piping and set to shut off when unsafe conditions occur. The device remains closed until manually reopened.

**manufactured gas.** A mixture of gaseous hydrocarbons produced from coal or oil; all gases made artificially or as byproducts; as distinguished from natural gas; a gas obtained by destructive distillation of coal, or by the thermal decomposition of oil, or by the reaction of steam passing through a bed of heated coal or coke. Examples are coal gases, coke oven gases, producer gas, blast furnace gas, blue (water) gas, carbureted water gas.

**march.** The border or limit of a mineral area leased to or owned by a mining company.

**margin.** The difference between the net system generating capability and system maximum load requirements including net schedule transfers with other systems.

**marginal well.** A well which is producing oil or gas at such a low rate that it may not pay for the drilling.

**marine band.** Roof shale overlying coal seams and being uncommonly high in the content of freshwater or marine shells. Also called mussel band.

**marketable natural gas.** Raw gas from which certain hydrocarbon and non-hydrocarbon compounds have been removed or partially removed by processing. Marketable natural gas is often referred to as pipeline gas, residue gas, or sales gas.

**market discount rate.** The rate of return that can be expected on the best investment.

**marsh.** A low-lying tract of soft, wet or periodically inundated land that provides an important ecosystem for a variety of plant and animal life.

**marsh gas.** Methane. If the decaying matter at the bottom of a marsh or pond is stirred, bubbles of methane rise to the surface, thus the name marsh gas. Chemical formula $CH_4$.

**maser.** Contracted version of microwave amplification by simulated emission of radiation. A class of amplifier from which the optical laser was developed.

**mass.** A basic property of matter, for everyday purposes identical with weight. But even when objects become weightless (for example, in earth orbit), they lose none of their mass—they still have inertia and momentum. For the nuclear physicist, mass and energy are interchangeable. There are two ways of determining the mass of a body. One is to weigh it and apply Newton's law. The other is to apply a force to it and to calculate the mass from the resulting acceleration.

**mass–energy equation.** The statement developed by Albert Einstein that "the mass of a body is a measure of its energy content", as an extension of his 1905 Special Theory of Relativity. The statement was subsequently verified experimentally by measurements of mass and energy in nuclear reactions. The equation, usually given as: $E = mc^2$, shows that when the energy of a body changes by an amount, E, no matter what form the energy takes, the mass, m, of the body will change by an amount equal to $E/c^2$. (The factor $c^2$, the square of the speed of light in a vacuum, may be regarded as the conversion factor relating units of mass and energy.) This equation predicted the possibility of releasing enormous amounts of energy (in the atomic bomb) by the conversion of mass to energy. It is also called the Einstein equation.

**mass materials.** Concrete, masonry or other heavy materials used to store thermal energy for heating or cooling.

**mass number.** The sum of the neutrons and protons in a nucleus. It is the nearest whole number to an atom's atomic weight. For instance, the mass number of uranium-235 is 235.

**mass spectrograph.** A device for detecting and analyzing isotopes. It separates nuclei that have different charge-to-mass ratios by passing the nuclei through electrical and magnetic fields.

**mast.** A drill derrick or tripod mounted on a drill unit, which can be raised to operating position by mechanical means.

**materials processing reactor.** A reactor employed for the purpose of changing the physical characteristics of materials by utilizing the reactor-generated ionizing radiation. Such characteristics may be color, strength, elasticity, and dielectric qualities.

**materials testing reactor.** A reactor employed for testing materials and reactor components in intense radiation fields.

**matter.** The substance of which a physical object is composed. All materials in the universe have the same inner nature, that is, they are composed of atoms, arranged in different ways; the specific atoms and the specific arrangements identify the various materials.

**mattress.** A blanket of brush or poles interwoven or otherwise lashed together and weighted with rocks and concrete blocks to hold it in place, that is used to cover an area subject to scour from currents.

**maximum actual operating pressure.** The maximum operating pressure existing in a piping system during a normal annual operating cycle.

**maximum credible accident.** The most serious reactor accident that can reasonably be imagined from any adverse combination of equipment malfunction, operating errors, and other foreseeable causes. The term is used to analyze the safety characteristics of a reactor. Reactors are designed to be safe even if a maximum credible accident should occur.

**maximum demand.** The greatest of all the demands under consideration during a specified period of time.

**maximum gas in storage.** The highest volumetric balance of total gas in storage during any storage cycle.

**maximum permissible concentration.** The amount of radioactive material in air, water, or food which might be expected to result in a maximum permissible dose to persons consuming it at a standard rate of intake. An obsolescent term.

**maximum permissible dose.** That dose of ionizing radiation established by competent authorities as an amount below which there is no reasonable expectation of risk to human health, and which at the same time is somewhat below the lowest level at which a definite hazard is believed to exist. An obsolescent term.

**maximum permissible dose equivalent (MPDE).** The largest dose equivalent received within a specified period which is permitted by a regulatory committee on the assumption that there is no appreciable probability of somatic or genetic injury. Different levels of MPDE may be set for different groups within a population.

**maxwell.** The unit of magnetic flux the centimeter-gram-second electromagnetic system. The maxwell is $10^{-8}$ weber.

**mean daily temperature.** Average of the minimum and maximum daily temperatures used to determine the number of degree days.

**mean free path.** The average distance traveled by a particle, atom, or molecule between collisions or interactions.

**mean life.** The average time during which an atom, an excited nucleus, a radionuclide, or a particle exists in a particular form.

**mechanical burner system.** A system which proportions air and gas and mechanically compresses the mixture for combustion purposes.

**mechanical equivalent of heat.** The conversion factor for transforming heat units into mechanical units of work. One British thermal unit equals 778 foot-pounds.

**mechanical rig.** A drilling rig whose source of power is one or more internal-combustion engines.

**mechanical turbulence.** The erratic movement of air caused by obstructions such as buildings.

**mechanics.** The study of how matter behaves under the influence of force.

**median lethal dose.** The absorbed dose which will kill 50 percent of a large population of a given species within a specified time.

**median lethal time.** The time required for the death of 50 percent of a large population of a given species as a consequence of receiving (or attributable to) a specific absorbed dose.

**medium-Btu gas.** Manufactured gas with value of 250 to 500 British thermal units per cubic foot.

**megaton energy.** The energy of a nuclear explosion which is equivalent to that of an explosion of one million tons (or 1000 kilotons) of TNT.

**megawatt.** A unit of power. A megawatt equals 1000 kilowatts, or one million watts.

**megawatt-day per ton.** A unit that expresses the burnup of nuclear fuel in a reactor; specifically, the number of megawatt-days of heat output per metric ton of fuel in the reactor.

**melting point.** That temperature at which a single, pure solid phase changes to a liquid or to a liquid plus another solid phase, upon the addition of heat at a specific pressure. Unless otherwise specified, melting points are usually stated in terms of 1 atmosphere pressure.

**mercaptans.** A group of organic chemical compounds containing distinctive offensive odors in small concentrations; added to natural or LP-gases to warn of leaks.

**mercurialism.** Chronic poisoning with mercury, as from excessive medication or industrial contacts with the mercury metal or its fumes.

**mercury.** A heavy metal, highly toxic if breathed or ingested. Mercury is residual in the environment, showing biological accumulation in all aquatic organisms, especially fish and shell fish.

**meridional wind.** The wind or wind component along the local meridian, as distinguished from the zonal wind (along the local parallel of latitude).

**mesh number (grit number).** The designation of size of an abrasive grain, derived from the openings per linear inch in the control sieving screen.

**meson.** One of a class of medium-mass, short-lived elementary particles with a mass between that of the electron and that of the proton.

**mesothermal deposit.** A mineral deposit formed at moderate temperatures (between 175 and 300°C) and moderate pressures, in and along fissures or other openings in rocks, by deposition at intermediate depths (4,000 to 12,000 feet) chiefly from hydrothermal fluids derived from consolidating intruding rocks. A mesothermal deposit differs from a mineral deposit formed in the deep veins and from one formed at shallow depths in its mineral composition and in the character of the alteration of the wall rock accompanying its formation.

**Metallurgical coal.** Coal with strong or moderately strong coking properties that contains no more than 8.0 percent ash and 1.25 percent sulfur, as mined or after conventional cleaning.

**metamorphic grade.** The grade or rank of metamorphism depends upon the extent to which the metamorphic rock differs from the original rock from which it was derived. If a shale is converted to a slate or phyllite, the metamorphism is low grade; if it is converted to a mica schist containing garnet and sillimanite, the metamorphism is high grade.

**metamorphism.** Any process by which consolidated rocks are altered in composition, texture, or internal structure by conditions and forces not resulting simply from burial and the weight of the subsequently accumulated overburden. Pressure, heat, and the introduction of new chemical substances are the principal causes of metamorphism, and the resulting changes, which generally include the development of new minerals, are a thermodynamic response to a greatly altered environment.

**metashale.** Shale altered by incipient metamorphic reconstitution but not recrystallized and without the development of partings or preferred mineral orientation.

**meter bar.** A metal bar for mounting a gas meter, having fittings at the ends of the bar for connecting the inlet and outlet connections of the meter and to which, in turn, the gas service line and house piping are connected.

**meter density.** The number of meters per unit of area or per unit length of distribution main.

**meter index (meter register).** That part of a meter which indicates the volume of gas passed through the meter.

**metering pump.** A portable, high-precision pump used to aid studies of rock and ground pressure changes. Small enough to be carried, the hand-operated device preloads high-pressure hydraulic cells, that, embedded in rock or concrete, measure the variations in load or pressure that accompany nearby excavations. The pump also meters, without leakage, the fluid—usually mercury, glycerin, or oil—that must be added to or withdrawn from a cell to obtain a desired pressure.

**meter manifold.** Gas piping between gas service line and meter.

**meter prover.** A device for testing the accuracy of a gas meter. A quantity of air is collected over water or oil in a calibrated cylindrical bell and then passed through the meter by allowing the bell to sink into the water or oil. A comparison of the measured amount of air passing through the meter and the amount registered on the meter dial gives a measure of meter accuracy.

**meter seal.** A metal wire or tape seal attached to a gas meter or a service stop in such a way as to prevent its being opened by an unauthorized person.

**meter set.** Includes the meter, meter bar, and connected pipe and fittings.

**meter stop.** A shutoff valve located on the inlet side of the meter. It may be integral with the meter bar.

**meter swivel.** The fitting that connects to the inlet and the outlet of a small gas meter.

**methanation.** A process of converting carbon monoxide and carbon dioxide present in synthetic gas to methane, using hydrogen, steam, heat, and appropriate catalysts.

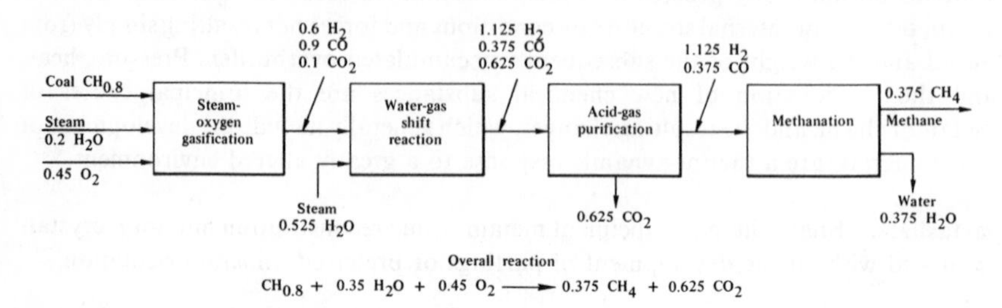


Production of pipeline gas by synthesis gas methanation. (*Courtesy Annual Review of Energy, Volume 1*)

**methane.** A colorless odorless flammable gaseous hydrocarbon that is a product of the decomposition of organic matter in marshes or mines or of the carbonization of coal. It is used as a fuel and as a raw material in chemical synthesis. Chemical formula $CH_4$. $CH_4$ is the prime constituent of natural gas. Methane can also be made by certain biomass conversion processes.

**methane indicator.** A portable instrument to determine the methane content in a given area.

**methanol.** A light volatile, flammable, poisonous, liquid alcohol, $CH_3OH$, formed in the destructive distillation of wood or made synthetically and used especially as a fuel, a solvent, an antifreeze, or a denaturant for ethyl alcohol, and in the synthesis of other chemicals. Methanol can be used as fuel for motor vehicles.

**methanol synthesis process.** A process in which synthesis gas from a high temperature (1316°C) entrained bed gasifier is passed through a shift converter where the hydrogen to carbon monoxide ratio is adjusted to 2:1, purified, and sent to a methanol converter. Crude methanol is distilled to remove water, higher alcohols, and other chemicals, to get chemical-grade methanol. A copper-zinc-chromium catalyst is used. Catalyst operating conditions are 260°C and 800 pounds per square inch. Low temperature fluid bed gasification is used for methane-methanol coproducts.

**methyl alcohol.** A poisonous liquid, and also known as methanol, which is the lowest member of the alcohol series. Also known as wood alcohol, since its principal source is the destructive distillation of wood. Chemical formula $CH_3OH$.

**metric ton.** A unit of mass and weight that equals 1000 kilograms or 2204.6 avoirdupois pounds.

**mho.** The practical unit of conductance equal to the reciprocal of the ohm.

**micrinoid.** A coal constituent similar to material derived from finely macerated vegetation.

**microgas survey.** A prospecting method which seeks to locate oil by the detection in soil samples of gases such as ethane, propane, and butane as evidence of leakage in the vicinity of oil pools. Methane is not significant as it is also formed by the decomposition of vegetable matter.

**microlog.** A resistivity log in borehole surveying obtained with a device consisting of closely spaced electrodes, the arrangement of which is basically the same but in miniature, as the normal and lateral devices in the regular electric survey. It is designed to measure the resistivity of a small volume of rock next to the borehole.

**microorganisms.** In geochemical prospecting, may be taken to include bacteria, algae, fungi, and any others of the relatively small forms of plant and animal life that inhabit soils and natural waters.

**microseism.** A very slight tremor or vibration of the earth's crust.

**microseismic movement.** Rather permanent, faint vibrations of the earth's crust, usually not exceeding 25 microns, caused by breakers on the coast or by storms far out at sea (up to 3000 kilometers from the coast). Synonym for microseism.

**microspectroscopy.** A method of identifying metallic constituents; it consists of drilling out the minute portion to be analyzed, flowing collodion over the resulting chips, and transferring the collodion together with the chips to a pure carbon electrode for analysis in a standard spectrographic arc.

**microsphere.** A small nuclear fuel particle that is coated with layers of graphite.

**microwave.** Electromagnetic radiation with wavelengths of a few centimeters. It falls between infrared and radio wavelengths on the electromagnetic spectrum.

**middle distillate.** One of the distillates obtained between kerosene and lubricating oil fractions in the refining processes. These distallates include light fuel oils and diesel fuel.

**migma.** A mush of partly fluid and partly solid rock material from which migmatite arises by consolidation. If the amount of its liquid portion becomes great enough, it will acquire mobility and may intrude into its surroundings in typical eruptive or intrusive fashion.

**migration.** The movement of oil, gas, or water through porous and permeable rock. Parallel (longitudinal) migration is movement parallel to the bedding plane. Transverse migration is movement across the bedding plane.

**migration area.** The sum of the slowing-down from fission to thermal energy and the diffusion area for thermal neutrons.

**migration of oil.** The movement of seepage of oil through rocks wherever they are sufficiently permeable to allow such passage.

**millidarcies.** A measure of permeability.

**milligram hours.** A measure of gamma-ray exposure expressed as the product of the equivalent radium content of the source in milligrams, and the time of exposure in hours.

**milling.** A process in the uranium fuel cycle where ore, which contains only 0.2 percent uranium oxide ($U_3O_8$), is converted into a compound called yellowcake, which contains 80 to 83 percent $U_3O_8$.

**million tons of coal equivalent.** A comparative unit of energy content widely used in the oil industry; equals 4.48 million barrels oil or 25.19 trillion cubic feet natural gas.

**mine.** An opening or excavation beneath the surface of the ground for the purpose of extracting minerals; a pit or excavation in the earth from which metallic ores or other mineral substances are taken by digging.

**mine acids.** Acids; usually a sulfuric acid formed by the action of water on sulfur left over from coal mining operations.

**mine bank.** An area of ore deposits that can be worked by excavation above the water level.

**mine cooling load.** The total amount of heat, sensible and latent, in British thermal units per hour, which must be removed by the air in the working areas.

**mine development.** Designates the operations involved in preparing a mine for ore extraction. These operations include tunneling, sinking, crosscutting, drifting, and raising.

**mine drainage.** Any water forming on or discharging from a mining operation.

**mined strata.** In mine subsidence, the strata lying vertically over the excavated area.

**mine dust.** Dust from rock drills, blasting, or handling rock. In the quantity inhaled by workers, dust may be classified as dangerous, harmless, and borderline, though the classification is purely arbitrary. Silica is a dangerous dust; bituminous coal dust is relatively harmless; and aluminum hydroxide is borderline.

**mine fan signal system.** A system which indicates by electric light, or electric audible system, or both, the slowing down or stopping of a mine ventilating fan.

**mine fires.** These very dangerous fires may arise from: spontaneous combustion, the ignition of timbers by gob fires, electric cable defects, or the heating and ignition of conveyor belts due to friction.

**mine heads.** In a mine ventilation system, the cumulative energy consumptions are called the mine heads. These heads are pressure differences determined in accordance with Bernoulli's principle.

**mine-mouth plant.** A stream-electric plant or coal gasification plant built close to a coal mine and usually associated with delivery of output via transmission lines or pipelines over long distances, as contrasted with plants located nearer load centers and at some distance from sources of fuel supply.

**mineral interests.** Mineral interests in land means all the minerals beneath the surface. Such interests are a part of the realty, and the estate in them is subject to the ordinary rules of law governing the title of real property.

**mineral tar.** A viscid variety of petroleum; tar derived from various bituminous minerals, such as coal, shale, and peat; shale tar.

**miner's inch.** The miner's inch of water does not represent a fixed and definite quantity, being measured generally by the arbitrary standard of the various ditch companies. Generally, however, it is accepted to mean the quantity of water that will escape from an aperture 1-inch square through a 2-inch plank, with a steady flow of water standing 6 inches above the top of the escape aperture, the quantity so discharged amounting to 2274 cubic feet in 24 hours.

**miner's weight.** The term used in a coal mining lease as the basis for the price per ton to be paid for mining. It is not a fixed, unvarying quantity of mine-run material, but is such a quantity of material as operators and miners may, from time to time, agree is necessary or sufficient to produce a ton of prepared coal.

**mine static head.** The energy consumed in the ventilation system to overcome all flow head losses. It includes all the decreases in total head (supplied from static head) which occur between the entrance and discharge of the system.

**mine velocity head.** The velocity head at the discharge of the system. Throughout the system, the velocity head changes with each change in duct area or number and is a function only of the velocity of air flow. It is not a head loss. The velocity head for the system must technically be counted a loss, because the kinetic energy of the air is discharged to the atmosphere and wasted. Therefore, it must be considered a loss to the system in determining overall energy loss.

**minimum critical mass.** For a specified fissile material, the minimum mass which can be made critical with no restriction as to geometrical arrangement, material composition, and moderating and reflecting media.

**minimum critical volume.** For a specified fissile material, the smallest volume of this material or of a mixture of this material and any other material that can be made critical with no restriction as to geometrical arrangement, material composition, and moderating and reflecting media.

**minimum energy dwelling (MED).** The minimum energy dwelling uses available energy-saving building techniques, advanced household appliances, and a solar/natural gas central energy system for space heating, cooling, and water heating. It is designed to save energy in building techniques and materials, site selection, and construction.

**mining.** The science, technique, and business of mineral discovery and exploitation. Strictly, the word connotes underground work directed to severance and treatment of ore or associated rock. Practically, it includes opencast work, quarrying, alluvial dredging, and combined operations including surface and underground attack and ore treatment.

**mining claim.** That portion of the public mineral lands which a miner, for mining purposes, takes and holds in accordance with mining laws.

**mining explosives.** High explosives used for mining and quarrying can be divided into four main classes: gelatins, semi-gelatins, nitroglycerin powders, and non-nitroglycerin explosives.

**mining hazards.** The dangers peculiar to the winning and working of coal and minerals including collapse of ground, explosion of released gas, inundation by water, spontaneous combustion, and inhalation of dust and poisonous gases.

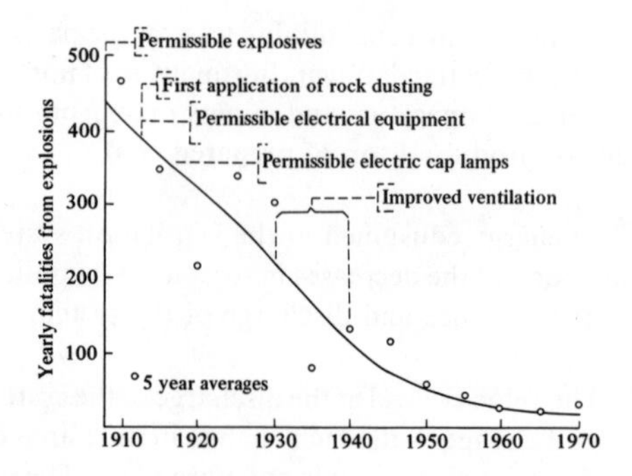


Fatalities from explosions in underground coal mines. (*Courtesy U.S. Council on Environmental Quality*)

**minor mine disaster.** Defined by the U.S. Bureau of Mines as any accident coming within one of the following categories: (a) a mine accident (not an explosion or fire) causing the death of less than five persons and considerable property damage; (b) a mine explosion or ignition causing injury to one or more persons or considerable property damage but no loss of life; (c) a mine explosion or ignition, resulting in the death of less than five persons; (d) a mine fire causing injury to one or more persons or considerable property damage but no loss of life; and (e) a mine fire resulting in the death of less than five persons.

**mirror.** A nuclear fusion device that confines the reaction within magnetic fields in such a way as to aid plasma confinement by particle reflection.

**misdirected solar radiation.** Concentrated and focused solar radiation resulting from the emergency or accidental defocusing of the heliostat field from the central receiver unit can be an invisible but serious danger. This misdirected radiation can cause fires, burns, serious glare problems, and possible eye damage. To protect against possible eye damage and glare problems, plant personnel could be required to wear protective goggles. Misdirected solar radiation from the heliostat field of central receiver solar thermal electric plants is perhaps the greatest potential safety hazard associated with solar thermal electric technology. Provisions for an "at rest" or "face down" position for the heliostats could alleviate other potential dangers resulting from misdirected solar radiation. Finally, both terrestrial and overhead "exclusion areas" could prevent potential receptors from being subjected to the hazards of this phenomenon.

**mist.** Liquid particles in air formed by condensation of vaporized liquids, and varying in size from 500 to 40 microns.

**mistral.** A strong, squally, cold, dry north wind that blows down the Rhone Valley south of Valence, France and into the Gulf of Lyons.

**mixed explosion.** One in which each ingredient, firedamp and coal dust, is presented below its lower limits, but in combination produces sufficient heat of combustion to propagate an explosion.

**mixed spectrum reactor.** A reactor having widely different neutron spectra in different parts of the core.

**MMBTU.** Million Btu's(British thermal units).

**mobile reactor.** A reactor that is mounted on or in a vehicle and is capable of operation while the vehicle is in motion.

**mobile source.** In air pollution, a moving source, such as an automobile.

**model.** A simplified description of a system, used as an aid to understanding the system. Models may be verbal or mathematical. Mathematical models are constructed from numerical values given to the components of the system and the relationships among components. If the model is sufficiently accurate to be approximately true, it can be used to predict the effect of changes within the system. Models may be constructed as initially-complex models, by starting with a replica and discarding non-essential details, or as initially-simple models, which begin with the most essential elements and add others as they are needed. Apart from mathematical models, models may be physical.

**moderator.** A substance used in a nuclear reactor in order to slow down or moderate fast neutrons. Such neutrons might otherwise escape from the reactor or be captured by unsuitable atoms. The moderator slows the neutrons down until they reach an appropriate speed for triggering further chain reactions.

**moderator control.** Reactor control by an adjustment of the properties, position, or quantity of moderator in such a way as to change the reactivity.

**modular integrated utility systems.** Small plants within housing developments or communities that provide all utility services.

**moil.** A tool for breaking and wedging out rock or coal.

**moisture bed.** The total moisture (percent) in a seam of coal before working.

**molded coal.** An artificial fuel made of charcoal refuse and coal tar, molded into cylinders, dried, and carbonized.

**mole.** A large diameter drill mounted on a movable framework and capable of tunneling holes of 5 to 30 feet in diameter.

**molecular weight.** The sum of the atomic weights of all the atoms in a molecule.

**molecule.** A group of atoms held together by chemical forces. The atoms in the molecule may be identical as in $H_2$, $S_2$, and $S_8$, or different as in $H_2O$ and $CO_2$. A molecule is the smallest unit of matter which can exist by itself and retain all its chemical properties.

**Molten Carbonate process.** A process in which coal and steam are fed into a molten bath of sodium carbonate which serves as a heat source and catalyst. The product gas can be upgraded to methane. Sulfur entering with the coal accumulates in the bath as sodium sulfide. Circulating melt carries char to a combustor where char is burned in oxygen or air which reheats the melt for the gasifier. A stream of melt is continuously withdrawn to purge the melt of ash and sulfur (in the form of hydrogen sulfide). Most of the sulfur in the coal is removed in this way. This two-vessel process is now being developed into one vessel to perform both gasification and combustion functions. The process gasifies coal at 999°C and burns char at 1038°C at a pressure of 420 pounds per square inch absolute.

**Molten-Iron process.** Using steam as a carrier, a mixture of coal and limestone is injected into a molten bath of iron. By injecting either air or oxygen into the molten bath, the carbon from the coal is oxidized to carbon monoxide. Injected steam dissociates to produce hydrogen and additional carbon monoxide. A low, medium, or high-Btu gas can be formed through this basic process. The limestone in the mixture absorbs sulfur and forms slag near the surface of the molten bath. The slag is removed through a slag port, is desulfurized, and the limestone is then recycled to the gasifier. Process conditions are 50 pounds per square inch gage and 1371°C.

**Molten Salt Breeder Reactor (MSBR).** A breeder reactor in which the fuel would be in the form of a molten salt of plutonium or uranium. It offers several technical advantages, but poses severe, unresolved engineering problems. The AEC's support for MSBR research terminated in June 1973.

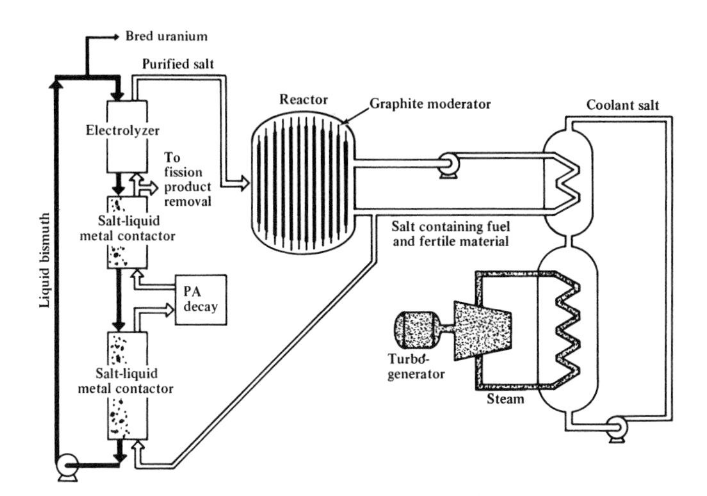


Flow diagram for a single-salt two-region molten salt breeder reactor. (*Courtesy Basic Nuclear Engineering*)

**Molten Salt process.** Air carries coal and sodium carbonate into a molten bath of sodium carbonate. Coal volatiles crack, producing a low-Btu off-gas. A stream of melt is continuously removed to purge the melt of sulfur and ash. With some modifications, a high-Btu gas could be produced via this basic reactor concept. Process conditions are 5 to 10 atmospheres and 316° to 371°C.

**momentum.** The measure of the motion of a body, determined by multiplying its mass by its velocity.

**monitor.** A gas sampler for continuous sampling of the atmosphere in a mine; an instrument that measures the level of ionizing radiation in an area; also a device which senses the presence of a flame, called a flame monitor.

**monitoring.** The taking of air samples to determine the amount of pollutants or radioactivity present in the environment.

**monochromator.** A device for producing monochromatic or narrow bands of radiant energy from the source.

**monsoon.** A name for seasonal winds. The primary cause is the annual variation of temperature over large land areas compared with neighboring ocean surfaces, causing an excess of pressure over the continents in winter and a deficit in summer; but other factors, such as the relief features of the land, have a considerable effect.

**mother lode.** The principal lode or vein passing through a district or particular section of country.

**motor gasoline production.** Total production of motor gasoline by refiners, measured at refinery outlet. Relatively small quantities of motor gasoline are produced at natural gas processing plants, but these quantities are not included.

**motor gasoline stocks.** Primary motor gasoline stocks held by gasoline producers. Stocks at natural gas processing plants are not included.

**motorized flue damper.** A device that reduces the amount of heat escaping up the chimney, by opening when the furnace is operating and closing when the furnace shuts off.

**mountain and valley winds.** A system of diurnal winds, prevailing in calm, clear weather, along the axis of a valley, blowing uphill and upvalley by day (anabatic wind), and downhill and downvalley by night (katabatic wind).

**mountain top removal.** A mining method wherein 100 percent of the overburden covering a coal seam is removed in order to recover 100 percent of the mineral. Excess spoil material is hauled to a nearby hollow to create a valley fill.

**moving annual total.** In the study of process costs (in large or in detail), a series of costs-per-unit observed and recorded at regular intervals (usually in monthly financial summaries cross-referenced to analyzed detail cost). Twelve months are covered and each month the new month's figures are added and those for the corresponding month of the previous year are removed. Therefore, like periods are always compared and seasonal fluctuations are smoothed out.

**muck.** Stone, dirt, and debris; also unconsolidated soils, sands, clays, and loams encountered in surface mining; generally, earth which can be severed and moved without preliminary blasting.

**mud.** Generally, any soil containing enough water to make it soft.

**mud analysis logging.** A continuous examination of the drilling fluid circulating in the well bore for the purpose of discovering evidence of oil or gas regardless of how small the quantities may be entrained in the fluid. When this service is utilized, a portable mud logging laboratory, which is incorporated in a trailer, is set up at the well. This method is widely used in drilling wildcat wells.

**mudcap.** A charge of dynamite, or other high explosive, fired in contact with the surface of a rock after being covered with a quantity of wet mud, wet earth, or sand—

no borehole being used. The confinement given the dynamite by the mud or other material permits part of the energy of the dynamite to be transmitted to the rock in the form of a blow.

**muddling off.** Commonly thought of as reduced productivity caused by the penetrating, sealing or plastering effect of a drilling fluid.

**mud drilling.** Drilling operations in which a mud-laden circulation fluid is used.

**mud flush.** To clear fragmented materials from a borehole by circulating a mud-laden fluid.

**mud pit.** A pit in which drilling mud is mixed, prepared, stored, or caught as it overflows from the drill-hole collar.

**mud swivel.** A modification of a water swivel designed for use when a mud-laden drill fluid is circulated in borehole-drilling operations.

**mud up.** The act or process of filling, choking, or clogging the waterways of a bit with consolidated drill cuttings. Also called sludging.

**mule.** A small car, or truck, attached to a rope and used to push cars up a slope or inclined path.

**mullock.** The accumulated waste or refuse rock about a mine.

**multigroup model.** A model which divides a neutron population into a finite number of neutron energy groups.

**multiple detectors.** Two or more seismic detectors whose combined output energy is fed into a single amplifier-recorder circuit. This technique is used to effect a cancellation of undesirable near-surface waves. Synonym for multiple geophones; multiple recording groups.

**Multiple Mineral Development act.** Law permitting oil and gas leases and mining claims on the same land, with the oil and gas lessee obtaining his rights under his lease and the mining claimant being accorded his rights to go patent, subject to the oil and gas lessee's interest.

**multiplication factor.** The ratio of the number of neutrons present in a reactor in any one neutron generation to that in the immediately preceding generation. Criticality is achieved when this ratio is equal to one.

**multiplying medium.** A medium in which a neutron-induced fission chain reaction can take place.

**multipurpose transmission line.** Employment of a transmission line for more than one function, such as regular transmission, wheeling, reserve capacity, and peak capacity usage.

**multishift working.** The working of two or three shifts per day on production, faces underground. Face machines and power supports for coal mines represent a heavy capital outlay and the aim is to make them productive as long as possible in the 24 hours.

**muon.** Contraction of mu-meson. An elementary particle, classed as a lepton (not as a meson), with 207 times the mass of an electron. It may have a single positive or negative charge.

**mush.** Soft and damp small coal; a coal which had been so crushed that it is unprofitable to mine.

**mutabilite.** A soft "corklike" bitumen of porous or resinous consistency. Partly soluble in organic solvents.

**mutation.** A permanent transmissible change in the characteristics of an offspring from those of its parents.

# N

**nacelle.** The housing portion of a wind electric conversion machine that contains the electric generating equipment.

**nadir.** The point of the celestial sphere that is vertically downward from the observer; the direct opposite of zenith; the lowest point.

**naked light.** Any light which is not so enclosed and protected as to preclude the ignition of an ambient firedamp-air mixture. Also called open light.

**naked light mine.** A nongassy coal mine where naked lights may be used by the miners. Such mines are exceptional and limited to small collieries operating near the outcrop of the seams.

**name plate rating.** A statement by the manufacturer of an energy system which gives the performance of the heating system under specific operating conditions; the full-load continuous rating of a generator, prime mover, or other electrical equipment under specified conditions as designated by the manufacturer.

**nano.** A prefix that divides a basic unit by one billion ($10^9$).

**Nansen bottle.** An oceanographic water sampling bottle made of a metal alloy which is little reactive with seawater, equipped with a rotary valve at each end so that when it is rotated at depth the valves close and lock shut, entrapping a water sample and setting the reversing thermometers.

**naphtha.** A volatile, colorless distillation product between gasoline and refined oil obtained from petroleum.

**naphtha gas.** Illuminating gas charged with the decomposed vapor of naptha.

**naphtha stripper.** A piece of equipment in which light hydrocarbon fractions are removed from naphtha for recover or sale.

**napthoid.** Liquid petroleumlike product found in cavities of igneous rocks and assumed to be a product of thermal distillation of bituminous substances contained in the rocks.

**narrow beam.** In beam attenuation measurements, a beam in which only the unscattered and small angle forward-scattered radiation reaches the detector.

**nascent state.** Just formed by a chemical reaction, and therefore very reactive. Nascent gases are probably in an atomic state.

**native.** Occurring in nature, either pure or uncombined with other substances; usually applied to metals, such as native mercury, native copper. Also used to describe any mineral occurring in nature in distinction from the corresponding substance formed artificially.

**native gas.** Gas originally in place in a particular underground structure as opposed to injected gas.

**National Environmental Policy Act of 1969 (NEPA).** The National Environmental Policy Act of 1969 (NEPA), implemented by Executive Order on March 5, 1970, and the guidelines of the Council on Environmental Quality of August 1, 1973, require that all agencies of the Federal government prepare detailed environmental statements on major Federal actions significantly affecting the quality of the human environment.

**National Environmental Research Parks (NERP).** National Environmental Research Parks are protected areas in which scientists can conduct long-term experiments to learn the impact of man's activities on the natural environment and on environments already altered by man such as abandoned farm fields and land that has been strip mined. Four Parks, totalling more than a million acres, have been established since 1972 when the first came into being. They are located on ERDA-owned land near Los Alamos, New Mexico; Richland, Washington; Idaho Falls, Idaho; and Aiken, South Carolina.

**National Uranium Resource Evaluation program (NURE).** DOE is carrying out a National Uranium Resource Evaluation program to better understand the long-range prospects for expanded domestic uranium supply for reactor development strategy and planning, and to assure adequate uranium supplies to fuel nuclear power growth.

**natural abundance.** Of a specified isotope of an element, the isotopic abundance in the element as found in nature.

**natural associated gas.** Natural gas existing in a free state in a reservoir containing oil, the gas being in contact with but not in solution in the oil of the reservoir.

**natural background radiation.** The amount of radiation present in the environment which is not the result of man's activities.

**natural circulation reactor.** A reactor in which the coolant (usually water) is made to circulate without pumping, but by natural convection resulting from the different densities of its cold and reactor-heated portions.

**natural draft.** Draft that is caused by a thermal upset in which temperature differences change the weight (pressure) of air.

**natural draining.** Any water course which has a clearly defined channel.

**natural gas.** Naturally occurring mixtures of hydrocarbon gases and vapors, found in porous geologic formations beneath the earth's surface, often in association with petroleum. The more important are methane, ethane, propane, butane, pentane, and hexane. The energy content of natural gas is usually taken as 1032 British thermal units per cubic foot.

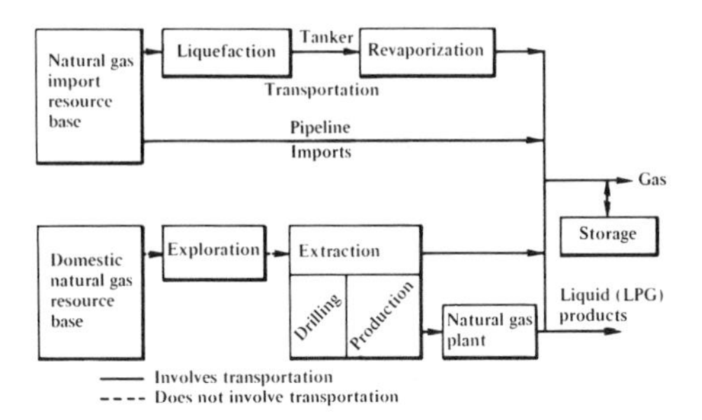


Natural gas resource development. (*Courtesy U.S. Council on Environmental Quality*)

**natural gas, associated-dissolved.** Gas occurring in the form of a gas cap associated with an oil zone, and in solution in the oil itself.

**natural gas design stress.** The estimated maximum tensile stress in the wall of the pipe in the circumferential orientation due to internal natural gas pressure that can be applied continuously with a high degree of certainty that failure of the pipe will not occur.

**natural gas distillate.** Material removed from natural gas at the "heavy end" portion; aliphatic compounds ranging from $C_4$ to $C_8$.

**natural gas fields.** An area containing one or more reservoirs of commercially valuable gas. (See Fig., p. 296)

**natural gas indicator.** Consists of a naphtha-burning safety lamp with a mirror attached to one side so that the action of the flame may be observed from above.

**natural gas liquids.** Liquid hydrocarbon mixtures which are gaseous at reservoir temperatures and pressures but are recoverable by condensation or absorption. Natural gasoline and liquefied petroleum gases fall in this category.

**natural gas marketed production.** Gross production of associated-dissolved gas from oil reservoirs plus gross production of gas from gas reservoirs less the sum of vented or flared gas, reinjected gas, and reduction in gaseous volume due to removal of light hydrocarbons, such as ethanes, propanes, butanes, pentanes and natural gasoline.

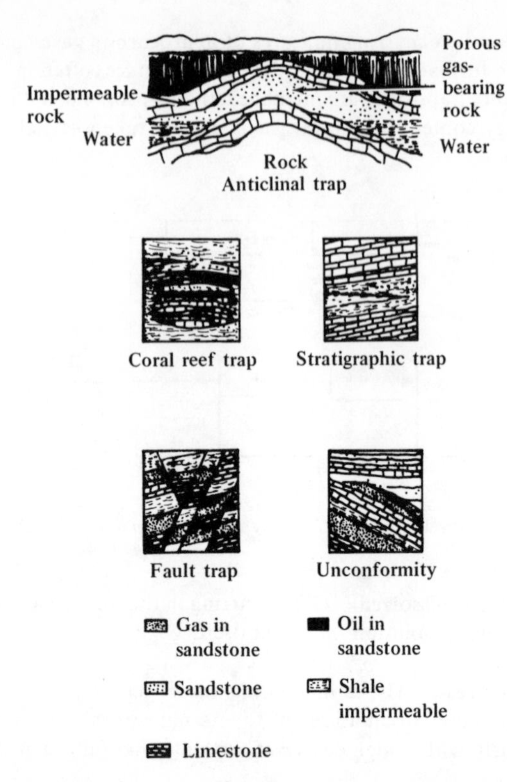


Types of natural gas reservoirs and entrapments. (*Courtesy Van Nostrand's Scientific Encyclopedia*)

**natural gasoline.** A mixture of liquid hydrocarbons extracted from natural gas and stabilized to obtain a liquid product suitable for blending with refinery gasoline.

**natural gas products.** Liquids (under atmospheric conditions), including natural gasoline, which are recovered by process of absorption, adsorption, compression, refrigeration, cycling, or a combination of such processes, from mixtures of hydrocarbon that existed in a reservoir.

**natural gas shrinkage.** The reduction in volume of wet natural gas due to the extraction of some of its constituents, such as hydrocarbon products, hydrogen sulphide, carbon dioxide, nitrogen, helium, and water vapor.

**natural pollutant.** A substance of natural origin that may be regarded as an environmental pollutant when present in excess (e.g. volcanic dust, sea salt particles, ozone formed photochemically or by lightning, products of forest fires).

**natural radioactivity.** Radioactivity of naturally occurring nuclides.

**natural uranium.** Uranium as found in nature, containing 0.7 percent of uranium-235, 99.3 percent of uranium-238 and a trace of uranium-234. It is also called normal uranium.

**Naval Petroleum Reserves.** Establishment of the public domain containing oil under the jurisdiction of the Department of the Navy. By law, petroleum cannot be taken from the Naval Petroleum Reserves other than for national defense purposes, and only with the expressed approval of the Congress. Oil-bearing lands identified as Naval Petroleum Reserves include: Elk Hills in California as Petroleum Reserve No. 1; Buena Vista Hills, adjacent to the southern boundary of Elk Hills, as Reserve No. 2; Teapot Dome in Wyoming as Reserve No. 3; Alaskan Prudhoe Bay Field as Reserve No. 4; Oil Shale Reserves Nos. 1, 2, and 3 (Nos. 1 and 3 in Colorado, No. 2 in Utah).

**naval reactors.** Improved naval nuclear propulsion plants and reactor cores developed in a wide range of power ratings for installation in naval vessels ranging in size from small submarines to large surface ships.

**necrosis.** Death of plant cells.

**needle coal.** Variety of lignite composed of a fibrous needlelike mass of vascular bundles of palm stems.

**needling.** To cut holes, notches, or ledges in a coal or rock surface to receive the ends of timber supports.

**negative gradient.** Descriptive of conditions in a layer where the temperature decreases with increasing depth.

**nephelometer.** An instrument which measures the scattering of light by determining the amount of light emitted at right angles to the original beam direction. Such devices are useful in studies of particles (size and amount) suspended in water.

**neptunium series.** The series of nuclides resulting from the radioactive decay of the man-made nuclide, neptunium-237. Many other man-made nuclides decay into this sequence. The end-product of the series is stable bismuth-209, which is the only nuclide in the series that occurs in nature.

**net calorific value.** Net heat of combustion. In the case of solid fuels and liquid fuels of low volatility, a lower value calculated from gross calorific values as the heat produced by combustion of unit quantity, at constant atmospheric pressure, under conditions such that all water in the products remains in the form of vapor.

**net electric system distribution.** The kilowatt-hours available for total electric system or company load. It is the sum of net generation by the system's own plants, purchased energy, and net interchange.

**net generating station capability.** The capability of a generating station as determined by test or operating experience less the power generated and used for auxiliaries and other station uses.

**net generation.** Total generation minus the kilowatt-hours consumed for station use.

**net plutonium.** The plutonium which is recovered after the irradiated fuel assemblies have been chemically processed.

**net radiometer.** An instrument for measuring all radiation components.

**net reserves.** The recoverable quantity of an energy resource that can be produced and delivered.

**network.** A system of transmission or distribution lines so cross-connected and operated as to permit multiple supply to any principal point on it.

**neutral flame.** A gas flame in which there is no excess of either fuel or oxygen.

**neutralization.** The process of adding an acid or alkali to a solution to make that solution neutral or inert.

**neutralization number.** A number indicating the acidity of an oil.

**neutral soil.** A soil in which the surface layer is neither acid nor alkaline in reaction; for most practical purposes, soil with a pH ranging from 6.6 through 7.3.

**neutrino.** An electrically neutral elementary particle with a negligible mass. It interacts very weakly with matter and hence is difficult to detect. It is produced in many nuclear reactions, for example, in beta decay, and has high penetrating power; neutrinos from the sun usually pass right through the earth.

**neutron.** An uncharged elementary particle with a mass that nearly equals that of the proton. The isolated neutron is unstable and decays with a half-life of about 13 minutes into an electron, a proton, and a neutron. Neutrons sustain the fission chain reaction in a nuclear reactor.

**neutron absorber.** A material with which neutrons interact significantly by reactions resulting in their disappearance as free particles; also an object with which neutrons interact significantly or predominantly by reactions resulting in their disappearance as free particles without production of other neutrons.

**neutron absorption.** A neutron interaction in which the incident neutron disappears as a free particle even when one or more neutrons are subsequently emitted accompanied by other particles.

**neutron activation analysis.** Activation analysis in which neutrons are the activating agent.

**neutron capture.** The process in which an atomic nucleus absorbs or captures a neutron. The probability that a given material will capture neutrons is measured by its neutron capture cross section, which depends on the energy of the neutrons and on the nature of the material.

**neutron chopper.** A device for periodically interrupting a beam of neutrons.

**neutron cross section.** A measure of the probability of a reaction of a neutron with a single atomic nucleus (microscopic cross-section $\sigma$); or with one of the nuclei in a unit volume of material (macroscopic cross-section $\Sigma$).

**neutron cycle.** The average energy, interaction, and migration history of neutrons in a reactor, beginning with fission and continuing until they have leaked out or have been absorbed.

**neutron density.** The number of neutrons per cubic centimeter.

**neutron diffusion.** A phenomenon in which neutrons in a medium tend, through a process of successive scattering collisions, to migrate from regions of high concentration to regions of low concentration.

**neutron economy.** The degree to which neutrons in as reactor are used for desired ends instead of being lost by leakage or nonproductive absorption. The ends may include propagating the chain reaction, converting fertile to fissionable material, producing isotopes, or research.

**neutron energy group.** One of a set of groups consisting of neutrons having energies within arbitrarily chosen intervals. Each group may be assigned effective values for the characteristics of the neutrons within the group.

**neutron flux.** A measure of the intensity of neutron radiation. It is the number of neutrons passing through one square centimeter of a given target in one second.

**neutron flux density.** The number of neutrons that enter a sphere of unit cross-sectional area per unit time; also called neutron flux.

**neutron log.** Identification of the fluid bearing zones of rocks, recorded by a nuclear device; also strip recording of the secondary radioactivity arising from the bombardment of the rocks around a borehole by neutrons from a source being caused to move through the borehole.

**neutron multiplication.** The production by a neutron of other neutrons in a medium containing fissionable material.

**neutron source.** An apparatus or a material emitting, or capable of emitting, neutrons.

**neutron yield per absorption.** The average number of primary fission neutrons (including delayed neutrons) emitted per neutron absorbed by a fissionable nuclide or by a nuclear fuel, as specified. It is a function of the energy of the absorbed neutrons.

**neutron yield per fission.** The average number of primary fission neutrons (including delayed neutrons) emitted per fission. It is a function of the energy of the absorbed neutrons.

**new gas.** Gas being made available for the first time by a contract of purchase and sale.

**new oil.** The volume of domestic crude petroleum produced from a property in a specific month which exceeds the base production control level for that property.

**Newton's law of gravitation.** Every particle in the universe attracts every other particle; the attraction between any two particles is proportional to their masses and inversely proportional to the square of the distance between them. Named after its discoverer Sir Isaac Newton (1642–1727), British physicist and mathematician.

**Newton's law of motion.** A body at rest remains at rest unless an external force acts on it; a body in motion continues to move uniformly and in a straight line unless an external force acts on it. An external force that acts on a body changes that body's momentum; the rate at which the momentum changes is proportional to the force; the change takes place in the direction of the force. To every action there is an equal and opposite reaction.

**nicking.** The cutting of a vertical groove in the seam to liberate the coal after it has been holed or undercut.

**nigritite.** A product of the coalification of nonfluid bitumens rich in carbon; insoluble or only slightly soluble in organic solvents.

**nip out.** The disappearance of a coal seam by the thickening of the adjoining strata which takes its place.

**nipple.** Tubular pipe fitting, usually threaded at both ends and under 12 inches in length.

**nitinol.** A nonmagnetic alloy having a wide range of use, including use in underwater demolition.

**nitinol heat engines.** A solar heat engine incorporating as its working fluid intermetallic nickel titanium compound, nitinol.

**nitric oxide.** A gas formed in great part from atmospheric nitrogen and oxygen when combustion takes place under high temperature and high pressure, as in internal combustion engines. Nitric oxide is not itself a pollutant; however, in the ambient air it converts to nitrogen dioxide, a major contributor to photochemical smog. Chemical formula NO.

**nitrogen.** An odorless, colorless, generally inert gas. It comprises 79 percent of earth's atmosphere in the free state. Chemical symbol N.

**nitrogen dioxide.** A compound produced by the oxidation of nitric oxide in the atmosphere; a major contributor to photochemical smog; also the most dangerous of mine gases produced by the incomplete detonation of some explosives. Chemical formula $NO_2$.

**nitrogen fixation.** The process of converting atmospheric (free) nitrogen to nitrogen compounds for use in either agricultural or industrial operations.

**nitrogenous wastes.** Wastes of animal or plant origin that contain a significant concentration of nitrogen.

**nitrogen oxides.** Compounds of nitrogen and oxygen which may be produced by the burning of fossil fuels. They are very harmful to health and a contributor to the formation of smogs. Chemical formula $NO^x$.

**nivenite.** A variety of uraninite high in uranium and containing 10 percent or more of the yttrium earths and 6.7 to 7.6 percent thoria.

**NLGI number.** The National Lubricating Grease Institute series of numbers (grades) classifying the consistency range of lubricating greases.

**noble gas.** The rare inert gases helium, neon, argon, krypton, xenon, and radon, which do not normally combine chemically with other elements.

**noble metals.** A metal with marked resistance to chemical reaction, particularly to oxidation and to solution by inorganic acids. The term is synonymous with precious metal.

**node.** A point, line, or surface in a standing wave system where some characteristic of the wave field has essentially zero amplitude. Nodes may be of several types, such as pressure or velocity.

**nodule.** Rounded masses of pyrite found deposited in coalbeds which range from a fraction of an inch to several feet in diameter.

**"no-flow" condition.** That condition obtained when the heat transfer fluid is not flowing through the collector array due to shut-down or malfunction and the collector is exposed to the amount of solar radiation that it would receive under normal operation conditions.

**nog.** Roof support formed of rectangular piles of logs; a block of wood wedged tightly into the cut in a coal seam after the coal cutter has passed and forming a temporary support for the coal and the roof above.

**noise equivalent level.** A scale which accounts for the acoustic energy received at a point from various noise events causing noise levels above some prescribed value.

**noise pollution.** Any undesired audible signal; any undesired sound.

**nominal capacity.** A figure expressed in tons per hour used in the title of the flowsheet and in general descriptions of the plant, applying to the plant as a whole and to the specific project under consideration. It may be taken as representing the approximate tonnage expected to be supplied to the plant during the hour of greatest load.

**nominal wall thickness.** Specified wall thickness of pipe without adding an allowance to compensate for the underthickness tolerances permitted in approved specifications.

**nonassociated natural gas.** Free natural gas neither in contact with nor dissolved in crude oil in the reservoir.

**nonbanded coal.** Coal that does not display a striated or banded appearance on the vertical face. It contains essentially no vitrain and consists of clarain or durain, or of material intermediate between the two.

**noncoincident demand.** The sum of two or more individual maximum demands, regardless of time of occurrence, within a specified period.

**noncoking.** A coal that does not form coke under normal conditions.

**noncoking coal.** A bituminous coal that burns freely without softening or displaying any appearance of incipient fusion. The percentage of volatile matter may be the same as for coking coal, but the residue is not a true coke.

**noncombustible.** Generally, a substance or gas that will not burn.

**nondestructive testing.** Method of examination that does not involve damaging or destroying the part being tested. Employed, for example, in order to detect cracks, erosion, or density changes in coal plant components, it includes such techniques as radiography, thermography, and ultrasonics.

**nonfirm gas.** Gas which is not required to be delivered or to be taken under the terms of a gas purchase contract.

**nonionic.** An electrically neutral particle.

**nonoperating interest.** A share in the mineral interest without operating burdens.

**Non-Proliferation Treaty (NPT).** Establishes the framework within which international safeguards operate to verify that nuclear materials are not diverted to nuclear weapons or other nuclear explosive devices.

**nonselective mining.** A low-cost method of mining combined with large-scale operations; in general, nonselective mining is applicable where the valuable sections of the deposits are numerous, irregular in occurrence, and separated by thin lenses of waste.

**nontabular deposits.** Mineral deposits of irregular shape.

**nontidal currents.** Permanent currents in the general circulatory system of the sea as well as temporary currents arising from meteorologic conditions such as winds, density and water temperature.

**nonutility generation.** Generation of electric power required in the conduct of the industrial and commercial operations of generating plants.

**nonweathering coal.** Coal having a weathering index of less than 5 percent, as defined by the U.S. Bureau of Mines standards.

**normal air.** A mixture of dry air and water vapor, varying from 0.1 percent to 3 percent by volume (usually over 1 percent in mines).

**normal gradient.** There are extensive areas of useful, low temperature geothermal heat at some depth under all regions of the country. Economical recovery and use of this low quality energy hinges on very significant improvements in drilling technology and energy extraction methods. Recovery of thermal energy from the normal gradient is a long-term project which could contribute significantly to United States energy supplies.

**normal gradient energy source.** Geothermal energy source; areas within the first 6 miles of the earth's crust which are heated (temperatures ranging from 16° C to 299° C) by a variety of sources.

**normal pressure.** Usually equal to the weight of a column of mercury 760 millimeters in height.

**normal radiation.** The component of solar radiation which is perpendicular to the absorbing surface.

**normal recovery capacity.** Amount of water in U.S. gallons raised 38° C per hour or per minute when calculated on a thermal efficiency of 70 percent, representing the water heated by a gas input of 1190 British thermal units per gallon.

**normal solution.** A solution made by dissolving one gram-equivalent weight of a substance in sufficient distilled water to make one liter of solution.

**normal temperature.** Normal temperature and pressure are taken as 0° C (273° absolute) and 30 inches (760 millimeters) of mercury pressure. Also called standard temperature.

**normal test pressures.** Those pressures specified for testing purposes at which adjustment of burner ratings and primary adjustments are made.

**normal uranium.** Natural uranium.

**norm system.** A system of classification and nomenclature for igneous rocks based on the norm of each rock. Only decomposed rocks of which accurate chemical analyses are available are classifiable.

**notching.** A method of excavating in a series of steps.

**Nottingham system.** A longwall method of working coal seams in which the trams run on a rail track along the face and are handloaded at the sides. The system can only be adopted in relatively thick seams where the trams can travel along the face without any roof ripping. The method is now largely replaced by face conveyors.

**noxious.** Causing or tending to cause injury, especially to health; for example, noxious gases.

**NTP.** Normal temperature and pressure; a temperature of 0° C and pressure of 760 millimeters of mercury.

**nuclear battery.** A radioisotopic generator; a device in which power is continuously derived from the emission of charged particles from a radionuclide. A potential is achieved by collection of the charged particles on a suitable grid or plate.

**nuclear chemistry.** The chemical aspects of nuclear science and the applications of nuclear science to chemistry.

**nuclear device.** A nuclear explosive used for peaceful purposes, tests, or experiments. The term is used to distinguish these explosives from nuclear weapons, which are packaged units ready for transportation or use by military forces.

**nuclear disintegration.** Transformation of a nucleus, possibly a compound nucleus, involving a splitting into more nuclei or the emission of particles.

**nuclear emulsion.** An ionization-sensitive material such as a photographic emulsion used for permanently recording the tracks of charged particles.

**nuclear energy.** The energy liberated by a nuclear reaction (fission or fusion) or by radioactive decay. When released in sufficient and controlled quantity, this heat energy may be converted to electrical energy.

**nuclear explosive.** An explosive based on fission or fusion of atomic nuclei.

**nuclear fission.** A nuclear reaction in which a heavy and unstable atomic nucleus (i.e., the nucleus of any element with atomic number 84 or greater) splits into approximately equal parts, emitting neutrons, radiation, and heat energy. The neutrons may trigger further fission and so set up a chain reaction.

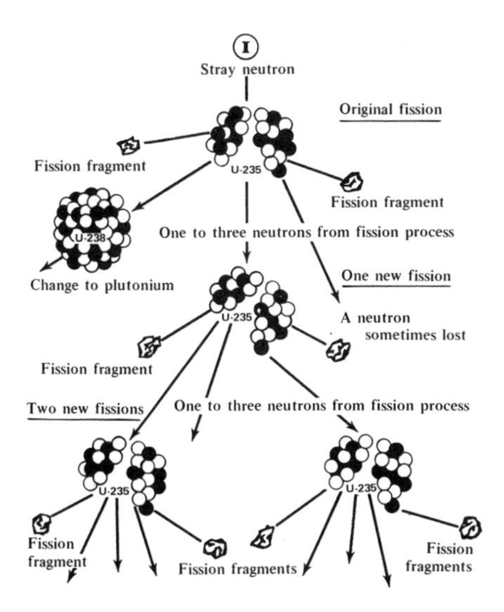


Nuclear fission chain reaction. (*Courtesy U.S. ERDA—69*)

**nuclear fuel.** Material containing fissionable materials of such composition that when placed in a nuclear reactor will support a self-sustaining fission chain reaction.

**nuclear fuel assemblies.** The fabricated form of fissionable material and other materials used in reactors to facilitate assembly of the fissionable material into the geometric lattice required to sustain a controlled fission process.

**nuclear fuel costs.** Nuclear fuel costs, given on a fuel cycle basis, accounts for the cost of each step in the fuel cycle, including interest charges.

**nuclear fuel cycle.** All the steps involved in the production, processing, use, reprocessing, and disposal of nuclear fuels. (See Fig., p. 306)

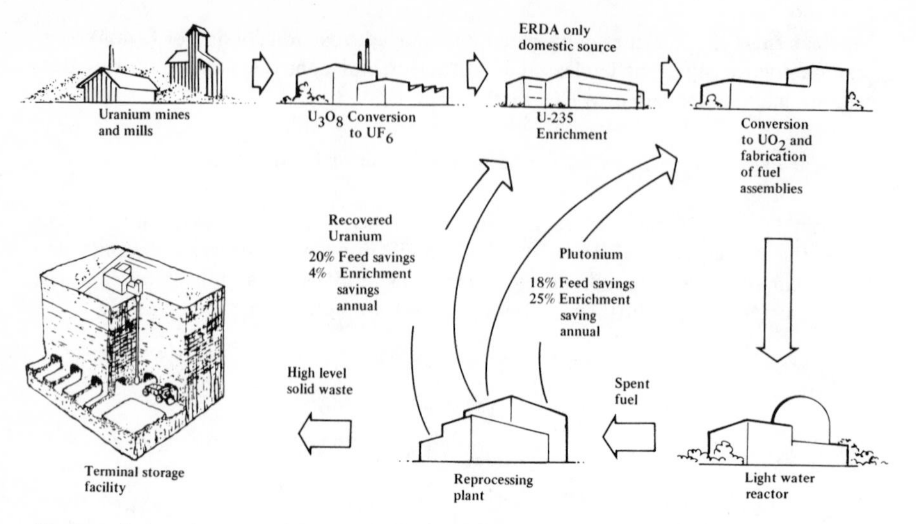


Nuclear fuel cycle. (*Courtesy U.S. Energy Research and Development Administration*)

**nuclear fuel reprocessing.** The processing of used or spent reactor fuel to recover the unused fissionable material and other valuable isotopes and place them in usable form.

**nuclear fusion.** A nuclear reaction in which two light atomic nuclei fuse, or combine, to form one heavier nucleus, emitting particles, radiation, and heat energy.

**nuclear fusion reaction.** Reaction between two light nuclei resulting in the production of at least one nuclear species heavier than either initial nucleus, together with excess energy.

**Nuclear Gasification process.** A process in which coal is slurried and dissolved in the presence of hydrogen. Liquid coal is further hydrogenated in subsequent steps, resulting in a desulfurized high-Btu gas. A portion of the product gas is cycled to a steam-methane reformer—a heat exchanger in the primary coolant loop of a high temperature gas-cooled reactor (HTGR). Endothermic steam-methane reforming produces hydrogen-rich gas and carbon dioxide. The hydrogen is then separated and cycled back to the coal hydrogenation part of the process. Liquid fuel products can also be formed by this fundamental process. Reformer pressures can be in excess of 700 pounds per square inch and temperatures can be 649° to 871°C.

**nuclear generating station.** An electric generating station in which the prime mover is a steam turbine, and the steam is generated in a reactor by heat from the fissioning of nuclear fuel.

**nuclear isobars.** Nuclides having the same mass number but different atomic numbers.

**nuclear isomers.** Nuclides having the same mass number and atomic number, but occupying different nuclear energy states.

**Nuclear Materials Management and Safeguards System (NMMSS).** NMMSS is a DOE-centralized, automatic data-processing system which provides periodic reports for management purposes to Government organizations in connection with nuclear materials inventory and financial management programs, nuclear material contract administration activities, and safeguards and security activities. NMMSS also provides nuclear material data to the Nuclear Regulatory Commission (NRC) for safeguards purposes related to licensee programs.

**nuclear poison.** A substance which can reduce reactivity.

**nuclear power.** Power released in exothermic nuclear reactions, which can be converted to mechanical or electrical power.

**nuclear powered generating capacity.** The rated electrical output of a turbine-generator utilizing a nuclear reactor as the heat-energy source for producing the steam which drives the turbine.

**nuclear power plant.** Any device, machine, or assembly that converts nuclear energy into some form of useful power, such as mechanical or electrical power. In a nuclear electric power plant, heat produced by a reactor is generally used to make steam to drive a turbine that in turn drives an electric generator.

**nuclear reaction.** A reaction involving a change in an atomic nucleus, such as fission, fusion, neutron capture, or radioactive decay, as distinct from a chemical reaction, which is limited to changes in the electron structure surrounding the nucleus.

**nuclear reaction energy.** The kinetic energy of the resultant particles and photons less the kinetic energy of the primary particles and photons for a nuclear reaction.

**nuclear reactor.** A device in which a fission chain reaction can be initiated, maintained, and controlled. Its essential component is a core with fissionable fuel. It usually has a moderator, a reflector, shielding, coolant, and control mechanisms. Sometimes called an atomic "furnace," it is the basic machine of nuclear energy. Also called reactor or pile. (See Fig., p. 308)

**nuclear rocket.** A rocket powered by an engine that obtains energy for heating a propellant fluid (such as hydrogen) from a nuclear reactor, rather than from chemical combustion.

**nuclear safeguards.** Include safeguard measures used at facilities handling nuclear materials, rigid standards for processing or storage of material, sophisticated systems for protecting material in transport or against acts of terrorists, and international safeguards. DOE provides IAEA with complete access to all safeguards research and

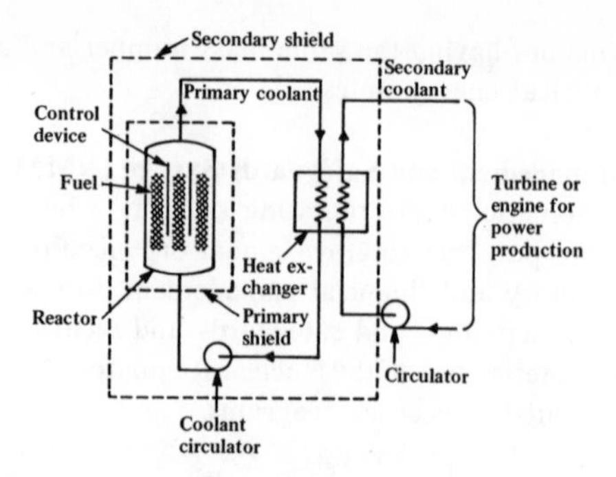


Diagram of a nuclear reactor power plant. (*Courtesy Basic Nuclear Engineering*)

development programs carried out in the United States. DOE also carries out detailed technical exchanges with other countries developing nuclear power to provide a basis for comparable protective systems.

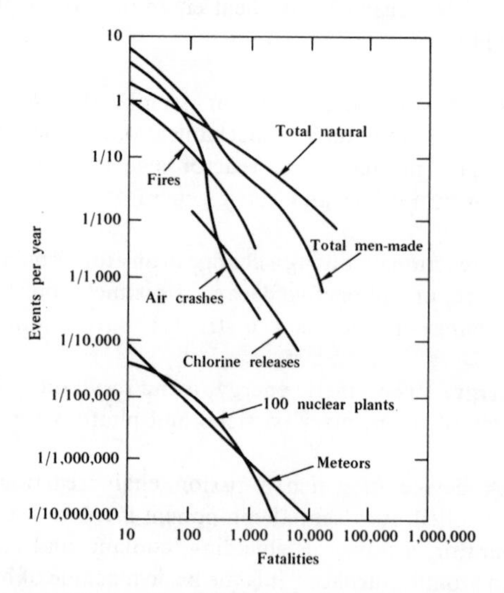


Estimated frequency of fatalities from various accidents. (*Courtesy Basic Nuclear Engineering*)

**nuclear superheating.** Superheating the steam produced in a reactor by using additional heat from a reactor. Two methods are commonly employed: recirculating the steam through the same core in which it is first produced (integral superheating) or passing the steam through a second and separate reactor. (See Fig., p. 309)

**nuclear trilinear chart.** A trilinear coordinated grid on which the nuclides have been plotted according to their mass, atomic, and neutron numbers.

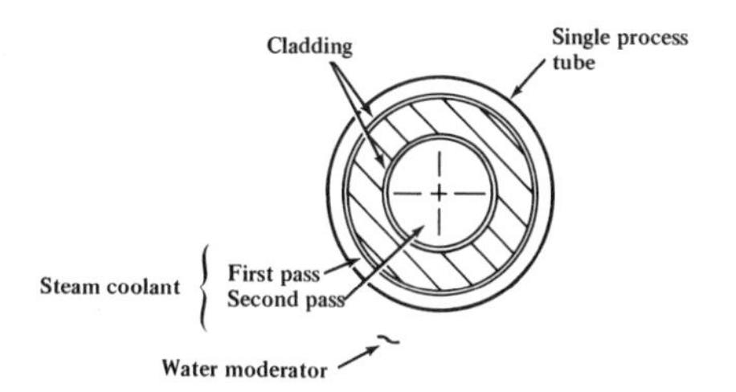


Fuel element design for use in a superheating reactor. (*Courtesy Basic Nuclear Engineering*)

**nuclear weapons.** A collective term for atomic bombs and hydrogen bombs; any weapons based on a nuclear explosive.

**nucleon.** A constituent of an atomic nucleus, that is, a proton or a neutron.

**nucleonics.** The science and technology of nuclear energy and its applications.

**nucleus.** The small, positively charged core of an atom. It is only about 1/10,000 the diameter of the atom but contains nearly all the atom's mass. All nuclei contain both protons and neutrons, except the nucleus of ordinary hydrogen, which consists of a single proton.

**nuclide.** A general term applicable to all atomic forms of the elements. The term is often erroneously used as a synonym for "isotope," which properly has a more limited definition. Whereas isotopes are the various forms of a single element (hence are a family of nuclides) and all have the same atomic number and number of protons, nuclides comprise all the isotopic forms of all the elements. Nuclides are distinguished by their atomic number, atomic mass, and energy state.

**nut coal.** A commercial term for sized coal (irrespective of size).

**nutrients.** Elements or compounds, such as carbon, oxygen, nitrogen, and phosphorous, which are essential as raw materials for organism growth and development.

**nutty slack.** Mixture of small coals, sized from two inches downward and probably of high ash content.

# O

**oakum.** A caulking filler used to fill extra wide cracks or as a backup for elastomeric caulking.

**OAPEC.** Acronym for Organization of Arab Petroleum Exporting Countries. It was founded in 1968 for cooperation in economic and petroleum affairs.

**observation well.** A well completed in any horizon which is used to obtain information relating to storage operations.

**observed-life table.** A table of plant experience relating survivors exposed to retirement at the beginning of each age interval to the actual retirements during each interval. The table may reflect all past experience or only a selected bank of years.

**occluded gases.** Gases which enter the mine atmosphere from pores, as feeders and blowers, and also from blasting operations. These gases pollute the mine air chiefly by the absorption of oxygen by the coal, and in addition by chemical combination of oxygen with carbonaceous matter. These gases include oxygen, nitrogen, carbon dioxide, and methane.

**ocean basin.** That part of the floor of the ocean that is more than about 600 feet below sea level.

**oceanic stratosphere.** The cold, deep layers of the ocean consisting of waters of polar or subpolar origin.

**oceanography.** The broad field of science pertaining to the sea, which includes the studies of the boundaries of the ocean, its bottom topography, the physics and chemistry of sea water, the characteristics of its motion, and marine biology.

**ocean temperature.** The mean surface temperature of both the Pacific and Atlantic Oceans is 17°C; that of the Indian Ocean is 18°C. Maximum temperatures are respectively, 32°C, 30°C, and 35°C.

**ocean thermal.** Providing power by harnessing the temperature differences between the surface waters and the ocean depths.

**ocean thermal conversion.** Process of using the warm ocean water (ocean's sunheated top layer) to vaporize a working fluid such as ammonia to run a turbine to produce electricity.

**Ocean Thermal Energy Conversion (OTEC) power plant.** The basic operating principles of an OTEC power plan have remained unchanged since d'Arsonval proposed them almost 100 years ago. The ocean is, for all practical purposes, an infinite heat source, converting and storing the incident solar energy from the sun in the form of warm surface water. The warm water is pumped through an evaporator containing a working fluid in a closed Rankine-cycle system. The vaporized working fluid drives a gas turbine which provides the plant's power. Having passed through the turbine, the vapor is condensed by colder water drawn up from deep in the ocean and then pumped back into the evaporator for reuse in the same cycle. No "fuel" of any kind is used; the enclosed working fluid simply is evaporated and condensed over and over by the warm surface and colder deep ocean water. Although OTEC operating principles are simple and well known, both the closed and open cycle systems pose complex engineering and cost problems. In both cases, the small temperature differentials (approximately 4°C vs. 538°C in coal-fired boilers) dictate that large quantities of water must be pumped, the pumping power being subtracted from the net power of the system. (See Fig. p. 312)

**ocean thermal gradients.** The temperature differences between deep and surface water.

**ocean tidal power.** Energy source producing electricity.

**octane.** A rating scale used to grade gasoline as to its antiknock properties; also any of several isomeric liquid paraffin hydrocarbons, $C_8H_{18}$; specially, normal octane, a colorless liquid boiling at 124.6°C found in petroleum.

| Octane rating systems | | | |
|---|---|---|---|
| Type of gasoline | Research method (traditional) | Averaged research and motor (new) | Gasoline symbol no. |
| Economy | 91 | 87 | 2 |
| Regular | 94 | 89 | 3 |
| Mid-premium | 96 | 91.5 | 4 |
| Premium | 100 | 95 | 5 |

Source: State of Maryland Comptroller's Office

Octane rating systems. (*Courtesy Help: The Useful Almanac 1976–1977*)

**octane number.** A number indicating the antiknock value of motor fuel.

**odorant.** Any material added to natural or LP gas in small concentrations to impart a distinctive odor. Odorants in common use include various mercaptans, organic sulfides, and blends of these.

**Office of Coal Research (OCR).** A bureau of the Department of the Interior established in 1960 to develop new and more efficient methods of mining, preparing, and utilizing coal. Now part of the Department of Energy.

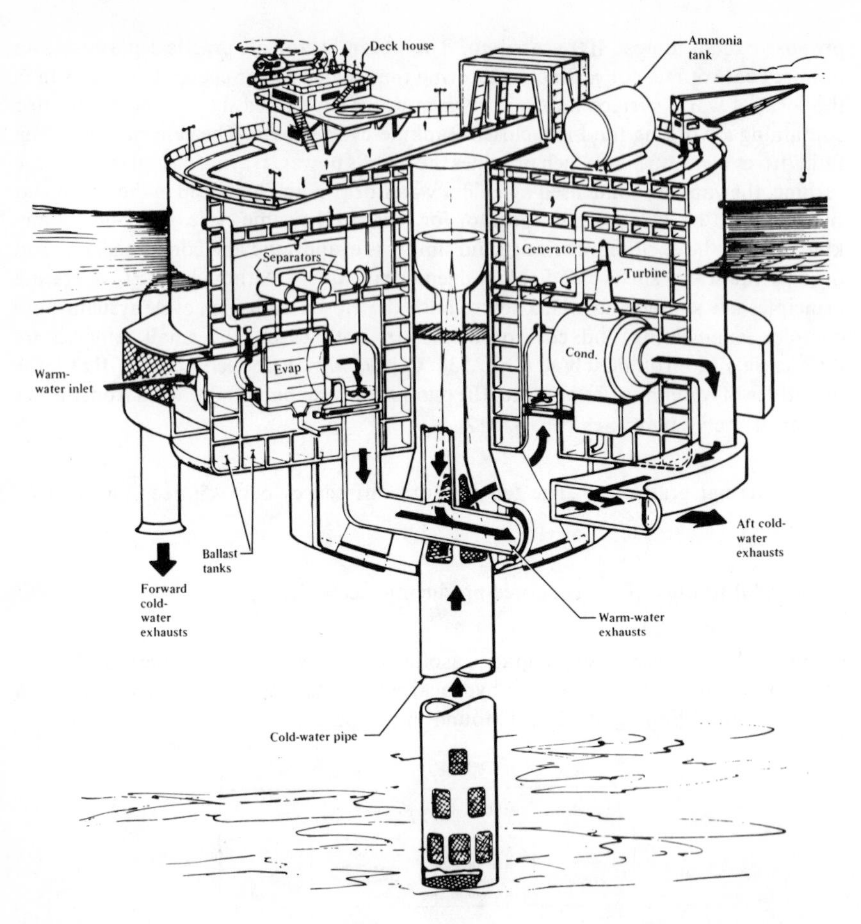


Ocean Thermal Energy Conversion (OTEC) power plant. (*Courtesy Popular Science, August 1975, p. 8*)

**off-peak service.** The period during a day, week, month, or year when the load being delivered by a gas system is not at or near the maximum volume delivered by that system for the corresponding period of time.

**offset line.** In surveying, a line established parallel to the main survey line, and usually not far from it.

**offset well.** A well drilled in the next location to another well according to the spacing rules of the state.

**Off Shore Windpower System (OWPS).** A system for generating electricity by wind turbines mounted on off-shore platforms.

**offshore winds.** Land breeze; winds blowing seaward from the coast.

**ohm.** The practical meter-kilogram-second (SI) unit of measurement of electrical resistance that equals the resistance of a circuit in which a potential difference of 1 volt produces a current of 1 ampere.

**oil-base mud.** A mud-laden drill-circulation medium in which an oil instead of water is used as the laden liquid.

**oil-bearing shale.** Shale impregnated with petroleum. Not to be confused with oil shale.

**oil burned for fuel.** Includes fuel oil, crude oil, diesel oil, and small amounts of tar and gasoline.

**oil burner.** A device in which oil fuel is vaporized and mixed with air for combustion.

**oil costs.** Average cost per barrel (42 gallons); dollars per barrel; includes fuel oil, crude and diesel oil, and small amounts of tar and gasoline.

**oildag.** Collodial dispersion of graphite in oil.

**oil derrick.** A towerlike frame used in boring oil wells to support and operate the various tools.

**oilfield.** An area underlain by one or more oil (or gas) pools which are adjacent or overlapping in plan view, and which are related to a single geological feature, either structural or stratigraphic. The size of an oilfield refers to the volume of recoverable oil (the "reserves") given either in barrels, or tons (depending on the gravity of the oil, 1 ton = 6.5 to 7.5 barrels).

**oilfield rotary.** The type and size of drilling machines used to rotary-drill boreholes in search of petroleum.

**oil fogging.** Spraying a fine mist of oil into a gas stream of a distribution system to avoid the drying effects of gas in certain distribution and utilization equipment.

**oil fuel costs.** Average cost per barrel (42 gallons).

**oil gage.** An instrument of the hydrometer type arranged for testing the specific gravity of oils.

**oil gas.** A gas of high calorific value obtained by the destructive distillation of high-boiling mineral oils and consisting chiefly of methane, ethylene, acetylene, benzene, and higher homologues.

**oil gas tar.** A tar produced by cracking oil vapors in the manufacture of oil gas.

**oil gravity.** The density of oil compared to the density of water as measured in degrees API (American Petroleum Institute). Oil with a low number is less valuable than oil with a high number. The high number indicates a high density.

**oil lease.** The permission granted by a landowner to a company to prospect and drill for oil and gas under his land and to produce them when found.

**oilless bearing.** Sleeve bearings of porous metal which depend solely on the porosity of the metal for oil storage.

**oil pocket.** A cavity in the rocks containing oil.

**oil pool.** An accumulation of oil in sedimentary rock that yields petroleum on drilling. The oil occurs in the pores of the rock and is not a pool or pond in the ordinary sense of these words.

**oil refinery.** A plant where petroleum is distilled and otherwise refined.

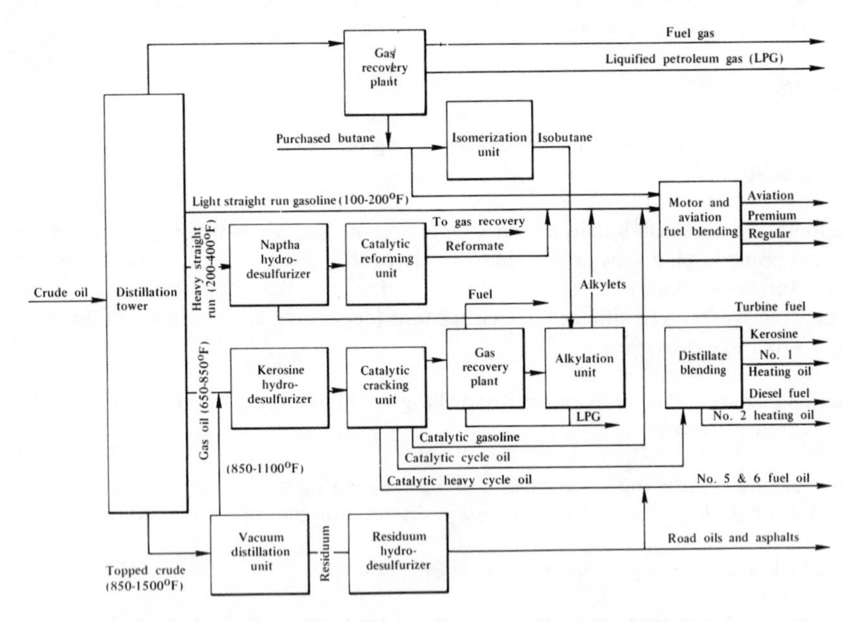


Oil refinery. *(Courtesy U.S. Council on Environmental Quality)*

**oil reforming.** Step in producing carbureted water gas in which a fraction of the carburetion oil is cracked to useful gas in the water-gas carburetor and superheater.

**oils.** A group of neutral liquids comprising three main classes: (1) fixed (fatty) oils from animal, vegetable, and marine sources, consisting chiefly of glycerides and esters of fatty acids; (2) mineral oils, derived from petroleum, coal, shale, etc., consisting of hydrocarbons; and (3) essential oils, volatile products, mainly hydrocarbons with characteristic odors, derived from certain plants.

**oil sands.** Sands which have been bonded with oil, for example, linseed oil. Such sands are particularly suitable for the production of large cores where high strength and considerable permeability are required.

**oil scrubbing.** The removal of certain impurities from manufactured or natural gas by passing the gas through an oil spray or bubbling the gas through an oil bath.

**oil seepage.** The slow leakage of petroleum oil from its underground accumulation.

**oil shale.** A very fine-grained sedimentary rock that contains enough organic matter (hydrocarbon) to yield 10 gallons or more oil per ton when properly processed. Some shales yield much more oil, and in the United States some thin layers of shale have been reported to yield 140 gallons of oil per ton. Most shales that are of commercial interest yield from 25 to 65 gallons of oil per ton. Some foreign deposits that contain shales yielding between 10 and 25 gallons per ton have been mined on a large scale. Many other organic-rich shales yeild less oil (1 to 10 gallons per ton), but these shales are so low grade that they usually are not called "oil shale."

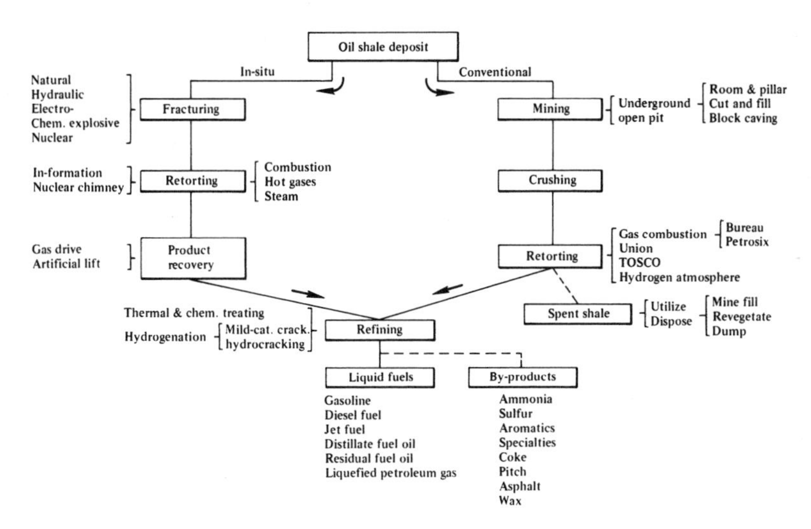


Processes for utilization of oil shade. (*Courtesy Annual Review of Energies, Volume 1*)

**oil shale lands.** Lands on or under which oil shale is present.

**oil shale retorting (in situ).** The process of extracting oil from shale by injecting intense heat into a deep well and then forcing it through the shale. (See Fig. p. 316)

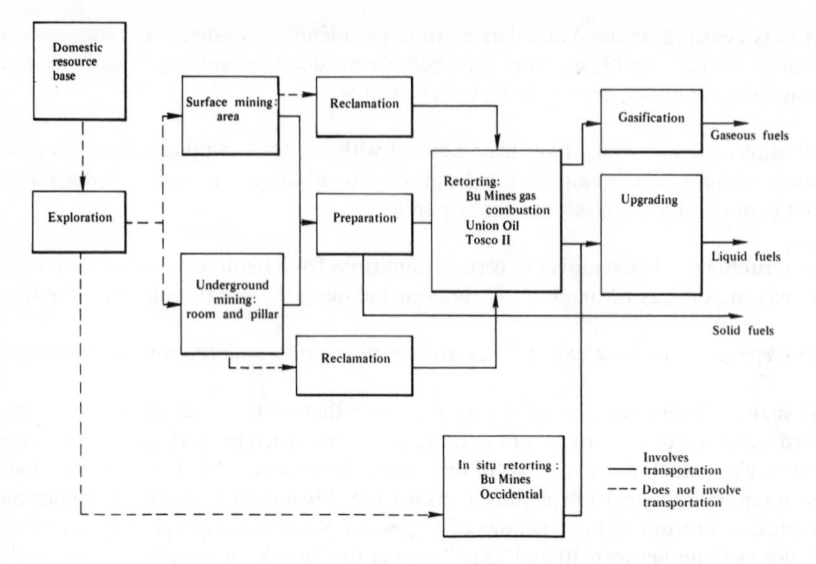


Oil shale resource development. (*Courtesy U.S. Council on Environmental Quality*)

**oil slick.** Oil, discharged naturally or by accident or design, floating on the surface of water as a discrete mass carried by wind, currents and tides.

**oil trap.** A geologist's term for a place where oil collects underground.

**oil well.** A dug or bored well, from which petroleum is obtained by pumping or by natural flow.

**oil-well casing.** Ordinary outside-coupled pipe used as borehole casing or drivepipe. Also called oilfield casing. (See Fig. p. 317)

**oil-well cement.** A hydraulic cement which sets at a slower rate than portland cement.

**oil zone.** A formation that contains capillary or supercapillary voids, or both, that are full of petroleum and will move under ordinary hydrostatic pressure.

**old oil.** Same as controlled crude oil.

**olefinic hydrocarbon.** A class of unsaturated hydrocarbons containing one or more double bonds and having the general chemical formula $C_nH_{2n}$.

**oligotrophic lakes.** Deep lakes that have a low supply of nutrients and thus contain little organic matter.

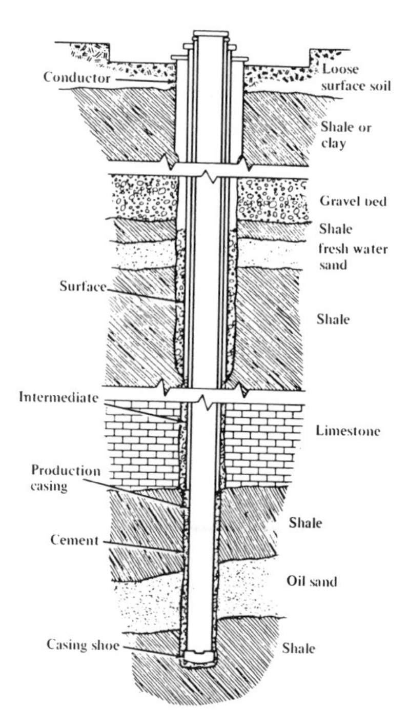


Oil well casing. (*Courtesy U.S. Council on Environmental Quality*)

**once-through cooling.** Cooling of an electric generating or industrial plant by pumping cold water from a river or lake through the condenser and then discharging the heated water back into the source.

**one-group model.** A model in which the neutron population is assigned a single effective energy.

**one-group theory.** A theory of neutron behavior in which all neutrons are assumed to belong to a single energy group. In this theory all neutrons born in fission are considered to be injected into this group, all absorptions are removals from the group, and reaction probabilities are taken as identical for all neutrons in the system.

**one mile population density index.** A number, roughly proportional to population density, applicable to a specific one-mile length of pipeline or main, and used in some cases to determine design and/or test requirements.

**on-orbit power generation.** Large (5 and 10 gigawatt electric ground output) orbital plants using solar cells for energy conversion have been proposed. Transmission of the generated power to earth is accomplished by microwave beam. The solar cell arrays are oriented by an attitude control system so as to face the sun, while the transmitting antenna constantly faces the ground receiving station. Consequently, the antenna must be mounted on gimbal joints. The frame structure joining the two large array areas is nonmetallic to allow passage of the microwave beam. To reduce weight and cost of cell arrays, plane mirrors would be arranged to effect concentration of incident sunlight. The concentration ratio would be limited to 2 in order to limit cell heating and the consequent reduction in cell efficiency.

**on-peak energy.** Energy supplied during periods of high system demands.

**onshore winds.** Sea breeze; winds blowing shoreward from the sea.

**on-site generation.** Generation of any electrical energy on a customer's property, with or without utilization of recoverable heat.

**opacity.** The property of reflecting light. For example, a window has zero opacity; a wall is 100 percent opaque.

**open burning.** Uncontrolled burning of wastes in an open area.

**opencast mining.** A coal mining method in which the overburden is removed, coal is extracted, and the overburden is replaced.

**opencut.** A method of excavation in which the working area is kept exposed.

**open-cycle reactor system.** A reactor system in which the coolant passes through the reactor core only once and is then discarded.

**open-flow test.** A test made to determine the volume of gas that will flow from a well in a given time when flowing unrestricted and open to the atmosphere. This is usually calculated from pressure tests of restricted flow.

**open hole.** Coal or other mine workings at the surface or outcrop. Also called opencast, opencut, open pit.

**open pit mining.** A form of operation designed to extract minerals that lie near the surface; a type of mining in which the overburden is removed from the product being mined and is dumped back after mining; may specifically refer to an area from which the overburden has been removed and has not been filled.

**open pressure.** The pressure on a gas well that has been open long enough for the pressure to stabilize.

**operating range.** The range of reactor power within which a reactor is designed to operate in steady-state condition.

**operative temperature.** That temperature of an imaginary environment in which, with equal wall (enclosing areas) and ambient air temperatures and some standard rate of air motion, the human body would lose the same amount of heat by radiation and convection as it would in some actual environment at unequal wall and air temperatures and for some other rate of air motion.

**Oppenheimer-Phillips process.** A nuclear reaction in which a low-energy deuteron approaches sufficiently close to a nucleus for that nucleus to strip the neutron from the deuteron but still repel the remaining proton.

**optimum air supply.** Volume of air delivered to a burner that will produce the maximum thermal efficiency under specific operating conditions.

**orange heat.** A division of the color scale, generally given as about 900°C.

**orbicular.** Containing spheroidal aggregates of megascopic crystals, generally in concentric shells composed of two or more of the constituent minerals; for example, the structure of some granular igneous rocks, such as corsite.

**orbit.** The region occupied by an electron as it moves about the nucleus of an atom.

**orbital electron capture.** A radioactive transformation in which the nucleus captures an orbital electron.

**organic.** Being, containing, or relating to carbon compounds, especially in which hydrogen is attached to carbon whether derived from living organisms or not. Usually distinguished from inorganic or mineral.

**organic-cooled reactor.** A reactor that uses organic chemicals, such as mixtures of polyphenyls (diphenyls and terphenyls), as coolant.

**organic petrochemicals.** Petrochemicals containing atoms of both hydrogen and carbon, tracing their origin to organic matter.

**organic sulfides.** A group of organic compounds containing a sulfur atom that is directly bonded between two carbon atoms. Some of the organic sulfides, such as dimethyl sulfide and thiophene, are considered to be suitable odorants.

**organic sulfur.** Carbon and sulfur compounds that are found in gas.

**Organization of Arab Petroleum Exporting Countries (OAPEC).** Founded in 1968 for cooperation in economic and petroleum affairs. Original members were Saudi Arabia, Kuwait, and Libya. In 1970, Abu Dhabi, Algeria, Bahrain, Dubai, and Qatar joined.

**Organization of Petroleum Exporting Countries (OPEC).** Founded in 1960 to unify and coordinate petroleum policies of the members. The members and the date of membership are: Abu Dhabi (1967); Algeria (1969); Indonesia (1962); Iran (1960);

Iraq (1960); Kuwait (1960); Libya (1962); Nigeria (1971); Qatar (1961); Saudi Arabia (1960); and Venezuela (1960). OPEC headquarters is in Vienna, Austria.

**orifice.** Opening; commonly used to apply to discs placed in pipelines or radiator valves to reduce the fluid flow to desired amount; the opening in an orifice cap, orifice spud, or other device whereby the flow of gas is limited and through which the gas is discharged.

**orifice cap (hood).** A movable fitting having an orifice which permits adjustment of the flow of gas by the changing of its position with respect to a fixed needle or other device.

**orifice meter.** A meter using the differential pressure across an orifice plate as a basis for determining volume flowing through the meter. Ordinarily, the differential pressure is charted.

**orifice plate.** A plate of noncorrosive material which can be fastened between flanges or in a special fitting, perpendicular to the axis of flow and having a concentric circular hole. The primary use is for the measurement of gas flow.

**orifice spud.** A removable plug or cap containing an orifice which permits adjustment of the flow of gas either by substitution of a spud with a different sized orifice or by motion of a needle with respect to it.

**original oil-in-place.** The estimated quantity of crude oil in known reservoirs prior to any production.

**orogenic.** Pertaining to the processes by which great elongated chains and ranges of mountains are formed.

**orphan banks.** Abandoned surface mines, operated prior to the enactment of reclamation laws that require additional reclamation.

**Orsat gas-analysis apparatus.** An instrument for analyzing flue gases; also used for analyzing mine air.

**osmotic energy conversion.** Osmotic energy conversion is an unusual concept for obtaining energy from salt water based on the principle of simple osmosis. If salt water and fresh water are placed in a tank, separated by a semi-permeable membrane resembling an extremely fine filter, the fresh water will flow through the membrane into the salt water. The membrane has pores big enough to let water molecules through, but too small to allow the passage of salt molecules. A "battery" could be built by stacking these osmotic membranes, alternating the two types, and filling the spaces in between with either fresh water or salt water in an alternating pattern. Again, less salty water could be used instead of absolutely fresh water. By placing electrodes at each end of the battery, the accumulated charges could be collected and converted into electric current. (See Fig. p. 321)

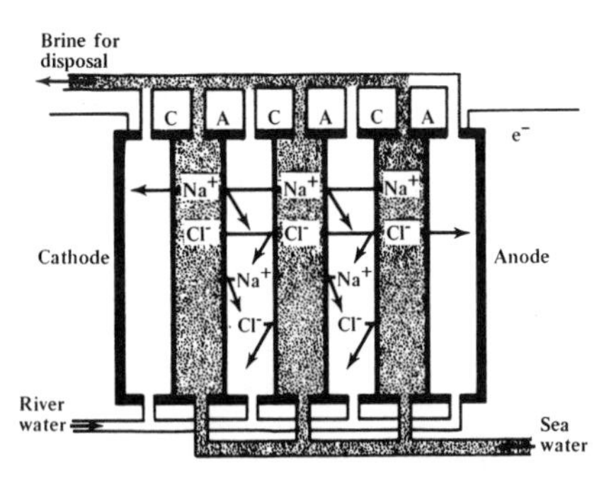


Diagram of a salt-water/fresh-water battery made with alternate osmionic cells of sea and fresh water separated by semipermeable membranes. (*Courtesy Popular Science, July 1975, p. 65*)

**Otto cycle engine.** The early form of internal combustion engine was named for Nikolaus Otto. The standard Otto piston burns a uniform fuel/air charge.

**outage.** Production time lost because of breakdowns or other unforeseen causes; the difference between the full-rated capacity and the actual contents of a barrel or tank; the loss of a volatile liquid such as gasoline ascribed to evaporation or pilferage. Also called shrinkage.

**outcrop.** A line on the earth's surface where a coal bed or other rock strata come to the surface. An outcrop does not necessarily require the visible presence of coal or other mineral at the surface but includes those places where they may be covered by a mantle of soil or other surface material.

**outer continental shelf.** The submerged lands extending from the outer limit of the historic territorial sea (typically 3 miles) to some undefined outer limit, usually a depth of 200 meters. In the United States, this is the portion of the shelf under federal jurisdiction.

**outfall.** A structure extending into a body of water for the purpose of discharging an effluent.

**outgassing.** The generation of vapors by materials usually during exposure to elevated temperature and/or reduced pressure.

**outpost well.** A well drilled to extend a known oil or gas pool. If successful in this objective, it is an extension well; if unsuccessful, it is a dry outpost.

**output.** The quantity of coal or mineral raised from a mine and expressed as being so many tons per shift, per week, or per year; also the power or product from a plant or prime mover in the specific form and for the specific purpose required.

**oven coke.** Coke produced by the carbonization of coal for the primary purpose of manufacturing coke.

**overburden.** Material of any nature, consolidated or unconsolidated, that overlies a deposit of useful materials, ores, or coal, especially those deposits that are mined from the surface by open cuts; also designates loose soil, sand, and gravel that lie above bedrock.

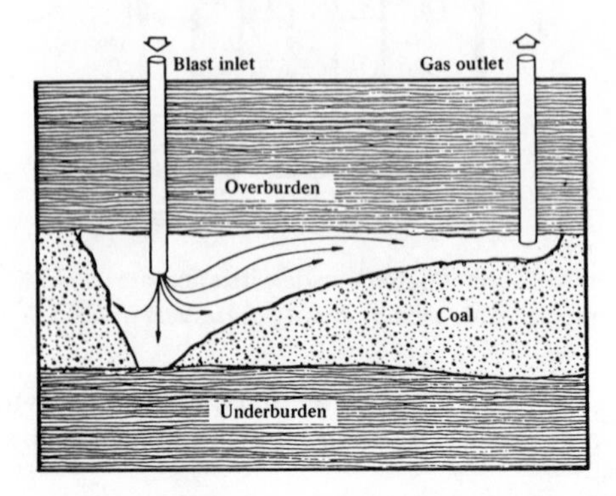


Overburden overlying coal deposits. (*Courtesy Van Nostrand's Scientific Encyclopedia*)

**overheating.** Heating a metal or alloy to such a high temperature that its properties are impaired. When the original properties cannot be restored by further heat treating, by mechanical working, or by a combination of working and heat treating, the overheating is known as burning.

**overload.** A load or weight in excess of the designed capacity, as applied to mechanical and electrical engineering plants, to loads on buildings and structures, and to excess loads on haulage ropes and engines.

**overloader.** A loading machine of the power-shovel type for quarry and opencast operations. It may be either pneumatic-tired or caterpillar-tracked. The bucket is filled, the machine retracted, and the bucket swung over to the discharge point.

**overpressure.** The transient pressure over and above atmospheric pressure caused by a shock wave from a nuclear explosion.

**over-the-shoulder handling.** A method of handling overburden whereby it is moved parallel to the highwall instead of at right angles to the wall as normally done.

**overwind.** To hoist the cage into or over the top of the headframe. In hoisting through a mine shaft, failure to bring the cage smoothly to rest at the proper unloading point at the surface can lead to serious accident unless preventive devices function effectively.

**ovonic.** A device for converting heat or sunlight directly to electricity. Invented by Standford F. Ovshinsky, it has a unique special glass composition which changes from an electrically nonconducting state to a semiconducting state.

**oxidant.** Any oxygen containing substance that reacts chemically in the air to produce new substances. Oxidants are the primary contributors to photochemical smog.

**oxidation.** A chemical reaction in which oxygen unites or combines with other elements. Organic matter is oxidized by the action of aerobic bacteria; thus, oxidation is used in waste water treatment to break down organic wastes.

**oxidation of coal.** The absorption of oxygen from the air by coal, particularly in the crushed state; this engenders heat which can result in fire. Ventilation, while dispersing the heat generated, supports oxidation which increases rapidly with a rise in temperature. Fresh air should not gain access to the coal.

**oxidation pond.** A man-made lake or pond in which organic wastes are reduced by bacterial action.

**oxygen.** A gas occurring in the atmosphere, of which it forms about 21 percent by volume. It is chemically very active and is necessary for combustion; the combination of the oxygen with other substances generally produces heat. Chemical symbol O.

**ozone.** A molecule containing three oxygen atoms ($O_3$). It occurs in minute quantities in the air near the earth's surface and in larger quantities in the stratosphere as a product of the action of ultraviolet light of short wavelengths on oxygen ($O_2$). At the top of the earth's atmosphere, it acts as a protective layer by absorbing ultraviolet radiation. It is also a major component of photochemical smog; it has a sharp, unpleasant odor and is an eye irritant.

# P

**package power reactor.** A small nuclear power plant designed to be crated in packages small enough to be conveniently transported to remote locations.

**packed tower.** An air pollution control device in which polluted air is forced upward through a tower packed with crushed rock or wood chips while a liquid is sprayed downward on the packing material. The pollutants in the air stream either dissolve or chemically react with the liquid.

**pair production.** The transformation of the kinetic energy of a high-energy photon or particle into mass, producing a particle and its antiparticle, such as an electron and positron.

**palingenesis.** The process of formation of new magma by the melting or fusion of country rocks with heat from another magma, with or without the addition of granitic material.

**PAMCO SRC process.** A process in which pulverized coal is mixed with a recycled coal-derived solvent, pumped through a preheater where hydrogen is mixed in and then sent to a dissolver. The coal is dissolved in solvent. The liquid product from the dissolver is degassed; hydrogen is recirculated and passed through a distillation unit to recover solvent and heavy product which is cooled and solidified to result in solvent refined coal. The PAMCO SRC has a heating value of 16,000 British thermal units per pound and a sulfur content of 0.5 to 0.9 percent by weight. Process operating conditions are 427° to 482°C and 1000 to 2000 pounds per square inch.

**panemone.** A vertical-axis wind collector capable of reacting to horizontal winds from any direction.

**panhandle formula.** A formula for calculating gas flow in large diameter pipelines, particularly at relatively high pressures and velocities.

**pans.** Peroxyacetyl nitrates, components of photochemical smog formed by the action of sunlight on the nitrogen oxides and hydrocarbons of air pollution.

**parabolic collector.** A device for collecting the sun's energy that utilizes a bowl or disk-shaped reflector. The reflector, which is often mirrored, concentrates the radiation, producing temperatures as high as 1649°C. (See Fig. p. 325)

**parabolic focusing collector.** A concentrating collector which focuses beam radiation with a parabolic reflector.

**parabolic mirror.** A solar energy collecting device.

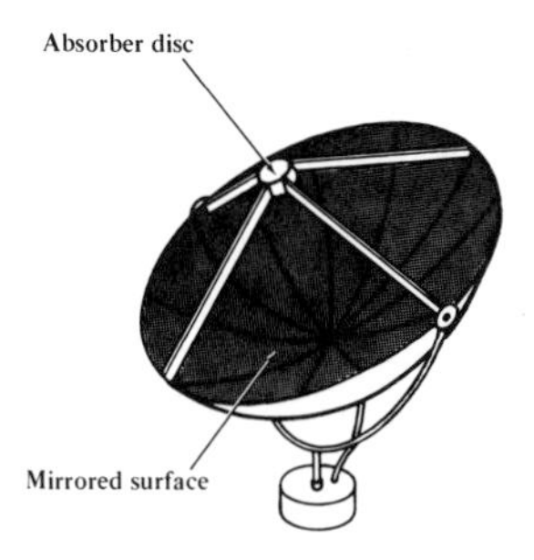


Parabolic solar collector. (*Courtesy Living with Natural Energy, Design for a Limited Planet*)

**parabolic reflector.** A device to obtain solar energy by concentrating sunlight into air-filled heat pipes.

**paraffin-base petroleum.** Crude oil which contains solid paraffin hydrocarbons and practically no asphalt.

**paraffin coal.** A light-colored bituminous coal used for the production of oil and paraffin.

**paramarginal resources.** Deposits not currently produced because the recovery is not economically feasible or because, although recovery is economically feasible, legal or political circumstances do not allow it.

**parasitic capture.** Any absorption (as in a reactor) of neutrons in reactions which do not cause further fission or the production of new fissionable material. In a reactor the process is undesirable.

**parent.** A radionuclide that upon radioactive decay or disintegration yields a specific nuclide, either directly or as a later member of a radioactive series.

**parent material.** The unconsolidated mass of rock material from which the soil profile develops.

**Parr's classification of coal.** The classification is based on the proximate analysis and calorific value of the ash-free, dry coal. The heating value of the raw coal is obtained, and from these data a table is drawn up at one end of which are the celluloses and woods of about 7000 British thermal units per pound. These figures are then plotted against the percentage of volatile matter in the unit of coal.

**partial decay constant.** For a radionuclide, the probability per unit time for the spontaneous decay of one of its nuclei by one of several possible modes of decay.

**particle.** A minute constituent of matter, generally one with a measurable mass. The primary particles involved in radioactivity are alpha particles, beta particles, neutrons, and protons.

**particle accelerator.** Machines which accelerate subatomic particles to such great velocities that as these particles strike atoms, and the nucleus of these atoms may be altered or split. Among those now in use are the cyclotron, the linear accelerator, and Van de Graaff generator, Proton synchroton, and the Bevatron.

**particle fluence.** At a given point in space, the number of particles or photons incident during a given time interval on a small sphere centered at that point divided by the cross-sectional area of that sphere. It is identical with the time integral of the particle flux density.

**particle flux density.** At a given point in space, the number of particles incident per unit time on a suitably small sphere centered at that point divided by the cross-sectional area of that sphere. It is identical with the product of the particle density and the average speed. This quantity may be also referred to as particle fluence rate.

**particulate loading.** The introduction of particulates into the ambient air.

**particulate matter.** Solid particles, such as the ash, which are released from combustion processes in exhaust gases at fossil-fuel plants.

**particulates.** Microscopic pieces of solids which emanate from a range of sources and are the most widespread of all substances that are usually considered air pollutants. Those between 1 and 10 microns are most numerous in the atmosphere, stemming from mechanical processes and including industrial dusts and ash.

**parting.** A layer of rock contained within, and lying roughly parallel to, a seam of coal. A parting has the effect of splitting a seam into two divisions or benches.

**Pascal's law.** States that the component of the pressure in a fluid in equilibrium that is due to forces externally applied is uniform throughout the body of fluid.

**passive solar design.** A structural design that makes use of the structural elements of a building, using no moving parts, to heat or cool spaces in the building.

**passive solar heating.** Solar heating of a building accomplished by architectural design without the aid of mechanical equipment.

**passive solar system.** A system that uses gravity, heat flows, or evaporation to operate without mechanical devices to collect and transfer energy, such as south-facing windows.

**PCBs.** Polychlorinated biphenyls, a group of organic compounds used in the manufacture of plastics. In the environment, PCBs are highly toxic to aquatic life. They persist in the environment for long periods of time and are biologically accumulative.

**pea coal.** In anthracite only, coal small enough to pass through a mesh 3/4- to 1/2-inch square, but too large to pass through a 3/8-inch mesh.

**peak.** The topmost point; summit; the tops of strip mine banks before grading.

**peak day.** The 24-hour day period of greatest total energy sendout.

**peaking capacity.** The generating capacity of facilities or equipment designed for use under extreme demand conditions.

**peaking turbines.** Supplemental generating capacity during times of high demand.

**peaking units.** Utility generating equipment used for intermittent maximum loads.

**peak load.** The maximum amount of power required and delivered during a stated period of time.

**peak load station.** A generating station which is operated to provide power during maximum load periods.

**peak shaving.** The use of fuels and equipment to generate or manufacture gas to supplement the normal supply of pipeline gas during periods of extremely high demand; also the process of supplying power from one extraneous source to help meet the peak demand on a system.

**peat.** Partially decomposed organic material formed in marshes and swamps. Stagnant ground water is necessary for peat formation to protect the residual plant material from decay.

**peat coal.** A natural product intermediate between peat and lignite; also an artificial fuel made by carbonizing peat.

**pebble bed.** A large bin of uniform size pebbles used for storing solar heat in solar air heating or cooling systems.

**pebble bed reactor.** A reactor in which the fissionable fuel (and sometimes also the moderator) is in the form of packed or randomly placed pellets, which are cooled by gas or liquid.

**pelagic.** Related to water of the sea as distinct from the sea bottom; related to sediment of the deep sea is distinct from that derived directly from the land.

**Pelton wheel.** An impulse water turbine with buckets which are bolted to its periphery and struck by a high velocity jet of water.

**penetrometer.** A simple device for measuring the penetrating power of a beam of x-rays or other penetrating radiation by comparing transmission through various absorbers.

**Pennsylvania oil.** Oil refined from crude petroleum produced in Western Pennsylvania, Southwestern New York, Eastern Ohio, and West Virginia.

**pennyweight.** One-twentieth troy ounce. Used in the United States and in England for the valuation of gold, silver, and jewels.

**penstock.** A pipe which transports water to a turbine for the production of hydroelectric energy.

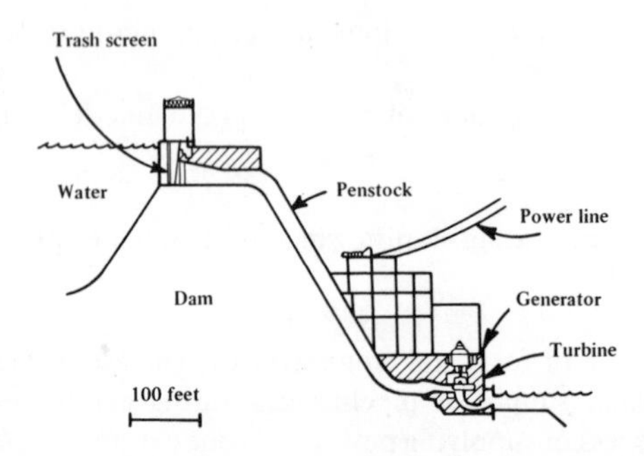


Components of a hydropower system. (*Courtesy U.S. Council on Environmental Quality*)

**percentage depth dose.** The ratio expressed as a percentage of the absorbed dose at any given depth within a body to the absorbed dose at some reference point of the body along the central ray. For x or gamma radiation, the location of the reference point depends on the energy of the incident radiation. It is at the surface for low energies or at the position of peak absorbed dose for high energies.

**percentage extraction.** The portion of a coal seam which is removed from the mine. The remainder may represent coal in pillars or coal which is too thin or inferior to mine or coal lost in mining. Shallow coal mines may extract only about 50 percent of the entire seam.

**percolation.** Movement, under hydrostatic pressure, of water through the pores or spaces of rock or soil.

**percolation rate.** The rate, usually expressed as a velocity, at which water moves through saturated granular material.

**period.** The time required for one cycle of a regularly repeated series of events. In a nuclear reactor, it is the time required for the power level to change by the factor 2.718, which is known as e (the base of natural logarithms.)

**periodic law.** A law in chemistry discovered by Mendeleev. The physical and chemical properties of the elements are periodic functions of their atomic weights.

**period meter.** An instrument for measuring the reactor time constant (reactor period).

**period range.** A range of reactor power levels over which the reactor time constant (reactor period) rather than the reactor power is of primary importance for reactor control.

**period (reactor).** The length of time in which the reactor power level increased by the factor e = 2.718.

**perlite.** An insulation material used for home insulation. It is usually installed by pouring the material into unfinished attic flour spaces. Perlite has the same insulation value as vermiculite.

**permafrost.** A permanently frozen layer of soil, subsoil, or other deposit occurring at variable depth below the earth's surface in arctic or subarctic regions.

Photo illustrating the permafrost along the route of the trans Alaska oil pipeline. (*Courtesy Alyeska Pipeline Service Company*)

**permeability.** A measurement of the ability of a rock to transmit fluid. The degree of permeability depends upon the size and shape of the pores and the size, shape, and extent of their interconnections.

**permissible.** Allowable; a machine or explosive is said to be permissible when it has been approved by the United States Bureau of Mines for use underground under prescribed conditions. All flameproof machinery is not permissible, but all permissible machinery is flameproof.

**permissible dose.** That amount of radiation which has been established as the maximum quantity that can be absorbed without undue risk to human health.

**permissible hydraulic fluids.** Commercially available, fire-resistant fluids developed by the oil industry in cooperation with the U.S. Bureau of Mines. They are water-in-oil emulsions, and can be substituted for flammable hydraulic fluids by users of large machinery, whether the equipment is operated underground or on the surface.

**personnel monitoring.** Determination by either physical or biological measurement of the amount of ionizing radiation to which an individual has been exposed, such as by measuring the darkening of a film badge or performing a radon breath analysis.

**petrochemicals.** Chemical compounds made with a petroleum hydrocarbon as one of their basic components; as, for example, ammonia, synthetic rubber, and carbon black.

**Petrocoal process.** A process in which crushed coal with recycle oil and hydrogen is fed to a hydropyrolytic reactor where coal is converted to asphaltenes. The liquid product, after the removal of solids, is subjected to catalytic hydrogenation at 427° to 454°C and 1500 pounds per square inch gage, and product oil suitable for refining is obtained.

**petroleum.** Material occurring naturally in the earth and composed predominantly of mixtures of chemical compounds of carbon and hydrogen with or without other nonmetallic elements, such as sulfur, oxygen, nitrogen, etc. Petroleum may contain or be composed of such compounds in the gaseous, liquid, and/or solid state, depending on the nature of these compounds and the existent conditions of temperature and pressure.

**petroleum coke.** The residue obtained by the distillation of petroleum. Primarily a fuel, it usually shows the following composition: 5 to 10 percent volatile and combustible matter, 90 to 95 percent fixed carbon, from a trace to 0.3 percent ash, and from 0.5 to 1 percent sulfur.

**petroleum gas.** Any of a number of flammable gases obtained by refining of petroleum and used both for organic synthesis and in liquefied form as fuels. Most important are the saturated compounds butane, isobutane, and propane and the unsaturated butylenes and propylene.

**petroleum naphtha.** A generic term applied to refined, partially refined, or unrefined petroleum products and liquid products of natural gas.

**petroleum refinery.** A plant that converts crude petroleum into the many petroleum fractions (asphalt, fuel, oil, gasoline, etc.). Usually this conversion is accomplished by fractional distillation.

**petroleum spirits.** A refined petroleum distillate with volatility, flash point, and other properties making it suitable as a thinner and solvent in paints, varnishes, and similar products.

**petroleum tar.** A viscous black or dark-brown product obtained in petroleum refining which will yield a substantial quantity of solid residue when partly evaporated or fractionally distilled.

**petropolitics.** Influence of oil exporting nations on the politics, especially the foreign policy, of a state.

**pF.** The numerical measure of energy with which water is held in the soil.

**pH.** Denotes the degree of acidity or alkalinity of a solution. At 25° C, pH of 7 is the neutral value. Acidity increases with decreasing values below 7, and alkalinity increases with increasing values above 7.

**phantom.** A volume of material approximating as closely as possible the density and effective atomic number of living tissue; used in biological experiments involving radiation.

**phenols.** A group of aromatic compounds having the hydroxyl group directly attached to the benzene ring. They give the reactions of alcohols, forming esters, ethers, and thiocompounds. Phenols are more reactive than the benzene hydrocarbons. They are present in coal tar and wood tar.

**phosphor.** A luminescent substance; a material capable of emitting light when stimulated by radiation.

**photochemical smog.** Smog produced by the action of sunlight on the pollutants emitted by automobiles. The major components of photochemical smog are ozone and peroxyacetyl nitrates.

**photoelectric cell.** A device which registers changes in the amount of light falling on it by converting it to electricity. Used to monitor processes and aid automatic control systems.

**photoelectric effect.** The complete absorption of a photon by an atom with the emission of an orbital electron (photoelectron).

**photofission.** Nuclear fission induced by photons.

**photogrammetry.** The process of making surveys or measurements by the aid of photography.

**photolysis.** The decomposition of molecules caused by the absorption of light energy.

**photometer.** An instrument for measuring the intensity of light, or for comparing intensities from two sources. The more accurate types are built around photoelectric cells.

**photomultiplier tube.** A device for measuring light at depths as great as 600 meters.

**photon.** A discrete quantity of electromagnetic energy. Photons have momentum but no mass or electrical charge.

**photoneutron.** A neutron released from a nucleus by a photonuclear reaction.

**photon propulsion.** Use of the directional emission of photons for propulsion.

**photonuclear reaction.** The interaction of a photon with a nucleus, usually with the release of a charged particle or neutron.

**photoproton.** A proton released from a nucleus by a photonuclear reaction.

**photosynthesis.** The process by which carbohydrates are compounded from carbon dioxide and water in the presence of sunlight and chlorophyll. Photosynthesis is the primary method for bioconversion of solar energy into forms more useful to man.

**photovoltaic cell.** A type of semiconductor in which the absorption of light energy creates a separation of electrical charges. The separation creates an electrical potential. The net effect is direct conversion of light, especially solar energy, into electricity. Typical materials used in the construction of photovoltaic cells are silicon, cadmium sulfide, and gallium arsenide.

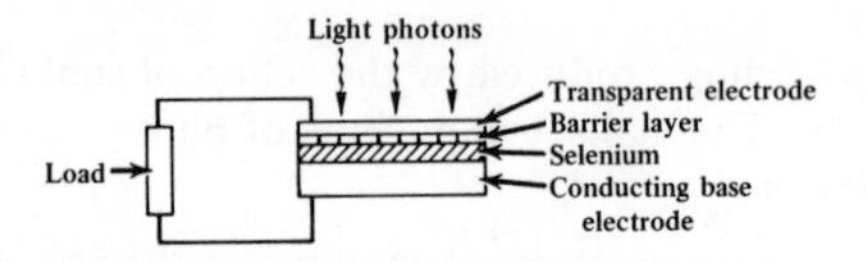


Diagram of a photovoltaic cell. (*Courtesy Van Nostrand's Scientific Encyclopedia*)

**photovoltaic concentrator arrays.** Systems that use lenses or mirrors to concentrate the sun's rays onto solar cells which convert sunlight directly to electricity. The concentrated sunlight increases the cells power output; thus, fewer cells are required to produce a given amount of electricity.

**photovoltaic conversion.** Transformation of solar radiation directly into electricity by means of a solid-state device such as the single-crystal silicon solar cell.

**photovoltaic process.** Process of converting light rays directly into electricity without going through the intermediate steps involving turbines and generators. In photovoltaics, solar cells convert sunlight into a stream of electrons easily converted into household current.

**pick mines.** Mines in which coal is cut with picks.

**piezoelectric.** Having the ability to develop surface electric charges when subjected to elastic deformation, and conversely.

**piezometer.** An instrument for measuring pressure head, usually consisting of a small pipe tapped into the side of a closed or open conduit and flush with the inside connected with a pressure gage, mercury, water column, or other device for the indicating pressure head.

**pig.** A heavily shielded container (usually lead) used to ship or store radioactive materials; also a flask having two or more tubulures to which smaller flasks may be attached, and used especially to collect fractions during fractional distillation; also a metal, barrel-shaped device covered with metal brushes and used to clean the internal surface of a pipeline.

**pile.** Old term for nuclear reactor. This name was used because the first reactor was built by piling up graphite blocks and natural uranium.

**pillar.** A solid mass of coal, rock or ore left standing to support a mine roof.

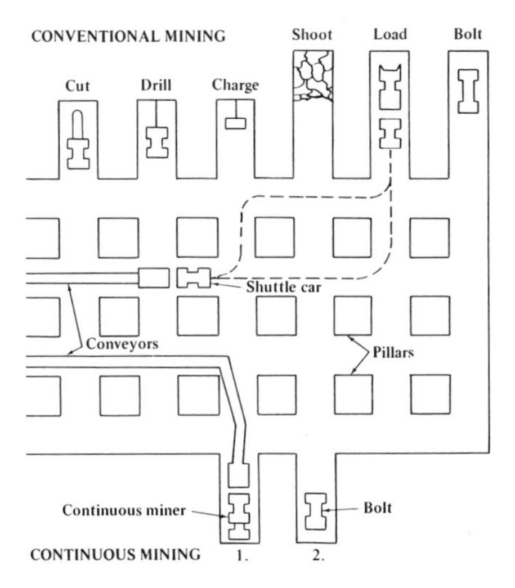


Pillar mining. (*Courtesy U.S. Energy Research and Development Administration, Division of Coal Conversion and Utilization*)

**pilot.** A small flame which is utilized to ignite the gas at the main burner(s).

**pilot plant.** A small-scale industrial process unit operated to test the application of a chemical or other manufacturing process under conditions that will yield information useful in the design and operation of full-scale manufacturing equipment.

**pinch effect.** In controlled fusion experiments, the effect obtained when an electric current, flowing through a column of plasma, produces a magnetic field that confines and compresses the plasma.

**pinch valve.** A valve used to pinch off or stop the flow of liquid, usually water, at a desired location.

**pion.** An elementary particle (contraction of pi-meson). The mass of a charged (positive or negative) pion is about 273 times that of an electron; that of an electrically neutral pion is 264 times that of an electron.

**pipe.** A tubular metal product, cast or wrought; any pipe or tubing used in the transportation of gas, including pipe-type holders.

**pipeline.** A line of pipe with pumping machinery and apparatus for conveying a liquid or gas; in the gas industry, the term applies to all parts of those physical facilities through which gas is moved in transportation, including pipe, valves, and other appurtenances attached to the pipe, compressor units, metering stations, regulator stations, delivery stations, holders, and fabricated assemblies.

**pipeline condensate.** A liquid containing lower boiling aliphatic and aromatic hydrocarbons which may be found in natural gas production, transmission, and distribution pipelines. Condensation to a liquid phase is induced by the higher pressure and lower temperature conditions in the pipeline.

**pipeline gas.** A methane-rich gas conforming to certain standards, including minimum water content, minimum inert gases, minimum hydrogen and carbon monoxide content, and 1000 pounds per square inch gage compression.

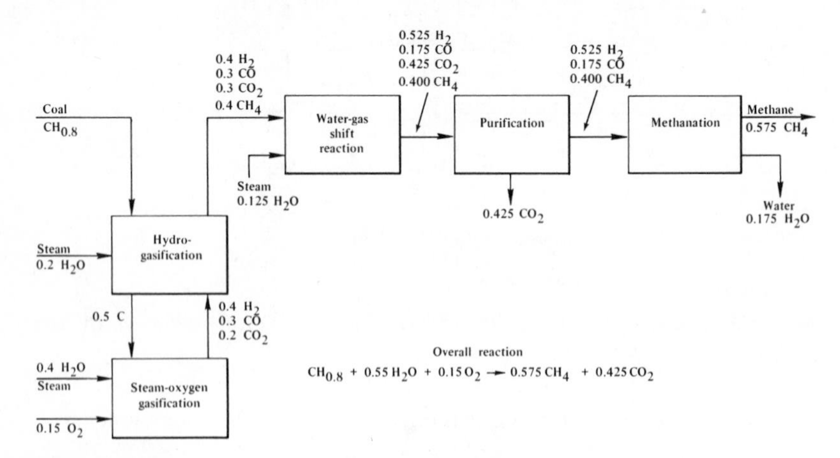


Production of pipeline gas by hydrogasification and methanation. (*Courtesy Annual Review of Energy, Volume 1*)

**pipeline peaking service.** Any service provided by a pipeline company to assist its customers in smoothing the fluctuations in their demand for gas.

**pipeline quality gas.** A synthetic natural gas very close to natural gas in heating value and burning characteristics.

**pipe-to-soil potential.** Electrical potential of pipe with reference to an electrode placed in the ground.

**piston.** The working part of a pump, hydraulic cylinder, or engine that moves back and forth in the cylinder, and is generally equipped with one or several rings or cups to control the passage of fluid. It ejects the fluid from the cylinder, as in a pump, or receives force from the fluid, which causes a reciprocating motion, as in an engine.

**pit.** A colliery; any area of open cut mining. May refer to only that part of the open cut mining area from which coal is being actively removed or to the entire contiguous mined area.

**pitch.** The angle at which a coal seam inclines below a horizontal line; the rise of a seam; also the solid or semisolid residue from the partial evaporation of tar. Primarily, pitch is a bitumen with extraneous matter such as free carbon, residual coke, etc.

**pitchblende.** A massive variety of uraninnite or uranium oxide, found in metallic veins. Contains 55 to 75 percent uraniumdioxide, up to 30 percent of uraniumtrioxide, and usually a little water and varying amounts of other elements. Thorium and the rare earths are generally absent.

**pitch coal.** A brittle, lustrous bituminous coal or lignite.

**pitch coke.** Coke made by the destructive distillation of coal-tar pitch.

**pith.** The soft part of a coal load.

**pithead.** The top of a mine shaft including the buildings, roads, tracks, plant, and machines around it.

**Pilot tube.** Two small, concentric tubes, one inside the other, that can be inserted into a pipe carrying fluids in order to measure the flow of liquid or gas in the pipe. The tubes are so arranged that one is affected by the total pressure of the fluid and the other by the static pressure existing in the pipe. The outside ends of the tubes are connected to either side of a manometer tube and the difference in pressure so measured is a function of the velocity of fluid flowing in the pipe ahead of the tube.

**pitting.** Formation of small depressions in a surface due to sand blasting, mechanical gouging, acid etching, or corrosion.

**pitting corrosion.** A form of extremely localized attack that results in penetration of the metallic component. It is one of the most destructive and insidious forms of corrosion because of the premature perforation of a component with only a small percent weight loss of the entire component, e.g., pitting of stainless steel condenser tubes in seawater service.

**placer deposit.** A deposit of clay, silt, sand, gravel, or some similar material deposited by running water, which contains particles of uranium, gold, or some other valuable mineral.

**plant factor.** The ratio of the average power load of an electric power plant to its rated capacity. Sometimes called capacity factor.

**plasma.** Gas comprising equal amounts of positively and negatively charged particles, so that in bulk it is electrically neutral; a fourth state of matter (solid, liquid, gas, plasma) capable of conducting magnetic force. The study of plasma motions is called magnetohydroynamics (MHD).

**plasma fuel.** A white-hot substance comprising the principal ingredient of suns and stars. Neither solid, liquid, nor gaseous, it could become the great energy-thrust provider of the future. The heavy hydrogen of the oceans may be the best source for this fusion-type power.

**plate-type heat exchanger.** An advanced heat exchanger design being developed for the Ocean Thermal Energy Conversion (OTEC) program which would have ammonia inside the structure and water outside of it, eliminating the need for a large heavy shell. The state-of-the-art heat exchangers consist of configurations of tubes inside large heavy shells.

**plenum.** A system of ventilation in which air is forced into a closed space, as a room or a caisson, so that the outward pressure of air in the space is slightly greater than the inward pressure from the outside, and leakage is outward instead of inward; a mode of ventilating a mine by forcing fresh air into it.

**plowshare.** The Atomic Energy Commission program of research and development on peaceful uses of nuclear explosives. The possible uses include large-scale excavation, such as for canals and harbors, crushing ore bodies, and producing heavy transuranic isotopes. (See Fig. p. 337)

**plug.** An external thread pipe fitting inserted into the open end of an internal thread pipe fitting to seal the end of a pipe; sealing a hole in a vessel, such as a pipe or tank, by inserting material in the hole and then securing it; also, refers to the material used to plug the hole.

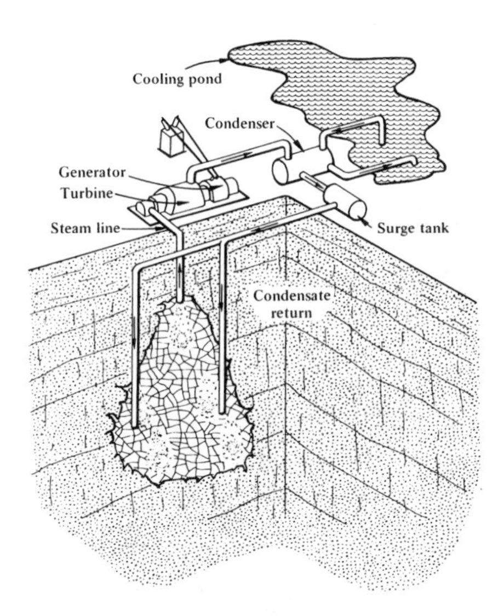


Plowshare concept of geothermal heat extraction. (*Courtesy U.S. Council on Environmental Quality*)

**plug back.** To seal off the bottom section of a well bore to prevent the inflow of fluid from that portion of the hole. This permits the inflow of oil and gas from the formation above the section so sealed off, without contamination of fluids below that depth.

**plume.** An elongated and mobile column or band as of smoke or exhaust gases.

**plutonium.** A heavy, radioactive, man-made, metallic element with atomic number 94. It occurs in nature in trace amounts only. Its most important isotope is fissionable plutonium-239, produced by neutron irradiation of uranium-238. It is used for reactor fuel and weapons. Chemical symbol Pu. (See Fig. p. 338)

**plutonium credit.** The value of plutonium (formed by conversion) in irradiated uranium.

**plutonium economy.** Processing plutonium produced in power reactors to reuse as fuel.

**pluvial.** Of a geologic change, resulting from the action of rain or sometimes from the fluvial action of rainwater flowing in stream channels.

**pneumoconiosis.** A lung disease caused by habitual inhalation of irritant mineral or metallic particles. It occurs in any work places where dust is prevalent. When the dust is from coal, the condition is often called black lung disease.

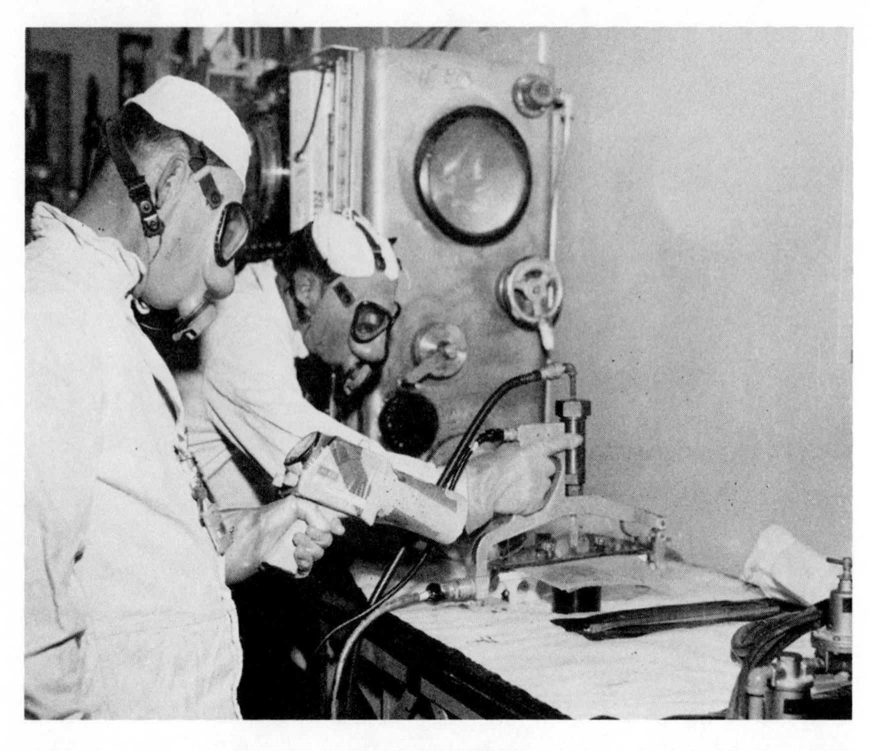

Packaging plutonium "button". Plaster envelope prevents spread of plutonium particles. Plutonium is toxic if inhaled. (*Courtesy U.S. Atomic Energy Commission*).

**point source.** A single point from which light emanates, such as the sun, the filament of an electric lamp, or other superheated metal; in air pollution, a stationary source of a large individual emission, generally of an industrial nature.

**poison.** Any material of high absorption cross section that absorbs neutrons unproductively and hence removes them from the fission chain reaction in a reactor, decreasing its reactivity.

**polar molecule.** A molecule in which the electrical charges of its constituent atoms or groups produce a dipole.

**polder.** A tract of low land reclaimed from the sea or other body of water by dikes

**pole miles.** Miles measured along the line of poles, structures, or towers carrying electric conductors regardless of the number of conductors or circuits carried.

**pollution.** The presence of matter or of energy whose nature, location, or quantity produces undesired environmental effects. (See Fig. p. 339)

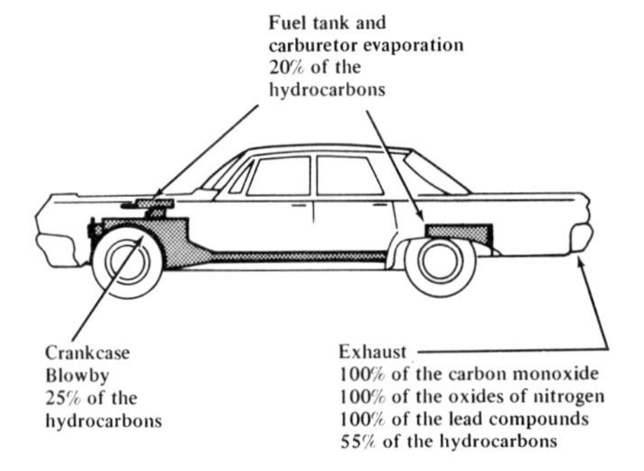


Approximate distribution of automobile pollutants. (*Courtesy National Science Teachers Association*)

**polymer.** A product of polymerization of normally gaseous olefin hydrocarbons to form high-octane hydrocarbons in the gasoline boiling range. Polymerization is the process of combining two or more simple molecules of the same type, called monomers, to form a single molecule having the same elements in the same proportion as in the original molecule, but having different molecular weights. The combination of two or more dissimilar molecules is known as copolymerization, and the product is called a copolymer.

**polyethylene.** A thin, transparent sheet of plastic used for moistureproofing floors and walls, as a covering for greenhouses, and as a cover material for solar collectors.

**polymerization.** Union of two or more molecules of a given structure to form a new compound with the same elemental proportions but with different properties and a higher molecular weight.

**polystyrene.** Panels of a highly flammable foamlike plastic used for house insulation.

**polyurethane.** Plastic that can be applied as a foam and used as an insulating barrier.

**polyvinyl chloride (PVC).** A common plastic building material that releases hydrochloric acid when burned.

**pond.** A body of water of limited size; a small lake.

**pool reactor.** A reactor in which the fuel elements are suspended in a pool of water that serves as the reflector, moderator, and coolant. Popularly called a swimming pool reactor, it is usually used for research and training.

**population density.** The number of persons per unit area (usually per square mile) who inhabit an area.

**porosity.** Voids in a reservoir rock available for storage of fluids; measured in percent of rock volume.

**port.** An opening in a research reactor through which objects are inserted for irradiation or from which beams of radiation emerge for experimental use.

**portal.** Any entrance to a mine.

**positive displacement meter.** An instrument which measures volume on the basis of filling and discharging gas in a chamber.

**positive displacement pump.** A pump that delivers a constant volume of fluid per cycle of operation at whatever pressure is necessary, within the design limits of the mechanism of the pump.

**positive drive.** A driving connection in two or more wheels or shafts that will turn them at approximately the same relative speeds under any conditions.

**positive gradient.** Layer of water where temperature increases with depth.

**positron.** An elementary particle with the mass of an electron but charged positively. It is the "antielectron". It is emitted in some radioactive disintegrations and is formed in pair production by the interaction of high-energy gamma rays with matter.

**posted field price.** Price for oil or gas in a given area, set by principal buyers.

**potential.** Electrical pressure, such as the electrical pressure normally existing between the conductors of a circuit on the terminals of a machine or apparatus; also a measure of the capacity of a well to produce oil or gas.

**potential difference.** The difference in pressure between any two points in an electrical circuit, measured in volts.

**potential energy.** The form of mechanical energy a body possesses by virtue of its position. If a body is being dropped from a higher to a lower position, the body is losing potential energy; if a body is being raised, it is gaining potential energy. The term also refers to "stored" energy; energy in any form not associated with motion, such as that stored in chemical or nuclear bonds.

**potential gradient.** An ascending or descending value of voltage related to a linear measurement, as a distance along the earth surface or ground.

**poundal.** Unit of force; the force needed to give a mass of one pound an acceleration of 1 foot per square second.

**pounds per square inch absolute (psia).** The measurement of pressure, whether hydraulic (liquids) or pneumatic (gases), made without including the effect of atmospheric pressure, which is 15 pounds per square inch at sea level.

**pour point.** The lowest temperature at which an oil can be poured.

**power.** Any form of energy available for doing any kind of work; specifically mechanical energy as distinguished from work done by hand; the rate at which work is done or energy expended; measured in units of energy per unit of time such as calories per second, and in units such as watts and horsepower.

**power density.** The rate of heat generated per unit volume of a nuclear reactor core; also the amount of power per unit of cross-sectional area of a windstream.

**power distillate.** The untreated kerosine condensates and still heavier distillates down to 28° Baume from mid-continent petroleum. Used as fuel in internal combustion engines.

**power factor.** The ratio of the mean actual power in an alternating-current circuit measure in watts to the apparent power measured in volt-amperes, being equal to the cosine of the phase difference between electromotive force and current.

**power from garbage and organic wastes.** Fuel obtained from the liquefaction of garbage and organic wastes; the direct incineration of municipal garbage as fuel.

**power gas.** A cheap gas (as Mond gas) made for producing power, especially for driving gas engines. (See Fig. p. 342)

**power per unit band.** The limit approached by the quotient obtained by dividing the power of the energy being transmitted by a given system, at a given time, and in a given frequency band; by the width of this band as the width of this band approaches zero.

**power reactor.** A reactor designed to produce useful nuclear power, as distinguished from reactors used primarily for research or for producing radiation or fissionable materials.

**power-shovel mining.** Power shovels are used for mining coal, iron ores, phosphate deposits, and copper ores. The shovels may be used either for mining or for stripping and removing the overburden or for both types of work, although at some coal mines the shovels used for stipping are considerably larger than those used for mining.

**power upon the air.** In coal mine ventilation, the horsepower applied is often known as the power upon the air. This may be the power exerted by a motive column due to natural causes or to a furnace, or it may be the power of a mechanical motor. The power upon the air is always measured in foot-pounds per minute.

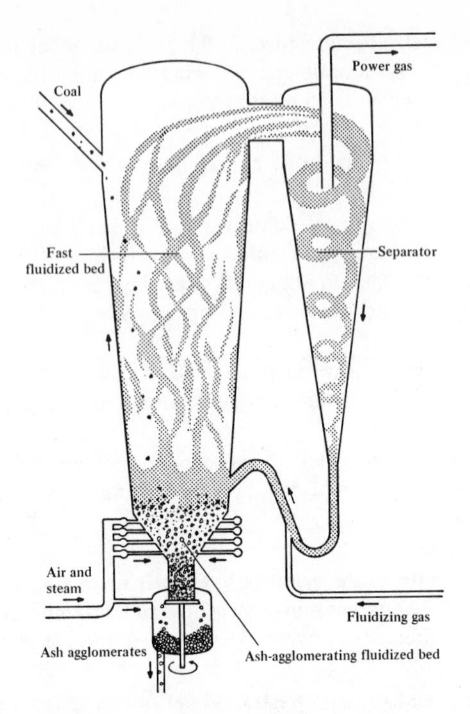


System for gasifying coal to produce power gas. The process incorporates a fluidized bed to partially oxidize coal at high pressures. (*Courtesy "Energy and the Future," Fig. 9*)

**ppm.** Parts per million. The unit commonly used to represent the degree of pollutant concentration where the concentrations are small. Larger concentrations are given in percentages. In air, ppm is usually a volume/volume ratio; in water, a weight/volume ratio.

**precipitate.** A solid that separates from a solution because of some chemical or physical change.

**precipitator.** A device to remove fine ash, tars, dusts, or smoke particles from flue gases or other gaseous streams; the device may employ mechanical, electrostatic, or chemical means, or a combination of these; in pollution control work, any of a number of air pollution control devices usually using mechanical/electrical means to collect particulates from an emission.

**precoat.** A thin coating which serves as a base.

**precursor.** Any radioactive nuclide which precedes that nuclide in a decay chain. The term is often restricted to the immediately preceding nuclide.

**preheat.** To heat beforehand.

**pressed distillate.** The oil left in petroleum refining after the paraffin has been separated from the paraffin distillate by cooling and pressing.

**pressed fuel.** An artificial fuel prepared from coal dust or waste coal, incorporated with other ingredients (such as tar), and compressed by molds into blocks or briquettes.

**pressure.** The load divided by the area over which it acts; the force per unit area applied to outside of a body, normally expressed as force per unit area in pounds per square inch (psi) or grams per square centimeter ($gm/cm^2$).

**pressure loader.** Device in which the rate of gas flow controls the operation of a pressure governor or regulator.

**pressure losses.** Losses in static or velocity pressure in a piping system due to friction, eddies, or leaks.

**pressure maintenance.** Injecting water or gas in a formation when natural oil reservoir pressures are insufficient.

**pressure-regulating station.** Equipment installed for the purpose of automatically reducing and regulating the pressure in the downstream pipeline or main to which it is connected. Included are piping auxiliary devices such as valves, control instruments, control lines, the enclosures, and ventilating equipment.

**pressure regulator.** A balanced valve device equipped with a diaphragm, used to control or restrict the flow of fluid or gas and designed such that the pressure of the fluid plus the force exerted by a spring or lever is sufficient to close the valve against the pressure exerted by the fluid or gas on the high-pressure side, thereby restricting its flow.

**pressure-relief station.** Equipment installed for the purpose of preventing the pressure on a pipeline or distribution system to which it is connected from exceeding by more than an established increment the maximum allowable operating pressure, by venting gas to the atmosphere whenever the pressure tends to rise too high.

**pressure and temperature relief valve.** A relief device activated by pressure and/or temperatures, commonly used on water heaters.

**pressure base.** An agreed reference pressure used to obtain uniform figures for gas flow and consumption.

**pressure control.** Maintenance of pressure, in all or part of a system, at a predetermined level or within a selected range.

**pressure-decline curve method.** A method of estimating nonassociated gas reserves in reserves which do not have a water drive.

**pressure differential.** Difference in pressure between any two points in a continuous system.

**pressure drop.** The loss in static pressure of the fluid (air, gas, or water) due to frictional obstruction in the pipe, valves, fittings, regulators, burners, appliances, or breeching.

**pressure gauge.** Instrument for measuring the pressure of a fluid.

**pressure limiting station.** Equipment installed for the purpose of preventing the pressure on a pipeline or distribution system from exceeding some maximum pressure, as determined by one or more regulating codes, by controlling or restricting the flow of gas when abnormal conditions develop.

**pressure-tube reactor.** A reactor in which the fuel elements are located inside tubes containing coolant circulating at high pressure. The tube assembly is surrounded by a tank containing the moderator at low pressure.

**pressure vessel.** A strong-walled container housing the core of most types of power reactors; it usually also contains moderator, reflector, thermal shield, and control rods.

**pressurized.** Any structure, area, or zone fitted with an arrangement that maintains nearly normal atmospheric pressure.

**pressurized reactor.** A reactor whose primary liquid coolant is maintained under such a pressure that no bulk boiling occurs.

**pressurized water reactor.** A power reactor in which heat is transferred from the core to a heat exchanger by water kept under high pressure to achieve high temperature without boiling in the primary system. Steam is generated in a secondary circuit. Many reactors producing electric power are pressurized water reactors. (See Fig. p. 345)

**pressurized water reserves.** Geologic formations beneath the Texas and Louisiana Gulf Coasts containing large volumes of hot brine under high pressure and possibly containing large amounts of entrapped natural gas. These deposits are capable of supplying both heat and mechanical energy, as well as potentially large amounts of dissolved methane.

**presulfide.** A step in the catalyst regeneration procedure which treats the catalyst with a sulfur-bearing material such as hydrogen sulfide or carbon bisulfide to convert the metallic constituents of the catalyst to the sulfide form in order to enhance its catalytic activity and stability.

**pretreatment.** In waste water treatment, any process used to reduce pollution load before the waste water is introduced into a main sewer system or delivered to a treatment plant for substantial reduction of the pollution load.

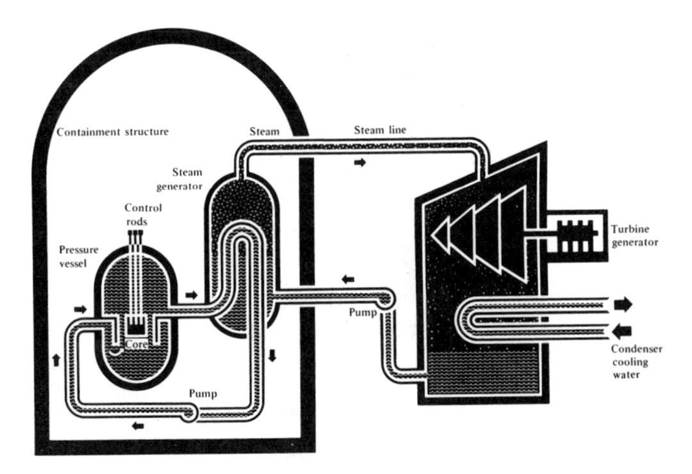


Structure of a pressurized water reactor. (*Courtesy U.S. Council on Environmental Quality*)

**Price-Anderson Act.** Provides for damages arising from a nuclear accident. It requires a utility which operates a nuclear power plant to purchase the maximum amount of liability insurance it can obtain. The federal government provides additional insurance to pay the public for damages from a nuclear accident.

**price tier.** The tiers refer to classes of crude oil production established for pricing purposes.

**prilling tower.** A tower that produces small solid agglomerates by spraying a liquid solution in the top and blowing air up from the bottom.

**primary air.** Air that is mixed with fuel before the mix reaches the ignition zone to enhance combustion.

**primary cell.** A cell which generates or makes its own electrical energy from the chemical action of its constituents. Examples of primary cells are the voltaic cell, Daniell cell, LeClanche cell, and the dry cell. (See Fig. p. 346)

**primary containment.** An enclosure which surrounds the nuclear reactor core and associated equipment for the purpose of minimizing the release of radioactive material in the event of a serious malfunction in the operation of the reactor.

**primary coolant.** A coolant used to remove heat from a primary source, such as a reactor core.

**primary distribution feeder.** An electric line supplying power to a distribution circuit.

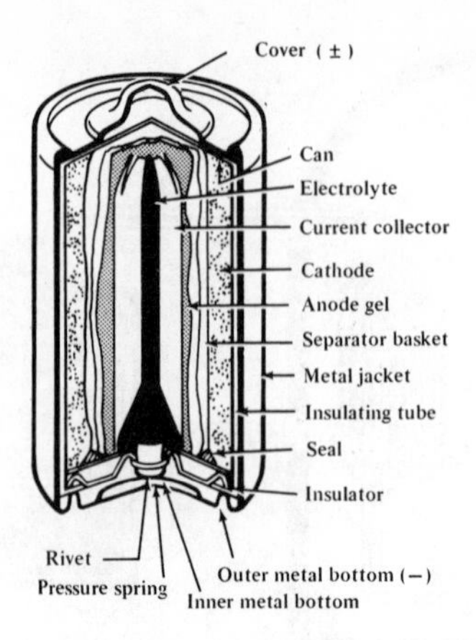


Primary alkaline cell. (*Courtesy Van Nostrand's Scientific Encyclopedia*)

**primary energy.** Energy in its naturally-occurring form (coal, oil, uranium) before conversion to end-use forms.

**primary fission yield.** The fraction of fissions giving rise to a particular nuclide before any beta or gamma decay has occurred.

**primary fuel.** Fuel consumed in original production (coal, oil, natural gas, for example) of energy as contrasted to a conversion of energy from one form to another.

**primary heating system.** A conventional oil, gas, or electric heating system as opposed to a solar heating system.

**primary stocks of refined products.** Stocks held at refineries, bulk terminals, and pipelines.

**primary system (nuclear power plant).** The nuclear portion of the plant, including the reactor, moderating, and cooling system. The generation of steam and electricity comprises the secondary system.

**primary voltage.** The voltage of the circuit supplying power to a transformer, as opposed to the output voltage or load-supply voltage which is called secondary voltage.

**prime mover.** A machine which converts fuel or other natural energy into mechanical power.

**priming.** The act of adding water to displace air, thereby promoting suction, as in a suction line of a pump; in a boiler, the excessive carry-over of fine water particles with the steam due to insufficient steam space, faulty boiler design, or faulty operating conditions.

**probable life.** The total expected service life for survivors at any given age.

**probable reserves.** A realistic assessment of the reserves that will be recovered from known oil or gas fields based on the estimated ultimate size and reservoir characteristics of such fields. Probable reserves include those reserves shown in the proved category.

**process gas.** Gas use for which alternate fuels are not technically feasible, such as in applications requiring precise temperature controls and precise flame characteristics.

**process heat.** Heat which is used in agricultural and industrial operations.

**process heat reactor.** A reactor that produces heat for use in manufacturing processes.

**process monitoring instrument.** Element in a process control system. Such instruments that are applicable to coal-processing systems include: on-line chemical analysis systems; acoustic leak detection; ultrasonic and acoustic transducer; temperature, pressure, and flow instrumentation; radiation monitoring equipment; digital computer-based data logging and plant-control systems; and plant protection systems.

**process unit.** A separate facility within a refinery, consisting of many types of equipment, such as heaters, fractionating columns, heat exchangers, vessels, and pumps, designed to accomplish a particular function within the refinery complex. For example, the crude processing unit is designed to separate the crude into several fractions, while the catalytic reforming unit is designed to convert a specific crude fraction into a usuable gasoline blending stock.

**process weight.** The total weight of all materials, including fuels, introduced into a manufacturing process. The process weight is used to calculate the allowable rate of emission of pollutant matter from the process.

**producer.** A producing well; also one who owns wells producing oil or gas.

**producer gas.** A combustible gas obtained by the partial combustion of coal or coke in air; consists mainly of carbon monoxide and nitrogen, with small proportions of hydrogen, methane, and carbon dioxide.

**producing horizon.** Rock from which oil or gas is produced.

**producing sand.** A rock stratum that contains recoverable oil or gas.

**producing zone.** The interval of rock actually producing oil or gas.

**production.** That which is produced or made; any tangible result of industrial or other labor.

**production reactor.** A reactor designed primarily for large-scale production of plutonium-239 by neutron irradiation of uranium-238; also a reactor used primarily for the production of radioactive isotopes.

**productivity.** The efficiency with which economic resources (men, material, and machines) are employed to produce goods and services; the value added to goods divided by the man-hours input.

**products of combustion.** Constituents resulting from the combustion of a fuel with the oxygen of the air including the inerts but excluding excess air.

**profile.** A vertical section of a soil showing the nature and sequence of its various zones.

**projection.** An estimation of probable future events.

**Project Independence.** A proposal by the US Government in January 1974, the aim of which was to develop America's own energy resources so that by 1980 the country would be self-sufficient in energy and invulnerable to pressures from overseas suppliers of oil. The Project required the relaxation of many restrictions imposed in earlier years to improve the quality of the environment.

**prompt criticality.** The state of a reactor when the fission chain reaction is sustained solely by prompt neutrons, that is, without the help of delayed neutrons.

**prompt gamma radiation.** Gamma radiation accompanying fission without measurable delay.

**prompt neutron fraction.** The ratio of the mean number of prompt neutrons per fission to the mean total number of neutrons (prompt plus delayed) per fission.

**prompt neutrons.** Neutrons that are emitted immediately following nuclear fission, as distinct from delayed neutrons, which are emitted for some time after fission has occurred. Prompt neutrons comprise more than 99 percent of fission neutrons.

**prompt radiation.** Radiation produced by the primary fission or fusion process, as distinguished from the radiation from fission products, their decay chains, and other later reactions.

**proof spirit.** Designates the alcoholic content of a liquid.

**propane.** A gaseous member of the paraffin series of hydrocarbons which when liquefied under pressure is one of the components of liquefied petroleum (LP) gas.

Contains approximately 2500 British thermal units per cubic foot (at 16°C and 30 inches mercury, and 91,740 British thermal units per gallon); used for domestic heating and for cooking. Chemical formula $C_3H_8$.

**proportional counter tube.** A gas-filled radiation-detection tube operated in that range of applied voltage in which the charge collected per isolated count is proportional to the charge liberated by the initial ionizing effect. The range of applied voltage depends upon the type and energy of the incident radiation.

**proportional meter.** A meter which measures automatically a proportional part of the volume flowing past a metering point.

**proportional region.** The range of applied voltage in a radiation counter tube in which the charge collected per ionizing event is proportional to the charge liberated by that event.

**proration.** The specified sharing of oil and/or gas production among the wells in a particular area.

**prospect.** A geographical area where exploration has shown that it contains sedimentary rocks and structure favorable for the presence of oil or gas; also to search for minerals or oil by looking for surface indications, by drilling boreholes, or both.

**prospecting.** Searching for new deposits; the removal of overburden, core drilling, construction of roads, or any other disturbance of the surface for the purpose of determining the location, quantity, or quality of the natural mineral deposit.

**protection.** Provisions to reduce exposure of persons to radiation; for example, protective barriers to reduce external radiation or measures to prevent inhalation of radioactive materials.

**protective action guide.** The absorbed dose of ionizing radiation to individuals in the general population which would warrant protective action following a contaminating event, such as a nuclear explosion.

**protective clothing.** Special clothing worn by a radiation worker to prevent contamination of his body or his personal clothing.

**protective survey.** An evaluation of the radiation hazards incidental to the production, use, or existence of radioactive materials or other sources of radiation under a specific set of conditions.

**protium.** The isotope of hydrogen having the mass number 1. Light hydrogen.

**proton.** An elementary particle with a single positive electrical charge and a mass approximately 1847 times that of the electron; the nucleus of an ordinary or light hydrogen atom. Protons are constituents of all nuclei. The atomic number of an atom is equal to the number of protons in its nucleus.

**proton synchrotron.** A type of particle accelerator for producing beams of very high energy protons in the billion electron volt range.

**prototype reactor.** A reactor that is the first of a series of the same basic design; sometimes also used to denote a reactor having the same essential features but of a smaller scale than the final series.

**proved reserves.** The estimated quantity of crude oil, natural gas, natural gas liquids, or sulfur which analysis or geological and engineering data demonstrates with reasonable certainty to be recoverable from known oil or gas fields under existing economic and operating conditions.

**proven acreage.** Land under which it is known that gas or oil exists in quantity and condition sufficient to support commercial production.

**province.** The largest unit used by the U.S. Geological Survey to define the areal extent of coal resources.

**proximate analysis.** The determination of the compounds contained in a mixture as distinguished from ultimate analysis, which is the determination of the elements contained in a compound; in the case of coal and coke, the determination of moisture, volatile matter, fixed carbon, and ash.

**Prudhoe Bay field.** The collection of reservoirs in Alaska's North Slope, proved in 1970, and currently in production. This includes reserves of 9.6 billion barrels of oil and 26 Tcf of associated-dissolved gas.

**P.S. detector tube.** A device for estimating the proportion of carbon monoxide in mine air. It consists of a glass tube containing a plug of silica gel impregnated with Pallado sulfite and enclosed between two plugs of purified silica gel which have no reaction with carbon monoxide.

**psia.** Pounds per square inch absolute.

**psig.** Pounds per square inch gage (gauge).

**psychrometer.** A device for measuring the vapor pressure and the relative humidity of the air or the quantity of moisture in the air, employing a wet bulb and a dry bulb thermometer.

**psychrometric.** Pertaining to the state of the atmosphere with reference to moisture.

**pulse.** An electrical signal arising from a single event of ionizing radiation.

**pulse amplifier.** An amplifier designed specifically to amplify the intermittent signals of a radiation-detection instrument, incorporating appropriate pulse-shaping characteristics.

**pulsed reactor.** A type of research with which repeated short, intense surges of power and radiation can be produced. The neutron flux during each surge is much higher than could be tolerated during a steady-state operation.

**pulse height.** The measure of the strength or signal amplitude of a pulse delivered by a detector; measured in volts.

**pulse height analyzer.** An electronic circuit which sorts and records pulses according to height or voltage.

**pulse height selector.** A circuit designed to select and pass voltage pulses in a certain range of amplitudes.

**pulverization.** The crushing or grinding of material into small pieces.

**pulverized fuel.** Finely ground coal or other combustible material, which can be burned as it issues from a suitable nozzle through which it is blown by compressed air.

**pump capacity.** The volume of fluid, at a specified pressure, that a pump can transfer or lift when powered by an engine or motor of any given horsepower.

**pumped-hydro storage.** A utility storage system in which off-peak power is used to pump water to an elevated reservoir for later reconversion to electricity by passage through turbines at peak-demand periods.

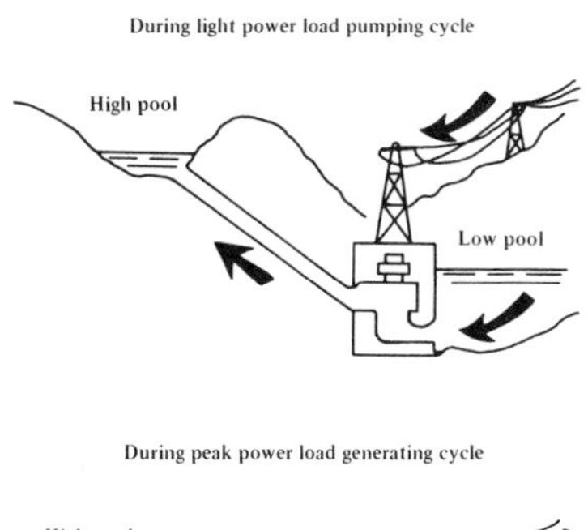


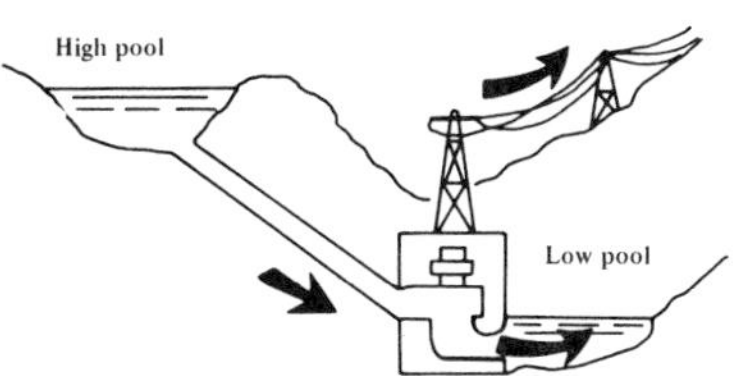


Pumped-storage operation. (*Courtesy U.S. Council on Environmental Quality*)

**pumped storage.** An arrangement in which water is pumped into a storage reservoir at a higher elevation when a surplus of electricity is being generated. During times of peak demand for electricity, this water is then used for the generation of electricity as in a hydroelectric power plant.

**pumpkin.** A reinforcing sleeve welded over a coupling.

**pump pressure.** The force per unit area or pressure against which the pump is acting to force a fluid to flow through a pipeline or drill string; also, the pressure imposed on the fluid ejected from a pump.

**purge.** To free a gas conduit of air, gas, or a mixture of gas and air.

**purge cycle.** As applied to electric pilot igniters, the period from the time of automatic closure of the main gas supply be the safety shutoff device to the time the electrical circuit is reenergized.

**purging.** The act of replacing the atmosphere within a container by an inert substance in such a manner as to prevent the formation of explosive mixtures.

**purification.** The process by which unwanted impurities, such as hydrogen sulfide, are removed from a gas mixture. Purification of gas is accomplished by two principal methods. The dry method in which the gas is passed through some purifying material such as iron oxide mixed with wood shavings, and the wet method in which the gas is brought in contact with some liquid containing an active purifying agent.

**Purox pyrolysis process.** Process for the pyrolysis of solid wastes using pure oxygen to produce a low-Btu gas that can be upgraded to a high-Btu gas.

**pushbutton coal mining.** A fully automatic and remotely controlled system of coal cutting, loading, and face conveying, including self-advancing roof support systems.

**pyranometer.** An instrument for measuring total (beam, diffuse, and reflected) solar radiation.

**pyrheliometer.** Instruments which measure the intensity of direct solar radiation.

**pyrite.** A common mineral that consists of iron disulfide ($FeS_2$) which has a pale brass-yellow color and metallic luster and is burned in making sulfur dioxide and sulfuric acid. Often called "fool's gold."

**pyrogenic.** Equivalent to igneous; formed through the effects of high temperature.

**pyrolysis.** Chemical decomposition by the action of heat; thermal decomposition of organic compounds in the absence of oxygen. (See Fig. p. 353)

**pyrometallurgical processing.** Fuel reprocessing based on reactions involving molten metals at high temperatures.

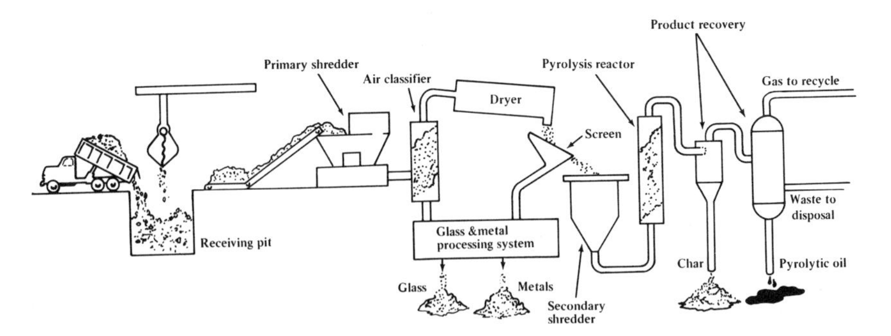


The Garrett pyrolysis system. (*Courtesy U.S. Council on Environmental Quality*)

**pyrometer.** An instrument for measuring high temperatures.

**pyroscope.** An instrument that, by a change in shape or size, indicates the temperature or, more correctly, the combined effect of time and temperature (called heat work).

# Q

**quad.** One quadrillion ($10^{15}$ or 1,000,000,000,000,000) Btus (British thermal units).

**quality factor.** The factor by which absorbed dose is to be multiplied to obtain a quantity that expresses on a common scale, for all ionizing radiations, the irradiation incurred by exposed persons.

**quantum.** Unit quantity of energy according to the quantum theory. It is equal to the product of the frequency of radiation of the energy and $6.6256 \times 10^{-27}$ erg-seconds. The photon carries a quantum of electromagnetic energy.

**quantum theory.** The statement according to Max Planck that energy is not emitted or absorbed continuously but in units or quanta. A corollary of this theory is that the energy of radiation is directly proportional to its frequency.

**quarry.** An open or surface working, usually for the extraction of building stone, as slate, limestone, etc.

**quench.** To cool suddenly; also to limit or stop the electrical discharge in an ionization detector.

**quenching.** Cooling by immersion in oil or water bath or spray; also applies to cooling in salt and molten metal baths or by means of an air blast; rapid cooling; the act of terminating an ionizing current by the use, for example, of an appropriate gas or vapor filling or by momentary reduction of the applied potential difference.

**quenching counter tube.** The process of terminating a discharge in a radiation counter tube by inhibiting continuous or multiple discharges caused by a single ionizing event.

**quench tank.** A water-filled tank used to cool incinerator residues.

**quitclaim.** Specifically a legal instrument by which some right, title, interest, or claim by one person in or to an estate held by himself or another is released to another. It is sometimes used as a simple but effective conveyance for making a grant of lands whether by way of release or as an original conveyance; also a document in which a mining company sells its surface rights but retains its mineral rights.

# *R*

**rabbit.** A device for moving a sample rapidly from one place (such as inside a research reactor) to another place (such as a radiochemistry laboratory).

**race.** An aqueduct or channel for conducting water to or from the place where it performs work.

**raceway.** Conduits, moldings, and other hollow material, often concealed, through which wires are run from one outlet to another.

**rad.** Radiation absorbed dose; the basic unit of absorbed dose of ionizing radiation. A dose of one rad means the absorption of 100 ergs of radiation energy per gram of absorbing material.

**radiant energy.** Energy that radiates or travels outward in all directions from its source.

**radiant heating.** Usually a system of heating by surfaces at higher than body temperatures whereby the rate of heat loss from human beings by radiation is controlled; the heat gained in a house or room by heat emitted from large, sun-warmed surfaces, such as walls or floors.

**radiant sky cooling system utilizing collectors.** A space conditioning system that circulates a fluid through the collectors during those hours which have the appropriate conditions for cooling (usually night and early morning). The cool liquid or gas is stored and cool air is transferred to the building interior during the cooling period.

**radiant systems.** A solar space conditioning system that transfers solar heated or climatically cooled fluid to a building component (floor or wall), thereby providing space conditioning to the building interior by radiation.

**radiation.** The emission and propagation of energy through matter or space by means of electromagnetic disturbances which display both wave-like and particle-like behavior; particles known as photons; streams of fastmoving alpha and beta particles, free neutrons, and cosmic radiation; also the energy so propagated. Nuclear radiation is that emitted from atomic nuclei in various nuclear reactions, including alpha, beta, and gamma radiation and neutrons.

**radiation absorbed dose (rad).** The basic unit of absorbed dose of ionizing radiation. A dose of 1 rad means the absorption of 100 ergs of radiation energy per gram of matter.

**radiation accidents.** Accidents resulting in the spread of radioactive material or in the exposure of individuals to radiation.

**radiation area.** Any accessible area in which the level of radiation is such that a major portion of an individual's body could receive in any one hour a dose in excess of 5 millirem, or in any 5 consecutive days a dose in excess of 150 millirem.

**radiation burn.** Radiation damage to the skin. Beta burns result from skin contact with or exposure to emitters of beta particles. Flash burns result from sudden thermal radiation.

**radiation chemistry.** The branch of chemistry that is concerned with the chemical effects, including decomposition, of energetic radiation or particles on matter.

**radiation chopper.** A device for periodically interrupting a flux or beam of radiation.

**radiation count.** A pulse that has been registered, corresponding either to an ionizing event or to an extraneous disturbance; also the number of pulses recorded in a specific period.

**radiation counter tube.** A radiation detector consisting of a gas-filled tube in which individual ionizing events give rise to discrete electrical pulses.

**radiation damage.** A general term for the alteration of properties of a material arising from exposure to ionizing radiation.

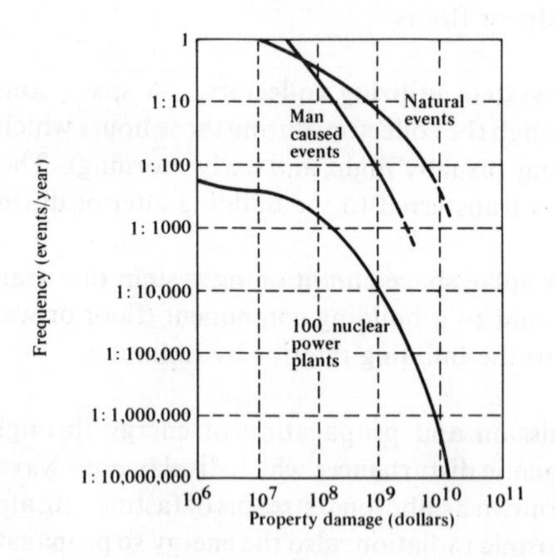


Average risk of accidents by various causes. (*Courtesy U.S. ERDA—69*)

**radiation detector.** A device for measuring the level of ionizing radiation (or quantity of radioactive material) and possibly giving warning when it exceeds a prescribed amount. It may also give quantitative information on dose or dose rate. Often called a geiger counter. (See Fig. p. 357)

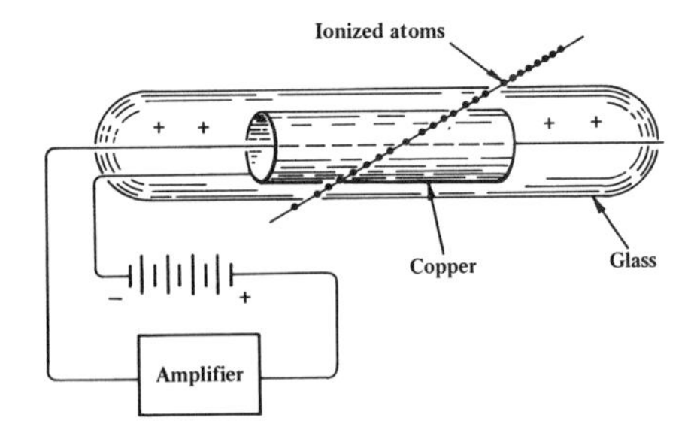


Geiger counter. (*Courtesy Van Nostrand's Scientific Encyclopedia*)

**radiation dose.** Accumulated exposure to radiation during a specified period of time.

| Dose (rems) | 0 to 100 | 100 to 200 | 200 to 600 | 600 to 1000 | 1000 to 5000 |
|---|---|---|---|---|---|
| Vomiting | None | 5% to 50% | 100% at 300 rems | 100% | 100% |
| Latent period | None | 3 hours | 2 hours | 1 hour | 30 minutes |
| Characteristic sign | | Leukopenia | Leukopenia<br>Purpura<br>Hemorrhage | Diarrhea<br>Fever | Convulsions<br>Tremor<br>Lethargy |
| Therapy | Reassurance | Reassurance | Blood transfusion<br>Antibiotics | Possible bone marrow trans-plant | Sedatives |
| Convalescent period | None | Several weeks | 1 to 12 months | Long | |
| Incidence of death | None | None | 0 to 80% | 80% to 100% | 90% to 100% |
| Time of death | | | 2 months | 2 months | 2 weeks |
| Cause of death | | | Hemorrhage<br>Infection | Hemorrhage<br>Infection | Circulatory collapse |

Effects of acute whole-body radiation. (*Courtesy Basic Nuclear Engineering*)

**radiation dosimetry.** The measurement of the amount of radiation delivered to a specific place or the amount of radiation that was absorbed there.

**radiation illness.** An acute organic disorder that follows exposure to relatively severe doses of ionizing radiation. It is characterized by nausea, vomiting, diarrhea, blood cell changes, and in later stages by hemorrhage and loss of hair.

**radiation monitoring.** Continuous or periodic determination of the amount of radiation present in a given area.

**radiation protection.** Legislation and regulations to protect the public and laboratory or industrial workers against radiation; also, measures to reduce exposure to radiation.

**radiation protection guide.** The officially determined radiation doses which should not be exceeded without careful consideration of the reasons for doing so; standards established by the Federal Radiation Council which are equivalent to what was formerly called the maximum permissible dose or maximum permissible exposure.

**radiation quantity.** A quantity characteristic of a particular radiation, which is capable of being measured.

**radiations.** Specific units or types of radiation.

**radiation sensor.** The component of a radiation detector which responds directly to ionizing radiation.

**radiation shielding.** Reduction of radiation by interposing a shield of absorbing material between any radioactive source and a person, laboratory area, or radiation-sensitive device.

**radiation source.** Usually a man-made, sealed source of radioactivity used in teletherapy and radiography as a power source for batteries or in various types of industrial gauges; also machines such as accelerators and radioisotopic generators, and natural radionuclides.

**radiation standards.** Exposure standards, permissible concentrations, rules for safe handling, regulations for transportation, regulations for industrial control of radiation, and control of radiation exposure by legislative means.

**radiation sterilization.** Use of radiation to cause a plant or animal to become incapable of reproduction; also the use of radiation to kill all forms of life (especially bacteria) in food, surgical sutures, etc.

**radiation survey.** An evaluation of the radiation hazard potential associated with any process or device involving radiation.

**radiation therapy.** Treatment of disease with any type of radiation; often called radiotherapy.

**radiation warning symbol.** An officially prescribed symbol (a magenta trefoil on a yellow background) which is always displayed when a radiation hazard exists.

**radiative capture.** A nuclear capture process whose prompt result is emission of electromagnetic radiation only, as when a nucleus captures a neutron and emits gamma rays.

**radiator.** A heating unit which transfers heat by radiation to objects within visible range, and by conduction to the surrounding air which in turn is circulated by natural convection.

**radioactive.** Exhibiting radioactivity or pertaining to radioactivity.

**radioactive chain.** A radioactive series.

**radioactive cloud.** A mass of air and vapor in the atmosphere carrying radioactive debris from a nuclear explosion.

**radioactive contamination.** Desposition of radioactive material in any place where it may harm persons, spoil experiments, or make products or equipment unsuitable or unsafe for some specific use; the presence of unwanted radioactive matter; also radioactive material found on the walls of vessels in used-fuel processing plants, or radioactive material that has leaked into a reactor coolant.

**radioactive cooling.** The reduction of the radioactivity of a material by radioactive decay.

**radioactive dating.** A technique for measuring the age of an object or sample of material by determining the ratios of various radioisotopes or products of radioactive decay it contains. For example, the ratio of carbon-14 to carbon-12 reveals the approximate age of bones, pieces of wood, or other archeological or geological specimens that contain carbon extracted from the air at the time of their origin.

**radioactive decay.** The spontaneous transformation of one nuclide into a different nuclide or into a different energy state of the same nuclide. The process results in a decrease with time of the number of the original radioactive atoms in a sample. It involves the emission from the nucleus of alpha particles, beta particles (or electrons), or gamma rays; or, the nuclear capture or ejection of orbital electrons; or fission.

**radioactive decay chain.** A series of nuclides in which each member transforms into the next through radioactive decay (not including spontaneous fission) until a stable nuclide has been formed.

**radioactive equilibrium.** That condition in which the activities of the members of a radioactive chain decrease exponentially in time with the half-life of the chain precursor. Such radioactive equilibrium is only possible when the half-life of the precursor is longer than that of any other chain member.

**radioactive half-life.** Time required for a radioactive substance to lose 50 percent of its activity by decay. Each radionuclide has a unique half-life.

**radioactive isotope.** A radioisotope.

**radioactive logging.** The logging process whereby a neutron source is lowered down the hole, followed by a recorder. When a hydrogen-bearing strata is located (which may be petroleum or water), the neutrons are absorbed. They disintegrate the hydrogen atoms, releasing alpha particles. The higher the alpha concentration, the higher the hydrogen concentration.

**radioactive series.** A succession of nuclides, each of which transforms by radioactive disintegration into the next until a stable nuclide results. The first member is called the parent, the intermediate members are called daughters, and the final, stable member is called the end product.

**radioactive source.** Any quantity of radioactive material which is intended for use as a source of ionizing radiation.

**radioactive standard.** A sample of radioactive material, usually with a long half-life, in which the number and type of radioactive atoms at a definite reference time is known. These are used in calibrating radiation measuring equipment or for comparing measurements in different laboratories.

**radioactive tracer.** A small quantity of radioactive isotope (either with carrier or carrier-free) used to follow biological, chemical, or other processes, by detection, determination, or localization of the radioactivity.

**radioactive waste.** Equipment and materials (from nuclear operations) which are radioactive and for which there is no further use. Wastes are generally classified as high-level (having radioactivity concentrations of hundreds to thousands of curies per gallon or cubic foot), low-level (in the range of 1 microcurie per gallon or cubic foot), or intermediate (between these extremes).

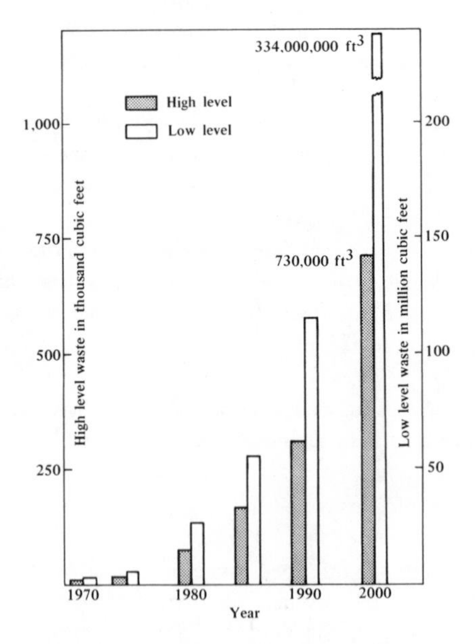


Annual production of radioactive waste. (*Courtesy National Science Teachers Association*)

**radioactivity.** The spontaneous decay or disintegration of an unstable atomic nucleus, usually accompanied by the emission of ionizing radiation.

**radioactivity concentration guide.** The concentration of radioactive material in an environment that would result in doses equal, over a period of time, to those in the Radiation Protection Guide. This Federal Radiation Council term replaces the former maximum permissible concentration.

**radiobiology.** The body of knowledge and the study of the principles, mechanisms, and effects of ionizing radiation on living matter.

**radiocarbon.** A radioactive isotope of carbon. It has a half-life of 5770 years.

**radiochemistry.** The body of knowledge and the study of the chemical properties and reactions of radioactive materials.

**radioecology.** The body of knowledge and the study of the effects of radiation on species of plants and animals in natural communities.

**radioelement.** An element containing one or more radioactive isotopes; a radioactive element.

**radiogenic.** Of radioactive origin; produced by radioactive transformation.

**radiographic inspection.** Method used to determine flaws in pipe or other metals by use of a machine which emits x-rays or gamma rays, which penetrate the metal and are transcribed onto film.

**radiography.** The use of ionizing radiation for the production of shadow images on a photographic emulsion. Some of the rays (gamma rays or x-rays) pass through the subject, while others are partially or completely absorbed by the more opaque parts of the subject and thus cast a shadow on the photographic film.

**radioisotope.** A radioactive isotope. An unstable isotope of an element that decays or disintegrates spontaneously, emitting radiation. More than 1300 natural and artificial radioisotopes have been identified.

**radioisotopic generator.** A small power generator that converts the heat released during radioactive decay directly into electricity. These generators generally produce only a few watts of electricity and use thermoelectric or thermionic converters; some also function as electrostatic converters to produce a small voltage; sometimes called an "atomic battery".

**radiology.** The science which deals with the use of all forms of ionizing radiation in the diagnosis and the treatment of disease.

**radioluminescence.** Visible light caused by radiations from radioactive substances; for example, the glow from luminous paint containing radium and crystals of zinc sulfide, which give off light when struck by alpha particles from the radium.

**radiolysis.** The dissociation (or decomposition) of molecules by radiation; for example, a small portion of water in a reactor core dissociates into hydrogen and oxygen during operation of the reactor.

**radiometric analysis.** An analytical technique for measuring, by radioactive tracer methods, elements which themselves are not radioactive.

**radiometric prospecting.** Finding minerals using a geiger counter or scintillometer that measures radioactivity.

**radiomimetic substances.** Chemical substances which cause biological effects similar to those caused by ionizing radiation.

**radiomutation.** A permanent, transmissible change in form, quality, or other characteristic of a cell or offspring from the characteristics of its parent, due to radiation exposure.

**radionuclide.** A radioactive nuclide.

**radioresistance.** A relative resistance of cells, tissues, organs, or organisms to the injurious action of radiation.

**radiosensitivity.** A relative susceptibility of cells, tissues, organs, or organisms to the injurious action of radiation.

**radiotherapy.** Radiation therapy.

**radium.** A radioactive metallic element with atomic number 88. As found in nature, the most common isotope has an atomic weight of 226. It occurs in minute quantities associated with uranium in pitchblende, carnotite, and other minerals; the uranium decays to radium in a series of alpha and beta emissions. By virtue of being an alpha- and gamma-emitter, radium is used as a source of luminescence and as a radiation source in medicine and radiography. Chemical symbol Ra.

**radon.** A radioactive element, one of the heaviest gases known. Its atomic number is 86, and its atomic weight is 222. It is a daughter of radium in the uranium radioactive series. Chemical symbol Rn.

**radon breath analysis.** Examination of exhaled air for the presence of radon to determine the presence and quantity of radium in the human body.

**radon gas.** A radioactive gaseous element formed by disintegration of uranium.

**raffinate stream.** In the solvent extraction process, the unneeded material in the output stream, as opposed to the "pregnant" stream containing the recovered valuable material such as uranium and plutonium.

**rake.** A train of mineral trucks; also a timber placed at an angle.

**Ralston's classification of coal.** A classification based on the percentage of carbon, hydrogen, and oxygen in the ash-, moisture-, sulfur-, and nitrogen-free coal. These figures are plotted on trilinear coordinates giving well-defined zones of bituminous coals, lignites, peats, etc.

**ramjet engine.** A simple type of jet propulsion system. The engine has a continuous inlet of air in its forward end, so that air is taken in and compressed only by the forward motion of the vehicle.

**ramp insertion of reactivity.** A linear increase of reactivity with time.

**random source of pollution.** Spillage of oil and hazardous chemicals on land or in the waters.

**Raney nickel catalyst.** Nickel sponge used as a catalyst in the hydrogenation of organic materials and the methanation of synthesis gas to methane.

**range.** In the nuclear field, the distance that a charged particle penetrates a material before it ceases to ionize.

**rangeability.** Ratio of maximum operating capacity to minimum operating capacity within a specified tolerance and operating condition.

**range-energy relation.** An expression giving the range of an energetic particle in matter as a function of the energy of the particle.

**range land.** Land that is more suitable for management by ecological rather than agronomic principles.

**rank.** A classification of coal according to percentage of fixed carbon and heat content. Higher ranked coal is presumed to have undergone more geological and chemical change than lower ranked coal.

**Rankine cycle.** The steam-Rankine cycle employing steam turbines has been the mainstay of utility thermal electric power generation for many years. The cycle, as developed over the years, is sophisticated and efficient. A typical cycle uses superheat, reheat, and regeneration. Heat exchange between flue gas and inlet air adds several percentage points to boiler efficiency in fossil-fueled plants. Modern steam Rankine systems operate at a cycle top temperature of about 800K with efficiencies of about 40 percent. All characteristics of this cycle are well suited to use in solar plants.
(See Fig. p. 364)

**Rankine-cycle solar air conditioner.** A small solar-powered turbine is used to run a conventional air conditioning unit. A working fluid is heated until it expands and

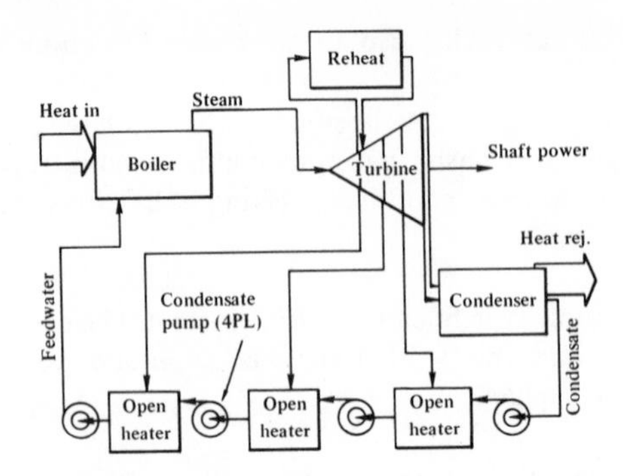


Schematic diagram of Rankine cycle. (*Courtesy Van Nostrand's Scientific Encyclopedia*)

turns a turbine. The turbine can either drive a conventional air conditioner or turn a generator and make electricity. Solar-heated hot water at 102°C enters the system from the solar panels. It flows through coils in three sealed tanks (vapor separator, boiler, and preheater), where the liquid refrigerant picks up heat and vaporizes. The Freon vapor at 93°C and 54 psi flows to the three-inch turbine. The Freon is cooled in a heat exchanger and condenser, and returns through the feed pump and gearbox back to the boiler where the cycle repeats. When there is no solar heat to run the turbine, the air conditioner can run on line power.

**Rankine engine.** A reversible heat engine.

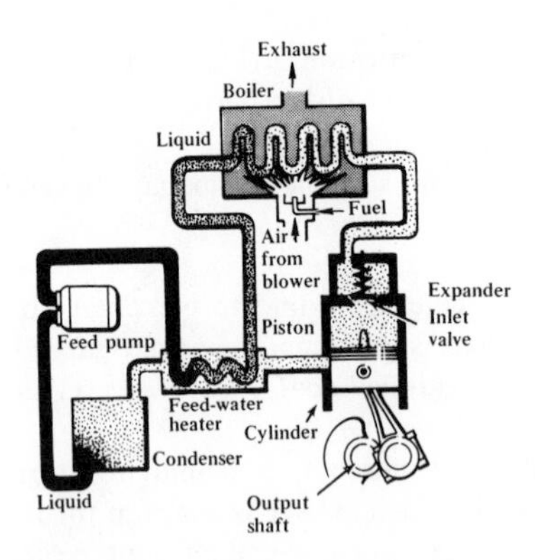


Diagram of a Rankine engine. (*Courtesy Popular Science, April 1976, p. 84*)

**Rankine scale.** A temperature scale in which the zero is absolute zero and the size of the degrees is that of the Fahrenheit scale; the temperatures in degrees Rankine are equivalent to degrees Fahrenheit plus 459.6.

**Raoult's law.** Principle stating that the vapor pressure of the solvent in a solution is proportional to the mole fraction of the solvent. This law has been of fundamental importance in the development of the theory of solutions.

**rare earths.** A group of 15 chemically similar metallic elements including Elements 57 through 71 on the Periodic Table of the Elements, also known as the Lanthanide Series.

**rash.** An impure and unmerchantable coal; a substance grading about halfway between a coal and a shale. It looks somewhat like an oil shale, which is characterized by a brown streak and leathery appearance with parting planes often smooth and polished. Rash is more flexible than oil shale and usually occurs in very thin flakes or sheets at the bottom, the top, or within the seam.

**rate base.** The value established by a regulator authority, upon which a utility is permitted to earn a specified rate of return.

**rated horsepower.** Theoretical horsepower of an engine based on dimensions and speed.

**rated load.** The kilowatt power output which can be delivered continuously at the rated output voltage; may also be designated as the 100 percent load or full-load rating of the unit.

**rated output capacity.** For a wind machine, it is the output power of the machine operating at the constant speed and output power corresponding to the rated wind speed.

**rated wind speed.** The lowest wind speed at which the rated output power of a wind machine is produced.

**rate of flow.** The volume or units of a material passing a given point in a system per unit of time.

**rating.** Limits placed on the operating conditions of a machine, apparatus, or device based on its design characteristics.

**ratio of specific heats.** For gases, the ratio of the specific heat at constant pressure to the specific heat at constant volume.

**raw coal.** Coal which has received no preparation other than possibly screening.

**raw fuel.** A fuel which is used in the form in which it is mined or obtained; for example, coal, lignite, peat, wood, mineral oil, natural gas.

**raw gas.** Natural gas in its natural state, existing in or produced from a field.

**Rayleigh wave.** A surface wave associated with the free boundary of a solid. The wave is of maximum intensity at the surface and diminishes quite rapidly as one proceeds into the solid.

**reaction rate.** The number of neutrons (or nuclei) undergoing a reaction per unit time.

**reactive power.** The portion of "apparent power" that does not work; for example, that portion supplied by generators or by electrostatic equipment, such as capacitors.

**reactivity.** A measure of the departure of a nuclear reactor from criticality. It is about equal to the effective multiplication factor minus one and is thus precisely zero at criticality. If there is excess reactivity (positive reactivity), the reactor is supercritical and its power will rise. Negative reactivity (subcriticality) will result in a decreasing power level.

**reactivity coefficient.** The change in reactivity caused by inserting a small amount of a substance in a reactor.

**reactivity oscillator.** Any device used to cause the reactivity of a system to vary periodically to gain information either about the reactor as a whole or about the absorbing sample.

**reactivity pressure coefficient.** The change in reactivity per unit change of pressure.

**reactor.** A device in which a fission chain reaction can be initiated, maintained, and controlled. Its essential component is a core with fissionable fuel. It is sometimes called an atomic pile. Also refers to a vessel in which coal-conversion reactions take place. Specific reactors may be found under alphabetical listing.

| Consequence | Largest Accident | Average Accident |
|---|---|---|
| Death | 92 | 0.05 |
| Acute illness | 200 | 0.1 |
| Property damage* | 1.7 | 0.5 |
| Probability (per reactor year) | 1 in 5 billion | 1 in 50,000 |

*In billions of 1973 dollars.

Reactor accidents and their consequences. (*Courtesy National Science Teachers Association*)

**reactor containment.** The prevention of release, even under the conditions of a reactor accident, of unacceptable quantities of radioactive material beyond a controlled area.

**reactor control.** The intentional variation of the reaction rate in a reactor or adjustment of reactivity to achieve or maintain a desired state of operation.

**reactor control system.** An association of equipment, assemblies, and materials used for the purpose of reactor control. Materials such as boron or cadmium are able to absorb neutrons and, by removing neutrons from the system, shut down a reactor, preventing new fissions from occurring. Common methods of introduction include the mechanical insertion of control rods into the core and the addition of liquid solutions of these neutron-absorbing elements to the water moderator. Most water reactors have both methods of control available.

**reactor core.** The part of a nuclear power plant which contains control rods and the fuel elements where fissioning occurs.

**reactor excursion.** Very rapid increase of reactor power above the normal operating level, either deliberately caused for experimental purposes, or accidental.

**reactor oscillator.** A device which produces periodic variations of reactivity by the oscillatory movement of an inserted material.

**reactor safety circuit instrumentation.** Instruments that monitor what is happening in the core. Improper signals concerning temperature, pressure, or other unwanted conditions will immediately shut down the reactor. Each safety system has one or more backup systems in case there is a failure in the primary system.

**reactor safety fuse.** A self-contained device designed to respond to excessive temperature or neutron flux density in a reactor and to act to reduce the reaction rate to a safe level.

**reactor vessel.** The principal vessel surrounding the reactor core.

**reagent.** Any material that causes a chemical reaction when added to a second substance.

**real calorific value.** The calorific value when determined by a calorimeter in the laboratory.

**real disposable income.** The figure, in dollars of constant value, obtained by subtracting corporate earnings not paid out as dividends, depreciation, and all taxes from gross national product and then adding welfare-type transfer payments and government interest payments. This figure is intended to represent the income which the public has available for making purchases.

**real gross national product.** Gross National Product in constant dollars.

**real income.** Current dollar income corrected for price level changes relative to some base period as measured by a suitable price index, or the Implicit GNP Deflator. Also termed "constant dollar" income.

**real power.** The rate of supply of energy, measured in kilowatts.

**reciprocating engine.** An apparatus which converts the energy in a fluid to mechanical energy by means of the expansion of the fluid (gas) against a piston. It normally includes a cylinder, closed by a piston connected by means of a connecting rod to a crankshaft; a valve mechanism admits and discharges fluid at appropriate times in the cycle.

**recirculated air.** Air returned from a space to be heated, conditioned, or cleaned then redistributed to the space.

**reclaimed oil.** Lubricating oil that is reprocessed for reuse.

**reclamation.** The process of reconverting mined land to its former or other productive uses; also the recovery of coal from a mine that has been abandoned because of fire, water, or other cause.

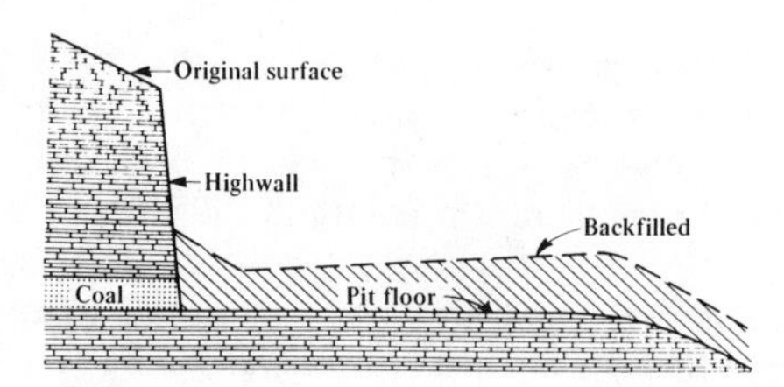


Reclamation by reshaping the spoil bank and partial backfilling. (*Courtesy U.S. Council on Environmental Quality*)

**recoverable heat.** That portion of thermal input to a prime mover that is not converted to mechanical power and can be reclaimed for utilization.

**recoverable resource.** That portion of a resource expected to be recovered by present-day techniques and under present economic conditions, including geologically expected but unconfirmed resources as well as identified reserves.

**recovered energy.** Reusing heat or energy that would otherwise be lost.

**rectification.** The process by which electric energy is transferred from an alternating-current circuit to a direct-current circuit.

**rectifier.** A device for converting alternating current to direct current; one use of this current is for external corrosion control of pipe and other metals; also the container or chamber in a cooling system where, typically, ammonia vapor is separated from water vapor.

**Rectisol process.** A process for the purification of coal-gasification gas based on the capability of cold methanol to absorb all gas impurities in a single step. Gas naphtha, unsaturated hydrocarbons, sulfur, hydrogen sulfide, and carbon dioxide are removed from the gas stream by the methanol at temperatures below 0°C.

**recuperator.** A continuous heat exchanger in which heat is conducted from the products of combustion to incoming air through flue walls.

**recycling.** The repetition of a particular process; the return of a stream or part of a stream to a previous process or location for additional recovery of the desired components; also the reuse of fissionable material after it has been recovered by chemical processing from spent or depleted reactor fuel, reenriched, and then refabricated into new fuel elements.

**red dog.** Material of a reddish color resulting from the combustion of shale and other mine waste in dumps on the surface; a gob pile after it has burned; also material generally used as a road-surfacing material.

**Redox battery.** An advanced battery system in early research stage under ERDA-supported programs, which uses liquid reactants such as metal salts dissolved in water and offers promise of longer-term (weekly) storage for utilities.

**reduced crude.** A residential product remaining after the removal by distillation or other means, of an appreciable quantity of the more volatile components of crude oil.

**reducer.** A fitting having a larger size at one end than the other and always threaded inside, unless specifically flanged or for some special joint.

**reducing gas.** Hydrogen, superheated steam, or any gas used as a reducing agent in redox reactions.

**refine.** To cleanse or purify by removing undesired components; to process a material to make it usable.

**refined products domestic demand.** A calculated value, computed as domestic production plus net imports (imports less exports), less the net increase in primary stocks.

**refined products imports.** Imports of motor gasoline, naphtha-type jet fuel, kerosene-type jet fuel, liquefied petroleum gases, kerosene, distillate fuel oil, residual fuel oil, petrochemical feedstocks, special naphthas, lubricants, waxes, and asphalt.

**refiner acquisition cost.** The cost to the refiner, including transportation and fees, of crude petroleum.

**refinery.** A device (usually a tower) or process which heats crude oil so that it separates into chemical components, which are then distilled off as more usuable substances. Simple structure components vaporize first. Typical crude fractions are: methane and ethane (the gasolines); propane and butane; kerosene, fuel oil, and lubricants; jelly paraffin, asphalt, and tar. The term is also applied to the plant in which the refinery process is carried out.

**refinery gas.** Any form or mixture of gas gathered in a refinery from the various stills.

**refinery gate.** The point at which oil or natural gas enters or leaves refinery facilities via pipeline, ship, truck, rail, or other transport mode.

**refinery pool.** An expression for the mixture obtained if all blending stocks for a given type of product were blended together in production ratio. Usually used in reference to motor gasoline octane rating.

**refining.** The separation of petroleum into distinct fractions to obtain useful products. The main fractions are gas, gasoline, naphtha, kerosine, gas oils, and residual fuel oil.

**reflected radiation.** Solar radiation reflected from a surface such as the ground (and, as used here, is ultimately incident on the collector surface).

**reflectance.** The ratio of radiation reflected from a surface to the total radiation incident on the surface.

**reflection.** The bounding back of light rays or ether rays as they strike a solid surface. In seismic prospecting, the returned energy (in wave form) from a shot which has been reflected from a velocity discontinuity back to a detector; the indication on a record of reflected energy.

**reflective loss.** The energy which strikes a surface and is not absorbed but reflected from it.

**reflectivity.** The ratio of radiant energy reflected by a body to that falling upon it.

**reflectometer.** A device for measuring the reflectances of light or other radiant energy.

**reflector.** A layer of material immediately surrounding a reactor core which scatters back or reflects into the core many neutrons that would otherwise escape. The returned neutrons can then cause more fissions and improve the neutron economy of the reactor. Common reflector materials are graphite, beryllium, and natural uranium.

**reflector control.** Reactor control by adjustment of the properties, position, or size of the reflector in such a way as to change the reactivity.

**reflux.** A distillate fraction having a certain boiling range which is introduced into fractionating equipment to bring about a cooling effect, resulting in more intimate contact between the vapors and thus in more efficient fractionation.

**reformer.** A device that processes fuel to be used in the fuel cell assembly. A single cell of the assembly generates roughly 1 volt of direct current and will create roughly 100 to 200 watts of electricity for each square foot of electrode cross-sectional area.

**reforming.** The cracking of petroleum naphtha or of straight-run gasoline of low octane number, usually to form gasoline containing lighter constituents and having a higher octane number.

**reforming processes.** A group of proprietary processes in which low-grade or low molecular weight hydrocarbons are catalytically reformed to higher grade or higher molecular weight materials; also applies to the endothermic reforming of methane for the production of hydrogen by the reaction of methane and steam in the presence of nickel catalysts.

**refraction.** A change of direction of a ray of light when it passes from one medium to another of different optical density.

**refractories.** Materials capable of withstanding extremely high temperatures and having relatively low thermal conductivities.

**refractory grate.** The assembly within or upon which refractory material is supported.

**refrigerant.** A substance which will absorb heat while vaporizing and whose boiling point and other properties make it useful as a medium for refrigeration.

**refrigeration capacity.** The rate of heat removal by a refrigerating system, usually expressed in British thermal units per hour or in tons.

**refrigeration cycle.** The full sequence of condensation and evaporation. The heat of evaporation is obtained from the material to be cooled.

**refrigeration ton.** The heat required to melt one short ton (2000 pounds) of ice in 24 hours. The latent heat of ice being 144 British thermal units per pound:

$$1 \text{ refrigeration ton} = \frac{144 \times 2000}{24} = 12{,}000 \text{ British thermal units}$$

**regenerative heating.** The process of utilizing heat which must be rejected in one part of the cycle to perform a useful function in another part of the cycle.

**regenerative reactor.** A converter or breeder reactor in which part or all of the fuel produced is used in the reactor.

**regenerative rockbed cooling system.** A space conditioning system that circulates air through rock storage during those hours which have the appropriate conditions for cooling (usually night and early morning). Water spray evaporate cooling of the rockbed is sometimes included as part of the system. The cool air is transferred to the building interior as required.

**regenerator.** A device in which the hot gases, usually waste combustion gases, pass through a set of chambers containing firebrick structures, to which the sensible heat is given up. The direction of hot gas flow is diverted periodically to another set of chambers and cold incoming combustion gas or air is preheated in the hot chambers.

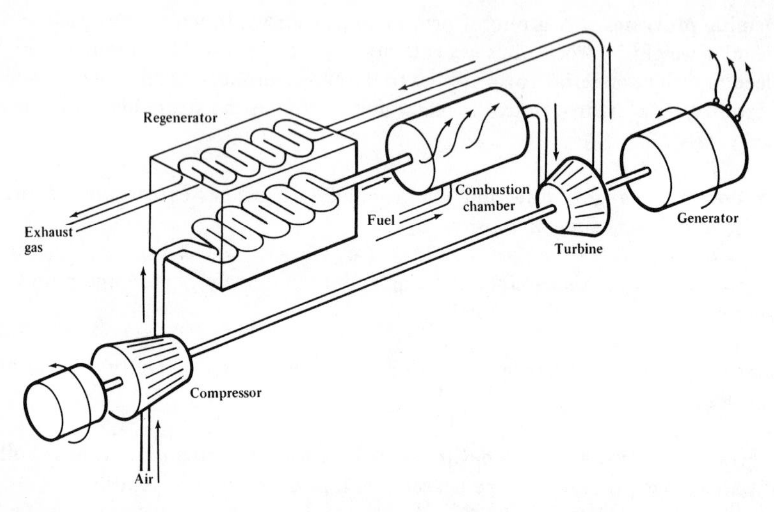


Diagram of a regenerative cycle gas turbine system. (*Courtesy U.S. Council on Environmental Quality*)

**registration.** The meter-dial positions of a gas meter index. The difference between two successive registrations indicates the volume of gas that has passed through the meter.

**regrading.** The movement of earth over a depression to change the shape of the land surface.

**regulating rod.** A reactor control rod used for making frequent fine adjustments in reactivity.

**regulator.** An opening in a wall or door in the return airway of a district to increase its resistance and reduce the volume of air flowing. It consists of a sliding shutter which can be adjusted to any proportion of the maximum aperture.

**regulator vent.** An atmospheric connection to the diaphragm of the regulator.

**rehabilitation.** Returning the land to a form in conformity with a prior land use plan, including a stable ecological state that does not contribute substantially to environmental deterioration.

**reheater load.** The amount of sensible heat in British thermal units per hour, restored to the air in reheating.

**Reid vapor pressure.** A test for gasolines; a measure of the vapor pressure of a sample at 38° C reported in pounds per square inch.

**reject.** The material extracted from the feed coal during cleaning for retreatment or discard; the stone or dirt discarded from a coal preparation plant, washery, or other process as of no value.

**relative biological effectiveness (RBE).** A factor used to compare the biological effectiveness of different types of ionizing radiation. It is the inverse ratio of the amount of absorbed radiation, required to produce a given effect, to a standard (or reference) radiation required to produce the same effect.

**relative humidity.** The amount of moisture contained by an atmosphere compared with the maximum amount that it could contain at the same temperature, expressed as a percentage.

**relative importance.** The average number of neutrons which must be added to a critical system to keep the chain reaction rate constant.

**relative wind.** Velocity of air with reference to a body in it.

**relaxation length.** The distance in a material through which nuclear radiation must pass in order to reduce (or relax) by nuclear interaction the intensity.

**released oil.** That portion of the base production control level for a property which is equal to the volume of new oil produced in that month and which may be sold above the ceiling price. The amount of released oil may not exceed the base production control level for that property.

**relief opening.** The opening provided in a draft hood to permit the ready escape to the atmosphere of the flue products from the draft hood.

**remaining reserves.** Those quantities of crude oil, natural gas, natural gas liquids, and sulfur as estimated under proved or probable reserves after deducting those quantities produced up to the respective date of the estimate.

**remote maintenance.** Maintenance of radioactive or contaminated equipment by means of a manipulator operated from a shielded position. (See Fig. p. 374)

**removal cross section.** An effective cross section ascribed to a material inserted between a fission neutron source and a thick hydrogenous medium.

**rending.** Breaking coal into lumps with a minimum of smalls.

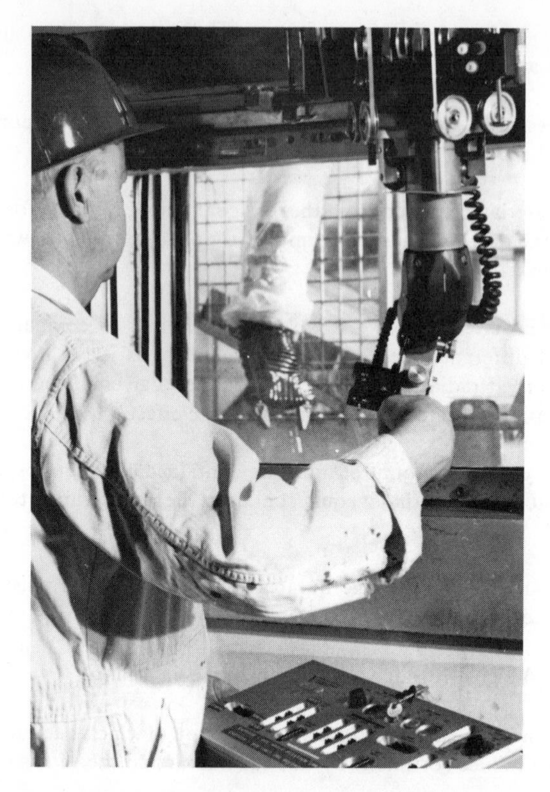

Photograph showing an operator using a remote manipulator from behind a lead-glass shielding window. (*Courtesy U.S. Atomic Energy Commission*)

**renewable resources.** Sources of energy that are regenerative or virtually inexhaustible such as solar energy or geothermal energy.

**repressuring.** Forcing natural gas under pressure into the oil reservoir in an attempt to increase the recovery of crude oil; also done with water or compressed air.

**reprocessing.** The processing of reactor fuel to recover the created plutonium-239 and the unused uranium-235, and to remove the unused fission products.

**required radiation.** The area of radiator surface required, based on the heat loss computation for the space to be heated.

**reradiation.** Reradiating heat after an object has received radiation or is otherwise heated.

**re-refined engine oil.** Waste oil collected from gas stations is re-refined in refinery plant to produce a satisfactory base stock oil. The process consists of first dehydrating and then treating the waste oil with acid to remove sludge and other contaminants. A clay slurry absorbs other undesirable elements in the oil and is then filtered out. An ashless base mineral oil remains, to which is added fresh additives and virgin oil to complete the process.

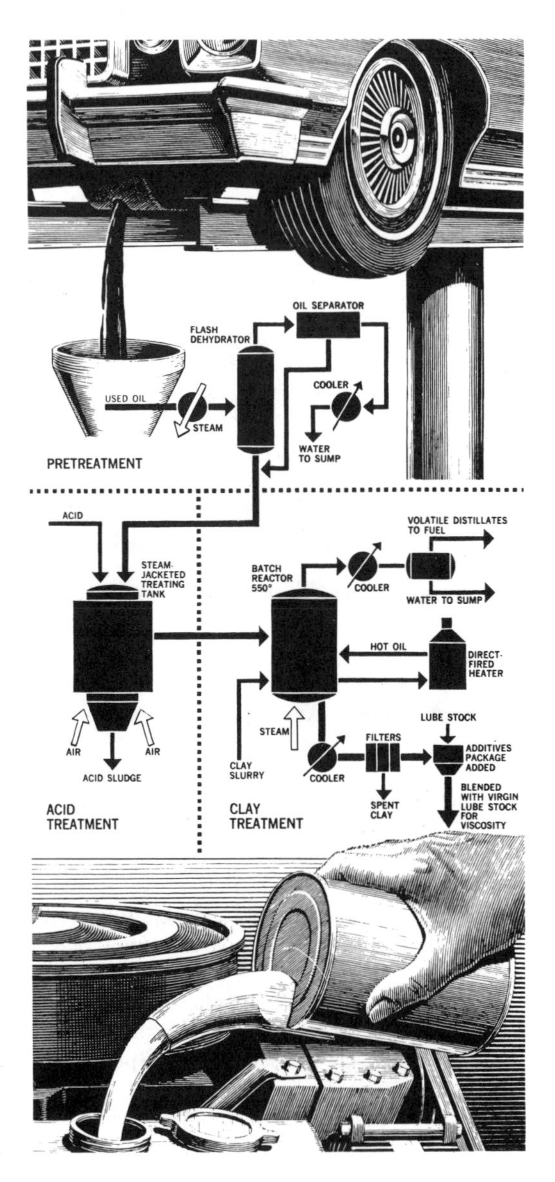


Re-refining waste oil collected from gas stations. (*Courtesy Popular Science, August 1976, p. 63*)

**research reactor.** A reactor primarily designed to supply neutrons or other ionizing radiation for experimental purposes, training, materials testing, and production of radioisotopes.

**reserve generating capacity.** Extra capacity maintained to generate power in the event of unusually high demand or a loss or scheduled outage of regular generating capacity.

**reserves.** Resources which are known in location, quantity, and quality and which are economically recoverable using currently available technologies.

**reservoir.** A space such as a pond, lake, tank, or basin which is used for storage and control of water; a natural underground container of liquids, such as oil or water, and gases.

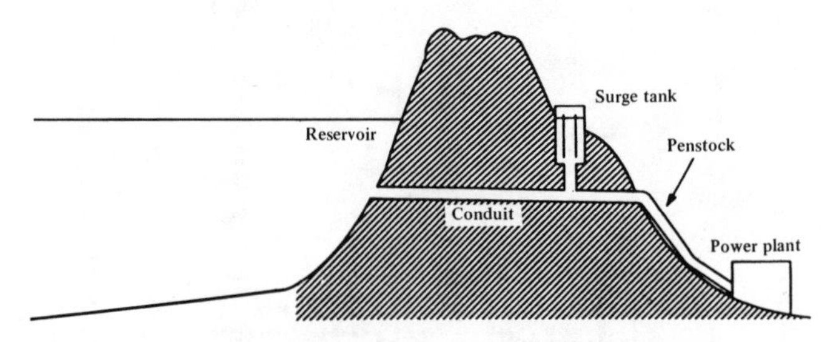


Characteristic design of a mountain reservoir hydroelectric plant. (*Courtesy Van Nostrand's Scientific Encyclopedia*)

**reservoir oil pressure.** The pressure within an oil pool sufficient to cause the oil to flow to the surface without pumping. Additional pressure may be created by injecting gas or compressed air as in repressuring.

**reservoir pressure.** The pressure existing at the level of the oil or gas productive formation in a well.

**residence time.** The time during which radioactive material remains in the atmosphere following the detonation of a nuclear explosive, usually expressed as a half-time, since the time for all material to leave the atmosphere is not well known.

**residual fuel oil.** The heavier oils that remain after the distillate fuel oils and lighter hydrocarbons are boiled off in refinery operations. Included are products known as ASTM grades Nos. 5 and 6 oil, heavy diesel oil, Navy Special Oil, Bunker C oil, and acid sludge and pitch used as refiner fuels. Residual fuel oil is used for the production of electric power, for heating, and for various industrial purposes.

**residual heat.** The total heat source remaining in a shutdown reactor, including after-heat.

**residual nuclear radiation.** Lingering radiation, or radiation emitted by radioactive material remaining after a nuclear explosion. It is arbitrarily designated as that emitted more than one minute after the explosion.

**residual power.** Radiation power released by decaying fission products in irradiated nuclear fuel after irradiation has ceased.

**residue.** That which remains after a part has been separated or otherwise treated.

**residue gas.** The natural gas remaining after the extraction of various liquid hydrocarbons.

**resistance.** The ability of all conductors of electricity to resist the flow of current, turning some of it into heat. Resistance depends on the cross section of the conductor (the smaller the cross section, the greater the resistance), and its temperature (the hotter the cross section, the greater its resistance).

**resistive cryogenic transmission cable.** New type of underground transmission cable being developed by the General Electric Research and Development Center in Schenectady, New York, which is capable of carrying larger blocks of electrical power than conventional overhead power lines.

**resonance.** The phenomenon whereby particles such as neutrons exhibit a very high interaction probability with nuclei at specific kinetic energies of the particles.

**resonance absorption.** Absorption of neutrons in the resonance energy region.

**resonance capture.** Radiative capture of neutrons in the resonance energy range.

**resonance detector.** An activation detector whose neutron cross section is characterized by large resonances.

**resonance energy.** The kinetic energy of an incident particle that excites an energy level in a compound nucleus.

**resonance escape probability.** The probability, in an infinite medium, that a neutron slowing down will traverse all or some specified portion of the range of resonance energies without being absorbed.

**resonance level.** An energy level in a compound nucleus which is excited in a nuclear reaction.

**resonance neutrons.** Neutrons having kinetic energy in the resonance energy range.

**resource recovery.** The process of obtaining materials or energy, particularly from solid wastes.

**resources.** The total estimated amount of a mineral, fuel, or energy source, whether or not discovered or currently technologically or economically extractable; quantities of an energy commodity that may be reasonably expected to exist in favorable geologic settings, but that have not yet been identified by drilling; also reserves and/or materials that have been identified, but cannot now be extracted because of economic or technological limitations, as well as economic or subeconomic materials that have not as yet been discovered.

**restoration.** The process of bringing back to a former condition.

**resue.** To mine or strip sufficient barren rock to expose a narrow but rich vein, which is then extracted in a clean condition.

**retaining wall.** A thick wall designed to resist the lateral pressure of earth behind it. Retaining walls are often necessary at mine sites in valleys to gain space for sidings.

**retort.** A vessel used for the distillation of volatile materials, as in the separation of some metals and the destructive distillation of coal; also a long semi-cylinder, now usually of fire clay or silica, for the manufacture of coal gas.

**retort carbon.** A very dense form of carbon produced by the deposition of carbon in the upper part of the retorts in the manufacture of coal gas.

**retort gas.** Gas resulting from the heating of coal in retorts, such as in the byproduct process of coke manufacture.

**retorting.** The heating of oil shale to drive out the oil and gas; in the sulfur industry, synonymous with sublimation. (See Fig. p. 379)

**retrofit.** The subsequent installation of equipment after initial construction is completed, such as the installation of solar collectors and other hardware onto an existing house to convert it for solar heating.

**return.** Any airway which carries the ventilating air out of the mine.

**return air.** Air returning to a heater or conditioner from the heated or conditioned space.

**reverse circulation.** The reverse of the normal course of drilling fluid circulation where the fluid returns to the surface through the drill pipe after being pumped down in the annular space.

**rib.** The termination of a coal face; the solid coal on the side of a gallery or long-wall face; a pillar or barrier of coal left for support.

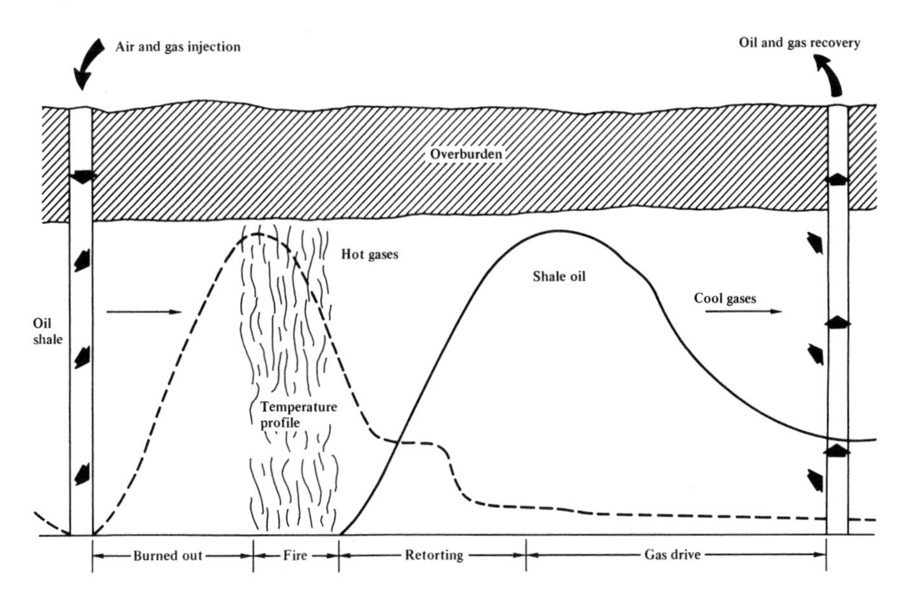


In-situ retorting. (*Courtesy U.S. Council on Environmental Quality*)

**rich mixture.** A gas-air mixture of which the air content is not sufficient for complete combustion.

**rich oil.** Oil that contains dissolved natural gasoline fractions.

**rider coal seam.** A "stray" coal seam usually above and divided from the main coal bed by rock, shale, or other strata material.

**riffle.** In mining, the lining of the bottom of a sluice, made of blocks or slats of wood or stones, arranged in such a manner that clinks are left between them.

**rifling.** Working coal which was left behind over the waste.

**rig.** A derrick, complete with enginehouse and other equipment necessary for operation, that is used for boring and afterwards pumping an oil well; an oil derrick. (See Fig. p. 380)

**rigging up.** The operation of getting the rig ready before the work of drilling is started, such as installing tools and machinery and establishing the water and fuel supply.

Offshore oil rig, off the Louisiana coast. (*Courtesy Bureau of Land Management*)

**right-of-way.** A strip of land, the use of which is acquired for the construction and operation of a pipeline or some other facility.

**rigid PVC.** Polyvinyl chloride or a polyvinyl chloride/acetate copolymer characterized by a relatively high degree of hardness, formulated with or without a small percentage of plasticizer.

**rig time.** The hours, days, etc., a drill rig is actually in use in actual drilling and other related borehole drilling operations.

**Ringelmann chart.** A series of illustrations ranging from light grey to black used to measure the opacity of smoke emitted from stacks and other sources. The shades of grey simulate various smoke densities and are assigned numbers ranging from one to five. Ringelmann No. 1 is equivalent to 20 percent density; No. 5 is 100 percent density. Ringelmann charts are used in the setting and enforcement of emission standards.

**ring-specimen.** A very short length of pipe cut for testing purposes, such as for the ring-tensile test.

**ring-tensile test.** A method of determining apparent tensile strength of plastic pipe by applying tensile forces in the hoop direction to a ring-specimen cut from pipe.

**Rio Blanco experiment.** The name of an experiment to stimulate production of natural gas by use of multiple nuclear explosions and to test the economic feasibility of future utilization of nuclear stimulation of an entire gas field. The test was made on May 16, 1973, near Meeker, in Rio Blanco County, Colorado.

**riparian rights.** Rights of a land owner to the water on or bordering his property, including the right to prevent diversion or misuse of upstream water.

**ripper.** A machine used to rip coal from the face.

**ripping.** The act of breaking, with a tractor-drawn ripper or long angled steel tooth, compacted soils or rock into pieces small enough to be economically excavated or moved by other equipment such as a scraper or dozer.

**riprap.** Heavy, irregular rock chunks used chiefly for river and harbor work, such as spillways at dams, shore protection, docks, and other similar construction for protection against the action of water.

**rise.** A vertical or inclined shaft from a lower to an upper level; upward inclination of a coal stratum; same as dip except taken in the upward direction versus the downward.

**riser.** General term for vertical runs of gas piping.

**Rittinger's law.** Principle stating that the energy required for reduction in particle size of a solid is directly proportional to the increase in surface area.

**river-clamp.** Long, heavy iron or concrete sleeves installed on a pipeline to prevent injury to pipe laid in a river bottom, and to weight the pipe.

**river mining.** Mining or excavating beds of existing rivers after deflecting their course, or by dredging without changing the flow of water.

**road octane.** A numerical value based upon the relative anti-knock performance of an automobile with a test gasoline as compared with specified reference fuels. Road octanes are determined by operating a car over a stretch of level road or on a chassis dynamometer under conditions simulating those encountered on the highway.

**road oil.** An asphaltic residual oil or a blend of such oil with distillates, which does not volatize readily and is used for dust laying or in the construction of various types of highways. Road oil is the most economic means of covering road stones containing hazardous asbestos.

**rocket.** A projectile driven by jet propulsion and containing its own fuel and propellants; may be powered either by nuclear reactors or by solid or liquid fuels.

**rocket fuel.** Propellant consisting of an oxidizer and a fuel, which react to give gaseous products and release energy; may be liquids or solids.

**rock-fill dam.** An earth dam built with any broken rock or similar material which is available.

**Rockgas process.** Process for the gasification of coal using the partial oxidation of coal in a molten sodium carbonate medium to produce a low-Btu fuel gas for consumption at the site of the gasification plant.

**rock pressure.** In petroleum geology, the pressure under which fluids such as water, oil, and gas are confined in rocks.

**Rockwell hardness test.** A method of determining the relative hardness of metals and case-hardened materials. The depth of penetration of a steel ball (for softer metals) or of a conical diamond point (for harder metals) is measured.

**rock wool.** A foam of natural mineral home insulation material, which may be installed in batts or blankets. Rock wool is fire and moisture resistant.

**rod.** A bar, the end of which is slotted, tapered, or screwed for the attachment of a drill bit; also a relatively long, slender body of material used in or in conjunction with a nuclear reactor, which may contain fuel, absorber, or material in which activation or transmutation is desired.

**rod dope.** Grease or other material used to protect or lubricate drill rods; also called gunk or rod grease.

**rod slap.** The impact of drill rods with the sides of a borehole, occurring when the rods are rotating.

**roentgen.** A unit of exposure to ionizing radiation. It is that amount of gamma or x-rays required to produce ions carrying one electrostatic unit of electrical charge (either positive or negative) in one cubic centimeter of dry air under standard conditions. Named after Wilhelm Roentgen, who discovered x-rays in 1895.

**roentgen equivalent man (rem).** The unit of absorbed radiation in biological matter. It is equal to the absorbed dose in rads multiplied by the relative biological effectiveness of the radiation.

**roentgen equivalent physical (rep).** An obsolete unit of radiation dosage superseded by the rad.

**roentgenography.** Radiography by means of x-rays.

**roentgen rays.** X-rays.

**roentgen therapy.** Radiation therapy with x-rays.

**roof bolting.** A system of roof support in mines.

**room and pillar.** An underground mining technique in which small areas of a coal or oil shale seam are removed and columns of the deposit are left in place to support the roof.

**room heater.** A self-contained, free-standing, nonrecessed, gas-burning, air-heating appliance intended for installation in the space being heated and not intended for duct connection.

**Röschen method.** A firedamp drainage method utilizing controlled drainage from the coal seams as they are being mined; a method, also known as the pack cavity method, devised to extract gas from the mined out areas of advancing longwall mining systems by leaving corridors or cavities at regular intervals in the pack.

**rotary bit.** The cutting tool attached to the lower end of the drill pipe of a rotary drilling rig.

**rotary boring.** A system of boring using hollow rods, with or without the production of rock cores. Rock penetration is achieved by the rotation of the cutting tool. It is the usual method in oil well boring with holes from 6 to 18 inches in diameter.

**rotary displacement meter.** An instrument which measures volume by means of rotating impellers, matching gears, or sliding vanes.

**rotary drill.** A machine which uses a revolving bit to bore out holes.

**rotary drilling.** The drilling method by which a hole is drilled by a rotating bit to which a downward force is applied. The bit is fastened to and rotated by the drilling string, which also provides a passageway through which the drilling fluid is circulated. New joints of drill pipe are added as drilling progresses. (See Fig. p. 384)

**rotary kiln.** A heated horizontal cylinder which rotates to dry coal.

**rotary-percussive drill.** A drilling machine which operates as a purely rotary machine to which is added a percussive action. This specially designed drilling bit gives a greater penetration rate and operates longer without deterioration of the cutting edges.

**rotary rig.** A machine used for drilling wells that employs a rotating tube attached to a bit for boring holes through rock.

**rotary scrubber.** A piece of equipment for removing impurities from gas by passing the gas over rotating surfaces or brushes that are partially immersed in liquid.

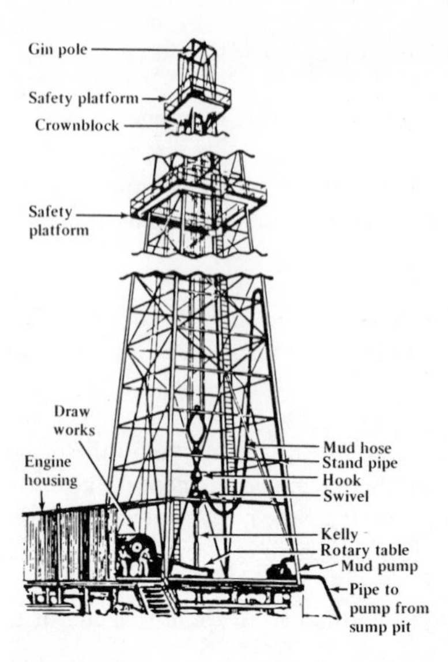


Typical rotary drilling rig. (*Courtesy Van Nostrand's Scientific Encyclopedia*)

**round coal.** Large coal, or large coal which has passed through the preparation plant and becomes more or less rounded.

**round trip.** The process of removing a drilling bit and drill pipe from a well and replacing them as required to change the bit.

**ROVAC system.** A no-Freon rotary-vane air cycle (ROVAC) heating and cooling system. (See Fig. p. 385)

**Rover.** A joint program of the Atomic Energy Commission and the National Aeronautics and Space Administration to develop a nuclear rocket for space flight.

**royalty.** The amount paid to the owner of mineral rights as payment for minerals removed. In gas and oil operations, the royalty is usually based on a percentage of the total gas or oil production.

**Rulison.** The name of an experiment which is used to stimulate production of natural gas by the use of a nuclear explosive to fracture impermeable rocks. Conducted in 1969.

**run.** The direction in which a vein lies; a length of pipe made of more than one piece of pipe; that portion of a fitting having its end in line or nearly so, as distinct from the branch or side opening, as of a tee; continuous production, operation of any furnace between major repairs; also a test made of a process or material.

**runaway reactor.** An increase in power or reactivity that cannot be controlled by the normal reactor control system, but might be terminated safely by the emergency shutdown system or a negative reactivity coefficient.

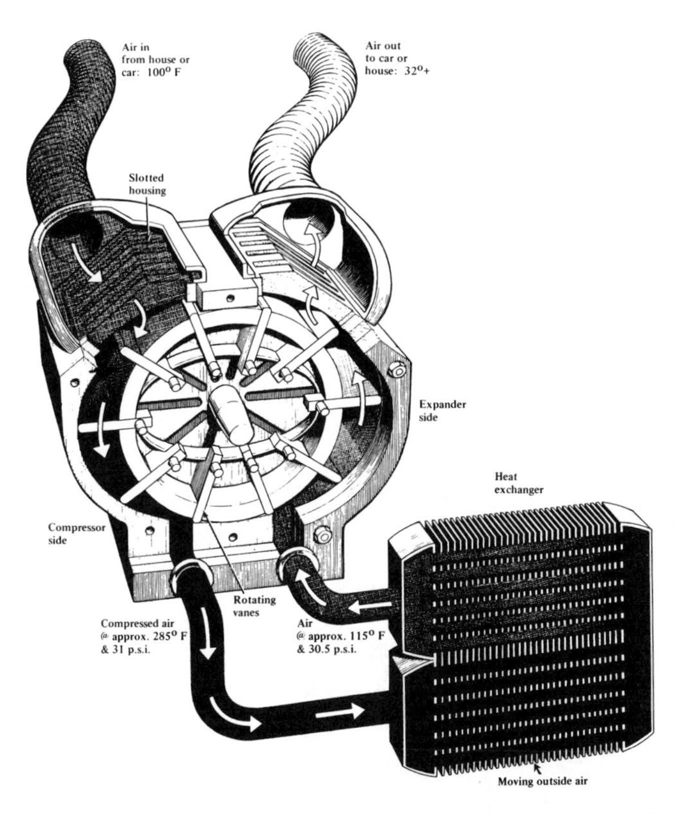


ROVAC air-conditioning system takes in air from house or car and compresses it, raising its temperature and pressure. The heated air is blown through the heat exchanger where the air is cooled. The cooler air enters the expander and cools to subfreezing on leaving the expander. (*Courtesy Popular Science, August 1976, p. 84*)

**running coal.** A term applied to bituminous coal on account of its tendency to soften and cohere when burning; also applies to a very friable coal which disintegrates and flows into the workings.

**runoff.** That portion of the rainfall not absorbed by the deep strata. It is utilized by vegetation or lost by evaporation, or it may find its way into streams as surface flow.

**run of river plant.** A hydroelectric power plant using the flow of the stream as it occurs and having little or no reservoir capacity for storage of water.

**rupture.** Deformation characterized by loss of cohesion (such as a failure in the pipeline for various reasons) where a complete loss to atmosphere of the gas or other media is sustained.

**rural windmills.** Wind power was once used widely on American farms but was replaced during the 1930's by inexpensive electricity and natural gas. Now, with rising fuel costs, wind power is being re-examined as a power resource for farms, rural homes, and remote rural regions. Old-fashioned windmills and multiple blades, still available from some manufacturers, cannot meet the power needs of most modern farms. They cannot be simply scaled up, because their large-heavy blades require tremendously strong towers to withstand strong winds. Tests began in 1977 on a spectrum of commercially available and prototype wind turbines, mostly in the 2 to 15 kilowatt range. (A "small windmill" is defined for this test program's purposes as one under 100 kilowatts.)

Traditional multi-bladed windmill (*1943 photo by Knell, U.S. Department of Agriculture*)

**ruthenium-106.** A radioactive fission fragment with a half-life of 369 days, not occurring in nature.

**R-value.** An insulation material resistant to temperature changes or heat flowing through it. The R-value required to adequately insulate a home will depend on such factors as the climate and the type of heating system. The larger the R-value is, the better the insulation. Adequate insulation in the attic and walls saves money on both heating and cooling bills. By the installation of adequate insulation, it has been estimated that 20 to 30 percent of the energy used to heat homes in winter and 10 percent required to cool them in summer can be conserved.

# S

**sack breakers.** Sacks of approximately 1 cubic foot, filled with dirt or sand and cement and used to prevent erosion or to form a barrier between pipelines and prevent coating or pipe damage when lowering in rocky trenches.

**sacrificial pile.** A mass of metal, usually scrap metal, used as an anode when a rectifier is used in cathodic protection; also, the magnesium and aluminum anodes used in cathodic protection which do not require an outside impressed voltage.

**saddle.** A fitted plate held in place (by clamps, straps, heat fusion, or welding) over a hole punched or drilled in a gas main, to which a branch line or service line connection is made.

**SAE viscosity number.** A system established by the Society of Automotive Engineers for classifying crankcase oils and automotive transmission and differential lubricants according to their viscosities.

**safety-control circuit.** A circuit wherein one or more safety controls in which failure due to grounding, opening, or shorting of any part of the circuit can cause unsafe operation of the valve or the controlled equipment.

**safety coupling.** A friction coupling adjusted to slip at a predetermined torque to protect the rest of the system from overload.

**safety engineering.** The planning, development, improvement, coordination, and evaluation of the safety component of integrated systems of men, materials, equipment, and environments to achieve optimum safety effectiveness in terms of both protection of people and protection of property.

**safety glass.** Glazing materials predominately inorganic in character which meet the appropriate requirements of the ANSI Standard and include laminated glass, tempered glass and wired glass.

**safety glazing materials.** Glazing materials so constructed, treated or combined with other materials as to minimize the likelihood of cutting and piercing injuries resulting from contact.

**safety member.** A control member which, singly or in concert with others, provides a reserve of negative reactivity for the purpose of emergency shutdown of a reactor.

**Safety Research Experiment Facilities (SAREF).** A project to develop safety information needed for use in commercial liquid metal fast breeder reactors.

**safety rod.** A standby control rod used to rapidly shut down a nuclear reactor in emergencies.

**safety shutoff device.** A device that will shut off the gas supply to the controlled burner if the source of ignition fails; device interrupting the flow of gas to the main burner only, or to the pilot and main burner under its supervision.

**safety shutoff valve.** A valve which automatically shuts off the supply of fuel through the functioning of a flame safeguard control or limiting device, which may interrupt the flow of fuel to the main burner only, or to the pilot and main burner.

**safety solvents.** Solvents free from fire or toxicity hazards and nondamaging to the surfaces or materials being cleaned.

**sag bolt.** Bolts installed at intersections to measure roof sag.

**salable output.** The total tonnage of clean coal produced at a mine as distinct from pithed output; the tonnage of coal as weighed after being cleaned and classified in the preparation plant.

**saline-alkali soil.** A soil having a combination of a harmful quantity of salts and either a high degree of alkalinity or a high amount of exchangeable sodium so distributed in the soil that the growth of most crop plants is less than normal.

**saline soil.** A soil containing excessive quantities of the neutral or nonalkaline soluble salts.

**salinity.** The total amount of solid material in grams contained in 1 kilogram of seawater when all the carbonate has been converted to oxide, the bromine and iodine replaced by chlorine, and all organic matter completely oxidized. Expressed as grams per kilogram of seawater or parts per thousand.

**salt.** A chemical compound formed when the hydrogen ion of an acid is replaced by a metal, or, together with water, when an acid reacts with a base. Salts are named according to the metal and the acid from which they are derived.

**salt dome.** A structure resulting from the upward movement of a salt mass, and with which oil and gas fields are frequently associated. Salt domes are now being converted to storage facilities for oil and natural gas, butane, propylene, and other hydrocarbon fuels.

**samarium poisoning.** The decrease in reactivity of a thermal reactor or chain reacting system caused by the presence of the fission product $^{149}Sm$.

**sand filter.** A filter for purifying domestic water, consisting of specially graded layers of aggregate and sand, through which the water flows slowly downwards. A similar type of filter with coarser sand is used for treating sewage effluent.

**sandstone.** A sedimentary rock composed predominantly of quartz grains or other noncarbonate mineral or rock detritus.

**sanitary landfilling.** An engineered method of solid waste disposal on land in a manner that protects the environment: waste is spread in thin layers, compacted to the smallest practical volume, and covered with soil at the end of each working day.

**sapropelic coal.** Coal of which the original plant material was transformed by putrefaction. Complete seams of sapropelic coals are rare, but layers or bands of varying thickness within seams are more frequent. This type of coal is not abundant and proves troublesome in cleaning processes.

**saturated hydrocarbon.** A chemical compound of carbon and hydrogen in which all the valence bonds of the carbon atoms not taken up with other carbon atoms are taken up with hydrogen atoms.

**saturated steam.** Steam at a temperature and pressure such that any lowering of temperature or increase in pressure will cause condensation.

**saturation.** Completely filled; a condition reached by a material, whether it be in solid, gaseous, or liquid state, which holds another material within itself in a given state in an amount such that no more of the material can be held within it in the same state; also the extent or degree to which the voids in rock contain oil, gas, or water—usually expressed in percent related to total void or bore space.

**Saybolt colorimeter.** An apparatus that has been universally adopted for determining the color of light oils.

**scaler.** An electronic instrument for rapid counting of radiation-induced pulses from geiger counters or other radiation detectors, by reducing the number of pulses entering the counter.

**scalping.** The removal by screen of undesirable fine materials from broken ore, stone, or gravel; the removal of vegetation before mining.

**scanning.** The sequential measurement of some quantity at a number of positions on an area or in a volume either continuously by mechanically moving a detector, or discontinuously by means of systematic electrical or mechanical switching. In the nuclear industry, this term applies to a method of determining the location and amount of radioactive isotopes within the body by measurements taken with instruments outside the body; usually the instrument, called a scanner, moves in a regular pattern over the area to be studied or over the whole body, and makes a visual record.

**scattering.** A process that changes a particle's trajectory. Scattering is caused by particle collisions with atoms, nuclei, and other particles or by interactions with fields of magnetic force. If the scattered particle's internal energy (as contrasted with its kinetic energy) is unchanged by the collision, elastic scattering prevails; if there is a change in the internal energy, the process is called inelastic scattering.

**Schlumberger logging.** An electric well logging technique originally devised by the Schlumberger brothers, in which electrical measurements are made and recorded at the surface, while a series of electrodes or coils is caused to traverse a borehole. The resulting curves can be used for purposes of geological correlation.

**Scholl's method.** A method for determining the uranium content in any uranium ore. The uranium is extracted with dilute nitric acid.

**scintillation counter.** An instrument that detects and measures ionizing radiation by counting the light flashes (scintillations) caused by radiation impinging on certain materials.

**scoria.** Slag; refuse of fused metals; clinker deposits characteristic of burned-out coal beds.

**scouring.** A wet cleaning or drycleaning process used in the manufacture of electrical porcelain.

**scram.** The sudden shutdown of a nuclear reactor, usually by rapid insertion of the safety rods. Emergencies or deviations from normal reactor operation cause the reactor operator or automatic control equipment to scram the reactor.

**scraper.** A rod for cleaning out shotholes prior to charging with explosives; a steel tractor-driven surface vehicle used for stripping and releveling topsoil and soft material at opencast pits.

**scraper ripper.** Strip-mine equipment that handles the jobs of breaking coal, loading, and hauling.

**scraper trap.** A fitting in either end of a pipeline with a shut-off valve and a door to insert or remove a pipeline scraper which is pushed through the pipeline to clean it and to increase flow efficiency.

**screening.** The separation of solid materials of different sizes by causing one component to remain on a surface provided with apertures through which the other component passes.

**scrub.** To remove certain constituents of a gas by passing it through scrubber equipment in which the gas is intimately mixed with a suitable liquid that absorbs or washes out the constituent to be removed from the gas.

**scrubber.** Special apparatus for cleaning waste gases with water or removing pollutants such as sulfur dioxides or particulate matter, from stack gas emissions by means of a liquid sorbent.

**Seacoke process.** Similar to the COED process, five fluidized bed pyrolyzers produce a syncrude (1.3 barrels/ton), char, and fuel gas. The operating pressure is 1 atmosphere. Process temperatures range from 316° to 871°C from the first to fifth stage.

**seam.** A stratum or bed of coal or other valuable mineral of any thickness; generally applied to large deposits of coal.

**seasonal gas.** Gas sold during certain periods of the year, either on a firm or on an interruptible basis.

**seat earth.** Floor of a coal seam; a bed representing old soil, usually containing abundant rootlets, underlying a coal seam; stratum underlying the valuable seam.

**second law efficiency.** The ratio of the minimum amount of work or energy necessary to accomplish a task to the actual amount used.

**secondary air.** The air for combustion externally supplied to the flame at the point of combustion.

**secondary blasting.** Reblasting oversize pieces of rock so as to reduce them to a size suitable for handling by the available excavators and crushers.

**secondary coal recovery.** Coal obtained after primary mining has been completed, by such methods as coal augers, push-button miner, and punch mining by underground machines.

**secondary coolant.** A coolant used to remove heat from the primary coolant circuit.

**secondary production.** Oil and gas obtained by the augmentation of reservoir energy; often by the injection of air, gas, or water into a production formation.

**secondary recovery.** The recovery obtained by any method whereby oil or gas is produced by augmenting the natural reservoir energy, as by injection of air, gas, or water into the production formation. (See Fig. p. 392)

**secondary system.** The steam- and electricity-generating (nonnuclear) portion of a nuclear power plant.

**second law of thermodynamics.** Heat cannot be changed directly into work at constant temperature by any cyclic process; one of the two "limit laws" which govern the conversion of energy, which describes the inevitable passage of some energy from a useful to a less useful form in any energy conversion.

**section.** A subdivision of anything, such as a portion of the working area of a mine.

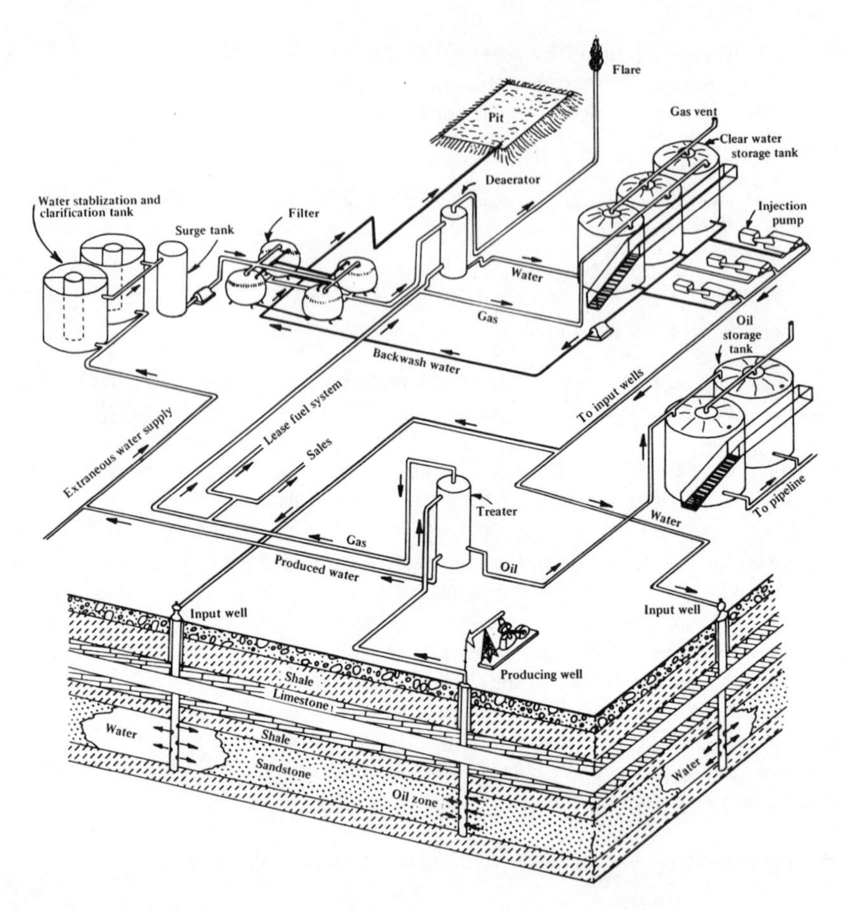


Waterflood secondary recovery system. (*Courtesy U.S. Council on Environmental Quality*)

**sediment.** Solid material, both mineral and organic, that is in suspension, is being transported, or has been moved from its site or origin by air, water, gravity, or ice, and has come to rest on the earth's suface either above or below sea level.

**sediment basin.** A reservoir for the retention of rock or other debris from a sediment-producing area.

**Seebeck effect.** Phenomenon first observed by Thomas J. Seebeck, involving the operation of a thermocouple. Describes the current flowing in a closed circuit due to an electromotive force which is generated when a temperature difference is maintained between the junction of two dissimilar metal conductors and another part of the circuit.

**seed core.** A reactor core which includes a relatively small volume of highly enriched uranium (the seed) surrounded by a much larger volume of natural uranium or thorium (the blanket). As a result of fissions in the seed, neutrons are supplied to the blanket where more fission takes place. In this way, the blanket is made to furnish a substantial fraction of the total power of the reactor.

**seep.** A spot where a fluid (as water, oil, or gas) contained in the ground oozes slowly to the surface and often forms a pool.

**seepage.** A quantity of a fluid that has seeped through a porous material such as soil; naturally occurring escape of crude oil, gas, or bitumen to the earth's surface; also failure that occurs through essentially microscopic breaks in the pipewall, frequently only at or near the test pressure.

**segment.** A portion of a river basin, the surface waters of which have common hydrologic characteristics.

**seismic.** Characteristics of, or produced by, earthquakes or earth vibration.

**seismic survey.** An exploration technique utilizing the variation in the rate of propagation of shock waves in layered media. It is used primarily to delineate subsurface geologic structures of possible economic importance. It is employed most frequently by the petroleum industry.

**seismograph.** A device for detecting vibrations in the earth; used in prospecting for probable oil-bearing structures.

**selective black paint.** A material more absorbent of the infrared long wavelengths of sunlight than nonselective black paint; used for coating the absorber plates in solar collectors.

**selective leaching.** The removal of one element from a solid alloy by corrosion processes. The most common example is the selective removal of zinc in brass alloys (dezincification), such as in power house heat exchangers using brass tubes with boiling water on one side and fuel combustion gases on the other.

**selective mining.** Mining to obtain a relatively high-grade mine product. It requires the use of expensive stoping systems.

**selective surface.** A special coating sometimes applied to the absorber plate in a solar collector. The selective surface absorbs most of the incoming solar energy and reradiates very little of it.

**Self-Agglomerating process.** A process for producing raw synthesis gas that could be upgraded to pipeline gas from coal. Two fluidized bed reactors are used. Coal is burned in a fluidized bed burner at a temperature approaching the ash fusion point of the feed coal. Off-gases from this burner should be sufficently clean of fly-ash to be expanded in an open cycle gas turbine. Ash agglomerates are transferred from the burner to the gasifier via a steam lift. Additional coal is fed to the gasifier. Superheated steam fluidizes the coal/ash mixture. The hot, inert ash supplies the heat of reaction for endothermic coal gasification reactions which can be the basis for producing methane.

**self-contained cooling unit.** A combination of apparatus for room cooling complete in one package; usually consists of compressor, evaporator, condenser, fan motor, and air filter. Requires connection to electric line.

**semiautomatic valve.** A valve that is opened manually and closed automatically, or vice versa.

**semiconductor.** A crystal system in which, though the electrons are ionically bound, a slight rise in temperature frees the valence atoms so that the system becomes a conductor; an example is germanium. Conduction of electricity proceeds in one direction only.

**semidiurnal.** Having a period or cycle of approximately one-half lunar day (12.42 solar hours). The tides and tidal currents are said to be semidiurnal when two flood periods and two ebb periods occur each lunar day.

**semidull coal.** A variety of banded coal containing from 21 to 40 percent of pure, bright ingredients (vitrain, clarain, and fusain), the remainder consisting of clarodurain and durain.

**semifusain.** A coal constituent transitional between vitrain and fusain. It displays gradual disappearance of cell structure, hardness, and yellowish color when observed in thin sections.

**semisplint coal.** Coal intermediate between durain coal and clarain coal (duroclarain); a banded coal containing 20 to 30 percent of opaque attritus and more than 5 percent anthraxylon; translucent humic matter, spores, pollens, and finely divided fusain are always present in small proportions.

**semiwater gas.** A mixture of carbon monoxide, carbon dioxide, hydrogen, and nitrogen obtained by passing a mixture of air and steam through incandescent coke. Its calorific value is low (about 125 British thermal units per cubic foot).

**sensible heat.** That heat which when added or subtracted results in a change of temperature.

**sensible heat storage.** A heat storage medium, typically water or gravel, in which the addition or removal of heat results in a temperature change only (as opposed to, say, a chemical reaction).

**separation energy.** The energy per unit mass required to achieve a specified degree of separation of a given mixture of two or more isotopic species. It can be either the theoretical minimum thermodynamic value or the actual value for a given process.

**separative work units.** A defined separation task expressed in terms of separative work units (SWU); a measure of the amount of physical effort required to separate the isotopes of uranium. The number of SWU is directly related to the amount of resources required physically to perform the separation. For example, the basic resources used to produce a given amount of enriched uranium at a specified level are the feed supplied and the enrichment services, or separative work effort, applied to that feed. SWU is generally expressed in kilogram units.

**separator.** A piece of equipment for separating one substance from another when the substances are intimately mixed, such as removing oil from water, oil from gas, ash from flue gas, or tramp iron from coal.

**sequence control.** An electrical method of control whereby once action has been initiated, a number of electrical circuits will automatically function in a prescribed order.

**service drip.** A liquid-collecting trap at the low point in a customer's gas service piping when the piping cannot be sloped back to the distribution main.

**service riser.** A vertical pipe, either inside or outside a foundation wall, running from the grade of the service pipe to the level of the meter.

**service shutoff.** Refers either to a service stop or to a meter stop used to cut off the supply of gas.

**service stop.** A plug-type valve located in the service line between the main and the building; often used synonymously with the meter stop, which is located within the building or immediately before the meter or regulator in outside settings.

**service stub.** A piece of pipe connected to a main and usually extended to the curb line for the addition of a service.

**service tee.** A tee in a customer's service piping with one leg closed and used for access to the service pipe in case of plugging with solids.

**set.** A timber frame for supporting the sides of an excavation shaft or tunnel; also, to point out the direction in which a current is flowing (for example when a current sets southward, the movement of water is toward the south).

**set casing.** To install steel pipe or casing in a well bore.

**settled production.** The production of an oil well that, apart from the normal progressive annual diminution will last a number of years.

**settling chamber.** In air pollution control, a device used to reduce the velocity of flue gases usually by means of baffles, promoting the settling of fly ash.

**seven sisters.** The seven sisters are the early primary oil barons—Exxon, Gulf, Texaco, Mobil, Socal, B.P., and Shell. They are larger and richer than most national governments and have dominated the world energy economy. They control significant energy sources such as gas, nuclear and coal.

| | | Energy industry | | | | |
|---|---|---|---|---|---|---|
| Petroleum company | Rank in assets | Gas | Oil shale | Coal | Uranium | Tar sands |
| Exxon | 1 | X | X | X | X | X |
| Texaco | 2 | X | X | X | X | — |
| Gulf | 3 | X | X | X | X | X |
| Mobil | 4 | X | X | — | X | — |
| Standard Oil of California | 5 | X | X | — | — | — |
| Standard Oil (Indiana) | 6 | X | X | — | X | X |
| Shell | 7 | X | X | X | X | X |
| Atlantic Richfield | 8 | X | X | X | X | X |
| Phillips Petroleum | 9 | X | X | — | X | X |
| Continental Oil | 10 | X | X | X | X | — |
| Sun Oil | 11 | X | X | X | X | X |
| Union Oil of California | 12 | X | X | — | X | — |
| Occidental | 13 | X | — | X | — | — |
| Cities Service | 14 | X | X | — | X | X |
| Getty | 15 | X | X | — | X | — |
| Standard Oil (Ohio) | 16 | X | X | X | X | — |
| Pennzoil United, Inc. | 17 | X | — | — | X | — |
| Signal | 18 | X | — | — | — | — |
| Marathon | 19 | X | X | — | — | — |
| Amerada-Hess | 20 | X | — | — | X | — |
| Ashland | 21 | X | X | X | X | — |
| Kerr-McGee | 22 | X | — | X | X | — |
| Superior Oil | 23 | X | X | — | — | — |
| Coastal States Gas Producing | 24 | X | — | — | — | — |
| Murphy Oil | 25 | X | — | — | — | — |

Petroleum company diversification in the energy industry. (*Courtesy National Science Teachers Association*)

**sewage gas.** A gas produced by the fermentation of sewage sludge; also marsh gas or firedamp.

**shading.** An effective way to keep a house cool in the summer is to shade it from the outside by selective planting of trees, installation of awnings, or initial house design.

**shading coefficient.** A multiplier determined by dividing the solar heat gain through a glazing system, under a specific set of conditions, by the solar gain through a single light of double strength sheet glass under the same set of conditions. The smaller the number the greater the reduction in solar heat gain.

**shadowing.** Placing an object in an electrolytic bath so as to alter the current distribution on the cathode.

**shaft.** An excavation of limited area compared with its depth. It is made for finding or mining ore or coal; raising water, ore, rock, or coal; hoisting and lowering men and material; or ventilating underground workings.

**shaft mine.** A mine in which the coal seam is reached by a vertical shaft which may vary in depth from less than 100 feet to several thousand feet; a mine in which the main entry or access is by means of a shaft.

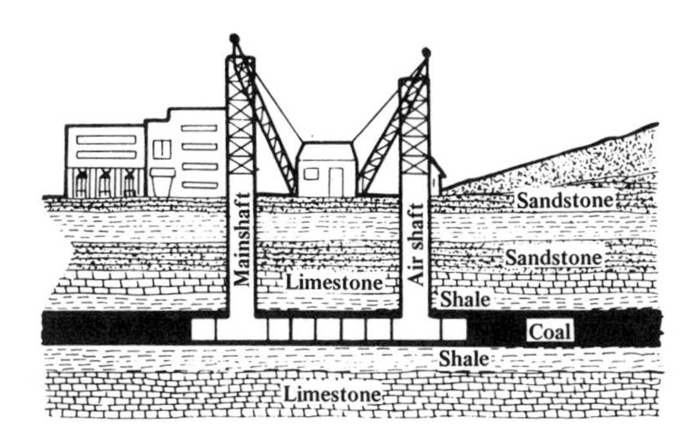


Shaft mines are used when coal lies roughly 100 feet below the surface. (*Courtesy Mining Enforcement and Safety Administration*)

**shake.** In a coal mine, a vertical crack in the seam and roof.

**shaker.** In drilling, a mechanically vibrated screen through which a returning drill fluid is passed to screen out larger chips, fragments, and drill cuttings before the drill fluid flows into the sump.

**shale.** Sedimentary or stratfied rock structure generally formed by the consolidation of clay or clay-like material; also one of the impurities associated with coal seams.

**shale oil.** A crude oil obtained from bituminous shales by submitting them to destructive distillation in special retorts.

**shale spirit.** The lower boiling fractions obtained in the refining of crude shale oil.

**sharp gas.** Gas when it is at its most explosive point.

**shear.** To make vertical cuts in a coal seam that has been undercut; also that type of force which causes two contiguous parts of the same body to slide relative to each other in a direction parallel to their plane of contact.

**shearing machine.** An electrically driven machine for making vertical cuts in coal.

**shell.** One of the series of concentric spheres, or orbits, at various distances from the nucleus, in which, according to atomic theory, electrons move around the nucleus of an atom. The shells are designated, in the order of increasing distance from the nucleus, as the k, l, m, n, o, p, and q shells. The number of electrons which each shell can contain is limited. Electrons in each shell have the same energy level and are further grouped into subshells.

**Sherwood.** The name of an Atomic Energy Commission program for research in controlled thermonuclear reactions.

**shield.** A body of material used to reduce the passage of radiation.

**shift conversion.** A process for the production of gas with a desired carbon monoxide content from crude gases derived from coal gasification. Carbon monoxide rich gas is saturated with steam and passed through a catalytic reactor where the carbon monoxide reacts with steam to produce hydrogen and carbon dioxide, the latter being subsequently removed in a wash plant.

**shift converter.** A device that converts carbon monoxide and water into hydrogen and carbon dioxide.

**shim rod.** A reactor control rod used in making infrequent coarse adjustments in reactivity, as in startup or shutdown.

**shock wave.** A pressure pulse in air, water, or earth, propagated from an explosion, which has two phases: in the first, or positive phase, the pressure rises sharply to a peak, then subsides to the normal pressure of the surrounding medium; in the second, or negative phase, the pressure falls below that of the medium, then returns. A shock wave in air usually is called a blast wave.

**shooting.** Exploding nitroglycerine or other high explosives in a hole to shatter the rock and increase the flow of oil; also, in seismographic work, the discharge of explosives to create vibrations in the earth's crust.

**shooting rights.** Permission to conduct geological and geophysical activity only, without the option to acquire lease acreage.

**shoring.** Timbers braced against a wall as a temporary support; also the timbering used to prevent a sliding earth adjoining an excavation.

**shortwall mining.** A method of mining whereby comparatively small areas are worked separately; a variation of longwall mining in which a continuous miner rather than a shearer is used on a shorter working face.

**shot.** An explosive charge.

**shovel.** Any bucket-equipped machine used for digging and loading earthy or fragmented rock materials.

**shroud.** A housing or jacket; also a structure used to concentrate or deflect a windstream.

**shutdown.** To make a reactor subcritical; the state of a reactor in a subcritical condition.

**shut-in.** Shutoff, so there is no flow; refers to a well, plant, pump, etc., when valves are closed at both inlet and outlet; in geology, a narrow gorge cut by a superposed stream across a ridge or hard rock between broad valleys of softer rock on each side of the ridge.

**shutoff valve.** Stops or valves readily accessible and operable by the consumer which are located in the piping system to shut off individual equipment, or between the meter and gas main to shut off the entire piping system.

**sial.** A layer of rock underlying all continents, that ranges from granitic at the top to gabbroic at the base. The thickness is variously placed at 30 to 35 kilometers.

**side slopes.** The slope of the sides of a canal, dam, or embankment.

**sidetracking.** The act or process of drilling past a broken drill or casing which has become permanently lodged in the hole. This operation is usually accomplished by use of a special tool known as a whip-stock.

**sidewall coring.** The taking of geological samples of the formation which constitutes the wall of the well bore. Another term in general use for this operation is "sidewall sampling."

**sidewall sampling.** The process of securing samples of formations from the sides of the borehole anywhere in the hole that has not been cased.

**Sierra Club.** Society devoted to conservation. Founded in 1892, the Club works in the US and in other countries to restore the quality of the natural environment and to maintain the integrity of ecosystems.

**sieve analysis.** The determination of the percentage of particles which will pass through screens of various sizes.

**sigma heat.** The sum of sensible heat and latent heat in a substance or fluid above a base point, usually 0° C.

**sigma pile.** In nuclear work, a large assemblage or "pile" of blocks of a pure material, within which measurements may be made of the absorption cross section and other nuclear parameters of the material.

**silica gel.** A desiccant; a hydroscopic material that readily absorbs substantial quantities of moisture and is used to reduce the relative humidity of air or gas.

**siltation.** The deposition or accumulation of fine particles suspended throughout a body of standing water or in some considerable portion of it.

**silviculture.** The technology of raising trees, or forest management.

**silvicultural farm.** An energy farm composed of trees.

**silver oxide cell battery.** A cell in which depolarization is accomplished by oxide of silver.

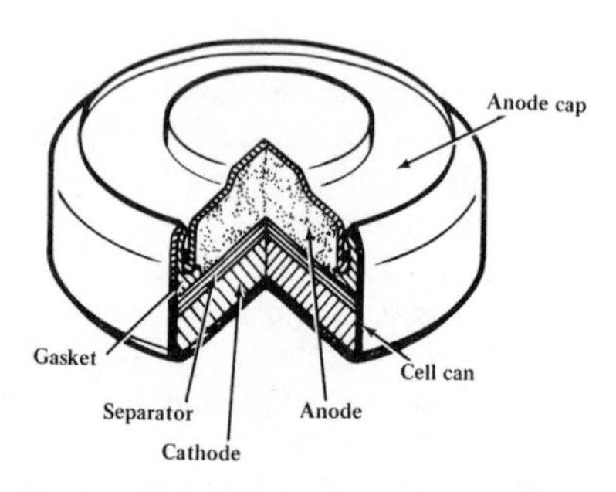


Cutaway of a silver oxide cell. (*Courtesy Van Nostrand's Scientific Encyclopedia*)

**sima.** The basic outer shell of the earth; under the continents it underlies the sial, but under the Pacific Ocean it directly underlies the oceanic water. Originally, the sima was considered basaltic in composition with a specific gravity of about 3.0. In recent years, it has been suggested that the sima is peridotitic in composition with a specific gravity of about 3.3.

**simple engine.** A reciprocating engine from which steam or compressed air is exhausted to atmosphere after expansion in one cylinder only.

**single-cycle reactor system.** A direct-cycle reactor system.

**sinking.** The process by which a shaft is driven; also a method employed in controlling oil spills by entrapping the oil droplets and sinking them to the bottom of the body of water.

**sintering.** The agglomeration of solids at temperatures below their melting point, usually as a consequence of heat and pressure.

**skidding the rig.** Moving a rig, with little or no dismantling of equipment, from the location of a lost or completed hole preparatory to starting a new one.

**skin dose.** The absorbed dose of radiation at a point on the skin.

**skylid.** Skylight or window in the roof that can be opened.

**skylight.** A window placed in the roof of a building; also a very poor quality of plate glass.

**slack.** Commonly used to describe the smaller sizes of coal passing through screen openings approximately 1 inch or less in diameter. Slack has a high ash content and is difficult to clean in the washery.

**slate.** A miner's term for any shale or slate accompanying coal; geologically, dense, fine textured, metamorphic rock.

**sleeve.** A pump-cylinder liner; tubular refractory shapes used to protect the metal rod that holds the stopper head in the valve assembly of a bottom-pouring ladle.

**slim hole.** Oil driller's term for diamond-drill borehole 5 inches or less in diameter; drill hole of the smallest practicable size.

**slip.** A joint or cleat in a coal seam; a small fault; a mass of spoil material that moves downward and outward to a lower elevation, generally caused by overloading of the downslope, freezing and thawing, or saturation of the fill.

**slop.** A name sometimes applied to processed oil which must be further processed to make it suitable for use.

**slope mine.** A mine opened by a slope or incline; a mine with an inclined opening used to reach the coal seam.

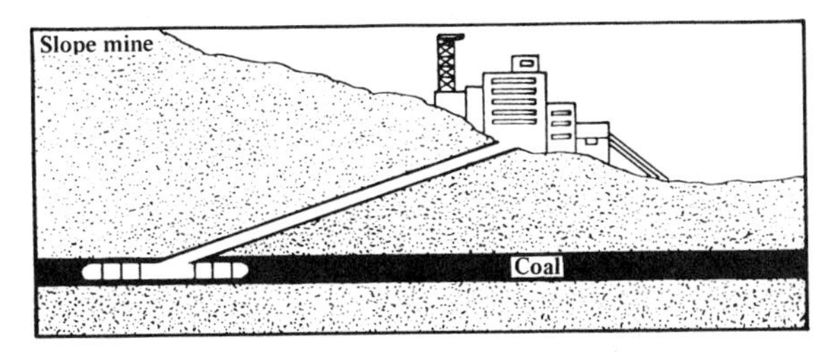


Slope mines are used when coal is readily accessible. (*Courtesy Mining Enforcement and Safety Administration*)

**slowing-down area.** One-sixth of the mean square displacement of neutrons in an infinite homogeneous medium from their points of origin to the points where they have been slowed down to a specified energy.

**slowing-down density.** The number of neutrons per unit volume and unit time which slow down past a given energy.

**slow neutron.** A low-energy neutron, sometimes called a thermal neutron. The energy of a slow neutron is about 0.025 electron volt, in contrast to the energy of a fast neutron, which may exceed 1,000 electron volts. Slow neutrons are very efficient in causing fission of uranium-235. Nuclear-power plants now in operation are designed to use slow neutrons to sustain the fission reaction.

**sludge.** The precipitate resulting from chemical treatment of water; refuse from a coal-washing plant. The term is also applied to the tar from the agitators in the chemical treatment of distillates.

**sludge abatement.** The control of the discharge into watercourses (or on adjacent land), of mineralized or impure water, or sludge, or mining debris.

**slug.** A section of heavy or dense fluid between two lighter fluids in a pipeline or other flow passage; in nuclear technology, a short, usually cylindrical fuel element.

**slug the pipe.** An operation performed before the drill pipe is hoisted in order to prevent crew members and tools from becoming covered with the drilling fluid. A quantity of very heavy mud is pumped into the top section of the drill pipe, which will cause the level of the fluid in the pipe to fall. When a stand of pipe is unscrewed, the drilling fluid will have been evacuated from it.

**slump.** The material that has slid down from high rock slopes; an en masse movement of material.

**slurry.** A suspension of pulverized solid in a liquid. Explosive slurries of ammonium nitrate, TNT, and water are used for blasting. Slurries of oil and coal or water and coal are used in coal processing and transportation.

**slurry reactor.** A reactor in which the fuel is a circulating suspension of fine particles in a liquid.

**slushing oil.** A nondrying oil used to coat metals, machine parts, etc., to prevent corrosion. It coats the metals very well and is easily removed when desired.

**small coal.** Thin seams of coal; coal with a top size less than 3 inches; slack; also called low coal.

**Smith-Putnam machine.** World's largest windmill, built in 1941 near Rutland, Vermont. The experimental wind turbine generator, named after its designer and builder, generated electricity and fed power into the commercial electrical network for 3½ years until structural damage occurred, and the system was not repaired due to wartime conditions.

**smog.** A mixture of smoke and fog; a fog made heavier and usually darker by smoke and chemical fumes; generally used as an equivalent of air pollution, particularly associated with oxidants.

**smoke.** Solid particles generated as a result of the incomplete combustion of materials containing carbon.

**SNAP program.** Acronym for Systems for Nuclear Auxiliary Power, an Atomic Energy Commission program to develop small auxiliary nuclear power sources for specialized space, land, and sea uses. Two approaches are employed: the first uses heat from radioisotope decay to produce electricity directly by thermoelectric or thermionic methods; the second uses heat from small reactors to produce electricity by thermoelectric or thermionic methods by turning a small turbine and electric generator.

**snowbird mine.** A mine that produces or ships only small quantities of coal, and operates only when coal is costly because of a scarcity or a shortage of cars for shipment.

**snow load.** The live load which must be included when designing a flat or low-pitched roof in temperate and cold climates.

**sodium.** The sixth most abundant element in the earth's crust and the most common alkali metal (Group IA); a soft electropositive metal. Chemical symbol Na.

**sodium-graphite reactor.** A reactor that uses liquid sodium as coolant and graphite as moderator.

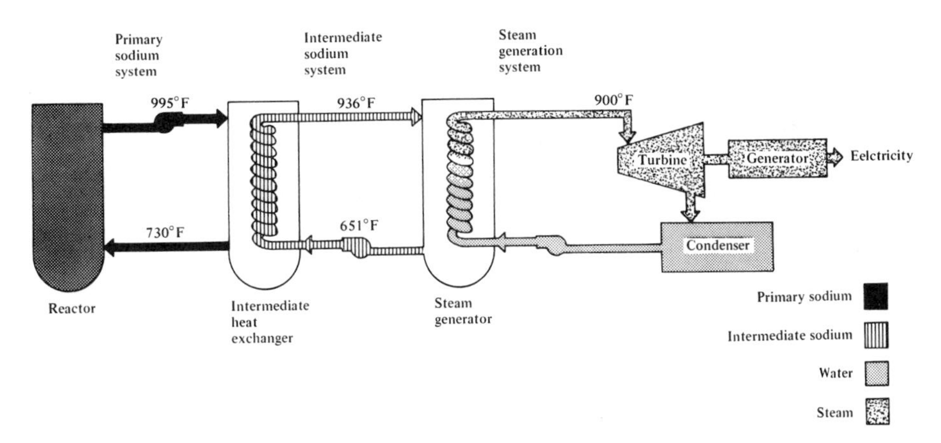


Heat transport system for sodium-cooled breeder power plants. (*Courtesy Breeder Reactor Corporation*)

**sodium-sulfur batteries.** An advanced battery system, consisting of molten sodium in the negative electrode and a mixture of sulfur and sodium polysulfide in the positive electrode. It has a solid ceramic electrolyte. The sodium-sulfur battery operates at a temperature of about 300 to 350° C. Engineering cells are presently delivering about

50 to 60 watt-hours per pound. Plans call for the sodium-sulfur battery to be ready foı testing in the Battery Energy Storage Test (BEST) facility in 1981.

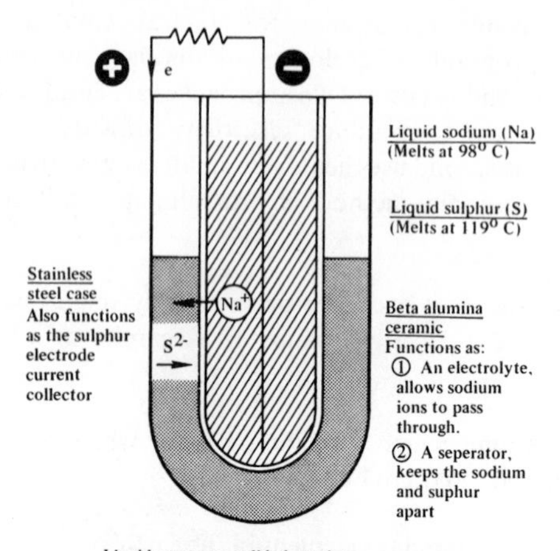
e
Liquid sodium (Na)
(Melts at 98° C)
Liquid sulphur (S)
(Melts at 119° C)
Stainless steel case
Also functions as the sulphur electrode current collector
Na+
S2-
Beta alumina ceramic
Functions as:
① An electrolyte, allows sodium ions to pass through.
② A seperator, keeps the sodium and suphur apart
Liquid reactants, solid electrolyte
Current is carried by sodium ions (Na ) which give up electrons (e) to the external circuit, pass through the solid electrolyte and react with sulphur (S)

Sodium-sulfur cell. (*Courtesy Electric Power Research Institute and the Applied Nucleonics Co., Inc.*)

**soft coal.** Bituminous coal as opposed to anthracite.

**soft fire.** A flame with a deficiency of air.

**soft radiation.** Ionizing radiation of long wavelength and low penetration.

**soft rays.** Beta particles or gamma rays having little penetration.

**soft rock.** Rock that can be removed by air-operated hammers, but cannot be handled economically by a pick.

**soil conserving crops.** Crops that prevent or retard erosion and maintain rather than deplete soil organic matter.

**soil horizon.** A layer of soil, approximately parallel to the surface, which differs from adjacent layers in chemical and physical properties.

**soil porosity.** The degree to which the soil mass is permeated with pores or cavities, expressed as the percentage of the whole volume of the soil unoccupied by solid particles.

**solar absorptance.** The ratio of the amount of solar radiation absorbed by a surface to the amount of radiation incident on it (for terrestrial applications usually calculated for Air Mass 2 characteristics).

**solar air heater.** Heaters that force across pipes containing sun-heated water and blow the air out through registers into rooms for heating. The faster the air is blown through a system, the lower the temperature of the air.

**solar altitude.** The angular elevation of the sun above the horizon.

**solar assisted absorption cooling system.** An air conditioning system that utilizes solar heat rather than conventional forms of energy as its primary power source for driving the refrigeration cycle.

**solar assisted heat pump system.** A heat pump that uses solar heated fluid to increase its coefficient of performance (COP).

**solar azimuth.** The horizontal angle measured between the sun and due south.

**solar balloon.** Using the sun's energy to heat air within an envelope (a balloon) to develop buoyancy.

**solar cell.** A device which converts radiant energy directly into electric energy by the photovoltaic process. Each cell produces a small potential difference, typically about 0.5 volts; an array of cells can provide a useful electric power capacity.

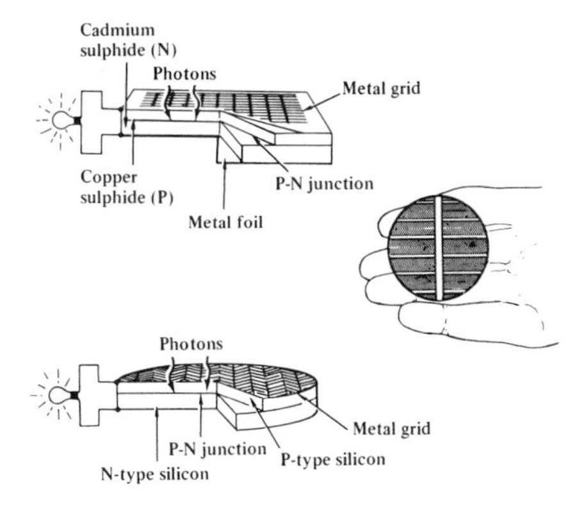


Solar cell components. (*Courtesy Ward Ritchie Press, Passadena, CA*)

**solar cell battery.** An electric cell that converts energy from the sun into electrical energy. (See Fig. p. 406)

**solar collection efficiency.** The ratio of the amount of energy collected to the total solar irradiation during the period under consideration.

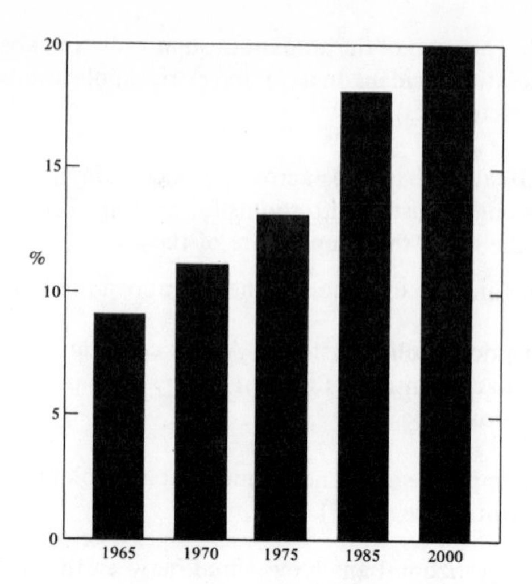


Solar battery efficiency. (*Courtesy The Coming Age of Solar Energy, D. S. Halacy, Harper and Row, New York, NY 10022*)

**solar collector.** A device used to gather and accumulate the sun's energy or solar radiation. Nearly all collectors have a layer of glass on top (some have plastic) to trap the heat once it passes into the collector. The medium used to transfer the heat to the rest of the system varies. Collectors using water to transport the heat have layers of honeycomb or corrugated or crimped-copper conductors and insulators. Air-transport collectors vary in their consistency. Another type of collector, the focusing collector, employs a concave mirror turned by a motor to follow the sun as the earth moves in order to maintain a focus on the sun. The basic function of the solar collector is to capture the sun's heat for household heating and cooling. (See Fig. p. 407)

**solar compass.** Synonym for sun compass.

**solar concentrator.** Reflector or lens designed to focus a large amount of sunshine into a small area, thus increasing the temperature.

**solar constant.** The average amount of solar radiation reaching the earth's atmosphere per minute. The solar constant is the equivalent of 1.94 calories of energy striking each square centimeter per minute or 430 British thermal units of energy striking each square foot per hour. The solar constant is measured on a plane perpendicular to the path of the radiation. It's value is 1.36 kilowatts per square meter.

**solar cooling.** Use of sunlight for cooling. Basically, solar cooling uses absorption equipment of the type employed in gas-burning refrigerators and air conditioners. Solar heat is applied to a solution, such as ammonia and salt water, to create a vapor that cools and condenses and absorbs heat. As the vapor cools an area, the heat it absorbs causes it to vaporize again, thus repeating the process.

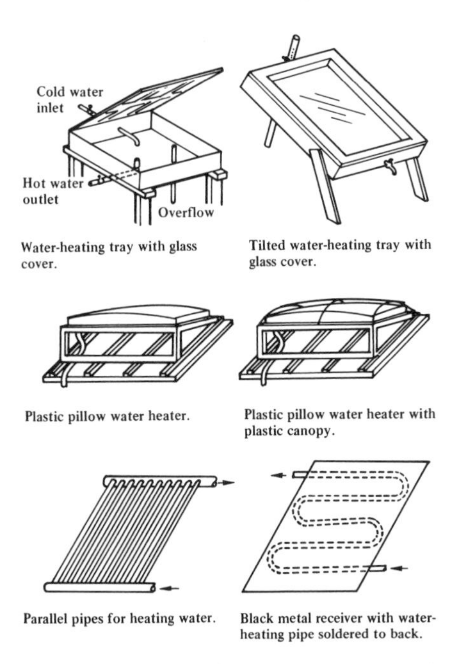


Water-heating tray with glass cover.

Tilted water-heating tray with glass cover.

Plastic pillow water heater.

Plastic pillow water heater with plastic canopy.

Parallel pipes for heating water.

Black metal receiver with water-heating pipe soldered to back.

Solar water heaters. (*Courtesy Direct Use of the Sun's Energy, Farrington Daniels, Random House, Inc., New York*)

**solar declination.** The angle of the sun north or south of the equatorial plane. It is plus if north of the plane and minus if south. The earth's tilted axis results in a day-by-day variation of this angle.

**solar electric conversion.** Technique in which energy from the sun is transformed into electricity by way of solar thermal, solar photovoltaic, ocean thermal, and wind conversion. (See Fig. p. 408)

**solar energy.** The energy transmitted from the sun in the form of electromagnetic radiation. Although the earth receives about one-half of one billionth of the total solar energy output, this amounts to about 420 trillion kilowatt-hours annually.

**solar energy equipment.** Those system components attached to premises in such a manner as to deliver heating, cooling or electricity by the system deriving its energy from the sun.

**Solar Energy Research Institute (SERI).** The Solar Energy Research, Development and Demonstration Act of 1974 called for the establishment of a Solar Energy Research Institute whose general mission would be to support DOE's solar energy program and foster the widespread use of all aspects of solar technology, including direct solar conversion (photovoltaics), solar heating and cooling, solar thermal power generation, wind, ocean thermal conversion, and biomass conversion.

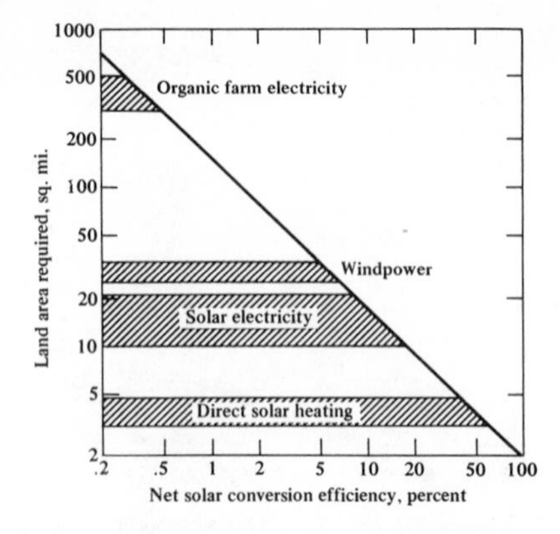


Land area required for 1000 megawatt electrical equivalent output as a function of solar conversion efficiency. (*Courtesy U.S. Council on Environmental Quality*)

**solar energy utilization.** Solar radiation may be utilized by photovoltaics, solar thermal electric, ocean thermal energy conversions, and wind energy conversion systems. These alternative energy sources utilize heliothermal, helioelectric, and heliochemical processes as additional means of producing energy.

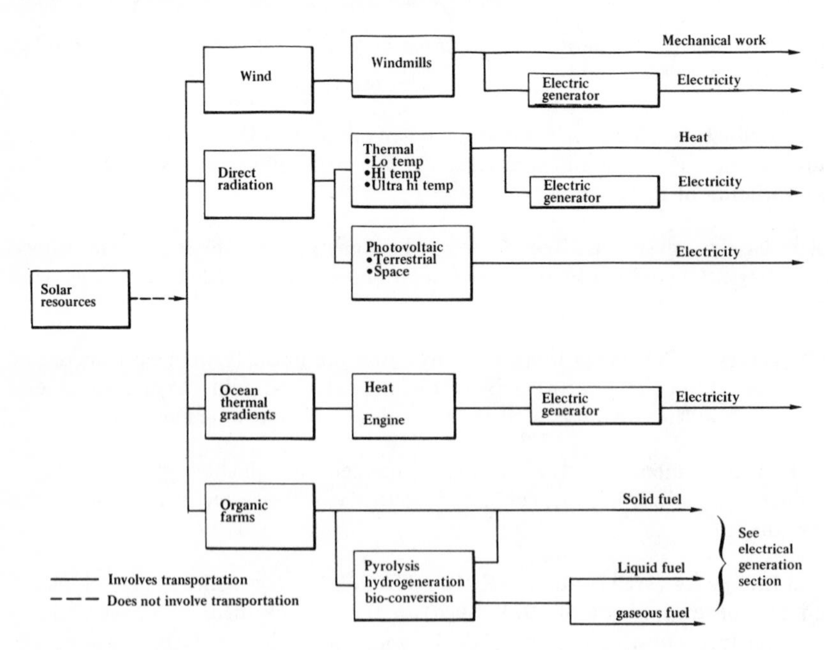


Solar energy resource development. (*Courtesy U.S. Council on Environmental Quality*)

**solar eyeball.** Type of solar concentrating collector developed in England, which uses Fresnel lens to focus on solar cell. As sun's rays move off focus, one of four gas reservoirs is heated. Gas expands, moving magnetic slug in curved pipe below. Since slug is held stationary by external magnet, it stays put while pipe and whole assembly move, bringing sun back into focus. "Eyeball" floats in a shallow tray of water, so movement is almost friction-free.

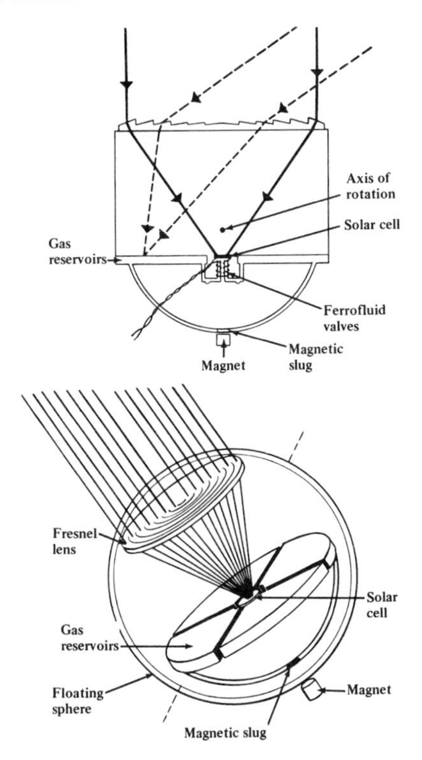


Solar "eyeball" collector. (*Courtesy Popular Science, October 1976, p. 98*)

**solar farm.** A suggested power utility, based in a desert and covering a considerable area, where large amounts of electrical energy would be generated from solar energy.

**solar furnace.** A particular type of image furnace; a unitized, self-contained, solar heating system; a device with large mirrors that focuses the sun's rays upon a small focal point to produce very high temperatures.

**solar heating and cooling.** The development, design, construction, and operation of systems that utilize and/or store the radiant energy of sunlight to provide comfort control and heated water for household, industrial, or agricultural use. (See Fig. p. 410)

**Solar Heating and Cooling Demonstration Act.** Federal legislation which became law in September 1974. This legislation establishes the necessary industrial capacity for the rapid expansion of the use of solar energy for heating and cooling. It calls for

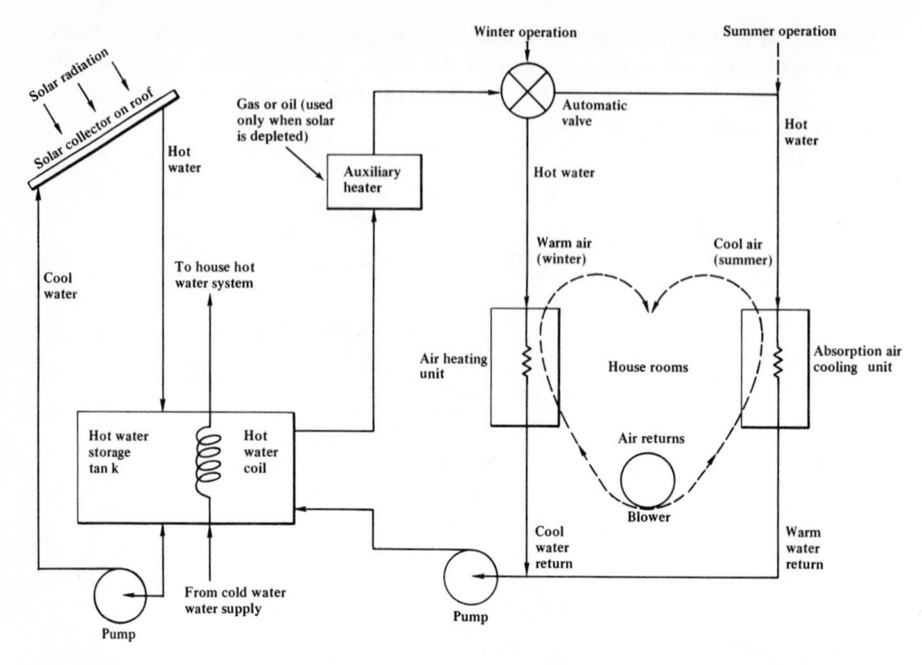


Residential heating and cooling with solar energy. (*Courtesy U.S. Council on Environmental Quality*)

demonstration of the practical use of solar energy on a large scale as follows: solar heating technology within 3 years, and combined solar heating and cooling technology within 5 years.

**solar heating and hot water systems.** The solar heating and hot water system provides two basic functions: (a) capturing the sun's radiant energy, converting it into heat energy, and storing this heat in an insulated energy storage tank; and (b) delivering the stored energy as needed to either the domestic hot water or heating system. The parts of the system which provide these two functions are referred to as the collection and delivery subsystems. The key component in the collection subsystem is the collector, whose basic function is to trap the sun's energy. The delivery subsystem is divided into two parts: one for providing heat to the hot water tank and another for providing heat to the building heating system. When either the domestic hot water or heating system requires heat, hot water from the energy storage tank is pumped to a heat exchanger in the domestic hot water tank or in the building ducts. (See Fig. p. 411)

**solar house.** A house equipped with glass areas and planned to utilize the sun's energy in heating. (See Fig. p. 411)

**solar irradiation.** The amount of radiation, both direct and diffuse, which can be received at any given location.

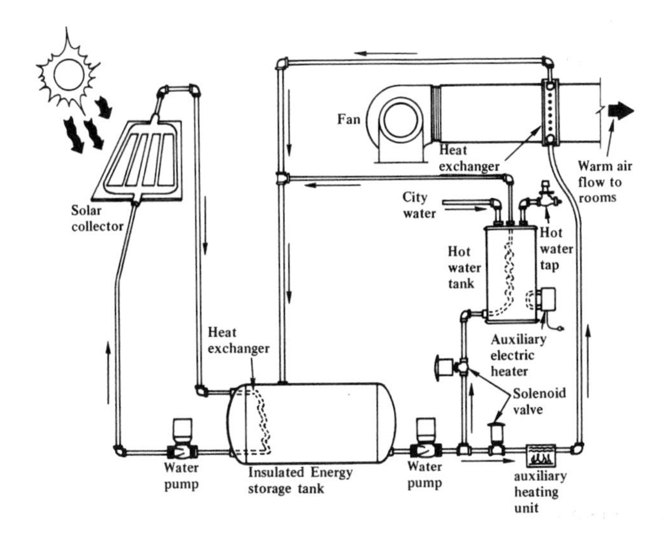


Schematic diagram of a solar heating and hot water system. (*Courtesy ERDA*)

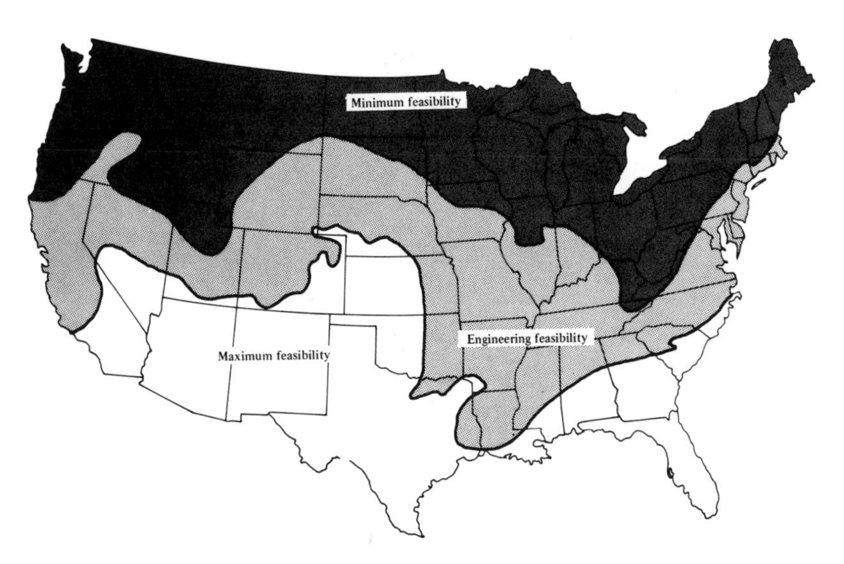


Regions of the different degrees of feasibility of heating houses by solar energy in the U.S. (*Courtesy National Science Teachers Association*)

**solarization.** An effect of strong sunlight (or artificial ultraviolet radiation) on some glasses, causing a change in their transparency. Glasses free from arsenic and of low soda and potash contents are less prone to this effect.

**solar photovoltaic conversion.** Conversion technique which involves sunlight passing through a flat solar cell to generate electricity directly.

**solar power.** Useful power derived from solar energy. Both steam and hot-air engines have been operated from solar energy. Large solar steam engines were built in California, Arizona, and Egypt between 1900 and 1914; however, not one of these engines has survived because of competition from the gasoline engine and the electric motor.

**solar power direct conversion systems.** The production of electricity from solar radiant energy by direct conversion devices such as an array of silicon cells. One direct conversion scheme involves the launching of a satellite-mounted array of such cells in synchronous orbit, permanently placing the cells over a preselected position on the earth's surface. Radiant energy would be converted to direct current, which in turn would be converted electronically into microwave energy. Microwave energy would be beamed to huge antennas located on the earth's surface beneath the satellite. The energy could then be converted to alternating current. At the present stage of development, the maximum efficiency of silicon cells is about 16 percent.

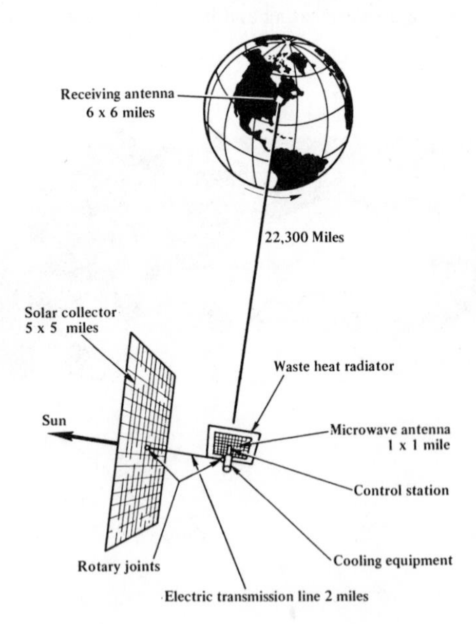


Satellite solar power station designed to produce $10^7$ kilowatts of electricity. (*Courtesy "Energy and the Future," Fig. 23*)

**solar powered irrigation.** Distributed collectors or small central-receiver systems are adaptable for pumping irrigation water in the Southwest, either by direct-coupled thermally driven heat engines or by generating electricity. A 25-horsepower solar turbine (nonelectric) pumping system is being constructed on a farm near Willard, New Mexico, for shallow-well irrigation.

**solar power thermal conversion systems.** Thermal conversion systems involving an extensive array of steel pipes coated with materials heated by the sun's rays. In one concept, nitrogen flowing through the pipes would gather the heat and transport it to molten salt. The molten salt can be heated to a temperature of about 538°C for production of steam, which would power conventional turbines at a projected efficiency of about 30 percent. The area required to supply energy to a 1000-megawatt power plant would be about 10 square miles of collection surface, plus a 300,000-gallon reservoir of molten salt. Energy storage would be necessary for nights and cloudy days.

**solar radiation.** The total electromagnetic radiation emitted by the sun.

**solar radiation intensity.** The amount of free energy outside the earth's atmosphere averaging 442 British thermal units per square foot-hour.

**solar rights.** An unresolved legal issue involving who owns the rights to the sun's rays.

**solar salt.** Salt obtained by solar evaporation of seawater.

**solar sea power.** The concept involving the use of temperature differences between the sun-heated surfaces of the ocean and the colder water deep under the surface to power heat engines. Since the water retains the heat of the sun, such plants, unlike other types of solar plants, could operate at night and during cloudy periods. The technology for such plants is envisioned to be large, extending a half mile or more under the water to reach the deep cold water. Since the temperature difference between this cold water and the surface waters is only in the range of 2°C, such a power plant would have a thermal efficiency less than one-tenth of the efficiency of a conventional, modern fossil-fueled plant. (See Fig. p. 414)

**solar simulators.** Equipment to simulate the solar flux for test purposes.

**solar space heaters.** A solar energy heating system which uses a black metal surface under glass on the roof of a building to absorb sunlight and create a "greenhouse" effect to heat the water or air which is circulated during the day and stored at night.

**solar spectrum.** The total distribution of electromagnetic radiation emitted from the sun, minus those wavelengths that are absorbed by the solar atmosphere.

**solar still.** A distillation apparatus consisting of an airtight space through which solar energy penetrates and is partially absorbed by brackish water in a basin at the bottom of the space. (See Fig. p. 414)

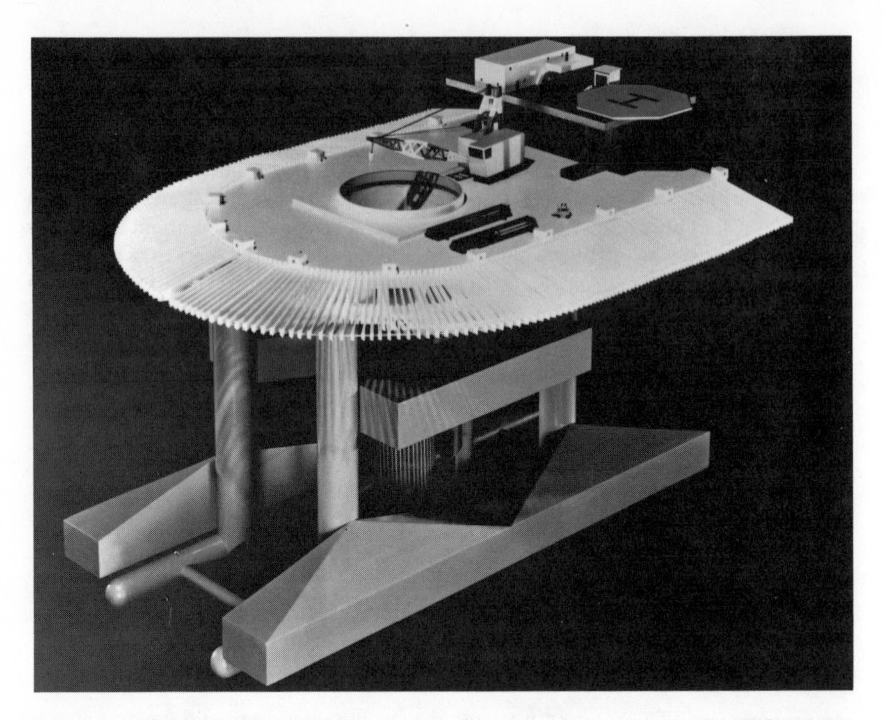

Froude model to scale of 10,000 kilowatt sea solar power plant. (*Courtesy Sea Solar Power, Inc.*)

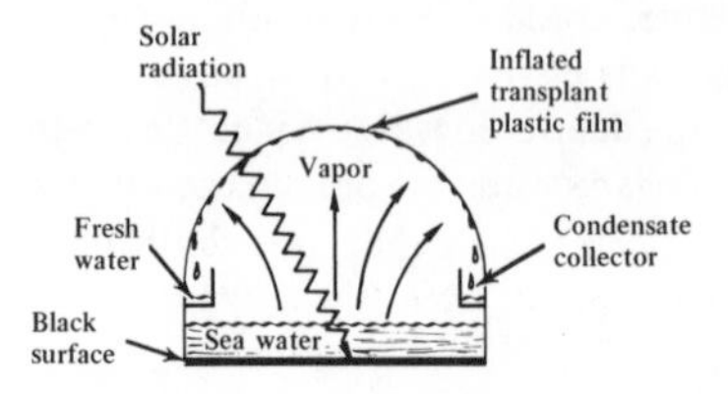


Solar still. (*Courtesy Van Nostrand's Scientific Encyclopedia*)

**solar system efficiency.** The amount of British thermal units lost from the time the sun's rays hit the solar collector to the moment they are used to heat the house or the water supply, compared with the original number coming in.

**solar thermal conversion.** The process of concentrating sunlight on a relatively small area to create the high temperatures needed to vaporize water or other fluids to drive a turbine for generation of electric power.

**solar thermal electric (STE).** The basic concept underlying solar thermal electric power generation is the utilization of solar radiation to heat a working fluid to a temperature high enough to be used (either directly or indirectly) to power a turbine which will, in turn, drive an electric power generator and thus produce electricity which can be integrated into existing electric power networks. To attain the high temperatures required for this application, concentrated and direct solar radiation is usually necessary (diffuse radiation cannot be focused), thus signifying the need for some sort of focusing collector configuration. Two methods presently exist for attaining these requisite temperatures in working fluid quantities large enough for electric power generation: the central receiver system and the distributed collector system.

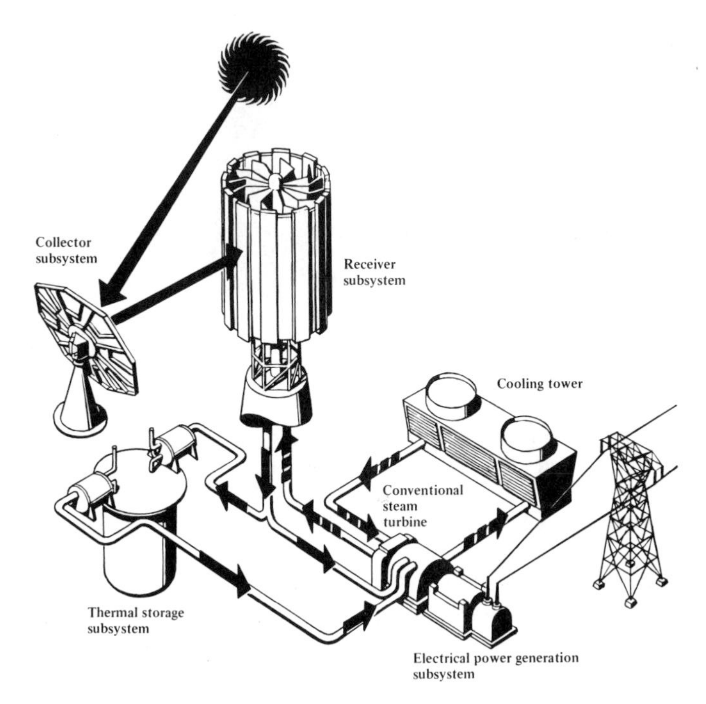


Typical central receiver of a solar thermal power plant. (*Courtesy Popular Science, October 1975, p. 90*)

**Solar Thermal Test Facility.** Facility being constructed by DOE at its Sandia Laboratories, New Mexico, for the purpose of testing boilers and other components that will be used in the pilot plants. (See Fig. p. 416)

**solar time.** Varying period, marked by successive crossings of the meridian by the sun; hour angle of the sun at the observer's point (apparent time) corrected to true time by use of the equation of time. Sundials show apparent solar time. Solar noon is that instant on any day at which time the sun reaches its maximum altitude for that day. Solar time is very rarely the same as local standard time in any locality.

Aerial view of DOE's Solar Thermal Test Facility at Sandia Laboratories. (*Courtesy U.S. Energy Research and Development Administration*)

**Solar Total Energy Systems.** Solar Total Energy Systems use the sun's heat to generate electricity and the leftover or residual heat for other purposes. Several technical approaches are being investigated as applicable to solar total energy systems. These include small central receivers, distributed collectors, and photovoltaics.

**solenoid.** An electromechanical device so designed that when the electric current is applied to a coil of wire wound around a cylinder, the electromotive force causes a bar or plunger inside the cylinder to move; a series of coils or wire carrying a current of electricity; a cylindrical coil used to produce a magnetic field.

**solenoid valve.** An automatic valve that is opened or closed by an electromagnet.

**solid fuels.** Any fuel that is a solid such as wood, peat, lignite, bituminous and anthracite coals of the natural variety, and the prepared variety such as pulverized coal, briquettes, charcoal, and coke. These are divided into two broad classes: naturally occurring solid fuels, and manufactured solid fuels.

**solidification.** The process of changing from a liquid or a gas to a solid.

**solid smokeless fuel.** Smokeless coal; a solid fuel, such as coke, which produces comparatively no smoke when burnt in an open grate. The gas industry produces certain brands of smokeless fuel such as Coalite, Rexco, Clean-Glow, and Phimax.

**solid waste disposal.** The ultimate disposition of refuse that cannot be salvaged or recycled.

**solstice.** Those times of the year when the sun is furthest north or south of the equator. In the northern hemisphere, it occurs about June 21 to begin summer and on about December 21 to begin winter.

**soluble oil.** A blend of mineral oil and emulsifiers which, when mixed with water, forms a dispersion for use as a cutting fluid.

**solution.** The change of matter from a solid or gaseous state into a liquid state by its combination with a liquid. All solutions are composed of a solvent (water or other fluid) and the substance dissolved is called the "solute". A true solution is homogeneous, as salt in water.

**solvent.** A substance used to dissolve another substance; that component of a solution which is present in excess, or whose physical state is the same as that of the solution.

**solvent extraction.** A method of extracting one or more substances from a mixture by treating a solution of the mixture with a solvent that will dissolve the required substances and leave the others. It is used in purifying certain fuels. In the coal industry, the term refers to the selective transfer of desired coal constituents from finely divided coal particles into a suitable solvent after intimate mixing, usually at high temperatures and pressure in the presence of hydrogen, with or without a catalyst, followed by phase separation. In uranium technology, the term refers to the selective transfer of metal salts from aqueous solutions or pulp, to immiscible organic liquid.

**solvent refined coal.** A process for treating coal to remove ash, sulfur, and other impurities. The end product contains about 16,000 British thermal units per pound, has an ash content of 0.1 percent, and a very low sulfur content of about 0.5 percent. The product is solid at room temperature, but can be liquefied by use of relatively low heat.

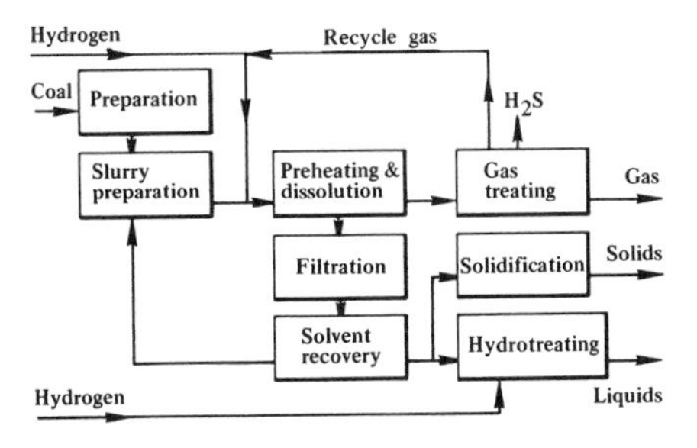


Solvent refined coal process. (*Courtesy U.S. Council on Environmental Quality*)

**Solvent Refined Lignite process.** Solvent extraction of lignite is produced from synthesis gas carbon monoxide and hydrogen under 1500 to 3000 pounds per square inch gage by vacuum flashing, producing solvent-refined lignite and light oils.

**solvent refining.** A method for removing ash and sulfur compounds from coal. The principal product is called solvent-refined coal; it has an energy content of about 16,000 British thermal units per pound and can be used as a substitute for coal in some applications.

**solvent refining of coal.** A technique for purifying coal. Pulverized raw coal is mixed with an aromatic solvent and reacted with hydrogen gas at high temperatures and pressures. This dissolves the coal and separates it from its ash, sulfur, oxygen, and water. The solvent is then removed, leaving a pitch-like product low in sulfur and ash and with a heat content improved by as much a 60 percent.

**somatic effects of radiation.** Effects of radiation limited to the exposed individual, as distinguished from genetic effects. Large radiation doses can be fatal. Smaller doses may make the individual noticeably ill, may produce temporary changes in blood-cell levels detectable only in the laboratory, or may produce no detectable effects whatever.

**soot.** A black substance, which consists essentially of carbon from the smoke of wood or coal (especially that which adheres to the inside of the chimney) and contains volatile products condensed from the combustion of the wood or coal, including certain ammonia salts.

**sorbent.** A material which extracts one or more substances present in an atmosphere or mixture of gases or liquids with which it is in contact, due to an affinity for such substances.

**sorption.** Any type of retention of a material at a surface. Sorption is basic to many processes used to remove gaseous and particulate pollutants from an emission and to clean up oil spills.

**source beds.** Rocks in which oil or gas has been generated.

**source material.** In atomic energy law any material, except special nuclear material, which contains 0.05 percent or more of uranium, thorium, or any combination of the two.

**source reactor.** A reactor specially designed to supply a stable flux of neutrons having a well-determined energy spectrum.

**sour crude.** Crude oil which (1) is corrosive when heated, (2) evolves significant amounts of hydrogen sulfide on distillation, or (3) produces light fractions which require sweetening. Sour crudes usually, but not necessarily, have high sulfur content. Examples are most West Texas and Middle East crudes.

**sour natural gas.** Natural gas which contains objectionable amounts of hydrogen sulfide and other sulfur compounds.

**space nuclear systems.** Nuclear electric power applications to space and terrestrial missions for the purpose of gaining solutions to energy-related problems.

**space velocity.** The volume of a fluid (usually measured at standard conditions) passing through a unit volume in a unit time; units are in reciprocal time.

**spallation.** A nuclear reaction induced by high-energy bombardment, in which several particles or fragments are simultaneously ejected from the nucleus.

**spark chamber.** An instrument for detecting and measuring the paths of elementary particles. It is analogous to the cloud chamber and bubble chamber. It consists of numerous electrically charged metal plates mounted in a parallel array. The spaces between the plates are filled with an inert gas, and any ionizing event causes sparks to jump between the plates along the radiation path through the chamber.

**sparse vitrain.** A field term to denote, in accordance with an arbitrary scale established for use in describing banded coal, a frequency of occurrence of vitrain bands comprising less than 15 percent of the total coal layer.

**special nuclear material.** In atomic energy law, this term refers to plutonium-239, uranium-233, uranium containing more than the natural abundance of uranium-235, or any material artificially enriched in any of these substances.

**special theory of relativity.** A theory developed by Albert Einstein that is of great importance in atomic and nuclear physics. It is especially useful in studies of objects moving with speeds approaching the speed of light. Two of the results of the theory with specific application in nuclear physics are statements that the mass of an object increases with its velocity and that mass and energy are equivalent.

**specific activity.** The radioactivity of a radioisotope of an element per unit weight of the element in a sample; the activity per unit mass of a pure radionuclide; the activity per unit weight of any sample of radioactive material.

**specific burnup.** The total energy released per unit mass of a nuclear fuel.

**specific energy.** In cutting or grinding, the energy expended or work done in removing a unit volume of work material, usually expressed as inch-pound per cubic inch or horsepower per minute per cubic inch.

**specific gamma-ray constant.** For a nuclide emitting gamma radiations, the product of exposure rate at a given distance from a point source of that nuclide and the square of that distance divided by the activity of the source, neglecting attenuation.

**specific heat.** The amount of heat needed to raise 1 gram of the substance through 1°C; also the ratio of the thermal capacity of a substance to that of water at 15°C. The specific heat of fluids varies with temperature and pressure.

**specific ionization.** The number of ion pairs formed per unit of distance along the track of an ion passing through matter.

**specific power.** The power generated in a nuclear reactor per unit mass of fuel, expressed in kilowatts of heat per kilogram of fuel.

**specific volume.** The volume of a unit weight of a substance at specific temperature and pressure conditions; the volume occupied by a unit of air, measured in cubic feet per pound.

**specific weight.** The weight of a unit volume of a substance, usually expressed as pounds weight per cubic foot.

**spectral shift reactor.** A reactor design in which a mixture of light water and heavy water is used as the moderator and coolant. The ratio of light to heavy water is varied to change (shift) the speed distribution (spectrum) of the neutrons in the reactor core. Since the probability of neutron capture varies with neutron velocity, a measure of reactor control is thus obtained.

**spectroscope.** An instrument in which collimated light is directed through a narrow slit onto either an optical prism or a diffraction grating, exhibiting its spectrum.

**spectrum.** A visual display, a photographic record, or a plot of the distribution of the intensity of a given type of radiation as a function of its wavelength, energy, frequency, momentum, mass, or any related quantity. (See Fig. p. 421)

**spent fuel.** Nuclear reactor fuel that has been used (irradiated) to the extent that it can no longer effectively sustain a chain reaction. (See Fig. p. 421)

**spent fuel reprocessing.** Reprocessing of spent nuclear reactor fuel for reuse as new fuel as well as on possible alternatives to reprocessing.

**spill.** The accidental release of a substance such as oil or radioactive material.

**spillway.** A passage for surplus water over or around a dam or other hydraulic structure.

**splent.** A hard variety of bituminous coal that ignites with difficulty because of its slaty structure, but makes a clear, hot fire.

**split coal.** A coalbed separated by a clay, shale, or sandstone parting which thickens so that both benches cannot be mined together.

**split system heating.** A combination of warm-air heating and radiator heating; also used for other combinations such as hot-water steam, steam-warm air, etc.

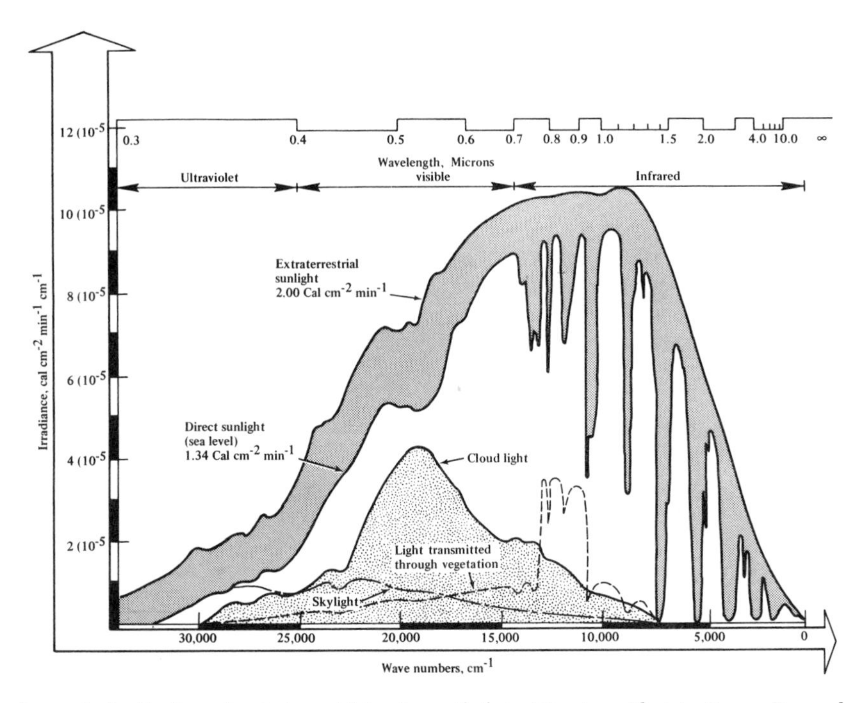


Spectral distribution of extraterrestrial solar radiation. (*Courtesy Electric Power Research Institute and the Applied Nucleonics Co., Inc.*)

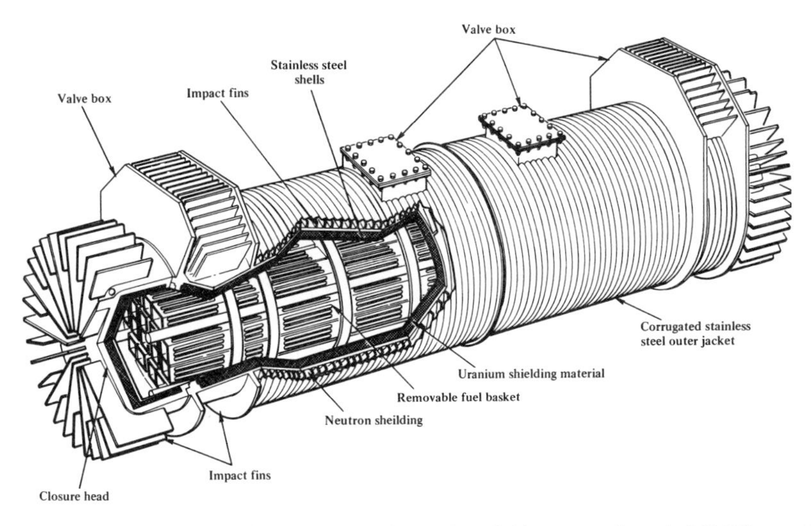


Cutaway view of the first multi-element spent fuel cask, weighing approximately 140,000 pounds and accomodating 7 pressurized water reactor fuel assemblies of 16 boiling water assemblies, each up to 180 inches long, with a burnup capacity of 35,000 megawatt days per metric ton of uranium. (*Courtesy General Electric*)

**spoil.** The overburden or noncoal material removed in gaining access to the coal or mineral material in surface mining; also called waste.

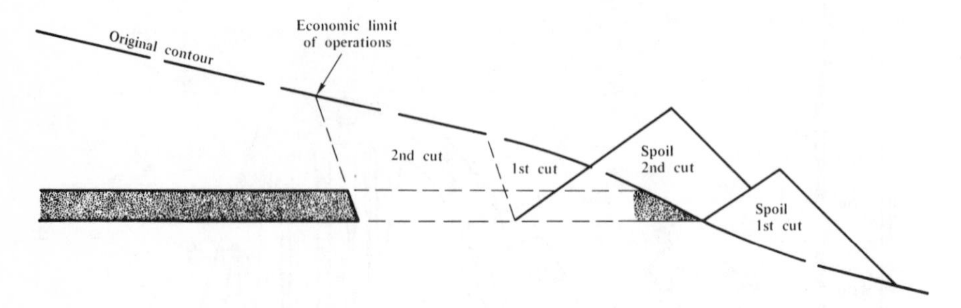


Spoil area in contour mining. (*Courtesy U.S. Energy Research and Development Administration, Division of Coal Conversion and Utilization*)

**spoil bank.** Area created by the deposited spoil or overburden material prior to backfilling; that part of the mine where the coal and other minerals that are not marketable are left.

**sponge.** A form of metal characterized by a porous condition, which results from the decomposition or reduction of a compound without fusion; a mixture of iron oxide and wood shavings for the removal of hydrogen sulfide from gas.

**spontaneous combustion.** The outbreak of fire in combustible material (such as oily rags or damp hay) that occurs without the direct application of a flame or a spark. It is usually caused by slow oxidation processes (such as atmospheric oxidation or bacterial fermentation) under conditions that do not permit the dissipation of heat.

**spontaneous fission.** Fission that occurs without an external stimulus. Several heavy isotopes decay mainly in this manner; examples: californium-252 and californium-254. The process occurs occasionally in all fissionable materials, including uranium-235.

**spontaneous nuclear reaction.** A nuclear reaction in which a nucleus spontaneously changes in mass, charge, or energy state.

**spouter.** An oil or gas well, the flow of which has not been controlled by the engineers.

**spud.** The small cap or plug, with an orifice through it, that admits gas into the mixing chamber of a burner; also a cable tool drill bit.

**spudding.** Refers to the act of hoisting the drill pipe and permitting it to fall freely so that the drill bit strikes the bottom of the well bore with considerable force. This is done to clean the bit of an accumulation of sticky shale which has slowed down the rate of penetration. Careless execution of this operation can result in kinks in the drill pipe and damaged bits.

**spudding-in.** The very beginning of drilling operations of a well.

**square foot of radiation.** The amount of heating surface in the form of radiators, convectors, unit heaters, or other devices, which will emit 240 British thermal units per hour.

**squeeze.** A crushing of coal with the roof moving nearer to the floor without breaking.

**squib.** A firing device that will burn with a flash which will ignite black powder.

**stabbing board.** A temporary platform erected in the derrick at an elevation of about 20 to 40 feet above the derrick floor, on which crew members work while casing is being run in a well.

**stability.** In petroleum products, the resistance to chemical change. Gum stability in gasoline means resistance to gum formation while in storage. Oxidation stability in lubricating oils and other products means resistance to oxidation to form sludge or gum in use.

**stabilization.** The addition of a gas to a gas to adjust its heat content to a specified value. Air is often used for the purpose of reducing heat content and LPG gases are used for the purpose of enriching or raising the heat content.

**stable.** Incapable of spontaneous change; not radioactive.

**stable isotope.** An isotope that does not undergo radioactive decay.

**stack.** A chimney or conduit for smoke; also a measure of fuel consisting of 108 cubic feet.

**stack effect.** The impulse of a heated gas to rise in a vertical passage such as a chimney, a small enclosure, or a stairwell.

**stack gas.** Gases resulting from combustion.

**stack gas cleaning.** Removing pollutants from combustion gases before those gases are emitted to the atmosphere.

**stack-gas desulfurization (scrubber).** Treating of stack gases to remove sulfur compounds.

**stack loss.** The flue gas loss; the sensible and latent heat lost up the chimney in the flue gas.

**stagnation.** Lack of wind in an air mass or lack of motion in water. Both cases tend to entrap and concentrate pollutants.

**standard cubic foot.** The volume of a gas at standard conditions of temperature and pressure. The American Gas Association uses moisture-free gas at 16° C and 30 inches of mercury (1.0037 atmospheres) as its standard conditions. The pressure standard is not universal in the gas industry; 14.7 pounds per square inch absolute (1.000 atmosphere) and 14.4 pounds per square inch absolute (0.980 atmosphere) are also used. The scientific community uses 0° C and 1 atmosphere as standard conditions.

**Standard Industrial Classification Manual.** A book, prepared and issued by the Office of Statistical Standards, United States Bureau of the Budget, to enable classification of business establishments by the type of activity in which they are engaged.

**standard metering base.** Standard conditions, plus agreed corrections, to which volumes are corrected for purposes of comparison and payment.

**standard service pressure.** The gas pressure that a utility undertakes to maintain on its domestic customers' meters.

**standing gas.** A body of firedamp known to exist in a mine, but not in circulation; it is sometimes fenced off.

**standpipe.** A relatively short length of pipe driven into the upper soil-like portion of the overburden as the first step of spudding-in a borehole; also a vertical pipe or reservoir for water used to secure a uniform pressure.

**staple.** A shaft that is smaller and shorter than the principal one and joins different levels; a small pit.

**start cart.** A heating device which supplies load pressure steam for fuel cell heat-up.

**static pressure.** The force per unit area acting on the surface of a solid boundary parallel to the flow. The static pressure is constant in all directions and is the component normal to the direction of flow; the force exerted per unit area by a gas or liquid measured at right angles to the direction of flow, or the pressure when no liquid is flowing.

**stationary source.** A pollution emitter that is fixed.

**station meter.** A meter of high capacity for measuring the output of a gas plant or pipeline delivery station.

**station use.** In the electrical industry, the kilowatt-hours used at the generating station for purposes essential to the operation of the station.

**STEAG-Combined Plant process.** A process being developed by STEAG Atiengesellschaft of West Germany. A Lurgi gasifier is used to produce high pressure, low-Btu gas which is scrubbed, expanded in an expansion turbine, burned in a pressurized boiler, and finally expanded again in a gas turbine. The pressurized boiler drives a steam turbine which, in combination with gas turbine power extraction, extracts a maximum of heat energy from the combustion power extraction, extracts a maximum of heat energy from the combustion and gasification of coal. The combined plant currently operating in Lupen, Germany, is too small to actually maximize the heat energy recovery from coal that a commercial size plant may be able to achieve. The 170-megawatt combined plant currently in operation is the design basis of an 800-megawatt plant in planning.

**steam.** Water in the form of a vapor. It is often the medium by which heat energy liberated from fuel by combustion is converted into mechanical work.

**steam accumulator.** A vessel to smooth out violent steam loads on a boiler. While the steam demand is relatively low, the accumulator is charged and acts as a buffer when sudden steam demands occur.

**steam coal.** Coal suitable for use under steam boilers; coal which is intermediate in rank between bituminous coal and anthracite.

**steam-electric plant.** A plant in which the prime movers (turbines) connected to the generators are driven by steam.

**steam engine.** A reciprocating engine, worked by the force of steam on the piston; the steam expands from the initial pressure to the exhaust pressure in a single stage.

**steam gas.** Highly superheated steam.

**steam generating station.** An electric generating station in which the prime mover is a steam turbine. (See Fig. p. 426)

**steam generator.** The equipment which uses a heat source to change water into steam.

**steam injection.** The injection of steam to reenergize oil reservoirs.

**Steam-Iron process.** A process being developed to supply hydrogen for the HYGAS coal gasification process.

**steam jet refrigeration.** A method of cooling involving the use of steam nozzles to reduce the pressure in a meter chamber so that the water boils at a low temperature. Since heat is drawn from the water, it is thus cooled.

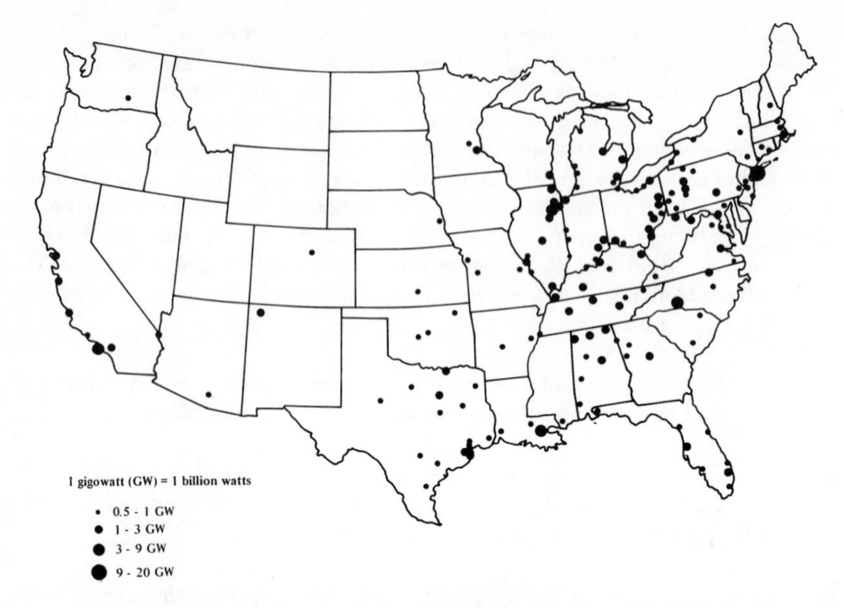


Major steam generating centers in the U.S. (1970). (*Courtesy U.S. ERDA—69*)

**steam or gas turbine.** An enclosed rotary type of prime mover in which heat energy in steam or gas is converted into mechanical energy by the force of a high velocity flow of steam or gas directed against successive rows of radial blades fastened to a central shaft.

**steam power.** Energy or power generated through the use of a steam engine.

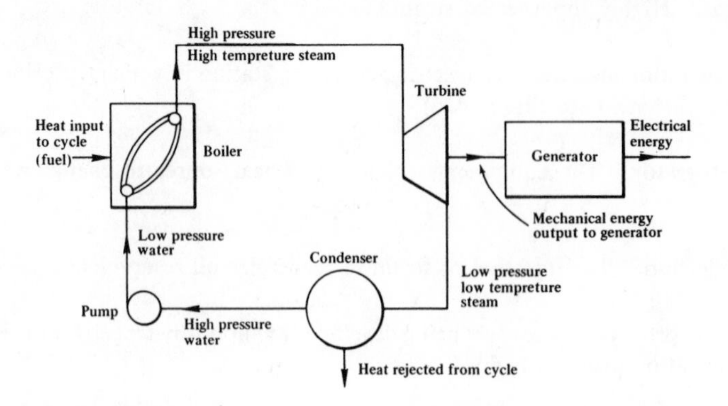


Simplified schematic of a steam power plant. (*Courtesy U.S. Council on Environmental Quality*)

**steam tracing.** A graphic recording of steam pressure and temperature done instrumentally.

**steam trap.** A device for allowing the passage of condensate, or air and condensate, while preventing the passage of steam.

**steam turbine.** A machine powered by high pressure steam and used to drive mechanical apparatus. It has a rotary motion in contrast to a reciprocating motion.

**Stefan-Boltzmann law.** The energy radiated in unit time by a black body, given as $E = K (T^4 - To^4)$, where T is the absolute temperature of the body, To the absolute temperature of the surroundings, and K is a constant.

**stemming.** The inert material (clay; water) used to fill a hole after the explosive charge has been placed.

**stepdown substation.** An assemblage of plant and equipment in one place, used to change electricity from a higher to a lower voltage.

**stepup substation.** An assemblage of plant and equipment in one place, used to change electricity from a lower to a higher voltage.

**sterilized coal.** The part of a coal seam which, for various reasons, is not mined.

**still.** An apparatus in which a substance is changed by heat into vapor, with or without chemical decomposition. The vapor is then liquefied in a condenser and collected in another part of the apparatus. Used generally for separating the more volatile parts of liquids and obtaining them in pure form.

**still-process.** German process for removing sulfur dioxide from flue gases with 70 to 80% efficiency by absorbing the sulfur dioxide with brown-coal filter ash, hydrated lime or limestone, at 100 to 1300°C, with a surplus 10% of surface water. The development of this process was interrupted and it has not been applied.

**Stirling hot-air engine.** The Stirling engine is expected to be clean burning and fuel efficient. It also can burn any type of fuel to generate the heat used in the engine. Like the current internal combustion engine, the Stirling has a cylinder and pistons. However, the fuel is burned outside of the cylinder (external combustion). The heat is used to expand a gas in the cylinder, which pushes down on the piston to produce the power stroke. The Stirling was invented in 1816 by a Scottish parson. The current basic design is the result of work done primarily by the N. V. Philips Company of The Netherlands. Problems which remain to be solved include the handling of the very large quantity of waste heat which must be removed by the radiator and the need for heater heads which can contain the high temperature and high pressure gas. (See Fig. p. 428)

**Stirred Fixed-Bed Gasifier process.** A pressurized, air-blown fixed-bed gasifier is being developed to produce a low-Btu gas from a wide variety of coals. A variable height stirrer facilitates the breaking-up and gasifying of caking coals. The gasifier is very similar to the Lurgi gasifier in operating principle. Process conditions are: a tem-

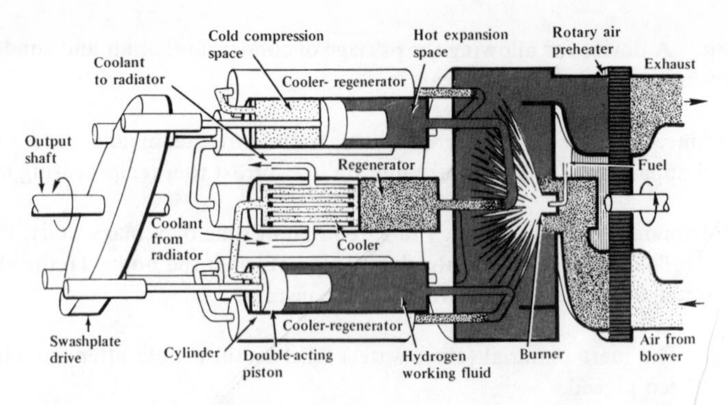


Stirling hot-air engine. (*Courtesy Popular Science, April 1976, p. 84*)

perature of 1260° C in the combustion zone and a pressure of 300 pounds per square inch gage.

**stock.** An accumulation of petroleum, or products in storage, awaiting transfer to the point of utilization; also, a bar used to support a windmill sail.

**stoker coal.** A screen size of coal specifically for use in automatic firing equipment.

**Stone Webster tonics process.** Process for removing fly ash and sulfur dioxide from flue gases by electrostatic precipitation and pre-scrubbing with water, followed by the removal of the sulfur dioxide by absorption in water, then contacting the absorbent with sodium bisulfate to produce sodium sulfate and sulfur dioxide gas. Caustic soda and sodium bisulfate are recovered. The process is used in the United States.

**stope.** An excavation from which ore has been excavated in a series of steps; usually applied to highly inclined or vertical veins. Each horizontal working is called a stope.

**stopping power.** A measure of the effect of a substance upon the kinetic energy of a charged particle passing through it.

**storage.** That component of a solar system that receives energy in the form of heat (or cold) and retains it for future use.

**storage battery.** A combination of secondary cells or accumulators which, once charged, may be used for a considerable time as a source of electric current.
(See Fig. p. 429)

**storage capacity.** The amount of energy that can be stored by a solar heating system to be used at a later time for space or water heating.

**storage cycle.** In the gas industry, a period commencing with an injection phase during which gas is stored, and ending with a subsequent withdrawal phase during which gas is removed.

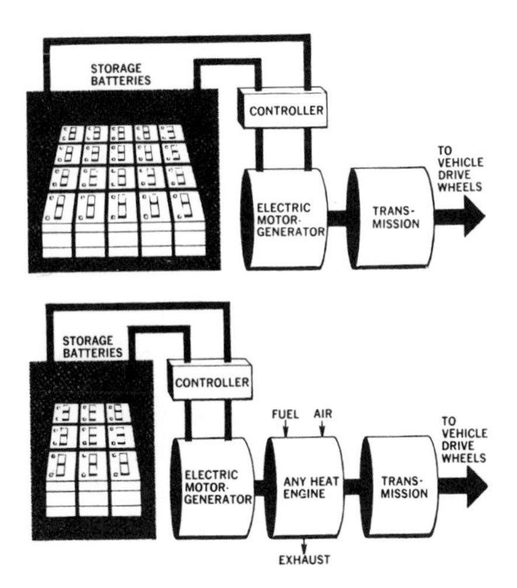


Use of storage batteries in electric and hybrid vehicles. (*Courtesy Popular Science, April 1976, p. 84*)

**storage holder.** A gas holder for storage of excess gas supply for use during times of excess demand.

**storage mains.** Those mains used primarily for injection and withdrawal of gas to and from underground storage.

**storage reservoir.** That part of the storage zone having a defined limit of porosity and/or permeability which can effectively accept, retain, and deliver liquids such as oil or water, and gases. (See Fig. p. 430)

**storage well.** A cased bore hole extending from the surface into the storage reservoir, which is used primarily for oil or gas input and/or output purposes.

**storage zone.** The geological name of that stratum in the earth's crust within which the storage reservoir is located.

**storm windows.** Glass or plastic applied over existing windows to save energy. Plastic can be self-installed for 50 cents per window each year. Single panes of glass can cover the window at a cost of 10 to 15 dollars per window. Triple-track glass combination storm windows cost 30 dollars each. These have screens and can be opened and closed easily. They are for installation on conventional double-hung windows.

**stove coal.** In anthracite only; two sizes of stove coal are made, large and small. Large stove coal, known as No. 3, passes through a 2¼ to 2 inch mesh and over a 1⅞ to 1½ inch mesh; small stove coal, known as No. 4, passes through a 1⅞ to 1⅜ inch mesh and over a 1⅛ to 1 inch mesh. Only one size of stove coal is usually made. It passes through a 2 inch square mesh and over a 1⅜ inch square mesh.

Underground gas-storage reservoirs in the U.S. (*Courtesy U.S. Council on Environmental Quality*)

**straightening vanes.** Round, square, or other shaped tubing installed axially inside the piping preceding an orifice or turbine meter to eliminate swirls and crosscurrents set up by the pipe fittings and valves.

**straight gas utility.** A company which derives the major portion of its total operating revenues from gas operations. A straight gas utility derives at least 95 percent of its total operating revenues from gas operations.

**straight-run bitumen.** A residual bitumen obtained as a result of the distillation of petroleum.

**strange particles.** A class of very short-lived elementary particles that decay more slowly than they are formed, indicating that the production process and decay process result from different fundamental reactions. They include K-mesons and hyperons.

**strata.** Artificial or natural sedimentary rock layers; plural of stratum.

**strata gases.** Gases occurring either in the mineral deposit itself or in adjacent or nearby formations. Their origin may be in a particular formation in which they were laid down or formed subsequently by chemical action, or they may occasionally migrate from other formations, frequently because of release of pressure with mining. Water flow and rock porosity and fissures also allow gas migration.

**stratified.** Formed or lying in beds, layers, or strata.

**stratified charge engine.** An engine in which the amount of charge, fuel plus air, is adjusted to engine conditions, directed to the area where it will burn best and fired at just the precise instant.

**stratigraphic trap.** A trap that results from variation in the lithology of the reservoir rock, a termination of the reservoir, or other interruption of its continuity.

**stratoscope.** An apparatus inserted in the drill hole which permits engineers to make a visual inspection of the strata.

**stray current.** Electrical current, normally direct current, from either natural or man-caused source, which could cause corrosion of structures such as pipelines, if not reduced properly.

**stream day.** A steady, 24-hour operation of a refinery unit.

**stream flow.** The quantity of water passing a given point in a stream or river in a given period of time.

**streamlines.** Hypothetical lines which show the velocity direction of the fluid stream at each point along the lines. A set of streamlines charts the flow pattern.

**stress.** The resultant force that resists change in the size or shape of a body acted on by external or internal forces.

**stress corrosion.** Stress corrosion cracking is cracking caused by the simultaneous presence of a tensile stress and a specific corrosion inducing medium. The metal or alloy surface will appear virtually unaffected while fine cracks progress through it. Some metals and alloys are susceptible to stress corrosion when stressed in the presence of contaminant commonly present in water, aqueous fluid, or in the atmosphere. Collector components may be subject to substantial residual stresses resulting from their fabrication. In addition, thermal cycling during operation may introduce applied stresses. Typical system component designs and operating conditions may be such that it is virtually impossible to avoid residual or applied stresses. The metals used in solar system components should be limited to those which are not susceptible to stress corrosion when in contact with the anticipated heat transfer liquid or atmospheric corrodents.

**stress crack.** Internal or external crack in a material caused by tensile or shear stresses less than that normally required for mechanical failure in air. The development of such cracks is frequently related to and accelerated by the environment to which the material is exposed.

**stress-rupture test.** A tension test performed at constant load and constant temperature, the load being held at such a level as to rupture; a method of testing plastic pipe to determine the hydrostatic strength by applying a constant internal pressure and observing time to failure.

**strike.** The course or bearing of the outcrop of an inclined bed or structure on a level surface; also the direction or bearing of a horizontal line in the plane of an inclined stratum, joint, fault, cleavage plane, or other structural plane. It is always perpendicular to the direction of the dip.

**stringer.** A small vein of mineralized rock; a narrow vein or irregular filament of mineral traversing a rock mass of different material; a thin layer of coal at the top of a bed, separating in places from the main coal by material similar to that comprising the roof.

**strip.** In coal mining, to remove the earth, rock, and other material from a seam of coal; in the gas industry, to remove light hydrocarbon fractions from gas for recovery and sale.

**strip mining.** A mining technique used when deposits of coal lie relatively near the surface (less than 100 feet). The overburden (the soil and rock above the coal) is stripped away, the coal removed, and the overburden from a trench dumped in the previous, parallel one. Strip mining is used primarily for coal, but may be used extensively in the mining of oil shale as well. (See Fig. p. 433)

**stripped atom.** One from which one or more electrons have been removed, rendering it ionically charged.

**stripper.** A pressure vessel in which the carbone dioxide and other heavy hydrocarbons are stripped from the liquid methanol by passing a clean stream of methane up through the methanol; also a nearly depleted well whose income barely exceeds operating cost of production.

**stripper production.** An oil well that provides slow but steady production.

**stripping ratio.** The unit amount of spoil or waste that must be removed to gain access to a similar unit amount of ore or mineral material.

**stroge.** The maximum distance a piston moves within a cylinder before the direction of its travel is reversed.

**strontium unit.** A measure of the concentration of strontium-90 in food and in the body (i.e., in bone). It is measured as the ratio of strontium to calcium, with which strontium becomes mixed in soil and living tissue. One strontium unit is one picocurie ($10^{-12}$ curie) of strontium-90 per gram of calcium.

**structural trap.** A reservoir, capable of holding oil or gas, formed from crustal movements in the earth that fold or fracture rock strata in such manner that oil or gas

A thick seam of coal exposed by stripping is being loaded into a hauler by a front end loader. (*Courtesy U.S. Energy Research and Development Administration*)

accumulating in the strata are sealed off and cannot escape. The most common structural traps are fault traps, anticlines, and salt domes.

**subanthracite.** Coal intermediate between anthracite and superbituminous coal.

**subaqueous mining.** Surface mining in which the material mined is removed from the bed of a natural body of water.

**subatomic particle.** Any of the constituent particles of an atom; an electron, a neutron, or a proton.

**subbituminous coal.** Ranking of soft coal generally having a heating value of 8,300 to 13,000 British thermal units per pound, high volatile matter, and ash.

**subcooling.** The cooling of a liquid to below its saturation temperature for the pressure under consideration.

**subcritical assembly.** A reactor consisting of a mass of fissionable material and moderator whose effective multiplication factor is less than one and hence cannot sustain a chain reaction.

**subcritical mass.** An amount of fissionable material of insufficient quantity or improper geometry to sustain a fission chain reaction.

**subcritical multiplication factor.** In a subcritical assembly containing a neutron source, the equilibrium ratio of the total number of neutrons resulting from fission and the source to the total number of neutrons which would exist in the assembly due to the source alone.

**subhydrous coal.** Coal of hydrogen content below average for the rank of coal; for example, coals containing a high proportion of fusain.

**sublimation.** The vaporization of a solid (especially when followed by the reverse change) without the intermediate formation of a liquid.

**sublime.** To cause to pass from the solid state to the vapor state by the action of heat, and to condense again to solid form.

**submarginal resource.** That portion of a resource that cannot presently be extracted economically. As the economic picture and technology change, some submarginal resources may become recoverable resources.

**submetering.** The practice of remetering purchased energy beyond the customer's utility meter, generally for distribution to building tenants through privately owned or rented meters.

**subsidence.** A sinking down of a part of the earth's crust due to underground excavation (often coal mines). If extensive, a cave-in can result.

**subsidence break.** A fracture in the rocks overlying a coal seam or mineral deposit as a result of its removal by mining operations. The subsidence break usually extends from the face upwards and backwards over the unworked area.

**subsoil.** A layer of the earth's mantle (regolith) grading into the soil above and into unmodified rock waste below, that is less oxidized and hydrated than the soil proper. It contains almost no organic matter, but is somewhat charged with and indurated by iron oxides and clay leached down from the overlying soil.

**substation.** An electrical installation containing generating or power-conversion equipment and associated electric equipment and parts, such as switchboards, switches, wiring, fuses, circuit breakers, compensators, and transformers.

**substitute natural gas (SNG).** A gas manufactured from carbonaceous material whose characteristics are substantially interchangeable with natural gas. The resultant gas is composed primarily of methane. At this time, SNG feedstocks are the light hydrocarbons, propane, butane, and the naphthas. Development of processes for production from heavier feedstocks and from coal is underway. Also called synthetic natural gas.

**subsurface waste disposal.** Waste disposal in which manufacturing wastes are disposed of in porous underground rock formations. Disposal wells should be at least 200 feet deeper than the deepest water-bearing formation and must be sealed with cement from top to bottom.

**subsurface water.** Water that exists below the surface of the solid earth; may be liquid, solid, or gaseous.

**suction anemometer.** A device that measures wind velocity by the degree of exhaustion caused by the blowing of the wind through or across a tube.

**suction burner system.** A system applying a vaccum to a combustion chamber to draw in the air and/or gas necessary to produce the desired combustible mixture.

**suction pressure.** The pressure of gas as it enters a compressor.

**sulfur dioxide.** One of several forms of sulfur in the air; an air pollutant generated principally from the combustion of fuels containing sulfur. A natural source of sulfur dioxide is volcanic gases. Chemical formula $SO_2$.

**sulfur minerals.** Element occurring naturally in association with volcanoes and hot springs, and in cap rocks in salt domes; extensively produced from pyrites and pyritic minerals, either directly or as a byproduct, also a byproduct in gas stripping. Main uses are as sulfuric acid, sulfur dioxide (paper making), and in vulcanizing compounds, fungicides, and insecticides. (See Fig. p. 436)

**sulfur oxides.** Compounds composed of sulfur and oxygen produced by the burning of sulfur and its compounds in coal, oil, and gas.

**sulfur smog.** A fog composed of smoke particles, sulfur oxides, and high humidity. The major cause of damage comes from the reaction of sulfur trioxide with the water to form sulfuric acid ($H_2SO_4$) droplets.

**summer valley.** The depression that occurs in the summer months in the daily load of an electric or gas distribution system, or a pipeline.

**sump.** A depression or tank that serves as a drain or receptacle for liquids for salvage or disposal; a pit or basin in which waste oil products are collected and stored; that part of a judd of coal which is extracted first.

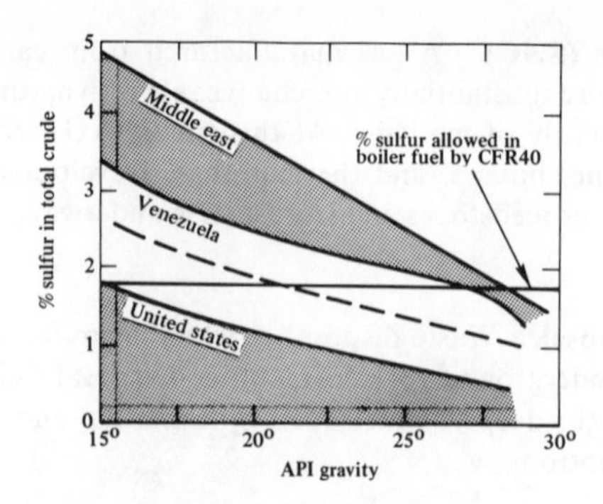


Sulfur content and API gravity of crude oils. (*Courtesy U.S. Council on Environmental Quality*)

**sun.** The luminous, celestial body around which the earth and other planets revolve, and from which they receive heat and light. Sunshine consists of a wide variety of electromagnetic waves. Its three main components are invisible heat waves, visible light rays of various colors, and invisible ultraviolet rays. Energy from the sun travels through space at a speed of 300,000 kilometers (or 186,000 miles) per second. Sunlight arriving at the edge of the earth's outer atmosphere carries energy at an approximately constant rate of 1.36 kilowatts per square meter (or 130 watts per square foot) of area covered. In terms of heat, this is equivalent to 428 British thermal units (Btu) per square foot per hour. The average amount of energy reaching the earth's surface is equivalent to about 177 watts per square meter (or 16.4 watts or 58.5 Btu per square foot) per hour. Measurements over an entire year (including night and day, cloudy and clear conditions, winter and summer) in various locations in the United States have shown that about 13 percent of the sun's original energy arrives at ground level. With an estimated $5 \times 10^9$ more years of life in its present state, the sun may be considered, for practical purposes, an inexhaustible source of fuel.

**Suncole.** Low-temperature coke used in making producer gas.

**sun compass.** A navigational compass that uses the sun and its calculated bearing to establish direction, especially in high altitudes.

**sun effect.** The quantity of heat from the sun tending to heat an enclosed space.

**sun observations.** In surveying, fixation of longitude and/or latitude of a station, or orientation of a survey line, by use of a theodolite, to relate the position of the sun, sideral time, and the location of the theodolite.

**sun pillar.** A vertical streak of light above the sun, usually seen at sunrise or sunset. It is caused by reflection from ice crystals.

**sunrise.** Defined in meteorology as the moment when the upper edge of the sun appears to rise above the apparent horizon on a clear day. See sunset.

**sunset.** In meteorological convention, the moment when the upper edge of the sun appears to fall below the apparent horizon on a clear day. Effects of refraction cause the apparent position to be about 34′ above the true position.

**sunshine.** The warmth and light given by the sun's rays; also a name of a soft grade paraffin wax with a low melting point, which can be burned in an ordinary miner's lamp with a nail (usually copper) in the wick and gives little smoke.

**sunstone.** A metaphorical name for fossil coal.

**sun tracking.** Following the sun with a solar collector to make the collector more effective.

**superanthracite.** Coal intermediate between anthracite and graphite.

**supercharger.** A blower that increases the intake pressure of an engine.

**superconductivity.** The abrupt and large increase in electrical conductivity exhibited by some metals as the temperature approaches absolute zero.

**supercritical flow.** Flow at velocities greater than one of the recognized critical values; Belanger's, Kennedy's, or Reynold's critical velocities.

**supercritical mass.** A mass of fuel whose effective multiplication factor is greater than one.

**supercritical reactor.** A reactor in which the effective multiplication factor is greater than one; consequently a reactor that is increasing its power level. If uncontrolled, a supercritical reactor will undergo an excursion.

**superficial velocity.** The linear velocity of a fluid flowing through a bed of solid particles calculated as if the particles were not present.

**superheating.** The heating of a vapor, particularly saturated (wet) steam, to a temperature much higher than the boiling point at the existing pressure. This is done in power plants to improve efficiency and to reduce condensation in the turbines.

**supertankers.** Extremely large oil tankers that can hold up to 4 million barrels (170 million gallons) of oil. (See Fig. p. 438)

**surface combustion.** Combustion of injected, properly proportioned fuel and air, on a surface or within a definite zone; when a mixture of air and gas or of air and oil vapor is forced through a porous wall, and ignited on the other side. Since necessary air is mixed with the gas, it will burn regardless of atmospheric conditions, even under water.

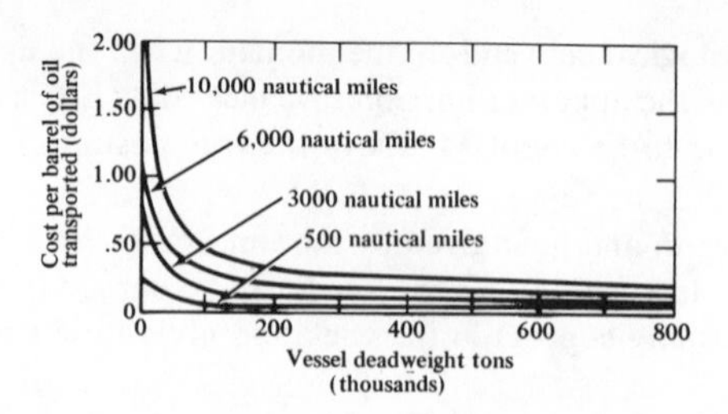


Costs of tanker transport. (*Courtesy U.S. Council on Environmental Quality*)

**surface conductance.** The heat transmitted from (or to) a surface to (or from) the fluid in contact with the surface in a unit of time per unit of surface area per degree temperature difference between the surface and the fluid.

**surface contamination.** The deposition and attachment of radioactive materials to a surface.

**surface mining.** The obtaining of coal from the outcroppings or by the removal of overburden from a seam of coal, as opposed to underground mining, or any mining at or near the surface. Also called strip mining, placer mining, opencast, opencut mining, open-lit mining.

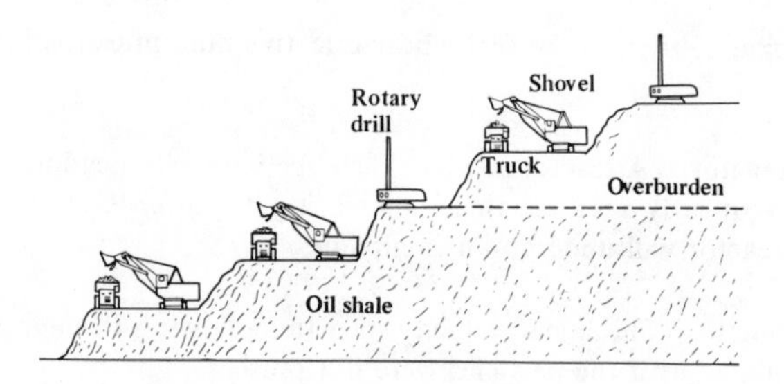


Drawing illustrating an oil shale surface mine. (*Courtesy U.S. Council on Environmental Quality*)

**surface pipe.** The first string of casing to be set in a well. The length varies in different areas from a few hundred feet to three or four thousand feet. Some states require a minimum length to protect fresh-water sands.

**surface water.** Water on the earth's surface exposed to the atmosphere as rivers, lakes, streams, oceans.

**surfactant.** An agent which modifies the physical, the electrical, or chemical characteristics of the surface of a solid, or surface tensions of solids or liquids. Used in flotation and in detergency.

**surveillance system.** A monitoring system used to determine environmental quality.

**survey meter.** Any portable radiation detection instrument especially adapted for surveying or inspecting an area to establish the existence and amount of radioactive material present.

**survival curve.** Curve obtained by plotting the number or percentage of organisms surviving at a given time against the dose of radiation, or the number surviving at different intervals after a particular dose of radiation.

**suspended solids.** Sediment which is in suspension in water but which will physically settle out under quiescent conditions (as differentiated from dissolved material); solids that can be separated from a liquid by filtration.

**sweat.** To gather surface moisture in beads as a result of condensation; the roof of a mine is said to sweat when drops of water are formed upon it, by condensation of steam formed by the heating of the waste materials.

**sweet.** Applied to oil and gas free of hydrogen sulfide.

**sweet crudes.** Low sulfur content crude oils.

**sweetening.** The process by which petroleum products are improved in odor and color by oxidizing or removing the sulfur-containing and unsaturated compounds.

**sweet natural gas.** Gas found in its natural state, containing small amounts of sulphur compounds.

**swimming pool reactor.** A pool reactor.

**switchgear.** A general term applying to switching, interrupting, controlling, metering, protective, and regulating devices; used primarily in connection with generation, transmission, distribution, and conversion of electric power.

**synchrocyclotron.** A cyclotron in which the frequency of the accelerating voltage is decreased with time so as to match exactly the slowing revolutions of the accelerated particles. The decrease in rate of acceleration of the particles results from the increase of mass with energy as predicted by the Special Theory of Relativity.

**synchronism.** The state when the phase difference between two or more periodic quantities is zero, or in phase.

**synchrotron.** An accelerator in which particles are accelerated around a circular path by radio-frequency electric fields. The magnetic guiding and focusing fields are increased synchronously to match the energy gained by the particles so that the orbit radius remains constant.

**syncline.** A fold in rocks in which the strata dip inward from both sides toward the axis.

**syncrude.** Synthetic crude oil; oil produced by the hydrogenation of coal or coal extracts which is similar to petroleum crude.

**synergism.** The action of two separate substances to produce an end effect which is greater than the sum of the effects of the two substances acting independently.

**syntectic.** Magmas produced by syntexis; also an isothermal reversible reaction in which a solid phase, on absorption of heat, is converted to two conjugate liquid phases.

**syntexis.** The formation of magma by direct melting of more than one kind of rock.

**Synthane process.** A coal gasification process being developed to produce pipeline quality gas. A fluidized bed gasifier is used to produce a medium-Btu gas which can be upgraded to a high-Btu gas. A pretreated or noncaking coal is fed into the top of the gasifier. The coal is devolatilized and gasified as it falls freely to the fluidized bed level. The bed is fluidized by a rising mixture of oxygen and steam fed into the bottom of the gasifier. Flue gas from the burned char would then have to be scrubbed to remove sulfur compounds. Unreacted char settles to the bottom of the reactor where it is removed. Gasifier conditions are 1000 pounds per square inch gage and 982° C.

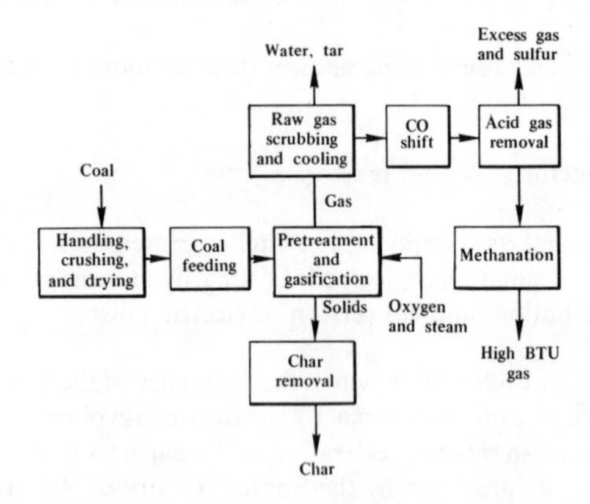


Block flow diagram of Synthane process. (*Courtesy Van Nostrand's Scientific Encyclopedia*)

**synthesis gas.** A mixture of carbon monoxide (CO) and hydrogen ($H_2$) containing small amounts of nitrogen, some carbon dioxide, and various trace impurities prepared for petrochemical synthesizing processes.

**synthetic natural gas (SNG).** A gaseous fuel manufactured from coal, containing almost pure methane, $CH_4$, and produced by a number of coal gasification schemes.

The basic chemical reaction is coal (C) + $2H_2 \longrightarrow CH_4$ ± heat. SNG contains 95 to 98 percent methane, and has an energy content of 980 to 1035 Btu per standard cubic foot, about the same as that of natural gas.

**synthetic oil.** Oil produced artificially as in the Bergius process or Fischer-Tropsch process.

**Synthoil process.** A hydrodesulfurization process in which coal is liquefied and sulfur removed as $H_2S$. Coal in a recycled oil slurry is mixed with hydrogen and fed into a fixed bed (packed bed) catalytic reactor. The catalyst is composed of pellets of cobalt molybdate on silic promoted alumina. The liquid product is fuel oil. Hydrogen consumption is from 3400 to 4375 standard cubic feet per barrel of product. Fuel oil yield is 3.2 to 3.4 barrels per ton of coal and has a heating value of 17,000 British thermal units per pound. Sulfur content is 0.4 to 0.2 percent by weight. Reactor operating conditions are 454°C and 4000 pounds per square inch gage.

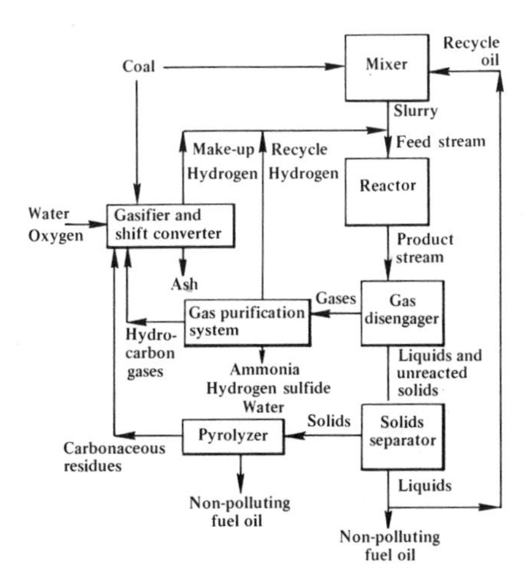


Synthoil process. (*Courtesy Van Nostrand's Scientific Encyclopedia*)

**system loss.** As applied in the electrical industry, the difference between the system net energy (power input and output), resulting from losses unaccounted for between the source of supply and the point of delivery.

# T

**tachometer.** An instrument of the direct-reading type, indicating the speed of a shaft or machine in revolutions per minute.

**tactite.** A general term for rocks of complex mineralogy formed by the contact metamorphism of limestone, dolomite, or other carbonate rocks into which foreign matter from the intruding magma has been introduced by hot solutions.

**tail gas.** The residue gas left after the completion of a treating process designed to remove certain liquids or liquefiable hydrocarbons.

**tailings.** The parts of any fluid material separated as refuse, or treated separately as inferior in quality or value; also mineral refuse from a milling operation usually deposited from a water medium.

**tailings dam.** One to which slurry is transported: the solids settle and the liquid is withdrawn.

**tall oil.** The oily mixture of rosin acids, fatty acids, and other materials obtained by acid treatment of the alkaline liquors from the digesting (pulping) of pine wood. Used in drying oils, in cutting oils, in core oils, in oil well drilling muds, in lubricants and greases, and in asphalt derivatives.

**tankage.** The process of storing liquids in a tank; the capacity of tanks; the price paid for tank storage of liquids.

**tank reactor.** A reactor in which the core is suspended in a closed tank, as distinct from an open pool reactor; commonly used as research and test reactors.

**tar.** The residue obtained by destructive distillation of materials such as coal, wood, or petroleum; also soft pitch or thickened petroleum found in cavities of some limestones.

**tar distillate.** A fraction in petroleum refining containing heavy oils and paraffin.

**target.** Material subjected to particle bombardment or irradiation in order to induce a nuclear reaction; also a nuclide that has been bombarded or irradiated.

**tariff.** A published volume of rate schedules and general terms and conditions under which a produce or service will be supplied.

**tar sand.** Sand impregnated with petroleum which dries up to viscous or solid bitumen; hydrocarbon bearing deposits distinguished from more conventional oil and gas reservoirs by the high viscosity of the hydrocarbon, which is not recoverable in its natural state through a well by ordinary oil production methods.

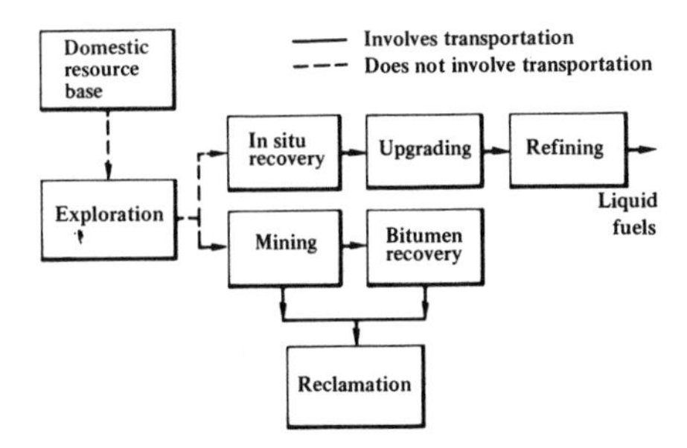


Tar sands resource development. (*Courtesy U.S. Council on Environmental Quality*)

**teaching reactor.** A research reactor or subcritical assembly.

**technology.** The application of knowledge for practical purposes; for example, engineering designs to convert solar energy into more useful forms of energy such as electricity or space heating.

**technological fix.** A solution to a problem based on technology, often used in a pejorative sense as an apparent, or simplistic technological solution to a complex human problem whose benefits may be only cosmetic.

**tectonic.** Pertaining to rock structures and topographic features resulting from deformation of the earth's crust; also earthquakes not caused by volcanic action, landslides, or collapse of caverns.

**tee.** A crossvein meeting a main vein without intersecting it; also a sleeve with a third opening in the side, usually at right angles, to allow a branch line to be connected to the main pipeline.

**telain.** Fragments of plant tissues, which are completely soaked with vitrain.

**telecontrolled power station.** A hydroelectric power station operated by remote control.

**telemetering.** Use of an electrical apparatus, transmitting data to a distant point, for indicating, recording, or integrating the values of a variable quantity.

**teletherapy.** Radiation treatment administered by using a source that is at a distance from the body, usually employing gamma-ray beams from radioisotope sources.

**temperature.** A degree of hotness or coldness measured on a definite scale; as distinguished from heat. Heat is a form of energy; temperature, a measurement of its thermal effects.

**temperature coefficient of reactivity.** The change in reactor reactivity (per degree of temperature) occurring when the operating temperature changes. The coefficient is said to be positive when an increase in temperature increases the reactivity, and negative when an increase in temperature decreases reactivity. Negative temperature coefficients are desirable because they help to prevent power excursions.

**temperature-compensated meter.** A meter in which the volume of gas is automatically corrected for variation in gas temperature.

**temperature-limiting device.** A device which automatically interrupts the gas flow to the burner when the temperature exceeds the limit set.

**tensile structure.** A structure formed by the stress or stiffening of its material.

**tephra.** A collective term for all clastic volcanic materials which, during an eruption, are ejected from a crater or from some other type of vent and transported through the air; includes volcanic dust, ash, cinders, lapilli, scoria, and pumice.

**ternary fission.** A rare type of nuclear fission in which three fission fragments are formed, one of which may be a light nucleus.

**tertiary recovery.** Use of heat and methods other than fluid injection to augment oil recovery (presumably occurring after secondary recovery).

**test bed plant.** Type of facility used in the development sequence of new technology; a plant, generally of intermediate size, designed to facilitate the introduction of experimental features for performance testing, and to provide for process changes and improvements as required.

**test pit.** Open excavations, dug by hand or by machine, large enough to permit a man to enter and examine formations in their natural condition.

**test reactor.** A reactor specially designed to test the behavior of materials and components under the neutron and gamma fluxes and the temperature conditions of an operating reactor.

**test weld.** The process of cutting out a portion of a weld in a pipeline for testing acceptability.

**test well.** One that determines not only the presence of petroleum oil, but also its commercial value, considering its abundance and accessibility; also an exploratory well for water.

**test year.** The 12-month period selected as the base for presenting data in a case or hearing before a regulatory agency.

**tetraethyllead (TEL).** A volatile lead compound which is added in concentrations up to 3 milliliters per gallon to motor and viation gasoline to increase the antiknock properties of the fuel. Formula, $Pb(C_2H_5)_4$.

**Tetrakis.** A light yellow liquid, chemically classified as a titanium ester, used in stopping water production in air- or gas-drilled wells.

**tetramethyllead (TML).** A highly volatile lead compound added to motor gasoline to reduce knock. May be used alone or in mixtures with TEL. Formula, $Pb(CH_3)_4$.

**Texaco Heavy Oil Gasification process.** Petroleum-based, heavy residual fuels are converted to hydrogen or gaseous fuels of several different heating values in this commercially-proven process. This partial oxidation process could also be used to gasify coal tars. The reactor is an entrained flow type.

**texture.** The character, arrangement, and mode of aggregation of the fragments, particles, or crystals that compose a rock.

**theodolite.** A survey instrument equipped with a sighting telescope.

**theoretical air requirement.** Volume of air necessary to insure the complete combustion of unit mass or volume of a fuel.

**therm.** A unit of heating value equivalent to 100,000 British thermal units; 1 therm equals 100 cubic feet of natural gas.

**thermal.** Hot or warm; also applied to springs which discharge water heated by nature.

**thermal batteries.** Type of battery that requires heat to activate. The electrolyte employed is a mixture of anhydrous salts which conduct electricity when melted. The heat source can be ignited by either an electric match or a mechanical primer. When the battery is ignited, the heat source evolves sufficient heat energy to melt the electrolyte and so permits the battery to deliver a considerable amount of electrical power. Widely used in rockets, missile systems, bomb fuses, and aircraft safety equipment.

**thermal boring.** Use of a high-temperature flame to fuse rock in drilling. Heat comes from ignition of kerosine with oxygen or other fuel system, at the bottom of a drill hole, and water with compressed air may be used to flush out the products.

**thermal breeder reactor.** A breeder reactor in which the fission chain reaction is sustained by thermal neutrons.

**thermal burn.** A burn of the skin or other organic material due to radiant heat, such as that produced by the detonation of a nuclear explosive.

**thermal capacity.** Heat required to raise the temperature of a body 1°C.

**thermal column.** A channel built into some research reactors to supply thermal neutrons for experimental purposes. It consists of a large body of moderator located adjacent to the core or reflector. Neutrons escaping from the reactor enter the thermal column where they are slowed down to thermal energies with velocities of about 2200 meters per second.

**thermal conduction.** The transfer of heat within a substance from points of higher to points of lower temperature.

**thermal conductivity.** The quantity of heat which will pass through a unit area of a material in unit time for a unit temperature difference between the material's surfaces under steady-state conditions.

**thermal conductor.** A material which readily transmits heat by means of conduction.

**thermal cracking.** A process in which the less volatile heating oil fractions are subjected to higher temperatures under increasing pressure, which causes the complex molecules in the gasoline range to break up into smaller ones, thus increasing gasoline quantity per barrel of crude oil.

**thermal diffusivity.** Thermal conductivity divided by the product of specific heat times unit weight.

**thermal dissociation.** The decomposition of a compound by the action of heat.

**thermal efficiency.** The ratio of the electric power produced by a power plant to the amount of heat produced by the fuel; a measure of the efficiency of heat with which the plant converts thermal to electrical energy; also, relating to heat, a percentage indicating the available British thermal unit input that is converted to useful purposes. (See Fig. p. 447)

**thermal energy.** Heat energy.

**thermal energy range.** Neutrons with energies less than 1 electron volt.

**thermal equilibrium.** In a system, the state at which there is no variation in temperature from one point to another.

**thermal factors (TF).** A rating given to aluminum thermalized windows and doors, based on standards and test procedures formulated by Architectural Aluminum Manufacturers Association. The number assigned to a window is based on temperature readings taken on various parts of the window using an inside temperature of 20°C, an outside temperature of −8°C, and a 15 mile per hour wind. The number

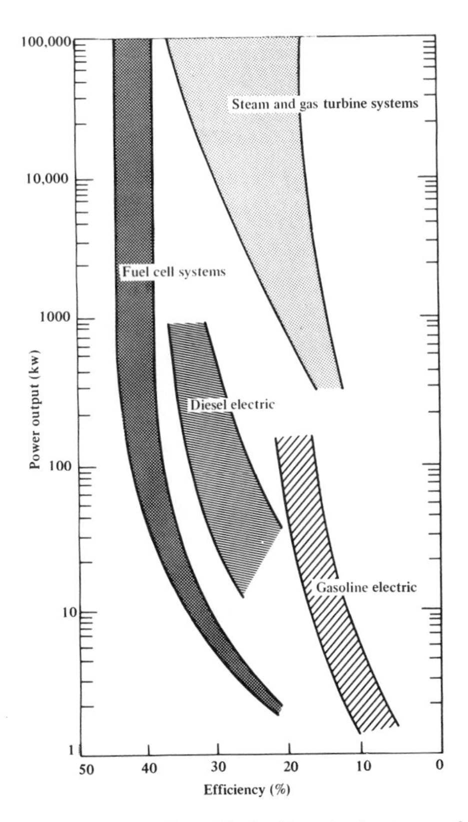


Thermal efficiency of electrical production of fuel cells and other types of generators. (*Courtesy "Energy and the Future," Fig. 36*)

takes into consideration both a window's heat-transfer characteristics as well as its resistance to condensate formation. The higher the TF number, the better the window will be. (See Fig. p. 448)

**thermal fission.** Fission cused by thermal neutrons.

**thermal gradient.** The rate at which temperature increases with depth below the earth's surface. A general average seems to be around 30° C increase per kilometer of depth, or 66° C per mile.

**thermal insulation.** A material having a relatively high resistance to heat transfer.

**thermalization.** The establishment of approximate thermal equilibrium between neutrons and their surroundings.

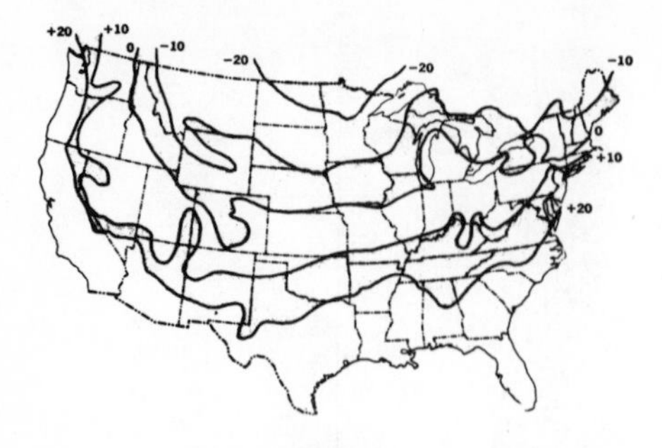


Minimum recommended thermal factors (TF)

| Exterior design temperature | Interior relative humidity | | | | | |
|---|---|---|---|---|---|---|
| | 15% | 20% | 25% | 30% | 35% | 40% |
| -20°F | 46 | 52 | 57 | 60 | --- | --- |
| -10°F | 39 | 46 | 52 | 57 | 60 | --- |
| 0°F | 30 | 39 | 45 | 52 | 57 | 62 |
| +10°F | 17 | 29 | 37 | 44 | 50 | 57 |
| +20°F | 0 | 16 | 25 | 34 | 40 | 48 |

Thermal factor numbers. (*Courtesy Popular Science, October 1975, p. 121*)

**thermal lag.** The time that it takes for the inside air temperature of an area to heat up or cool to that of the surrounding outside temperature.

**thermally actuated valve.** An automatic valve which utilizes the heat generated by the resistance of an electrical component in opening or closing the valve.

**thermal neutron (slow neutron).** A neutron in thermal equilibrium with its surrounding medium. Thermal neutrons are those that have been slowed down by a moderator to an average speed of about 2200 meters per second (at room temperature) from the much higher initial speeds they had when expelled by fission. This velocity is similar to that of gas molecules at ordinary temperatures.

**thermal pollution.** The discharge of heated effluents into natural waters at a temperature detrimental to existent ecosystems.

**thermal power.** Energy generated through the use of heat energy.

**thermal power plant.** Any electric power plant which operates by generating heat and converting the heat to electricity.

**thermal radiation.** Electromagnetic radiation emitted from the fireball produced by a nuclear explosion. Thirty-five percent of the total energy of a nuclear explosion is emitted in the form of thermal radiation, as light, ultraviolet and infrared radiation.

**thermal reactor.** A nuclear reactor in which the fission chain reaction is sustained primarily by thermal neutrons. Most reactors are thermal reactors.

**thermal recovery.** A petroleum recovery process that utilizes heat (rather than water or gas) to thin viscous oil in an underground formation and allows it to flow more readily toward wells through which it can be brought to the surface.

**thermal separation.** The separation of minerals by heat. The method is used, for example, to remove impurities from rock salt.

**thermal shield.** A layer or layers of high density material located within a reactor pressure vessel or between the vessel and the biological shield to reduce radiation heating in the vessel and the biological shield.

**thermal spike.** A momentary zone of high temperature produced along the track of a high-energy particle.

**thermal spring.** A spring that brings warm or hot water to the surface.

**thermal structure.** The temperature variation with depth of sea water.

**thermal utilization.** The probability in an infinite medium that a thermal neutron will be absorbed in the fissile material.

**thermal water.** The mineral-charged water that issues from a hot spring or geyser.

**thermal wind.** The increase in geostrophic wind with height due to horizontal temperature gradients. Its magnitude is about 3.6 km per second (7.2 knots) per km per 1°C per 100 km.

**thermionic conversion.** The conversion of heat into electricity by boiling electrons from a hot metal surface and condensing them on a cooler surface. No moving parts are required.

**thermistor.** A temperature measuring device that employs an electrical resistor made of material whose resistance varies sharply in a known manner with the temperature; commonly used for oceanographic temperature measurements because of their percentage response to unit temperature changes and their great sensitivity.

**thermochemical conversion process.** Any process which transforms an intial set of chemical reagents into a different product set of chemicals involving the application or deletion of heat energy.

**thermochemistry.** The study of the heat changes accompanying chemical reactions.

**thermocouple.** A device consisting essentially of two conductors made of different metals, joined at both ends, producing a loop in which an electric current will flow when there is a difference in temperature between the two junctions.

**thermodynamics.** A study of the transformation of energy into other manifested forms and of their practical application. The three laws are: first law, conservation of energy (energy may be transformed in an isolated system, but its total is constant);

second law, heat cannot be changed directly into work at constant temperature by a cyclic process; third law, heat capacity and entropy of every crystalline solid becomes zero at absolute zero (0° K).

**thermodynamics, law of.** The First Law of Thermodynamics states that energy can neither be created nor destroyed. The Second Law of Thermodynamics states that when a free exchange of heat takes place between two bodies, the heat is always transferred from the warmer to the cooler body.

**thermoelectric conversion.** The conversion of heat into electricity by the use of thermocouples.

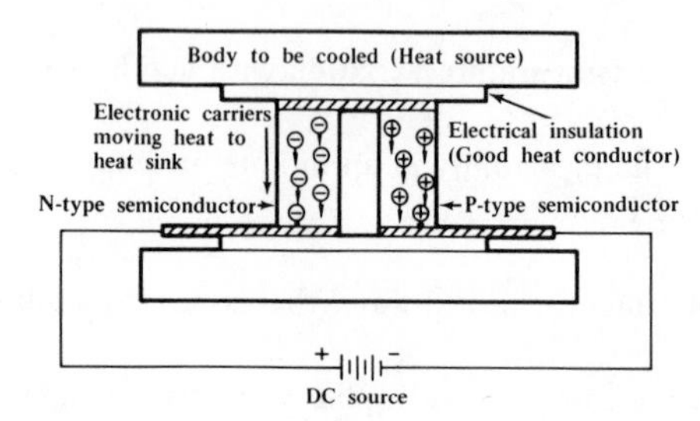


Cross section of a thermoelectric cooler. (*Courtesy Van Nostrand's Scientific Encyclopedia*)

**thermoelectricity.** The flow of electrical current when there is a difference in temperature between the junction of two different metal conductors joined together in a closed circuit.

**thermoelectric metals.** Metals or alloys used in thermocouples for measuring high temperatures.

**thermonuclear fusion.** Source of energy available from hydrogen isotopes in seawater.

**thermonuclear reaction.** A reaction in which very high temperatures bring about the fusion of two light nuclei to form the nucleus of a heavier atom, releasing a large amount of energy. In a hydrogen bomb, the high temperature to initiate the thermonuclear reaction is produced by a preliminary fission reaction.

**thermoplastic.** A quality which allows a material to repeatedly soften when heated and harden when cooled. Typical of the thermoplastics family are the styrene polymers and copolymers, acrylics, cellulosics, polyethylenes, vinyls, nylons, and the various fluorocarbon materials.

**thermoset.** A material that will undergo or has undergone a chemical reaction by the action of heat, catalysts, ultraviolet light, etc., leading to a relatively infusible state. Typical of the plastics in the thermosetting family are the aminos (melamine and urea), most polyesters, alkyds, epoxies, and phenolics.

**thermostat.** The thermostat is the brain of a heating system. It must swiftly detect minute changes in room temperature and turn the heat source on and off accordingly. Moreover, it has to anticipate "overrun" of the heating system—the period when the blower of a warm-air system continues to operate after the burner has shut down.

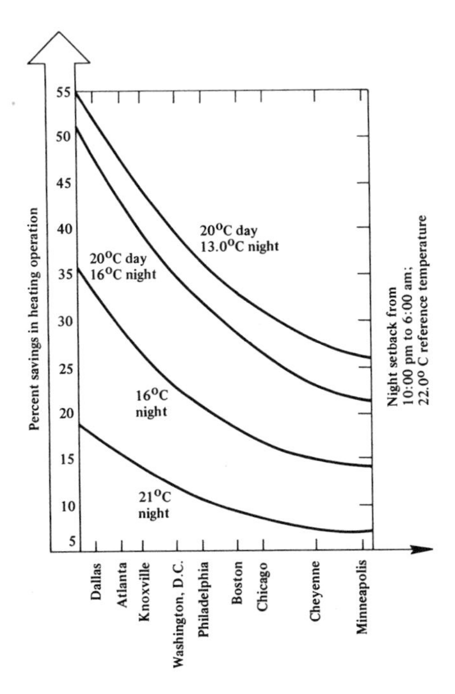


Estimated energy savings from lower thermostat settings. (*Courtesy Electric Power Research Institute and Applied Nucleonics Co., Inc.*)

**thermosyphon.** The principle that makes water circulate automatically between a solar collector and a storage tank above it, gradually increasing its temperature. A solar heating system that uses natural convection to transport heat from the collector to storage, by appropriately locating the storage in relation to the collector.

**thermosyphon water collector.** A flat-plate solar collector that circulates water (through the collector) by means of natural convection.

**theta pinch.** A nuclear fusion device employing high plasma density, strong magnetic fields, and field stabilization feedback to product confinement. The device can be toroidal or possibly linear (if equipped with magnetic mirrors).

**thick seam.** In general, a coal seam over 4 feet in thickness.

**Thomason solar house.** A series of solar heated homes constructed by Harry H. Thomason in the 1950's and early 1960's in the Washington, D.C. area. Solar-generated heat is stored in a combination water-rock storage bin. The Thomason solar collector converts approximately 55 percent of the available sunlight into heat.

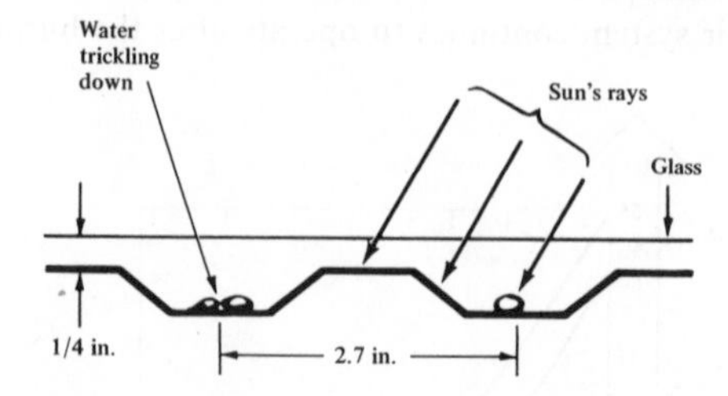


Thomason design of a solar panel. (*Courtesy Ward Ritchie Press, Passadena, CA*)

**thorium.** A radioactive, silvery-white, metallic element in Group III of the periodic system. Used in the preparation of gas mantles, as an alloying element in magnesium, to coat tungsten wire and as a nuclear fuel. New interest may be shown in thorium as a nuclear fuel material since plutonium security problems emphasized.

**thorium series.** The series of nuclides resulting from the radioactive decay of thorium-232. Many man-made nuclides decay into this sequence. The end product of this sequence in nature is lead-208.

**three-phase current.** Alternating current in which three separate pulses are present, identical in frequency and voltage, but separated 120 degrees.

**threshold dose.** The minimum dose of radiation that will produce a detectable biological effect.

**threshold limit values.** Threshold limit values refer to airborne concentrations of substances and represent conditions under which it is believed that nearly all workers may be repeatedly exposed day after day without adverse effect. Threshold limit values refer to time-weighted concentrations for a 7- or 8-hour workday and 40-hour workweek.

**throttle valve.** A valve designed to regulate the supply of a fluid (as steam or gas and air) to an engine; also a valve used in space-cooling equipment which expands the fluid in the system to produce a cooling effect.

**through coal.** Coal as it is mined, that is, large and small mixed together; run-of-mine coal; also coal after passage through a screen of stated size.

**throughput.** Quantity of material (ore or selected fraction) passed through the mill or a section thereof in a given time or at a given rate.

**throw.** A fault; a dislocation; the amount of vertical displacement up or down produced by a fault; in heating or air conditioning, the distance air will carry, measured along the axis of an air stream from the supply opening to the position in the stream at which air motion reduces to 50 feet per minute.

**tidal energy.** The creation of electric power from the movement of the sea.

**tidal power.** Mechanical power, which may be converted to electrical power, generated by the rise and fall of ocean tides. The possibilities of utilizing tidal power have been studied for many generations, but the only feasible schemes devised so far are based on the use of one or more tidal basins, separated from the sea by dams or barrages, and of hydraulic turbines through which water passes on its way between the basins and the sea. The world's largest tidal power plant is located on the estuary of the River Rance, in Brittany, France, completed in the 1960s, which generates 544 kwh per year. The disadvantages of tidal power generation are the very high capital cost of the dams or barrages, possible ecological disturbance, but the major limitation imposed by the tidal cycle can be overcome by generating energy from both the filling and emptying of the basins to provide base-load power, and by using base-load power at times of low demand to pump water to higher reservoirs, from which it can be released to meet peak load demands.

**tide.** Rise and fall of the surface of the sea due to the gravitational pull of the moon, generally taking place twice daily. In the open sea, this rise may not exceed 2 feet, whereas in the shallow seas bordering continents it may be more than 20 feet, and in narrow tidal estuaries from 40 to 50 feet. Since the moon travels in its own orbit in the same direction of the earth, a period of about 24 hours, 25 minutes will elapse between successive occasions when the moon is vertically above a given meridian. The interval between successive high tides will therefore be about 12½ hours.

**time-of-flight spectrometer.** A device for separating and sorting neutrons (or other particles) into categories of similar energy, measured by the time it takes the particles to travel a known distance.

**timetable.** A schedule showing a planned order or sequence. In ERDA's timetable, the near term covers the period from today through 1985, the midterm includes the period from 1986 to 2000, and the long term starts in 2001.

**tipple.** Originally the place where the mine cars were tipped and emptied of their coal, and still used in that sense, although now more generally applied to the surface structures of a mine, including the preparation plant and loading tracks.

**tip-speed ratio.** The trim aerodynamic blades of a wind generator extract more power from the wind at their tips than they do near the hub of the blade. This is expressed mathematically as the tip-speed ratio, which is the difference between the rotational speed of the tip of the blade and the actual velocity of the wind.

**tissue equivalent ionization chamber.** An onization chamber in which the materials of the walls, electrodes, and gas are selected to produce ionization essentially equivalent to that characteristic of the tissue being simulated.

**tissue-equivalent material.** Material that absorbs and scatters ionizing radiation to the same degree as a particular biological tissue.

**TNT equivalent.** A measure of the energy released in the detonation of a nuclear explosive expressed in terms of the weight of TNT (the chemical explosive, trinitrotoluene) which would release the same amount of energy when exploded. It is usually expressed in kilotons or megatons. The TNT equivalence relationship is based on the fact that 1 ton of TNT releases 1 billion ($10^9$) calories of energy.

**toe.** The base of the coal in an opencast mine; the burden of material between the bottom of the borehole and the free face.

**tokamak.** Type of nuclear fusion system employing a doughnut-shaped, magnetically confined plasma; acronym of Russian words for toroidal magnetic chamber.

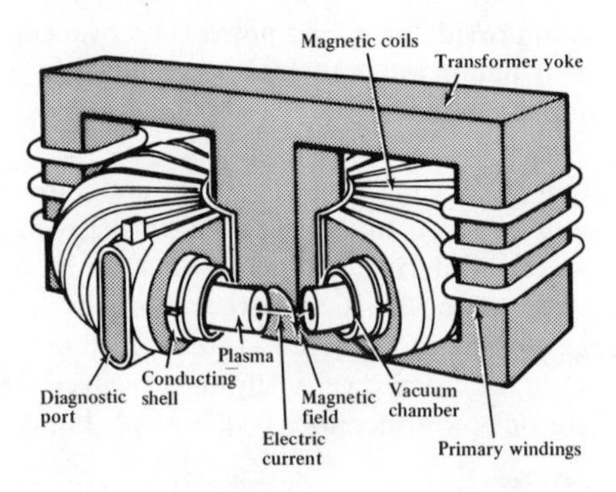


Tokamak fusion machine showing the hot plasma confined by helical magnetic field lines (only one shown). (*Courtesy Basic Nuclear Engineering*)

**Tokamak Fusion Test Reactor (TFTR)** The nation's first fusion device designed to use deuterium and tritium, the isotopes of hydrogen which will be used to fuel fusion power plants. It will be built at Princeton University. The purpose of the TFTR is to study the physics of burning plasmas and to gain engineering experience in the design, construction, and operation of a deuterium-tritium burning tokamak employing engineering features and systems needed in an experimental power reactor.

**toll enrichment.** An arrangement whereby privately owned uranium could be enriched in uranium-235 content in government facilities upon payment of a service charge by the owners.

**toluene.** A colorless liquid which occurs in coal tar and wood tar. Used as a solvent and as an intermediate for its derivatives. Symbol $C_6H_5CH_3$.

**ton.** Any of various units of weight for large quantities: (a) a unit that equals 20 short hundredweight of 2000 pounds, used chiefly in the United States, Canada, and the

Republic and South Africa; aso called a short ton; (b) a unit that equals 20 long hundredweight or 2240 pounds, used chiefly in England; also called a long ton; (c) a unit of internal capacity for ships that equals 100 cubic feet, also called a register ton; (d) a unit that approximately equals the volume of a long ton weight of sea water; used in reckoning the displacement of ships and it equals 35 cubic feet; also called a displacement ton; and (e) a unit of volume for cargo freight usually considered to be 40 cubic feet; also called a freight ton, a measurement ton.

**ton of cooling.** The extraction of 200 British thermal units per minute, 12,000 British thermal units per hour, or 288,000 British thermal units per day of 24 hours. The term is derived from the amount of heat energy required to convert a ton of water into ice at 0°C during a 24-hour period.

**topped crude.** A residual product remaining after the removal, by distillation or other processing means, of an appreciable quantity of the more volatile components of crude petroleum.

**topping cycle.** A means to increse thermal efficiency of a steam-electric power plant by increasing temperatures and interposing a device, such as a supercritical gas turbine, between the heat source and the conventional steam-turbine generator part of the plant to convert some of the additional heat energy into electricity.

**tornado-type wind turbine.** An advanced wind energy concept which would consist of a tall cylindrical tower with an open top, slotted side openings, and guide vanes to create a swirling tornado-like vortex flow. Outside air would be allowed to rush into the base of the tower and would be drawn upward through the vortex's low pressure core. Spinning rotor blades would drive a generator. (See Fig. p. 456)

**toroid.** A doughnut-shaped surface such as that of the magnetic container confining the plasma in a tokamak fusion power plant. (See Fig. p. 456)

**torr.** The pressure exerted per square centimeter by a column of mercury 1 millimeter high at a temperature of 0°C where the acceleration of gravity is 980.665 centimeters per second.

**TOSCOAL process.** A process for producing solid char and fuel oil. In this process crushed coal is preheated by the dilute phase fluid bed technique and fed to a pyrolysis drum where it is heated with hot circulated ceramic balls. The char product is passed through a revolving screen to separate ceramic balls, unreacted coal or char and vapors. Pyrolytic vapors are condensed and fractionated. Ceramic balls are heated and recirculated. Operating temperatures are 427° to 538°C. (See Fig. p. 457)

**total cooling effect.** The difference between the total heat content of the airstream mixture entering an air conditioner per hour and the total heat of the mixture leaving per hour.

**total cooling load.** The sum of the sensible and latent heat components.

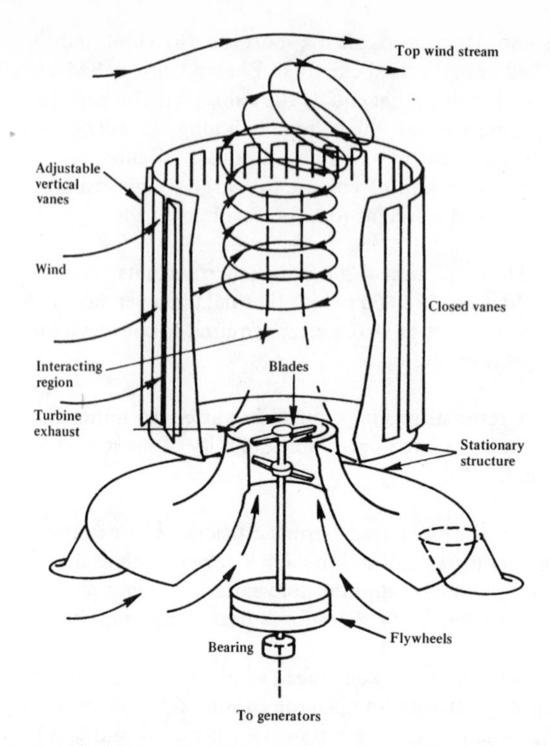


Tornado-type wind turbine. The vortex tower would create a high-velocity swirling vortex of air with a low-pressure core resembling the center of a tornado. Outside air would rush into the core, spinning rotor blades that would drive a turbine to produce electricity. (*Courtesy U.S. Energy Research and Development Administration*)

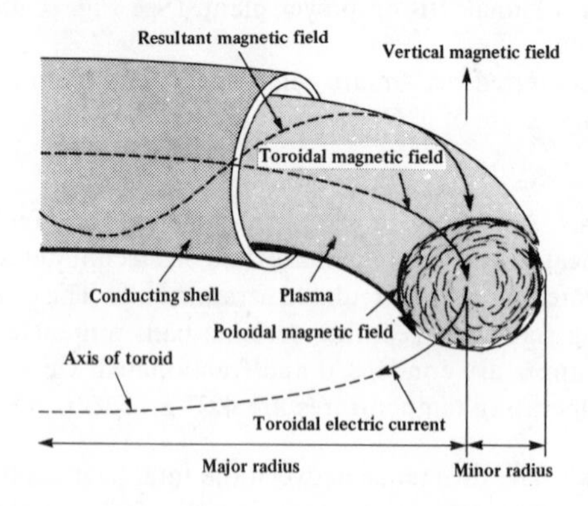


Tokamak toroidal plasma ring. (*Courtesy Basic Nuclear Engineering*)

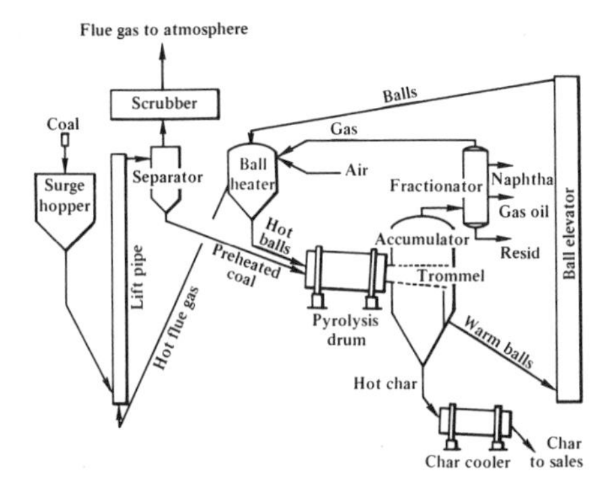


*Toscoal* process for pyrolysis of coal. (*Courtesy Van Nostrand's Scientific Encyclopedia*)

**total depth.** The greatest depth reached by a well bore.

**total energy.** The total energy at any point in a moving fluid consists of the sum of the internal static, velocity, and potential energies at that point.

**total energy house.** A house designed to conserve energy.

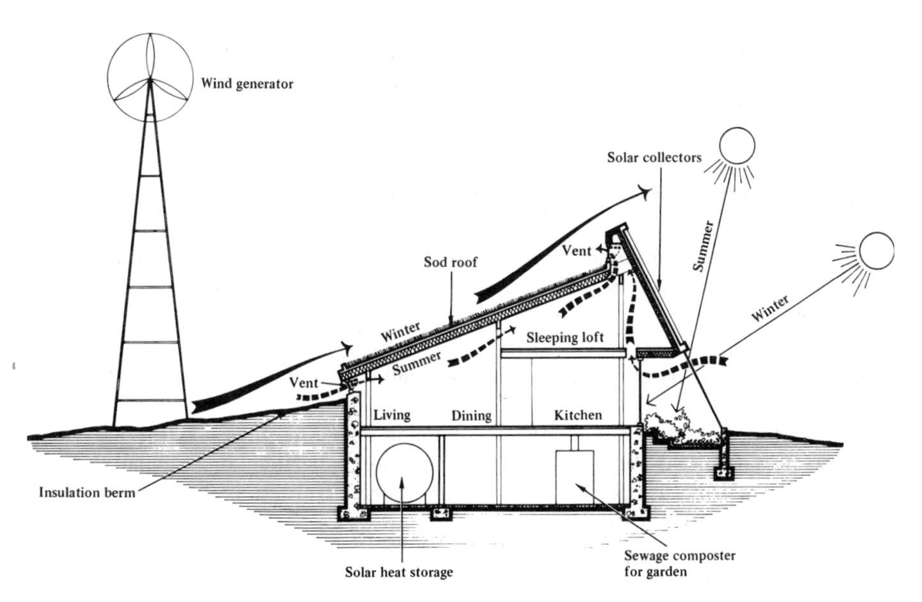


Total energy house is partially buried in the ground to utilize the earth's natural insulating properties. (*Courtesy Living with Natural Energy, Design for a Limited Planet*)

**total energy system.** A packaged energy system of high efficiency, utilizing gas-fired turbines or engines that produce electrical energy and utilize exhaust heat in applications such as heating and cooling.

**total gross energy consumption.** Total energy inputs into the economy, including coal, petroleum, natural gas, and the electricity generated by hydroelectric, nuclear, and geothermal power plants. Gross consumption includes conversion losses by the electric power sector.

**total input gas.** The volume of extraneous gas injected into a storage reservoir during a given period of time.

**total ionization.** The total number of ion pairs produced in any way by a directly ionizing particle.

**total net energy consumption.** Inputs into the final consuming sectors (household and commercial, industrial, and transportation) and consisting of direct fuels and electricity distributed from the elecric power sector. Conversion losses in the electric sector constitute the difference between net and gross energy.

**total output gas.** The volume of gas withdrawn from a storage reservoir during a given period of time.

**total pressure.** The algebraic sum of the static pressure and velocity pressure at any particular point.

**tower scrubber.** A vertical vessel filled with plates or suitable packing over which scrubbing liquid flows while the gas to be purified flows upward through the liquid, separating entrained liquids or solids from the gas.

**town's gas.** Gas manufactured from coal for use in cities for illumination and heating; usually a mixture of coal gas and carbureted water gas.

**toxic fluids.** Gases or liquids which are poisonous, irritating and/or suffocating, as classified in the Hazardous Substances Act, Code of Federal Regulations, Title 16, Part 1500.

**toxicity.** The quality or degree of being poisonous or harmful to plant or animal life.

**toxic pollutants.** A combination of pollutants including disease-carrying agents which, after discharge and upon exposure, ingestion, inhalation, or assimilation into any organism can cause death or disease, mutations, deformities, or malfunctions in such organisms or their offspring.

**trace metals.** Metals found in small quantities or traces, usually due to their insolubility.

**tracer.** An element or compound that has been made radioactive so that it can be followed easily in industrial and biological processes. Radiation emitted by the tracer (radioisotope) pinpoints its location.

**tracer gas.** A gas introduced in small quantities into the main body of the air to determine either the air current or the leakage paths in a ventillation system.

**trade winds.** The wind system which blows from the subtropical highs toward the equatorial trough; a major component of the general circulation of the atmosphere.

**training reactor.** A reactor used primarily for training in reactor operation and instruction in reactor behavior.

**tramp iron.** Stray metal objects such as coal-cutter picks or bolts, which have become mixed with coal or ore. To remove this iron before it damages the ore-handling machine, various types of magnets are widely used.

**Trans Alaska pipeline.** An 800-mile oil pipeline, across the state of Alaska, constructed by the Alyeska Pipeline Service Company of Anchorage, Alaska, for the purposes of making the 9.6 billion barrel oil reserves at Prudhoe Bay, Alaska, available to U.S. industry and consumers. At capacity, a total of two million barrels a day will be transported.

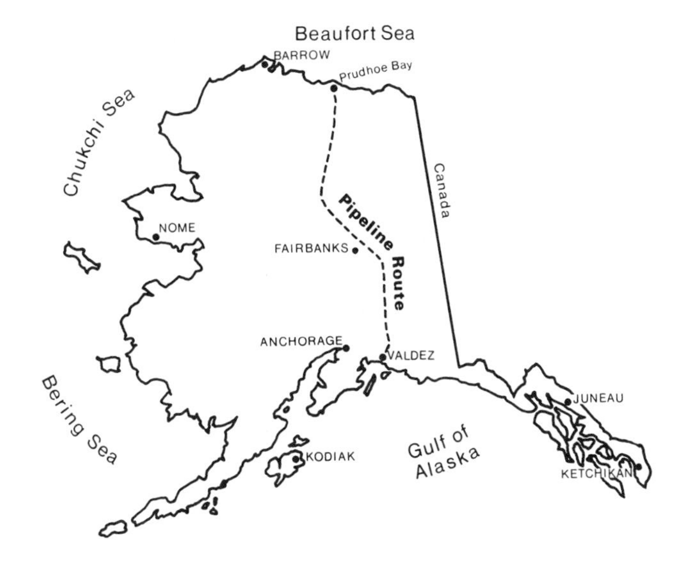


Trans Alaska pipeline route. (*Courtesy Alyeska Pipeline Service Company*)

**transducer.** A device for converting energy from one form to another; a device which measures quantities in a system (pressure, current, voltage) and converts them into related or proportional units.

**transfer admittance.** The transfer admittance of a network made up of an energy source and an energy load connected by a transducer is the quotient obtained by dividing the phasor representing the source current of the source by the phasor representing the load voltage of the load.

**transformer.** A device which, through electromagnetic induction but without use of moving parts, transforms alternating or intermittent electric energy in one circuit into energy of similar type in another circuit, commonly with altered values of voltage and current.

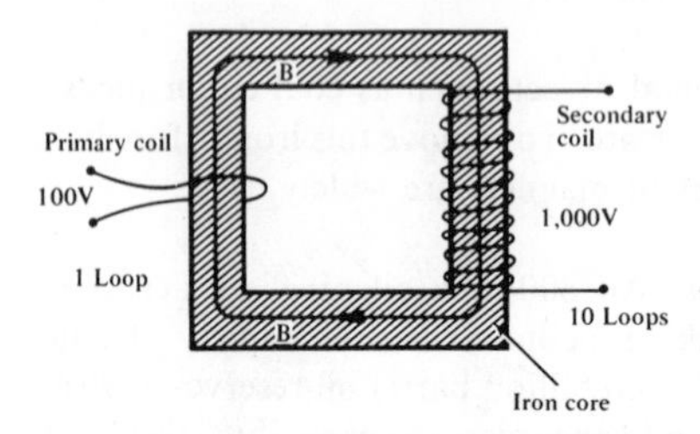


Model of a transformer. (*Courtesy National Science Teachers Association*)

**transformer oil.** A special type of oil of high dielectric strength, forming the cooling medium of electric power transformers.

**transistor.** A small, long-lasting device for amplifying electrical signal. It is dependent on the special qualities of germanium and silicon crystals as semi-conductors.

**transit.** A surveying instrument with the telescope mounted so that it can measure horizontal and vertical angles. Also called a transit theodolite.

**transmission level.** The transmission level of the energy at any point in an energy transmission system is the rate of flow of that energy as expressed in terms of (a) a specified reference rate of flow and of (b) the transmission loss by which the actual rate of flow must be reduced to equal the reference rate.

**transmission line.** A cable or wire through which electric power is moved from electric power generating plants to areas of use by consumers. It is usually operated at greater than 250,000 volts. (See Fig. p. 461)

**transmittance.** The ratio of the radiation passing through a material to the radiation incident on the upper surface of that material.

**transmittance-absorptance product.** The product of the transmittance of the transparent collector cover(s) and the absorptance of the collector plate.

**transmutation.** The transformation of one element into another by a nuclear reaction or series of reactions; for example, the transmutation of uranium-238 into plutonium-239.

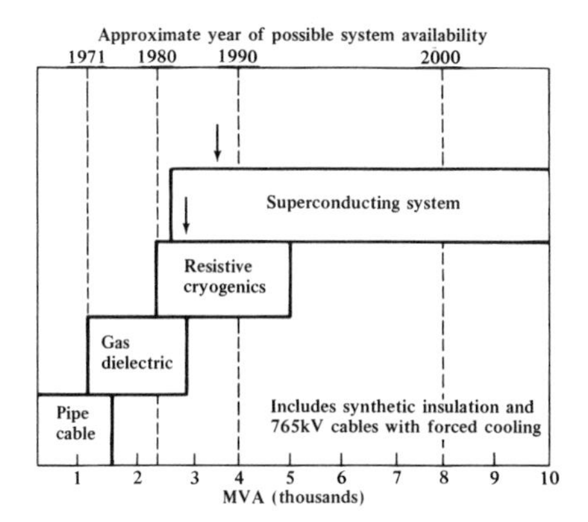


Comparative capabilities of underground transmission systems utilizing new technology. (*Courtesy Van Nostrand's Scientific Encyclopedia*)

**transpiration.** The process by which water vapor escapes from a living plant and enters the atmosphere.

**transplutonium element.** An element above plutonium in the Periodic Table; that is, one with an atomic number greater than 94.

**transuranic element.** An element above uranium in the Periodic Table; that is, with an atomic number greater than 92. All 11 transuranic elements are produced artificially and are radioactive. They are: neptunium, plutonium, americium, cirium, berkelium, californium, einsteinium, fermium, mendelevium, nobelium, and lawrencium.

**transuranium element.** A transuranic element.

**trap.** A natural subsurface petroleum reservoir; a device for removing liquids or solids from a gaseous stream; a low spot in a pipeline or main.

**trap pressure.** Pressure held at the trap. Also, oil and gas separator.

**treating.** The process of improving the quality of petroleum products with chemicals.

**triage.** The process of determining which casualties (from a large number of persons exposed to heavy radiation) need urgent treatment, which ones are well enough to go untreated, and which ones are beyond hope of benefit from treatment. Used in medical aspects of civil defense.

**trickle-down collectors.** "Wet" solar collector where water runs down the surface of the absorber plate and collects in a gutter at the bottom for transfer to the heat-storage tank.

**tricresyl phosphate (TCP).** Colorless to yellow liquid used as a gasoline and lubricant additive and plasticizer. Formula, $PO(OC_6H_4CH_3)_3$.

**trillion cubic feet.** A unit of measure commonly used for natural gas; equivalent to 39.3 million tons of coal, or 184 million barrels of oil.

**tritium.** A radioactive isotope of hydrogen with two neutrons and one proton in the nucleus. It is man-made and is heavier than deuterium (heavy hydrogen). Tritium is used in industrial thickness gauges and as a label in experiments in chemistry and biology. Its nucleus is a triton.

**triton.** The nucleus of a tritium atom.

**Trombe wall.** A passive solar heating system devised by Felix Trombe and Jacques Michel that combines the solar collector and heat storage in one, south-facing wall unit. The system consists of a thick concrete wall painted black on its outer face. Sheets of glass are placed in front of this wall with an airspace between. Air from the rooms of the building passes through openings at the foot of the wall and enters the airspace where it is heated by the sun. As it warms, the air rises up the air cavity by natural convection, and passes back into the building interior again through a second series of openings at the top of the wall. In order to arrest the flow of warm air into the building in summer, the openings in the wall are blocked by shutters.

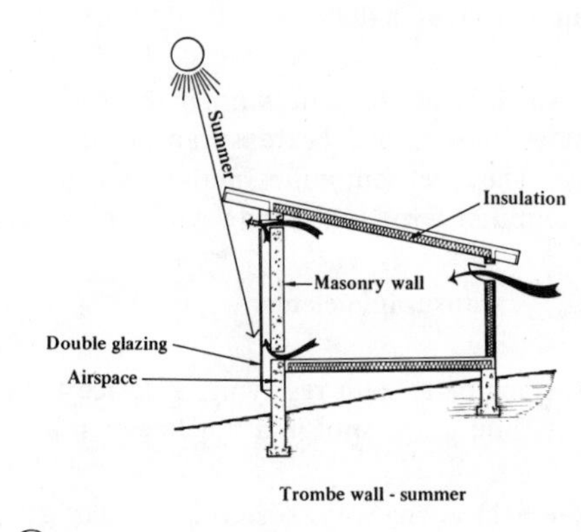


Trombe wall - summer

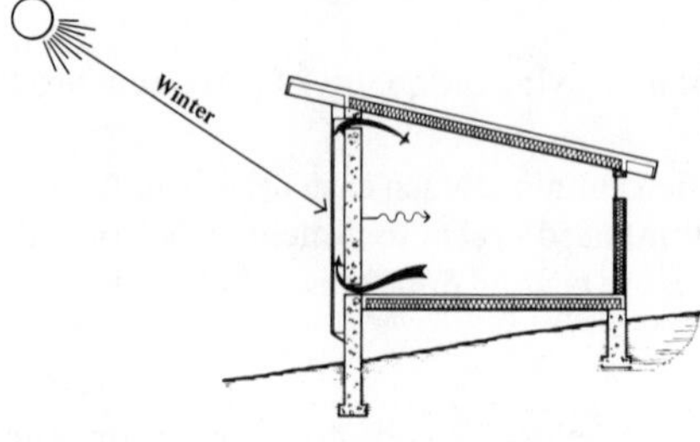


Trombe wall - winter

Trombe wall. (*Courtesy Living with Natural Energy, Design for a Limited Planet*)

**trommel screen.** A cylindrical or conical revolving screen used for screening or sizing substances such as rock, ore, or coal.

**tropic tide.** Tides occurring approximately every two weeks when the effect of the moon's maximum declination north and south of the equator is greatest.

**troposphere.** The portion of the atmosphere which extends out 7 to 10 miles from the earth's surface, and in which, generally, temperature decreases rapidly with altitude, clouds form, and convection is active.

**tunnel excavation.** Excavation carried out completely underground and limited in width, and in depth or height.

**turbidimeter.** An instrument for determining the concentration of particles in a suspension in terms of the proportion of light absorbed from a transmitted beam.

**turbidity.** The condition of having the transparence or translucence disturbed, as when sediment in water is stirred up, or when dust, haze, clouds, etc. appear in the atmosphere because of wind or vertical currents.

**turbine.** A machine that converts the energy in a stream of fluid into mechanical energy by passing the stream through a system of fixed and/or moving fan-like blades, causing them to rotate. Turbines have wide uses in large-scale power generation (usually employing steam-driven turbines), small-scale power generation (e.g., Pelton wheel), jet aircraft propulsion (gas turbines deriving power from a stream of heated air), marine engines, etc.

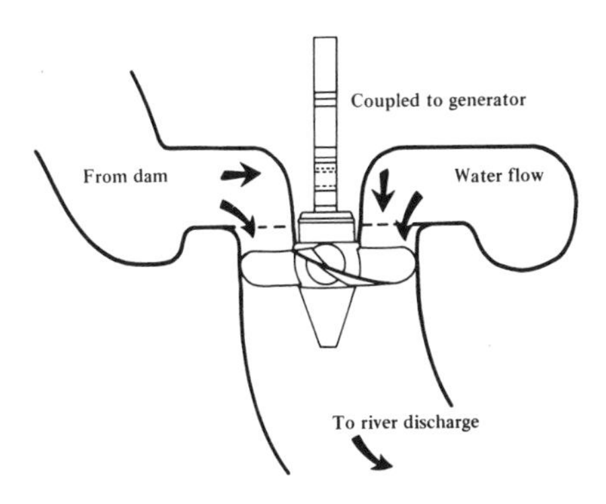


Reaction turbine. (*Courtesy U.S. Council on Environmental Quality*)

**turbine generator.** A combination of a turbine and an electric generator.

**turboblower.** A blower in which the rotating part is equipped with blades that rotate between stationary blades attached to the housing. The respective sets of blades are set at an angle such that, as the rotor turns, gases are pushed through the blades and again by the next set of rotating blades, and finally discharged from the casing at the opposite end from which they entered.

**turbojet engine.** A thermal jet in which air is compressed by a rotating compressor, heated by fuel combustion at compressor pressure, released through a gas turbine which drives the compressor, and finally ejected at high velocity through the rearward exhaust nozzle.

**turboprop engine.** A gas turbine engine designed to produce thrust by means of a propeller.

**turnaround efficiency.** The resulting efficiency when energy is converted from one form or state to another form or state, and then reconverted to the original form or state.

**two-cycle engine.** An engine in which only two strokes of the pistons, corresponding to one revolution of the crankshaft, are required to complete the cycle. In this cycle an explosion occurs on each downward stroke of the piston; the fresh charge is admitted and the exhaust gases expelled at or near the end of the stroke.

**two-group model.** A multigroup model in which the neutrons are divided into two energy groups, usually a thermal group and a fast or epithermal group containing all neutrons having energies above thermal.

**two-outlet heater.** A water heater typically having one outlet going to the piping for a domestic hot water system, and the other outlet going through a pipe to a large storage tank.

**two-valve burner system.** A system using separate control of air and gas, both of which are under pressure. The valves controlling the air and gas flows may or may not be interlocked.

# U

**U-Gas process.** A fluidized bed gasifier process which produces a low-Btu gas from a wide variety of coals. Pretreated coal or noncaking coal is fed into the gasifier where it is gasified directly with steam and air or oxygen. By carefully selecting the steam to air (or oxygen) ratio, ash will agglomerate and can be separated by weight from the fluidized bed. Process conditions are 300-350 pounds per square inch gage and 1038°C in the gasifier.

**ultimate analysis.** The determination of the elements contained in a compound, i.e., carbon, hydrogen, oxygen, nitrogen, sulphur, and ash.

**ultimate $CO_2$.** The maximum theoretical percentage of flue gas $CO_2$ that it is possible to produce from the complete combustion of a fuel with the chemically correct fuel-air ratio.

**ultimate recoverable reserves.** The total quantity of crude oil, natural gas, natural gas liquids, or sulfur estimated to be ultimately producible from an oil or gas field as determined by an analysis of current and engineering data. This includes any quantities already produced up to the respective date of the estimate. Also called ultimate production.

**ultimate reservoir capacity.** The total estimated volume of gas that could be contained in an underground storage reservoir when developed to its maximum design pressure.

**ultimate waste disposal.** As applied in the nuclear field, a two-step operation that comprises the preparation of radioactive waste for final and permanent disposal and the actual placing of the product at the final site.

**ultrahigh voltages.** Voltages greater than 765,000 volts.

**ultraviolet light.** Black light; invisible light rays from the portion of the spectrum that lies beyond the violet on the shorter wavelength side. Used to induce chemical reactions and produce fluorescence.

**ultraviolet radiation.** Electromagnetic radiation whose wavelength is shorter than that for visible light but longer than that for x-rays; radiant energy of wavelengths from 0.1- to 0.4-micron.

**underclay.** A bed of clay, in some cases highly siliceous, in many others highly aluminous, occurring immediately beneath a coal seam and representing the soil in which the trees of the Carboniferous swamp forests were rooted. Also called seat earth.

**underground coal gasification.** A process for producing synthetic gas from coal in natural, underground deposits, 100 or more feet below the surface.

**underground fires.** There are two types of underground fires: those which involve exposed surfaces and are known as open, freely burning fires; and those which may be wholly or partly concealed, and are invariably caused by spontaneous heating of the coal itself, known as gob fires.

**underground storage.** The utilization of subsurface facilities for storing gas which has been transferred from its original location for the primary purposes of conservation, fuller utilization of pipeline facilities, and more effective and economic delivery to markets. The facilities are usually natural geological reservoirs such as salt domes, depleted oil or gas fields, or waterbearing sands sealed on the top by an impermeable cap rock. The facilities may be man-made or natural caverns.

**unfinished oils.** Petroleum oils or a mixture or combination of such oils, or any component or components of such oils, which are to be further processed.

**unit coal.** Applied to coal prepared as for analysis, and being the pure coal substance considered altogether apart from extraneous or adventitious material (moisture and mineral impurities) which may by accident or through natural causes have become associated with the combustible organic substance of the coal; the differentiation between the noncoal substnce of a sample being analyzed and the coal itself. It is expressed by the formula: Unit Coal $= 1.00 - (W + 1.08A + 0.55S)$ where $W =$ water, $A =$ ash, and $S =$ sulfur.

**unit heater.** A forced convection heating device of two types: an assembly of encased heating surface with fan and motor (or turbine) for connection to a source of steam or hot water; or an assembly same as the above plus a fuel burner for connection to a source of oil or gas (or supplied with coal).

**unitization.** Joint operation of several leases usually for reasons of economy or conservation.

**unit train.** A system developed by the Baltimore & Ohio Railroad for delivering coal more efficiently. A string of cars with distinctive markings and loaded to "full visible capacity," is operated without stops along the way for cars to be cut in and out. In this way, the customer receives his coal quickly and the empty car is scheduled back to the coal fields.

**universal coal cutter.** A coal cutter with a jib capable of cutting at any height or angle. It may be mounted on crawler trackers.

**unproven area.** An area in which it has not been established by drilling operations whether oil and/or gas may be found in commercial quantities.

**unsaturated compounds.** Any compound having more than one bond between two adjacent atoms, usually carbon atoms, and capable of adding other atoms at that point to reduce it to a single bond.

**unsaturated hydrocarbon.** Chemcial compounds of carbon and hydrogen in which all the valence bonds of the carbon atoms are not taken up with hydrogen atoms.

**unsaturates.** Hydrocarbon compounds of such molecular structure that they readily pick up additional hydrogen atoms. Olefins and diolefins, which occur in cracking, are of this type.

**unscattering.** Any scattering collision in which the scattered neutron gains kinetic energy. It is important only in the thermal energy range.

**unscreened coal.** Coal for which no size limits are specified; run-of-mine coal.

**upcast.** An upward current of air passing through a shaft, or the like.

**unstable.** Readily decomposed; liable to spontaneous combustion or oxidation.

**upgrade.** To increase the commercial value of a coal by appropriate treatment.

**upstream.** From a reference point, any point located nearer the origin of flow, that is, before the reference point is reached.

**upwind.** On the same side as the direction from which the wind is blowing.

**uranium.** A radioactive element with the atomic number 92 and, as found in natural ores, an average atomic weight of approximately 238. The two principal natural isotopes are uranium-235, which is fissionable, and uranium-238, which is fertile. Natural uranium also includes a minute amount of uranium-234. Uranium is the basic raw material of nuclear energy. Chemical symbol U. (See Fig. p. 468)

**uranium enrichment.** The development, design, construction, and operation of systems, processes, and components to permit isotopic separation and enrichment of the isotope uranium-235 in uranium for use as nuclear fuel. The technology includes such processes as gaseous diffusion, centrifugation, and advanced systems involving lasers and aeronozzles. (See Figs. pp. 470, 471)

**uranium hexafluoride.** A gaseous compound of uranium; used in the diffusion process of enrichment. Chemical formula $UF_6$.

**uranium milling.** The process of crushing, grinding, and chemically treating uranium ore in order to remove the uranium oxide. Uranium ores typically contain on the order of 0.1 percent uranium.

Photograph of a uranium mine. (*Courtesy U.S. Energy Research and Development Administration*)

**uranium oxide.** The important oxides of uranium are $UO_2$, $U_3O_8$, and $UO_3$. The dioxide ($UO_2$) is used as a nuclear fuel element. Triuranium octoxide ($U_3O_8$), or Yellowcake, is the international standard for the form in which uranium concentrate is marketed. Uranium trioxide ($UO_3$) is an intermediate product in the refining of uranium.

**uranium resources.** The increasing commitment to the use of nuclear power worldwide has created a large demand for a uranium fuel supply. Currently, the U.S. Government estimates a domestic uranium resource base of about 3.7 million tons triuranium octoxide ($U_3O_8$). This includes only about 700,000 tons of reserves that have been defined by drilling. The additional 3 million tons of potential resources reflect geologic judgments on mineral trends and limited investigations. It is obvious the projected demand of 1,300,000 tons of $U_3O_8$ is considerably higher than the known reserves and must rely on probable potential resources. (See Figs. pp. 469, 472)

**uranium separation.** Application of new laser technologies for uranium enrichment and isotope separation. (See Fig. p. 470)

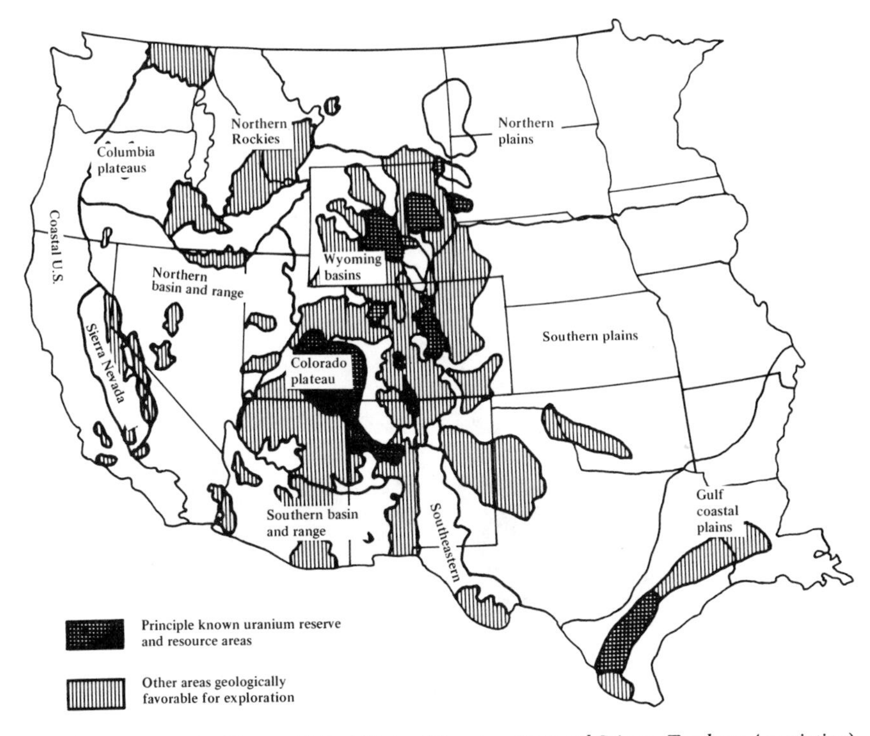


Uranium resources in Western United States. (*Courtesy National Science Teachers Association*)

**uranium series.** The series of nuclides resulting from the radioactive decay of uranium-238, also known as the uranium-radium series. The end product of the series is lead-206. Many man-made nuclides decay into this sequence.

**uranium tetrafluoride.** A solid green compound called green salt; an intermediate product in the production of uranium hexafluoride. Chemical formula $UF_4$.

**uranium trioxide.** An intermediate product in the refining of uranium, also called orange oxide. Chemical formula $UO_3$.

**uranium-235.** The readily fissionable isotope of uranium; concentrated from natural uranium by gaseous diffusion, by centrifugation, or by electromagnetic methods. (See Fig. p. 471)

**uranium-238.** The abundant, naturally occurring isotope of uranium; naturally occurring uranium consists of 99.29 percent uranium-238 and 0.71 percent uranium-235. It is nonfissionable but will capture neutrons in a nuclear reactor to produce plutonium-239, a nuclide which can substitute for uranium-235 as a nuclear fuel or a nuclear explosive. (See Fig. p. 471)

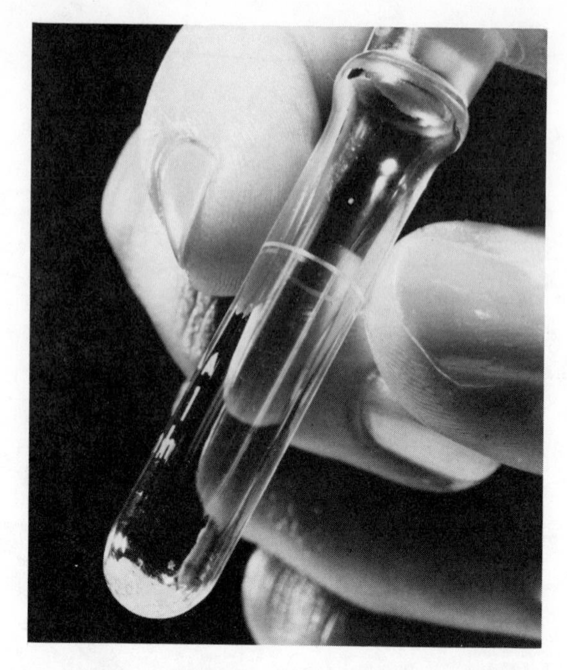

Photo shows, at the bottom of the test tube, about 4 milligrams of uranium enriched to about 3 percent by laser experiments conducted at Lawrence Livermore Laboratory of the University of California. (*Courtesy Lawrence Livermore Laboratory*)

**ureaformaldehyde foam.** An effective insulation which is foamed in place during installation and used to insulate finished frame walls and unfinished attic floor areas. It requires special care in quality controls to avoid problems.

**use charge.** An annual rental charge assessed by the Atomic Energy Commission for inventories of enriched fissionable material.

**useful energy gain.** The energy collected by a solar collector which is not lost to the surroundings and can ultimately be used for space or water heating.

**useful solar heat.** Heat delivered by a solar system.

**utility electric meter.** Device for measuring the amount of electricity used; the unit of measurement is the kilowatt hour. One kilowatt hour is the equivalent of 1000 watts of electricity used for one hour. Most electric meters have a set of dials read in multiples of 10, clockwise or counter-clockwise. The dial on the far right indicates tens of kilowatt-hours; the next one, hundreds of kilowatt-hours; the next, thousands of kilowatt-hours. When the pointer is between two numbers, the lower of the two numbers is read. (See Fig. p. 472)

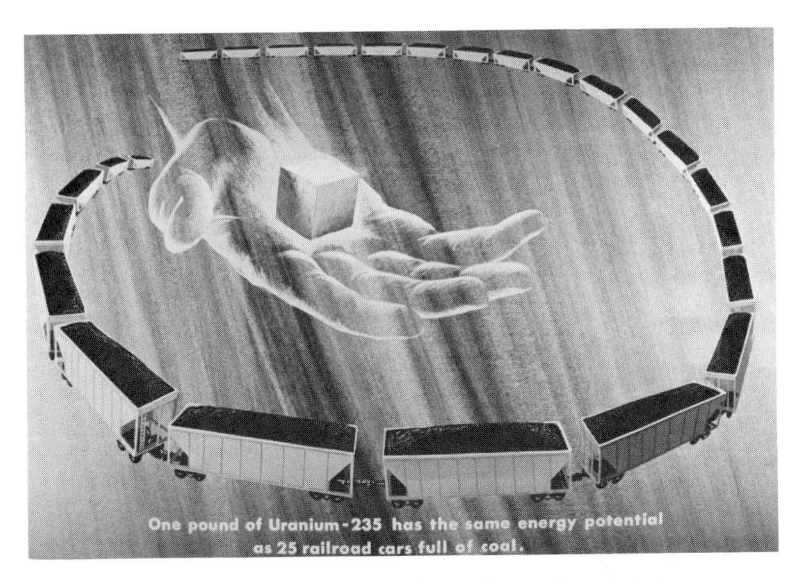


Uranium-235. (*Courtesy U.S. Atomic Energy Commission*)

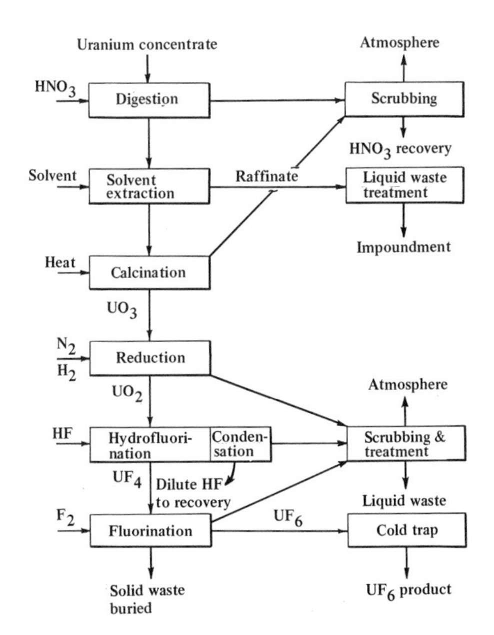


Uranium ($UF_6$) production by wet solvent extraction-flourination process. (*Courtesy U.S. Council on Environmental Quality*)

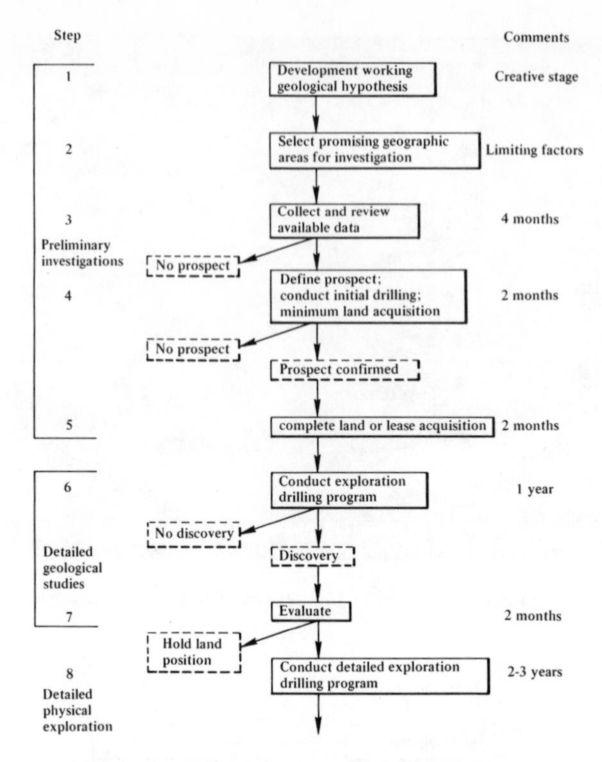


Uranium exploration. (*Courtesy U.S. Council on Environmental Quality*)

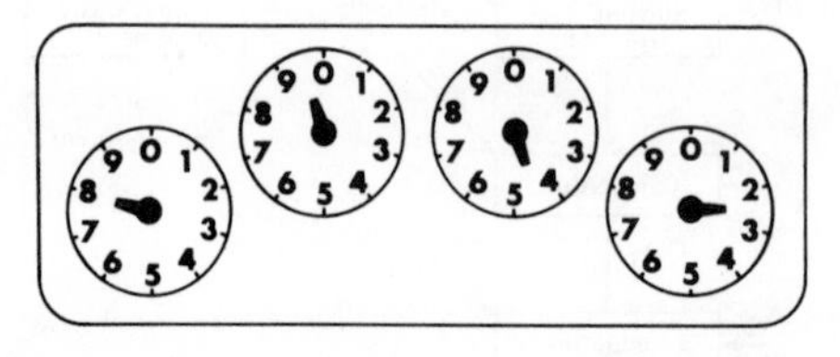

Dials showing typical reading of an electric meter. (*Courtesy Help: The Useful Almanac 1976–77*)

**utility gases.** Natural gas, manufactured gas, synthetic gas, liquefied petroleum gas-air mixture, or mixtures of any of these gases.

**utilization factor.** The ratio of the maximum demand of a system, or part of a system, to the rated capacity of the system, or part of the system, under consideration.

**U-value.** A unit of heat transmission used in heat loss calculations for buildings and defined as: British thermal units transmitted per square foot per hour per degrees Fahrenheit difference in air temperature between the two faces of the wall under consideration.

# V

**vacuum.** A pressure less than atmospheric pressure, measured either from the base of zero pressure or from the base of atmospheric pressure; also a space devoid of matter.

**vacuum distillation.** Distillation under reduced pressure, which sufficiently reduces the boiling temperature of the distilled material to prevent decomposition or cracking.

**vacuum-evaporated film.** A film generally formed on a sheet or plate by electrical evaporation of a metal or alloy in an evacuated chamber.

**vacuum filtration.** The separation of solids from liquids by passing the mixture through a filter where, on one side, a partial vacuum is created to increase the rate of filtration. It may be used to extract fine coal from the suspension.

**vacuum-relieving device.** A device to automatically admit air or gas into space at a pressure below atmospheric.

**vadose.** Applied to seepage waters occurring below the surface and above the water table.

**valence.** A property of atoms determining the number of atoms with which they can combine in chemical reactions.

**valley** Any hollow or low-lying tract of ground between hills or mountains, usually traversed by streams or rivers which receive natural drainage from the surrounding high ground.

**value added.** Calculated by taking the value of shipments of a firm, subtracting costs of materials, fuel, electricity, resales, and miscellaneous receipts, and then adding inventory increase. The sum of value added for industrial firms give industrial value added.

**valve.** Any contrivance inserted in a pipe or tube containing a lid, cover, ball, or slide that can be opened or closed to control the flow or supply of liquids, gases, or other shifting material through a passage.

**valve box.** A housing around an underground valve to allow access to the valve and to protect the valve from mechanical damage or the effects of weather.

**valve chamber.** The space in a gas dry-meter containing the slide valves and mechanism for their operation.

**valve control.** A fuel-air ratio control system that operates by means of mechanical linkage of related valves, common in industrial combustion systems.

**valve seat.** The stationary portion of the valve which, when in contact with the movable portion, stops flow completely.

**Van Allen radiation belts.** Several belts of ionizing radiation extending from a few hundred miles to a few thousand miles above the earth's surface. The radiation consists of protons and electrons which originate mostly in the sun and are trapped by the earth's magnetic field. Powerful doughnut-shaped zone of radiation 1000 to 3000 miles above the surface of the earth and parallel with the equator.

**Van de Graaff generator.** An electrostatic machine in which electrically charged particles are sprayed on a moving belt and carried by it to build up a high potential on an insulated terminal. Charged particles are then accelerated along a discharge path through a vacuum tube by the potential difference between the insulated terminal and the opposite end of the machine. A Van de Graaff accelerator is often used to inject particles into larger accelerators. Named after R. S. Van de Graaff, who invented the device in 1931.

**van der Waals' adsorption.** Physical, as distinct from chemical, cohesion; normal adhesive forces between molecules characterized by relatively low heats of adsorption.

**vapor.** The gaseous state of a substance; a gaseous substance that is at a temperature below its critical temperature and therefore liquefiable by pressure alone.

**vapor barriers.** A vapor barrier will prevent water vapor from condensing and collecting in home insulation. Vapor barriers can be mylar or aluminum foil.

**vapor density.** The relative density of a gas or vapor as compared with some specific standard, such as hydrogen.

**vapor-dominated convective hydrothermal resources.** Geothermal systems which produce superheated steam containing very small amounts of water.

**vaporization.** The act or process of changing a substance from the liquid to the gaseous state. One of three basic contributing factors to air pollution; the others are attrition and combustion.

**vaporizer.** A heat exchanger used to return liquid natural gas to a gaseous form and then to continue to heat the gas to a temperature at which it can be sent into the distribution system.

**vapor lock.** The formation of vapor in a feed line by vapors generated from the fuel, resulting in an interruption of the flow of the liquid fuel.

**vapor plume.** The stack effluent consisting of flue gas made visible by condensed water droplets or mist.

**vapor pressure.** The pressure at which a liquid and its vapors are in equilibrium at a definite temperature. If the vapor pressure reaches the prevailing atmospheric pressure, the liquid boils.

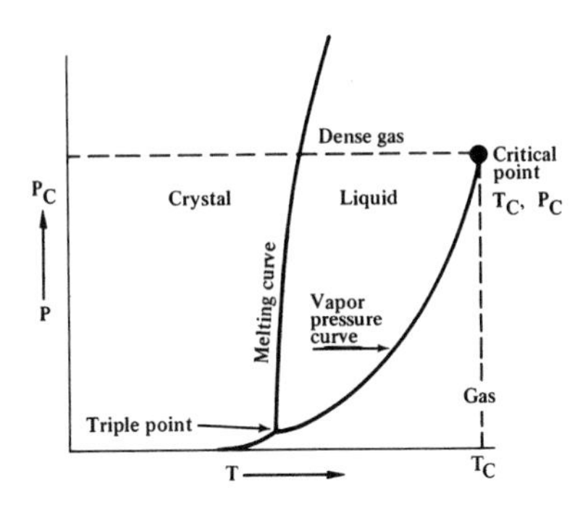


Universal phase diagram. (*Courtesy Van Nostrand's Scientific Encyclopedia*)

**vapor suppression.** A safety system that can be incorporated in the design of structures housing water reactors. In this system, the space surrounding the reactor is vented into pools of water open to the outside air. If surges of hot vapors should be released from the reactor in an accident, their energy (pressure) would be dissipated in the pools of water. Gases not condensed would be scrubbed clean of radioactive particles by the bubbling. Another system uses a suppression pool in a separate pressure vessel that can be vented through a stack.

**vapor system.** A steam heating system operating at pressure very near that of the atmosphere.

**variometer.** A geophysical device for measuring or recording variations in terrestrial magnetism; a variable inductance provided with a scale.

**veering wind.** A clockwise change in wind direction.

**vegetation survey.** Leakage surveys made for the purpose of finding leaks in underground gas piping by observing vegetation.

**vein.** A zone or belt of mineralized rock lying within boundaries clearly separating it from neighborhood rock; includes all deposits of mineral matter found through a mineralized zone or belt coming from the same source, impressed with the same forms, and appearing to have been created by the same processes. Typically, a vein is long, deep, and relatively narrow.

**vein bitumen.** Synonym for asphaltite.

**velocity.** Rate of motion in a given direction.

**velocity pressure.** The pressure exerted by a moving fluid in the direction of its motion. It is the difference between the total pressure and the static pressure.

**vent.** An opening in a tank or other piece of equipment, sealed to prevent escape of material within the equipment at normal pressures, but so arranged that it automatically opens to relieve excessive pressure in the equipment; also the relief opening in a pressure regulator, normally open to the atmosphere.

**vent connector.** The portion of the venting system which connects the gas appliance to the gas vent or chimney.

**vented recessed heater.** A self-contained, vented appliance complete with grilles or equivalent, designed for incorporation in or permanent attachment to a wall, floor, ceiling, or partition, and furnishing heated air circulated by gravity or by a fan directly into the space to be heated, through openings in the casing.

**vent gas.** Products of combustion from gas appliances plus excess air plus dilution air in the gas vent or chimney above the draft hood or draft regulator.

**ventilation.** The atmospheric air that is purposely allowed to enter an interior space to cool or freshen it; the principal air-conditioning process concerned with control of air circulation.

**Venturi meter.** A trademark for a form of the Venturi tube arranged to measure the flow of a liquid in pipes. Small tubes are attached to the Venturi tube at the throat and at the point where the liquid enters the converging entrance. The difference in pressure heads is shown on some form of manometer, and, from this difference and a knowledge of the diameters of the tubes, the quantity of flow is determined.

**Venturi scrub.** A method for cleaning particulates from stack gases which consists of water being injected into a high-speed gas flow. The particulates are removed with the water.

**Venuri throat.** A tube tapered down to a lesser diameter and then expanding gradually to its original diameter. Pressure-measuring taps are provided at the entrance and at the constricted throat for determining pressure differential through the tube used for metering.

**Venturi tube.** A closed conduit which is gradually contracted to a throat causing a reduction of pressure head by which the velocity through the throat may be determined.

**verifier.** In gas testing, an apparatus by which the amount of gas required to produce a flame of a given size is measured.

**vertical-axis rotor-type wind turbines.** A wind energy turbine which is designed to accept wind from any direction and thus need not be turned into the wind. One type of vertical-axis wind turbine is the Darrieus rotor, named after its inventor G. J. M. Darrieus of France in the 1920's. DOE has been operating a small Darrieus and is building a 50-foot (17 meter) model for tests.

**virgin coal.** An area of coal which is in place (in situ) and unimpaired by mining activities.

**virgin neutrons.** Neutrons from any source, before they make a collision.

**virgin stock.** Oil processed from crude oil which contains no cracked material. Also called straight-run stock.

**vermiculite.** A loose-fill, mineral-insulation material that can be poured into small areas for complete coverage.

**vertical axis wind turbines.** The vertical-axis wind turbine can accept wind from any direction and therefore does not have to be turned into the wind. One type of vertical-axis wind turbine, the Darrieus rotor, is considered to be a potential major competitor to the propeller-type system. It was invented by G.J.M. Darrieus of France in the 1920's. With its eggbeater-shaped blades, the Darrieus rotor has relatively high power output per given rotor weight and cost.

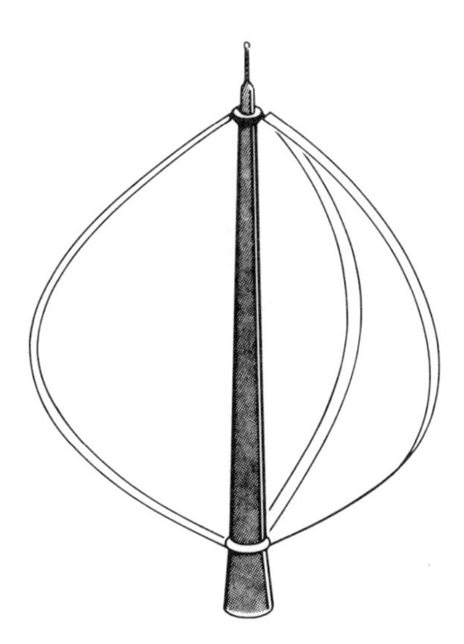

Vertical axis wind turbine. (*Courtesy Wind Machines*)

**violet cell.** A new form of silicon solar cell which converts more sunlight to electricity from the violet/ultraviolet range of the light spectrum than conventional solar cells. The violet cell was developed by COMSAT's Dr. Lindmoyer.

**visbreaking.** Lowering or breaking the viscosity of residuum by cracking at relatively low temperatures.

**viscosimeter.** An instrument used to measure the viscosity of liquids.

**viscosity.** Any resistance to deformation that involves dissipation of energy by internal friction; in general, resistance to flow; that property of semi-fluids, fluids, and gases by virtue of which they resist an instantaneous change of shape or arrangement of molecules.

**visible radiation.** Radiant energy of wavelengths from 0.4 to 0.76 micron; the visible spectrum.

**visible spectrum.** That portion of the electromagnetic spectrum, the waves of which normally produce upon the human eye color sensations of red, orange, yellow, green, blue, violet or their intermediate hues, or of white light if the rays are combined.

**vital capacity.** The term for the greatest volume of air that a man can expel from his lungs after full inspiration; it is the greatest volume of air that can be moved in and out of the lungs in a single breath. The average man's vital capacity is between 4 and 5 liters.

**Vitasul.** A trade name for a chemical additive to diesel fuel which reduces considerably the carbon monoxide danger from diesel locomotive exhausts.

**vitrain.** Term introduced by M. C. Stopes in 1919 to designate the macroscopically recognizable very bright bands of coals, usually a few millimeters (3 to 5) in width. In many coals the vitrain is permeated with numerous fine cracks at right angles to stratification, and consequently breaks cubically, with conchoidal surfaces. In other coals the vitrain is crossed by only occasional perpendicular cracks. In the macroscopic description of seams, only the bands of vitrain having a thickness of several millimeters are usually noted. Examination with the microscope shows vitrain to consist of microlithotypes very rich in vitrinite. After clarain, vitrain is the most widely distributed and common macroscopic constituent of humic coals.

**vitrinization.** The process in coalification that results in the formation of vitrain.

**void coefficient.** A rate of change in the reactivity of a water reactor system resulting from a formation of steam bubbles as the power level and temperature increase.

**volatile.** Capable of being readily evaporated at a relatively low temperature.

**volatile fluxes.** The volatile constituents of a magma.

**volatile matter.** Those products, exclusive of moisture, that are liberated by a material as gas and vapor, determined by definite prescribed methods which may vary according to the nature of the material. In the case of coal and coke, the methods employed are those prescribed in ASTM Designation D271.

**volt.** The practical meter-kilogram-second (SI) unit of electrical potential difference and electromotive force that equals the difference of potential between two points in a conducting wire carrying a constant current of 1 ampere when the power dissipated between these two points equals 1 watt. It equals the potential difference across a resistance of 1 ohm when 1 ampere of current is flowing through it.

**Volta, Alessandro.** Alessandro Volta, physicist and pioneer of electrical science. Volta invented the cell, pile, and battery that bear his name. His name was also given to the volt.

**voltage.** The amount of electromotive force, measured in volts, that exists between two points.

**voltage regulation.** This is the change in output voltage which occurs when the load is reduced from rated value to zero with the values of all other quantities remaining unchanged.

**voltaic cell.** A cell consisting of two electrodes and one or more electrolytes which, when connected in a closed circuit, will give out electrical energy.

**volt-ampere.** A unit of electric measurement equal to the product of a volt and an ampere.

**volume dose.** The product of absorbed dose and the volume of the absorbing mass.

**vortex.** A flow of a fluid in which the steamlines form concentric circles. There are two kinds of vortex, forced and free, depending on whether torque is applied externally, e.g., a forced vortex can be made in a liquid contained in a vessel and stirred with a paddle; a free vortex occurs when the liquid is allowed to leave the cylinder through a small hole in the bottom, in which case the force is provided by gravity acting on the fluid. Vortices occur when a solid body moves through a fluid unless the body is designed (e.g., in an aerofoil) to avoid creating them.

**wadding.** Paper or cloth placed over an explosive in a hole.

**wagon.** An underground coal car; any vehicle for carrying coal or debris.

**wagon drill.** A drilling machine mounted on a light, wheeled carriage.

**walk.** To deviate from the intended course, such as a borehole that is following a course deviating from its intended direction.

**walking beam.** The beam used to impart a reciprocating movement to the drilling column in percussive drilling.

**wall.** The side of a drift or of an entry; the face of a longwall working or stall; also a rib of solid coal between two rooms.

**wall coal.** The middle division of three in a seam, the other two being termed top coal and ground coal.

**wall friction.** The drag created in the flow of a liquid or gas because of contact with the wall surface of its conductor, such as the inside surface of a pipe or drill rod or the annular space between a drill string and the walls of a borehole.

**wall orientation.** The primary direction of an exterior vertical surface of a building facing within 45° of one of the directions (i.e., north, south, east, and west).

**wallplate.** A horizontal timber supported by posts resting on sills and extending lengthwise on each side of the tunnel. The roof supports rest on these wallplates.

**wall rock.** The country rock immediately adjacent to a vein.

**wandering coal.** A coal seam that exists only over a small area; an irregular seam of coal.

**Wankel engine.** An internal combustion rotary engine that has a rounded triangular rotor functioning as a piston and rotating in a space in the engine and that has only two major moving parts.

**warm spring.** A thermal spring.

**warrant.** A general term for the clay floors of coal seams, particularly when hard and tough.

**warrenite.** A general term for gaseous and liquid bitumens consisting of a mixture of paraffins, isoparaffins, etc.

**washability.** Coal properties determining the amenability of a coal to improvement in quality by cleaning.

**washbox.** In coal preparation, the jig box in which coal is stratified and separated into fractions (heavier below and lighter above).

**washed coal.** Coal from which impurities have been removed by treatment in a liquid medium.

**washed gases.** Purified coal gas from which the chemicals benzene and naphthalene have been extracted by scrubbing with oil.

**washer.** An apparatus for the wet cleaning of coal; also an apparatus in which gases are washed.

**washer cooler.** A washer, in the form of a tall tower, in which the washing liquid is sprayed in at the top, collected in the bottom of the tower, and then cooled and recycled through the tower. This serves the dual purpose of washing the gas free of impurities and cooling the gas.

**washout.** A channel cut into or through a coal seam at some time during or after the formation of the seam and filled generally with sandstone, more rarely with shale.

**wash plant.** A plant in which coal is freed from some of its ash and other impurities, such as inorganic surfur, in order to produce a higher quality fuel. The separation may be accompanied by a number of processes that generally take advantage of the differences in density between pure coal and any contaminating ingredients. A waste product consisting of various amounts of ash, sulfur, water, and coal is rejected from a wash plant.

**waste disposal.** The disposition of radioactive waste without specific provision for recovery.

**waste drainage.** The controlled leakage of air through waste to insure that large concentrations of mine gases do not accumulate in that waste.

**waste heat.** Heat which is at temperatures very close to the ambient and hence is not valuable for production of power and is discharged to the environment.

**waste-heat boiler.** One which uses heat of exit gases from furnaces to produce steam or to heat water.

**waste lubrication.** Consists of packing oil-soaked waste for use in lubricating railway cars.

**water-air heat exchanger.** A device in which air is either heated or cooled by flowing water.

**water boiler.** A research reactor whose core consists of a small metal tank filled with uranium fuel in an aqueous solution. Heat is removed by a cooling coil in the core. Not to be confused with boiling water reactor.

**water budget.** A budget of the incoming and outgoing water from a region, including rainfall, evaporation, runoff, seepage, and with perhaps special attention to ablation (evaporation) of snow, evapotranspiration from vegetation, dew, or other special aspects relevant for special interests (e.g., agriculture).

**water conservation.** The protection, management, and use of water resources in such a way as to sustain maximum benefits to people, agriculture, industry, commerce, and other segments of the national economy.

Aerial view of Lahontan Dam on the Carson River in Nevada. The Lahontan Power Plant, below the dam, has a capacity of 2400 kilowatts. (*Courtesv Bureau of Reclamation*)

**water-cooling tower.** A device for the evaporative cooling of water by contact with air.

**water drive.** A method of making a high-level oil well nearing exhaustion to continue to produce by pouring water into abandoned low-level wells in hydraulic communication with the producing horizon.

**waterflooding.** A secondary-recovery operation for oil fields in which water is injected into a petroleum reservoir to bring more oil to the surface.

**water gas.** A mixture of gases produced by forcing steam through a very hot coke or coal. It is a mixture of carbon monoxide and hydrogen with small amounts of nitrogen and carbon dioxide and is sometimes used as a fuel for heating and cooking. It burns with a blue flame.

**water heater.** An appliance for supplying hot water for domestic or commercial purposes other than space heating.

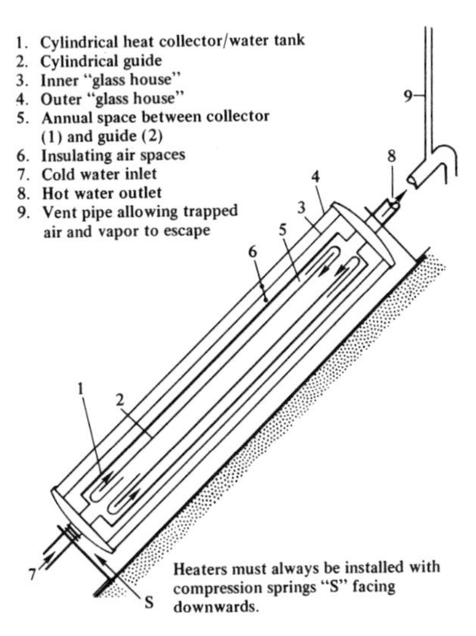


Construction of a water heater. (*Courtesy Ward Ritchie Press, Pasadena, CA*)

**water hyacinth.** A freshwater plant of tropical regions that has become a noxious weed in many parts of the world. Highly prolific, and reproducing mainly vegetatively, it can double its numbers in 8 to 10 days in water at a temperature of 10°C or more, provided nutrients are present. It was introduced to North America in 1884 by visitors to the New Orleans Cotton Exposition, who brought specimens from Venezuela. NASA has experimented with the ulitization of water hyacinth to provide fertilizer and feeding stuffs for livestock. May be candidate for energy biomass plant.

**water injection.** Technique for reducing the formation, and so the emission, of pollutants from internal combustion engines. Research has shown that water injection is the most efficient way of reducing pollutant emissions, but it requires vehicles to carry water tanks as large as fuel tanks, the water may freeze, and conventional anti-freeze compounds produce exhaust pollutants, and there may be long-term corrosion in the engine.

**water pollution.** The addition of sewage, industrial wastes, or other harmful or objectionable material to water in concentrations or in quantities sufficient to result in measurable degradation of water quality.

**waterpower.** The power of water derived from its gravity or its momentum as applied or applicable to the driving of machinery.

**water quality criteria.** The levels of pollutants that affect the suitability of water for a given use. Generally, water use classification includes: public water supply, recreation, propagation of fish and other aquatic life, agricultural use, and industrial use.

**water quality standard.** A plan for water quality management containing four major elements: the use (recreation, drinking water, fish and wildlife propagation, industrial or agricultural) to be made of the water; criteria to protect those uses: implementation plans (for needed industrial-municipal waste treatment improvements) and enforcement plans; and an antidegradation statement to protect existing high quality waters.

**water rate.** The weight of dry steam consumed by a steam engine for each horsepower per hour. The result is stated in either indicated horsepower or brake horsepower.

**water shutoff.** The sealing off of saltwater-bearing foundations from oil-bearing zones to prevent harmful underground water pollution.

**water soluble oils.** Oils having the property of forming permanent emulsions or almost clear solutions with water.

**water-to-carbon ratio.** The ratio by weight of the amount of water to carbon compounds in a gas (vapor) stream.

**water turbine.** A prime mover coupled to an alternator and using a purely rotary motion to generate an alternating current. The main types of water turbines are: the Pelton wheel for high heads, the Francis turbine for low to medium heads, and the Kaplan turbine for a wide range of heads.

**water wheel.** A wheel designed to be turned by the impact of flowing water; used to drive machinery, raise water, etc.

**watt.** The absolute meter-kilogram-second (SI) unit of power that equals 1 absolute joule per second. It is analogous to horsepower or foot-pounds per minute of mechanical power. One horsepower is equivalent to approximately 746 watts.

**watt-hour.** The total amount of energy used in 1 hour by a device that uses 1 watt of power for continuous operation. Electrical energy is commonly sold by the kilowatt hour (1000 watt-hours).

**wattmeter.** An instrument for measuring electric power in watts, the unit of electrical energy, volts times amperes.

**wavefront.** In seismology, the surface of equal time elapse from the point of detonation to the position of the resulting outgoing signal at any given time after the charge has been detonated; in a more restricted sense, the surface along which phase is constant at a given instant.

**wavelength.** The distance between similar points on successive waves. That of visible light varies between 300 angstrom units (violet) and 7600 angstrom units (red).

**wave power.** Using the sea's energy in advancing waves to generate power to drive a turbine.

**weathering.** The group of processes, such as chemical action of air and rainwater and of plants and bacteria and the mechanical action of changes in temperature, whereby rocks, on exposure to the weather, change in character, decay, and finally crumble into soil.

**weatherstripping.** The use of thin spring metal, rolled vinyl, or foam rubber tape to reduce drafts through cracks or window edges.

**web.** The slice or thickness of coal (usually restricted to thin or medium slices of coal) taken by a cutter loader when cutting along the face. The thickness of web varies from a few inches to about 6 feet.

**wedge pyrometer.** An instrument for the approximate measurement of high temperatures. It depends on a wedge of colored glass, the position of which is adjusted until the source of heat is no longer visible when viewed through the glass; movement to the wedge operates a scale calibrated in temperatures.

**weeper.** A hole in the ceiling of an underground aqueduct to let water from above drain through; a hole in a retaining wall to permit the escape of water from behind; a small feeder of water; also called weep hole.

**weight dropping.** A seismic technique by which energy can be sent downward into the earth without the necessity of drilling shotholes. The technique involves lifting a weight and permitting it to fall and strike the ground. The waves from the impact are then recorded. In areas where drilling is difficult or unduly expensive, this technique may be highly advantageous.

**weir.** An obstruction placed across a stream to divert the water through a desired channel.

**well.** A shaft or hole sunk into the earth to obtain oil, gas, etc. Also used as a synonym for borehole or drill hole. Wells are classified as oil wells, gas wells, dry holes, stratigraphic tests, or service wells.

**well abandoning.** The act of abandoning an oil site when the oil reservoir is depleted. The hole is plugged with cement, salvageable equipment is removed, and the site is regarded.

**well bleeding.** Allowing a bore hole to vent to the atmosphere for the purpose either of clearing it of impurities or of testing it.

**wellbore.** The hole made by the drilling bit.

**well casing.** Steel pipe inserted (and sometimes cemented) into a gas or oil well, intermittently as the well is drilled, to line the well to eliminate ground caving and water infiltration, and to prevent gas and/or oil from escaping or leaking from the native reservoir into other formations.

**well completion.** Final sealing off of a drilled well after the drilling apparatus is removed from the borehole.

**well core.** A sample of rock penetrated in a well or other borehole obtained by use of a hollow bit that cuts a circular channel around a central column or core.

**wellhead.** The equipment used to maintain surface control of a well. It is formed of the casing head, tubing head, and Christmas tree. Also refers to various parameters as they exist at the wellhead: wellhead pressure, wellhead price of oil, etc. Also applies to oil or gas brought to the surface and ready for transportation to refinery or ship or pipeline. Wellhead costs usually refer to the cost to bring the oil or gas to the surface and do not include the costs of transportation, refining, distribution, or profit.

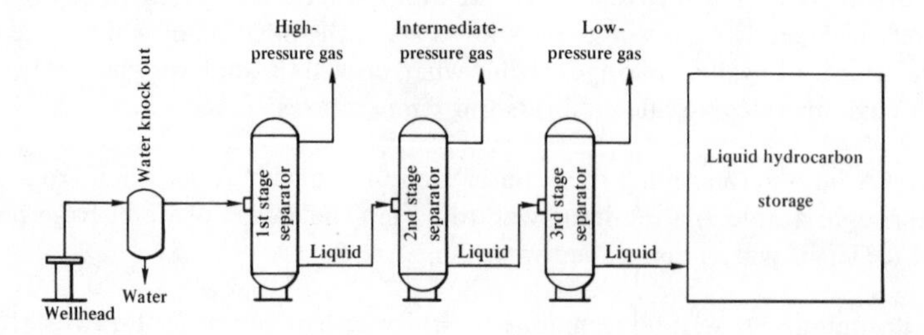


Three-stage wellhead separation unit. (*Courtesy U.S. Council on Environmental Quality*)

**well logging.** A widely used geophysical technique which involves probing the earth with instruments lowered into boreholes and recording their readings at the surface. Among rock properties currently being logged are electrical resistivity, self-potential, gamma-ray generation (both natural and in response to neutron bombardment), density, magnetic susceptibility, and acoustic velocity.

**Wellman-Galusha process.** A fixed-bed gasifier process, which produces low-Btu gas from primarily mildly caking or noncaking coal. It operates much like the Lurgi gasifier. The stirred fixed-bed gasifier being developed is based on the Wellman-Galusha gasifier design. The Wellman-Galusha operates at near atmospheric pressures.

**well pressure.** The natural pressure of the oil or gas in a well. It is often several hundred pounds per square inch and sufficient to cause the oil to rise to the surface. The well pressure is not related to the depth of the oil deposit below the surface.

**well rig.** An assemblage of all mechanisms, including power motors, necessary to drilling, casing, and finishing a tube well. Also called well drill.

**well shooting.** The firing of a charge of nitroglycerin or other high explosive in the bottom of a well to increase the flow of water, oil, or gas.

**Welsbach mantle lamp.** A type of lamp in which the flame impinges on a knitted cup or mantle saturated with certain chemical compounds that are heated to incandescence and emit a bright, white light.

**welt.** Large-scale topographic elevation, elongate in shape, with relatively steep sides. Generally parallel to continental coasts, welts may rise above water level to form islands, island chains, or even mountain ranges.

**westerlies.** Any winds with components from the west; also the dominant west-to-east motion of the atmosphere, centered over the middle latitudes of both hemispheres.

**wet cleaning.** A coal-cleaning method involving the use of washers plus the equipment necessary to dewater and heat-dry the coal; a method generally used when cleaning the coarser sizes of coal.

**wet critically.** Reactor criticality achieved with a coolant.

**wet gas.** Natural gas containing oil vapors. It occurs with or immediately above the oil. Also called casinghead gas.

**wet gas wells.** Gas wells that produce a gas from which gasoline can be obtained but which do not produce crude oil.

**wetlands.** Swamps or marshes, especially as areas preserved for wildlife.

**wet mining.** A mining method in which water is sprayed into the air at all points where dust is liable to be formed, and no attempt is made to prevent the air from picking up the moisture. It therefore soon becomes saturated and remains so throughout the ventilation circuits.

**wet natural gas.** Natural gas containing readily condensible gasoline that may be extracted in quantities sufficient to warrant the installation of a plant.

**wet scrubber.** An absorption tower used to remove polluted gases from a waste gas stream by contact with a liquid (e.g., hydrogen chloride being absorbed in water).

**wetting agent.** A substance that lowers the surface tension of water and thus enables it to mix more readily; a chemical promoting adhesion of a liquid (usually water) to a solid surface; wetting agents such as bone glue and sodium carboxymethyl-cellulose may be used for binding coal dust on mined roadways.

**wheelbase.** The distance between the points of contact of the front and back wheels of any vehicle with the rails or other surface upon which they travel; the distance between the leading and trailing axles of a vehicle.

**wheeling.** Transmission of electricity by a utility over its lines for another utility; also includes the receipt from and delivery to another system of like amounts (but not necessarily the same) energy.

**white damp.** Carbon monoxide (CO) gas that may be present in the afterdamp of a gas- or coal-dust explosion or in the gases given off by a mine fire; also one of the constituents of the gases produced by blasting. An important constituent of illuminating gas, it supports combustion and is very poisonous. It is colorless, odorless, and tasteless.

**white heat.** A common division of the color scale, generally given as about 1540°C.

**whole body counter.** A device used to identify and measure the radiation in the body (body burden) of human beings and animals; it uses heavy shielding to keep out background radiation, and ultrasensitive scintillation detectors and electronic equipment.

**Wigner effect.** In reactor operation, the change in physical properties of graphite resulting from the displacement of lattice atoms by high-energy neutrons and other energetic particles.

**wildcatter.** One who drills wells in the hope of finding gas or oil in territory not known to be a gas or oil field; also one who locates a mining claim far from where one has been discovered or developed.

**wildcat well.** An exploratory well drilled for oil or gas on a geologic structure or in an environment that has never produced.

**wild gas.** Blast-furnace gas that does not burn steadily or properly.

**wild well.** A well flowing while out of control.

**wind chill.** A measure of the quantity of heat that the atmosphere is capable of absorbing within an hour from an exposed surface. The term is based on a scale that correlates temperature and wind force. The term "wind chill" also used to describe the relative discomfort resulting from combinations of wind and temperature.

**wind chill factor.** That factor, dependent on wind velocity, which indicates the chilling effect on humans, of any given temperature. The general formula for heat loss $H$ is

$$H = (A + B\sqrt{V + C_V})\Delta t$$

where $H$ is the heat loss (wind chill) in kg cal/m$^2$/hr, $v$ is the wind speed in meters per second, $t$ is the difference in degrees Celsius between neutral skin temperature of 33° and air temperature, and constants $A$, $B$, and $C$ are, respectively, 10.45, 10.00, and −1.00. Values of $A$, $B$, and $C$ vary widely in formulas presented by different investigators. The above formula measures the cooling power of the wind and temperature in complete shade and does not consider the gain of heat from incoming radiation, either direct or diffuse.

**wind classification.** A system designed to emphasise the forces mainly responsible for the characteristics of wind. Geostrophic winds have a balance between pressure gradient and the coriolis force and blow along the isobars. Gradient winds are a modification in which the curvature of the isobars and of the flow is important, as in a cyclone. Isallobaric winds are caused by rapidly changing pressure patterns. Katabatic winds (cold downslope) and anabatic (warm upslope) winds are shallow and local, as are antitriptic (friction dominated) winds which are exemplified by katabatic ravine winds and cold outflows from storms. Land and sea breezes are antitriptic but the coriolis force becomes inmportant with time. The thermal wind is a geostrophic component. The ageostrophic component is due to acceleration or friction. See berg winds, moutain waves.

**wind concentrator.** A device or structure that concentrates a windstream.

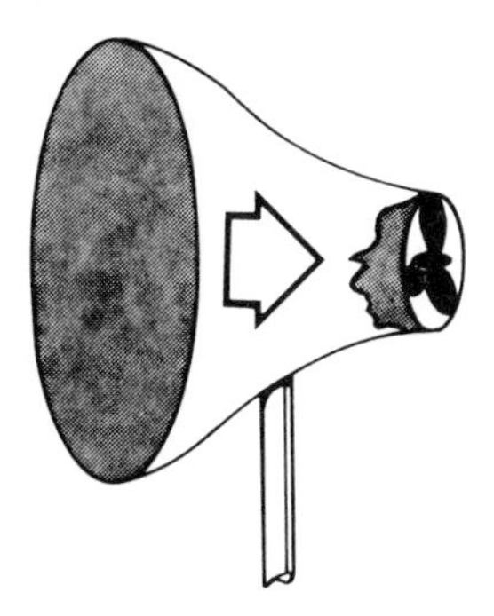

Wind concentrator. (*Courtesy Wind Machines*)

**wind-driven current.** The ocean current which develops as a result of the changes in density distribution caused by the wind drift.

**wind energy.** A form of solar energy, since winds are caused by variations in the amount of heat the sun sends to different parts of the earth. Electricity is produced when a windmill catches the wind and revolves, rotating a turbine which powers an electric generator. A major drawback of wind energy is the large amount of wind needed to generate the 6000 watts of electricity used by the average home. As with sunlight-produced energy, some backup resource is needed.

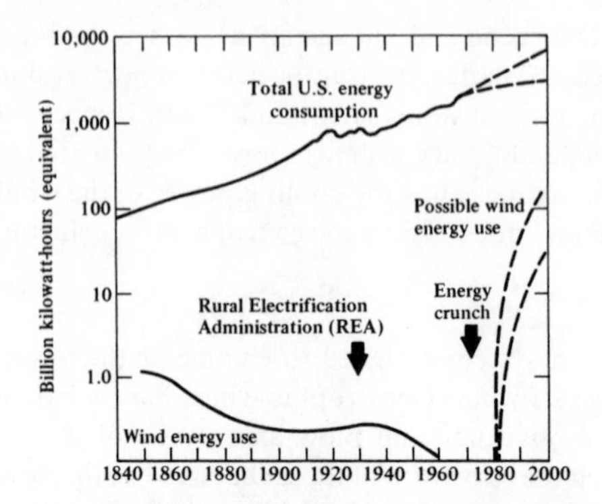


Comparison of wind energy and total U.S. energy consumption. (*Courtesy Wind Machines*)

**wind energy conversion (WEC).** A wide variety of machines have been designed to covert wind energy into useful power. The large scale WEC systems currently being developed utilize a horizontal axis design. This design will require siting in high average wind speed areas. They will require average wind speeds of 18 mph (29 km/hr) in order to be economically viable. (See Fig. p. 491)

**wind erosion.** The removal of material from the land or from buildings by the action of the wind. The mechanisms include straightforward picking up of dust by the airflow, and the dislodging or abrasion of surface material by the impact of particles already airborne.

**wind gage.** An anemometer for testing the velocity of the air in mines.

**wind generators.** Devices to extract energy from the wind to be used to generate electricity directly. (See Fig. p. 492)

**winding.** The operation of hoisting coal, ore, men, or materials in a shaft. The conventional system is to employ two cages actuated by a drum-type winding engine with steel ropes attached at either end of the drum, one over and the other under it, so that as one cage ascends the other descends, and they arrive at pit top and bottom simultaneously.

**wind load rating.** A specification used to indicate the resistance of a derrick to the force of wind. The typical wind resistance of derricks is 75 miles per hour with pipe standing in the derrick and 115 miles per hour and more with no pipe standing in the derrick.

Photograph of a 125-foot-tall, 100-kilowatt windmill developed for ERDA by NASA's Lewis Research Center in Cleveland, Ohio. (NASA Photo)

**wind measurement.** The measurement of wind may be achieved by the direct tracking of balloons, optically or by radar. Doppler radar may be used to determine the velocity of rain or other airborne objects, their horizontal motion being attributed to wind. The displacement of clouds as seen by satellite photography may be used, provided clouds can be satisfactorily identified as moving with the wind.

**windmill.** Machine with a rotor that is moved slowly by the wind to produce mechanical power, used originally to mill grain and pump water.

**windmill anemometer.** An anemometer in which a windmill is driven by the air stream, and its rotation transmitted through gearing to dials or other recording mechanisms. In some instruments the rotating vanes and dials are in the same plane (both vertical), and in others the dial is horizontal. In the windmill type, the operation of air measurement involves readings of the dials at the beginning and end of a measured period, and a watch or clock is required. Windmill instruments may be fitted with an extension handle providing a form of remote control, and may be used to measure the air speed in an otherwise inaccessible spot.

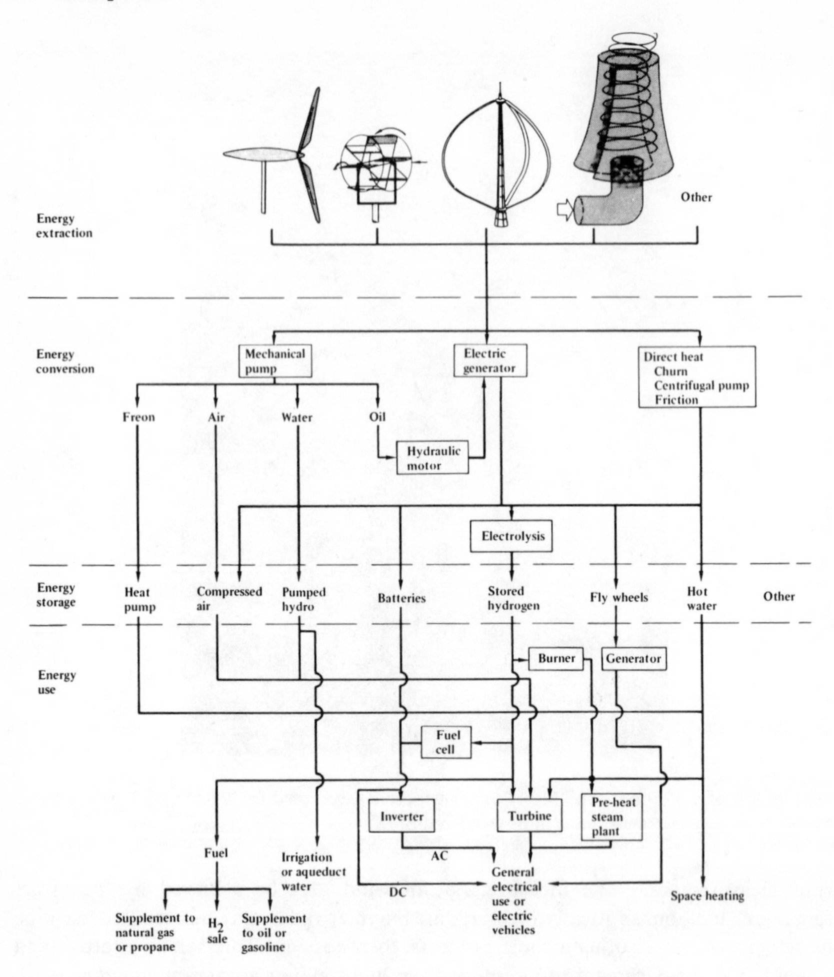


Energy from wind machines. (*Courtesy Wind Machines*)

**wind power.** A power source derived from the use of windmills which convert the wind's energy into electricity. The power contained in a moving wind stream is proportional to the first power of both the air density and the area of the wind stream and proportional to the third power of the wind velocity. (See Fig. p. 493)

**wind power plants.** Wind turbines supplying electric power to a grid.

**wind pressure.** The pressure on a structure due to wind, which increases with wind velocity approximately in accordance with the formula $p = 0.003v^2$, where p is the pressure in pounds per square feet of area affected and v is the wind velocity in miles per hour.

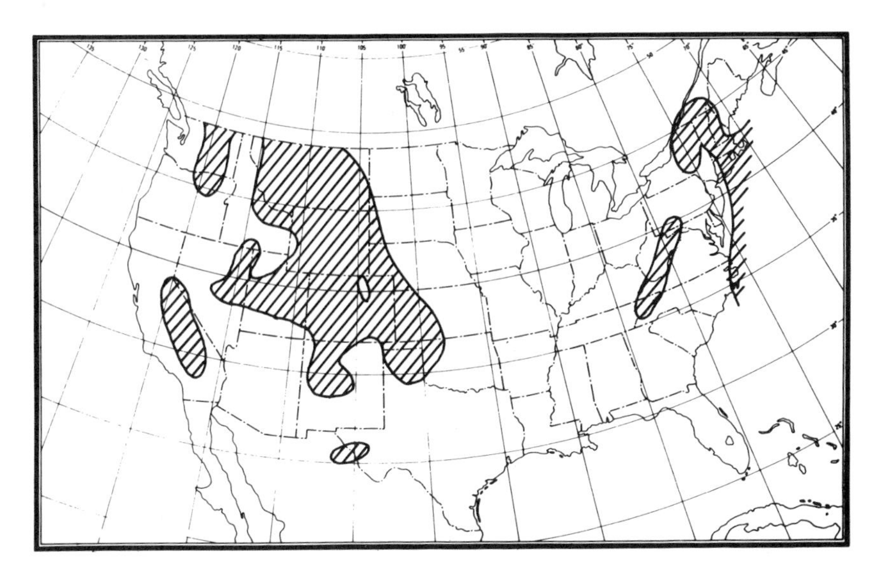

Areas in the U.S. where annual average wind speeds exceed 18 miles per hour at 150 feet elevation above ground level. (*Courtesy Wind Machines*)

**wind profile.** The variation of wind characteristics (e.g., mean speed, direction, turbulence level) with altitude.

**wind rose.** Diagrams designed to show the prevailing wind direction.

**wind shadow thermals.** The strength of the wind and the terrain determine the intensity of turbulence. Where the wind is reduced (e.g., on the lee side of a hill) the depth of air warmed by the ground heated in sunshine is less and the maximum temperature reached greater. Glider pilots find that intense thermal upcurrents rise from such places from time to time. A special case is a field of ripe wheat: the temperature within the crop is higher than in a green crop, and a gust of wind bending down the stalks releases a body of very warm air.

**wind shear.** In general, the rate of change of the wind vector with distance in a direction perpendicular to the wind direction. Usually, the term implies the change of wind with height, and near the ground it is often used to designate the vertical gradient of horizontal velocity.

**wind turbine generator.** A wind machine, powered by a rotating blade or propeller, that drives an electric generator. (See Fig. p. 494)

**wind variation.** The wind varies in time and place. In the wake of a building, tree or cliff it may fluctuate by 50 percent or more in a very few seconds and vary in direction by 180°. With the passage of storms and with time of day and season it varies over

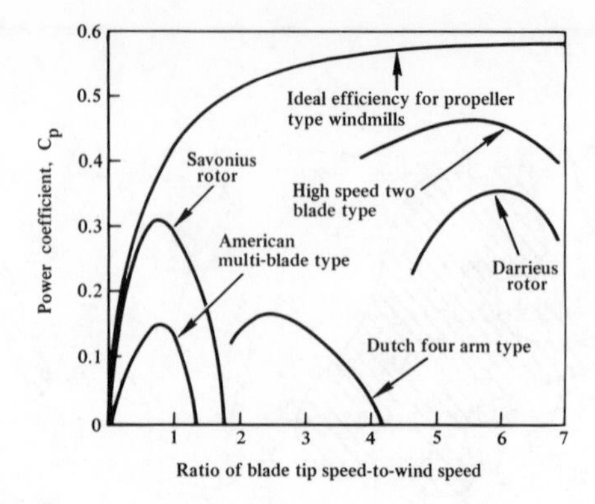


Comparison of typical performance of wind machines. (*Courtesy Wind Machines*)

minutes, hours, days, and months. Even one year may differ markedly from another at the same place. These variations have to be taken into account in defining "the wind," its average or mean, and in determining the exposure of the measuring instrument.

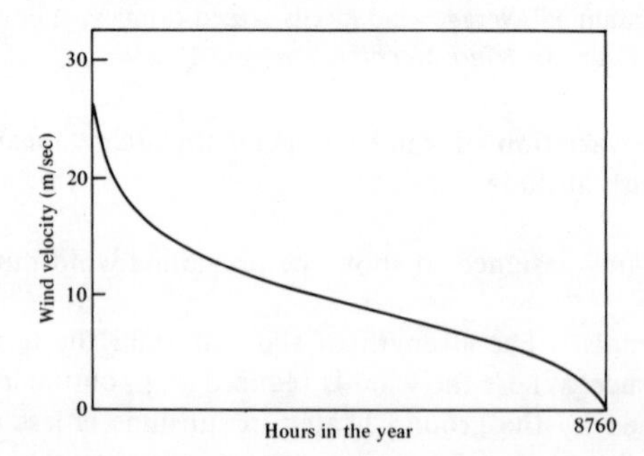


Wind velocity duration curve. (*Courtesy Van Nostrand's Scientific Encyclopedia*)

**Windworks.** Small, effective electric-wind generators, built by the members of a Wisconsin commune called "Windworks." Some wind plants designed by Hans Meyer cost less than $200 to build.

**Winkler process.** A fluidized bed gasifier process, which produces a low-Btu gas or a medium-Btu gas from a wide variety of coals. Coal is auger fed into the gasifier where it is gasified directly with steam and oxygen. Reactor conditions are atmospheric pressure and 816° to 982°C.

**winning.** The excavation, loading, and removal of coal or ore from the ground; winning follows development. The term also refers to that portion of a coalfield laid out for working.

**winter oil.** A heavy railway-car and engine oil which has a solidifying point below −29°C.

**winter peak.** The greatest load on an electric system during the demand interval in the winter or heating season.

**winze.** An interior mine shaft; a vertical or inclined opening or excavation connecting two levels in a mine.

**withdrawn gas.** Gas taken out of storage.

**Wobbe index.** A number indicating interchangeability of fuel gases. It is obtained by dividing the heating value of a gas by the square root of its specific gravity.

**wood-burning stoves.** Five of the basic stove types include: (a) simple box stove, (b) ashley airtight stove, (c) base-burning (Riteway) stove, (d) down-draft (Vermont) airtight stove, and (e) front-end (Scandinavian Jotul) combustion stove.

**wood gas.** Gas produced during production of charcoal by heating wood in the absence of air and usually used as a fuel at the production site.

**work.** Energy transferred from one body to another in such a way that a difference in temperature is not directly involved. Work is measured in terms of the amount of a force multiplied by the distance it has traveled. The practical unit is the foot-pound (force in pounds weight multiplied by distance in feet).

**workable.** A coal seam or ore body of such thickness, grade, and depth, as to make it a good prospect for development.

**working fluid.** Fluid in electrical generation plants that is heated by the energy source and then expands through the turbine without leaving the system.

**working pressure.** Normal operating gauge pressure in a device or system.

**workover.** Performance of one or more of a variety of remedial operations on a producing well with the hope of restoring or increasing production.

**worth.** As applied in the nuclear field, the reactivity attributable to a specified component, material, portion of material, or void, in a nuclear reactor or chain-reacting system.

# X

**xenon.** A very heavy, inert gas used in specialized electric lamps, present in air at about 0.05 parts per million. Chemical symbol Xe.

**xenon effect.** The decrease in the reactivity of a thermal reactor or chain-reacting system after a long period of power operation, due to the presence of large numbers of xenon-135 atoms generated in the fission products.

**x-ray.** A penetrating form of electromagnetic radiation emitted either when the inner orbital electrons of an excited atom return to their normal state (these are characteristic x-rays), or when a metal target is bombarded with high speed electrons (these are bremsstrahlung). X-rays are always nonnuclear in origin.

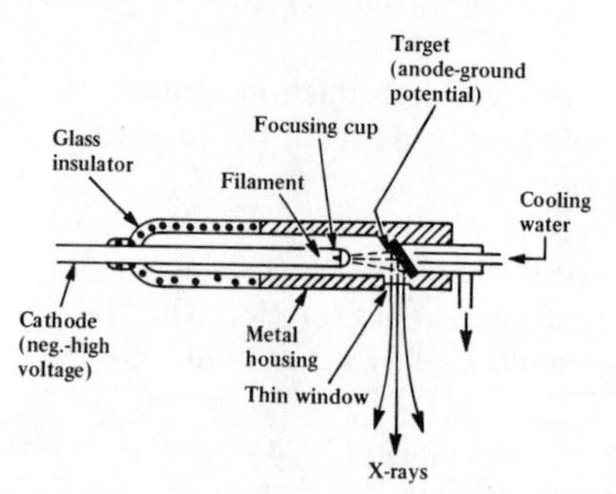


High-voltage, high-vacuum x-ray tube. (*Courtesy Van Nostrand's Scientific Encyclopedia*)

**xyloid coal.** Brown coal or lignite mostly derived from wood.

**xylovitrain.** Those coal constituents derived from lignified plant tissues and from which all structure has disappeared; also structureless vitrain.

**year.** Measure of time. Solar year (successive intervals between transits of first point of Aries) is 365.2422 mean solar days; civil year is 365.2425; and sidereal year 365.2564 days.

**yellowcake.** Product of the milling process in the uranium fuel cycle that contains 80 to 83 percent uranium oxide.

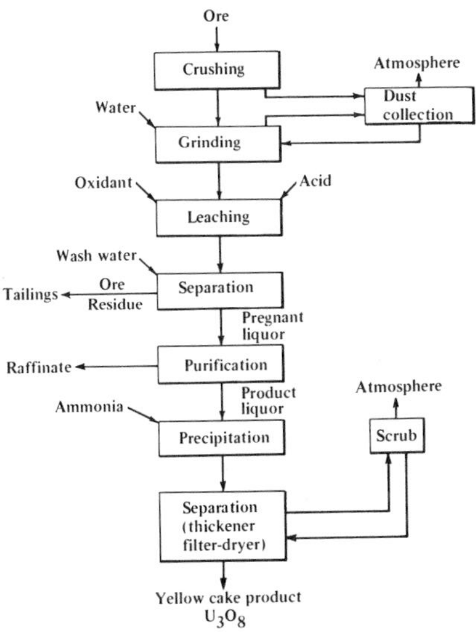


Milling plant for production of yellowcake. (*Courtesy U.S. Council on Environmental Quality*)

**yellow heat.** A division of the color scale, generally given as about 1090°C.

**yield.** The total energy released in a nuclear explosion. It is usually expressed in equivalent tons of TNT (the quantity of TNT required to produce a corresponding amount of energy). Low yield is generally considered to be less than 20 kilotons; low intermediate yield from 20 to 200 kilotons; and intermediate yield from 200 kilotons to 1 megaton. There is no standardized term to cover yields from 1 megaton upward.

**yield point.** The stress at which a material exceeds its elastic limit. Below this stress the material will recover its original size on removal of the stress; above this stress, it will not.

# Z

**Zeiss konimeter.** A portable dust-sampling instrument.

**zenith.** Point in celestial sphere directly above observer.

**zeolitic catalyst.** Since the early 1960s, modern cracking catalysts have contained a silica-alumina crystalline structured material called zeolite. This zeolite is commonly called a molecular sieve. The admixture of a molecular sieve with the base clay matrix imparts desirable cracking selectivities.

**zero gas.** Gas at atmospheric pressure.

**zero gas governor.** A gas pressure regulating device common to industrial combustion systems, used for controlling and maintaining varying inlet gas pressures to outlet atmospheric pressure.

**zero group.** The group of inert gases having a valence of 0 in the periodic system.

**zero-point energy.** Energy remaining in a substance at the absolute 0 of temperature.

**zero-power reactor.** An experimental reactor operated at such low power levels that a coolant is not needed and little radioactivity is produced.

**zinc.** A lustrous, bluish-white metallic element in group II of the periodic system. Chemical symbol Zn.

**zinc-bromine batteries.** An advanced battery system in early research stage. It is being developed under an ERDA-supported program.

**zinc-chlorine battery.** The zinc-chlorine battery operates at ambient temperatures and has an aqueous electrolyte. The reactants are zinc and chlorine gas. A unique feature of this system is the method used to store chlorine when the battery is charged. The chlorine gas is chilled in the presence of water, which is part of the electrolyte, to form an ice, known as chlorine hydrate. This solid form of storage provides added safety in case of accidental exposure. To discharge the battery, the chlorine ice is heated to revaporize the chlorine which is then consumed by reaction at one of the electrodes.

**Zinc Chloride Catalysts process.** A process for producing fuel oil and low-Btu gas. Coal is pulverized and slurried in the recycle oil. The slurry is fed to the reactor where hydrogen and molten zinc chloride are added at high reactor temperature and pressure. One-step hydrocracking takes place in the molten salt reactor. The distillate product is separated from gases. The spent catalyst is regenerated in the fluidized bed combustor and recycled. Fuel oil and low-Btu gas are produced. Reactor operating conditions are 371° to 441°C and 1500 to 3500 pounds per square inch absolute.

**zinc oxide.** An infusible white solid used in preparation of synthetic (or substitute) natural gas to absorb sulfur from naphtha.

**zircaloy.** An alloy of 98 percent Zr, 1.5 percent Sn, 0.35 percent Fe-Cr-Ni, and 0.15 percent O which optimizes the requirements of a low neutron cross section, reliable corrosion resistance in very hot water, and the strength required for cladding uranium fuel used in light water reactors.

**zone heat.** A central heating system which is arranged so that different temperatures can be maintained in two or more areas of the building being heated.

**zone refining.** A method used in refining germanium and silicon to produce the ultrapure elements used in making transistors.

# CONVERSION FACTORS

| MULTIPLY | BY | TO OBTAIN |
|---|---|---|
| atmospheres | 10,333 | kgs per sq meter |
| atmospheres | 14.70 | pounds per sq inch |
| Bars | $9.870 \times 10^{-1}$ | atmospheres |
| Bars | $10^{6}$ | dynes per sq cm |
| Bars | $1.020 \times 10^{4}$ | kgs per square meter |
| Bars | $1.450 \times 10$ | pounds per sq inch |
| British thermal units | 0.2530 | kilogram-calories |
| British thermal units | 777.5 | foot-pounds |
| British thermal units | $3.927 \times 10^{-4}$ | horsepower-hours |
| British thermal units | 1054 | joules |
| British thermal units | 107.5 | kilogram-meters |
| British thermal units | $2.928 \times 10^{-4}$ | kilowatt-hours |
| Btu per min | 0.02356 | horsepower |
| Btu per min | 0.01757 | kilowatts |
| centimeters | 0.3937 | inches |
| centimeters | 0.01 | meters |
| centimeters of mercury | 0.1934 | pounds per sq inch |
| centimeters per second | 1.969 | feet per minute |
| centimeters per second | 0.036 | kilometers per hour |
| centimeters per second | 0.02237 | miles per hour |
| cubic centimeters | $10^{-6}$ | cubic meters |
| cubic centimeters | $10^{-3}$ | liters |
| cubic feet | 1728 | cubic inches |
| cubic feet | 0.02832 | cubic meters |
| cubic feet | 7.481 | gallons |
| cubic feet | 28.32 | liters |
| cubic feet per minute | 0.1247 | gallons per sec |
| cubic feet per minute | 0.4720 | liters per sec |
| cubic inches | $5.787 \times 10^{-4}$ | cubic feet |
| cubic meters | 35.31 | cubic feet |
| cubic meters | 264.2 | gallons |
| cubic meters | $10^{3}$ | liters |
| decimeters | 0.1 | meters |
| degrees (angle) | 60 | minutes |
| degrees (angle) | 0.01745 | radians |
| dekameters | 10 | meters |
| ergs | $9.486 \times 10^{-11}$ | British thermal units |
| ergs | 1 | dyne-centimeters |
| ergs | $7.376 \times 10^{-8}$ | foot-pounds |
| ergs | $10^{-7}$ | joules |
| ergs | $2.390 \times 10^{-11}$ | kilogram-calories |
| ergs | $1.020 \times 10^{-8}$ | kilogram-meters |

| MULTIPLY | BY | TO OBTAIN |
|---|---|---|
| ergs per second | $1.341 \times 10^{-10}$ | horsepower |
| ergs per second | $10^{-10}$ | kilowatts |
| feet | 0.3048 | meters |
| feet per second | 18.29 | meters per minute |
| foot-pounds | $1.28 \times 10^{-3}$ | British thermal units |
| foot-pounds | $1.356 \times 10^{7}$ | ergs |
| foot-pounds | $5.050 \times 10^{-7}$ | horsepower-hours |
| foot-pounds | $3.241 \times 10^{-4}$ | kilogram-calories |
| foot-pounds | 0.1383 | kilogram-meters |
| foot-pounds | $3.766 \times 10^{-7}$ | kilowatt-hours |
| foot-pounds per minute | $1.286 \times 10^{-3}$ | Btu per minute |
| foot-pounds per minute | 0.01667 | foot-pounds per sec |
| foot-pounds per minute | $3.030 \times 10^{-5}$ | horsepower |
| foot-pounds per minute | $3.241 \times 10^{-4}$ | kg, calories per minute |
| foot-pounds per minute | $2.260 \times 10^{-5}$ | kilowatts |
| foot-pounds per second | $7.712 \times 10^{-2}$ | Btu per minute |
| foot-pounds per second | $1.818 \times 10^{-3}$ | horsepower |
| foot-pounds per second | $1.945 \times 10^{-2}$ | kg. calories per min |
| foot-pounds per second | $1.356 \times 10^{-3}$ | kilowatts |
| gallons | 3785 | cubic centimeters |
| gallons | 0.1337 | cubic feet |
| gallons | 231 | cubic inches |
| gallons | $3.785 \times 10^{-3}$ | cubic meters |
| gallons | $4.951 \times 10^{-3}$ | cubic yards |
| gallons | 3.785 | liters |
| gallons | 8 | pints (liq) |
| gallons | 4 | quarts (liq) |
| gallons per minute | $2.228 \times 10^{-3}$ | cubic feet per second |
| gallons per minute | 0.06308 | liters per second |
| grams | $10^{-3}$ | kilograms |
| grams | $10^{3}$ | milligrams |
| grams | 0.03527 | ounces |
| grams | 0.03215 | ounces (troy) |
| grams | $2.205 \times 10^{-3}$ | pounds |
| grams centimeters | $9.297 \times 10^{-8}$ | British thermal units |
| grams per cc | 62.43 | pounds per cubic foot |
| grams per cc | 0.03613 | pounds per cubic inch |
| grams per cc | $3.405 \times 10^{-7}$ | pounds per mil-foot |
| horse-power | 42.44 | Btu per min |
| horse-power | 33,000 | foot-pounds per min |
| horse-power | 550 | foot-pounds per sec |
| horse-power | 1.014 | horsepower (metric) |
| horse-power | 10.70 | kg. calories per min |
| horse-power | 0.7457 | kilowatts |
| horse-power | 745.7 | watts |

| MULTIPLY | BY | TO OBTAIN |
|---|---|---|
| horse-power-hours | 2547 | British thermal units |
| horse-power-hours | $1.98 \times 10^6$ | foot-pounds |
| horse-power-hours | 641.7 | kilogram-calories |
| horse-power-hours | $2.737 \times 10^5$ | kilogram-meters |
| horse-power-hours | 0.7457 | kilowatt-hours |
| inches of mercury | 0.03342 | atmospheres |
| inches of mercury | 1.133 | feet of water |
| inches of mercury | 345.3 | kgs per square meter |
| inches of mercury | 70.73 | pounds per square ft |
| inches of mercury | 0.4912 | pounds per square in |
| inches of water | 0.002458 | atmospheres |
| inches of water | 0.07355 | inches of mercury |
| inches of water | 25.40 | kgs per square meter |
| inches of water | 0.5781 | ounces per square in |
| inches of water | 5.204 | pounds per square ft |
| inches of water | 0.03613 | pounds per square in |
| kilograms | $10^3$ | grams |
| kilograms | 2.2046 | pounds |
| kilograms | $1.102 \times 10^{-3}$ | tons (short) |
| kilogram-calories | 3.968 | British thermal units |
| kilogram-calories | 3086 | foot-pounds |
| kilogram-calories | $1.558 \times 10^{-3}$ | horsepower-hours |
| kilogram-calories | 4183 | joules |
| kilogram-calories | 426.6 | kilogram meters |
| kilometers | $10^5$ | centimeters |
| kilometers | 3281 | feet |
| kilometers | $10^3$ | meters |
| kilometers | 0.6214 | miles |
| kilometers | 1093.6 | yards |
| kilometers per hour | 27.78 | centimeters per sec |
| kilometers per hour | 54.68 | feet per minute |
| kilometers per hour | 0.9113 | feet per second |
| kilometers per hour | 0.5396 | knots per hour |
| kilometers per hour | 16.67 | meters per minute |
| kilometers per hour | 0.6214 | miles per hour |
| kms per hour per sec | 27.78 | cms per sec per sec |
| kms per hour per sec | 0.9113 | ft per sec per sec |
| kms per hour per sec | 0.2778 | meters per sec per sec |
| kms per hour per sec | 0.6214 | miles per hr per sec |
| kilometers per min | 60 | kilometers per hour |
| kilowatts | 56.92 | Btu per min |
| kilowatts | $4.425 \times 10^4$ | foot-pounds per min |
| kilowatts | 737.6 | foot-pounds per sec |
| kilowatts | 1.341 | horsepower |
| kilowatts | 14.34 | kg. calories per min |

| MULTIPLY | BY | TO OBTAIN |
|---|---|---|
| kilowatts | $10^3$ | watts |
| kilowatt-hours | 3,412 | British thermal units |
| kilowatt-hours | $2.655 \times 10^6$ | foot-pounds |
| kilowatt-hours | 1.341 | horsepower-hours |
| kilowatt-hours | $3.6 \times 10^6$ | joules |
| kilowatt-hours | 860.5 | kilograms-calories |
| kilowatt-hours | $3.671 \times 10^5$ | kilogram-meters |
| meters | 100 | centimeters |
| meters | 3.2808 | feet |
| meters | 39.37 | inches |
| meters | $10^{-3}$ | kilometers |
| meters | $10^3$ | millimeters |
| meters | 1.0936 | yards |
| meter-kilograms | $9.807 \times 10^7$ | centimeter-dynes |
| meter-kilograms | $10^5$ | centimeter-grams |
| meter-kilograms | 7.233 | pound-feet |
| meters per minute | 1.667 | centimeters per sec |
| meters per minute | 3.281 | feet per minute |
| meters per minute | 0.05468 | feet per second |
| meters per minute | 0.06 | kilometers per hour |
| meters per minute | 0.03728 | miles per hour |
| meters per second | 196.8 | feet per minute |
| meters per second | 3.281 | feet per second |
| meters per second | 3.6 | kilometers per hour |
| meters per second | 0.06 | kilometers per min |
| meters per second | 2.237 | miles per hour |
| meters per second | 0.03728 | miles per minute |
| meters per sec per sec | 3.281 | feet per sec per sec |
| meters per sec per sec | 3.6 | kms per hour per sec |
| meters per sec per sec | 2.237 | miles per hour per sec |
| miles | $1.609 \times 10^5$ | centimeters |
| miles | 5280 | feet |
| miles | 1.6093 | kilometers |
| miles | 1760 | yards |
| miles per minute | 88 | feet per second |
| miles per minute | 1.6093 | kilometers per minute |
| miles per minute | 0.8684 | knots per minute |
| miles per minute | 60 | miles per hour |
| ounces | 8 | drams |
| ounces | 437.5 | grains |
| ounces | 28.35 | grams |
| ounces | 0.0625 | pounds |
| ounces per square inch | 0.0625 | pounds per sq inch |
| pounds | 444,823 | dynes |
| pounds | 7000 | grains |

| MULTIPLY | BY | TO OBTAIN |
|---|---|---|
| pounds | 453.6 | grams |
| pounds of water | 0.01602 | cubic feet |
| pounds of water | 27.68 | cubic inches |
| pounds of water | 0.1198 | gallons |
| Quadrants (angle) | 90 | degrees |
| Quadrants (angle) | 5400 | minutes |
| Quadrants (angle) | 1.571 | radians |
| Radians | 57.30 | degrees |
| Radians | 3438 | minutes |
| radians per second | 57.30 | degrees per second |
| radians per second | 0.1592 | revolutions per second |
| revolutions | 360 | degrees |
| revolutions | 4 | quadrants |
| revolutions | 6.283 | radians |
| revolutions per minute | 6 | degrees per second |
| square feet | 144 | square inches |
| square feet | 0.09290 | square meters |
| square feet | $3.587 \times 10^{-8}$ | aquare miles |
| square feet | 1/9 | square yards |
| square meters | $2.471 \times 10^{-4}$ | acres |
| square meters | 10.764 | square feet |
| square meters | $3.861 \times 10^{-7}$ | square miles |
| square meters | 1.196 | square yards |
| square miles | 640 | acres |
| square miles | $2.788 \times 10^{7}$ | square feet |
| square miles | 2.590 | square kilometers |
| square miles | $3.098 \times 11^{6}$ | square yards |
| square yards | $2.066 \times 10^{-4}$ | acres |
| square yards | 9 | square feet |
| temp (degs C) + 273 | 1 | abs temp (degs K) |
| temp (degs C) + 17.8 | 1.8 | temp (degs F) |
| temp (degs F) + 460 | 1 | abs temp (degs R) |
| temp (degs F) − 32 | 5/9 | temp (degs C) |
| tons (long) | 1016 | kilograms |
| tons (long) | 2240 | pounds |
| tons (metric) | $10^{3}$ | kilograms |
| tons (metric) | 2205 | pounds |
| tons (short) | 907.2 | kilograms |
| tons (short) | 2000 | pounds |
| tons (short) per sq ft | 9765 | kgs per square meter |
| tons (short) per sq ft | 13.89 | pounds per sq inch |
| tons (short) per sq in | $1.406 \times 10^{6}$ | kgs per square meter |
| tons (short) per sq in | 2000 | pounds per sq inch |

# GLOSSARY

**ACES** Annual Cycle Energy System
**ACS** American Chemical Society
**ADP** Atmospheric Dew Point
**AEC** Atomic Energy Commission
**AGA** American Gas Association
**AHAM** Association of Home Appliance Manufacturers
**AIMME** American Institute of Mining and Metallurgical Engineers
**AMP** American Melting Point
**ANSI** American National Standards Institute
**ASHRAE** American Society of Heating, Refrigerating, and Air Conditioning Engineers

**BB** Billions of Barrels
**BCF/CD** Billion Standard Cubic Feet per Calendar Day
**BCF/Y** Billion Standard Cubic Feet per Year
**BNL** Brookhaven National Laboratory
**BOD** Biochemical Oxygen Demand
**BOM** Bureau of Mines
**BWR** Boilingr Water Reactor

**CARI** Council of Air Conditioning and Refrigeration Industry
**CEQ** Council on Environmental Quality
**CIP** Cascade Improvement Program
**CLS** Class Life System
**CNG** Compressed Natural Gas
**COED** Char-Oil Energy Development
**COP** Coefficient of Performance
**CRBR** Clinch River Breeder Reactor
**CS** DOE Conservation and Solar Applications Division
**CRG** Catalytic Rich Gas
**CUP** Cascade Uprating Program

**DOE** Department of Energy
**DPU** Department of Public Utilities

**EBR** Experimental Breeder Reactor
**ECCS** Emergency Core Cooling System
**ECS** Environmental Control Systems
**ECPA** The Energy Conservation and Production Act
**E Cube** Energy Conservation Utilizing Better Engineering
**EEI** Edison Electric Institute
**EIS** Environmental Impact Statement
**EIA** DOE Energy Information Administration Division
**EPA** Environmental Protection Agency
**EPCA** The Energy Policy and Conservation Act

**FFTF** Fast Flux Test Facility
**FOB** Free On Board
**FPC** Federal Power Commission
**FRC** Future Requirements Committee
**FY** Fiscal Year

**GAMA** Gas Appliance Manufacturers Association
**GATE** Group to Advance Total Energy
**GCBR** Gas Cooled Fast Breeder Reactor
**GDP** Gross Domestic Product
**GNP** Gross National Product

**HEU** Highly Enriched Uranium
**HPFL** High Performance Fuels Laboratory
**HTGR** High-Temperature Gas-Cooled Reactor

**HVAC** Heating, Ventilating, and Air Conditioning

**ICOP** Imported Crude Oil Processing
**INGAA** Interstate Natural Gas Association of America
**IOCC** Interstate Oil Compact Commission
**IPAA** Independent Petroleum Association of America

**LDC** Less developed Countries
**LLL** Lawrence Livermore Laboratory
**LMFBR** Liquid Metal Fast Breeder Reactor
**LNG** Liquefied Natural Gas
**LOFT** Loss of Fluid Test (LOFT) Facility. Also Loss of Fluid Test (LOFT) Program
**LPA** Liquid Petroleum Air
**LPG** Liquefied Petroleum Gas
**LPGA** Liquefied Petroleum Gas Association
**LWBR** Light Water Breeder Reactor
**LWR** Light-Water Reactor

**MAOP** Maximum Allowable Operating Pressure
**MBA** Material Balance Area
**MB/CD** Thousands Barrels per Calender Day
**MB/D** Thousand Barrels per day
**MMBD** Millions of Barrels per day
**MMST/YR** Millions of Short Tons per Year
**MED** Minimum Energy Dwelling
**MHD** Magnetohydrodynamics
**MOU** Memorandums of Understanding
**MRG** Methane Rich Gas
**MSBR** Molten Salt Breeder Reactor
**MSL** Mean Sea Level
**MST** Mean Solar Time. Also, Mountain Standard Time

**NARUC** National Association of Regulatory Utility Commissioners
**NASA** National Aeronautics and Space Administration
**NBS** National Bureau of Standards
**NCA** National Coal Association
**NEC** National Electric Code
**NEIC** National Energy Information Center
**NEPA** National Environmental Policy Act
**NGL** Natural Gas Liquids
**NLPGA** National LP-Gas Association
**NMMSS** Nuclear Materials Management and Safeguards System
**NMRAS** Nuclear Materials Reports and Analysis System
**NPC** National Petroleum Council
**NPT** Non-Proliferation Treaty
**NTIS** National Technical Information Service

**OAPEC** Organization of Arab Petroleum Exporting Countries
**OCR** Office of Coal Research
**OCS** Outer Continental Shelf
**OEP** Office of Emergency Planning
**OIP** Oil-In-Place
**OPEC** Organization of Petroleum Exporting Countries
**OPS** Office of Pipeline Safety, a branch of the Department of Transportation
**ORMAK** Oak Ridge Tokamak
**OSHA** Occupational Safety and Health Act
**OTEC** Ocean Thermal Energy Conversion
**OWPS** Offshore Windpower System

**PGA** Purchased Gas Adjustment
**PLBR** Prototype Large Breeder Reactor

**PUC** Public Utility Commissions
**PUD** Public Utility District
**PWR** Pressurized Water Reactor

**RDT&E** Research, Development, Test, and Evaluation

**SEFOR** Southwest Experimental Fast Oxide Reactor
**SERI** Solar Energy Research Institute
**SIC** Standard Industrial Classification
**SMYS** Specified Minimum Yield Strength
**SNG** Synthetic Natural Gas. Also: Supplemental Natural Gas, Substitute Natural Gas
**SNM** Special Nuclear Material(s)
**SPE** Society of Petroleum Engineers
**SPQ** Synthetic Pipeline Quality Gas
**SWU** Selective Work Units
**SYD** Sum-Of-The Years Digits

**TARGET** Team to Advance Research for Gas Energy Transformation
**TEEC** Total Energy and Environmental Conditioning
**TERA** Total Energy Resource Analysis
**TFTR** Tokamak Fusion Test Reactor
**TIME** Total Industry Marketing Effort
**TSC** Thermal Stress Cracking
**TVA** Tennessee Valley Authority

**UL** Underwriters' Laboratories, Inc.
**UMWA** United Mine Workers of America
**USGS** United States Geological Survey

**VLCC** Very Large Crude Carriers

**WECS** Wind Energy Conversion System

# BIBLIOGRAPHY

## ENERGY—GENERAL


*_A National Plan for Energy Research, Development and Demonstration: Creating Energy Choices for the Future._ Vol. 1: The Plan. Washington: Energy Research and Development Administration, 1976.

Ayers, Robert U., and McKenna. _Alternatives to the Internal Combustion Engine._ Baltimore: Johns Hopkins Press, 1972.

Brubaker, Sterling. _In Command of Tomorrow: Resource and Environmental Strategies for Americans._ Baltimore: Johns Hopkins Press, 1975.

*Clark, Wilson. _Energy for Survival—The Alternative to Extinction._ New York: Anchor Book, Anchor Press/Doubleday, 1975.

Congressional Quarterly, Inc. _Energy Crisis in America._ Washington, 1973.

*Considine, Douglas M. (ed.). _Van Nostrand's Scientific Encyclopedia._ 5th ed. New York: Van Nostrand Reinhold Company, 1976.

*_Criteria for Energy Storage R&D._ Washington: The National Research Council, National Academy of Sciences, 1976.

Eaton, William W. _Energy Storage._ Washington: Energy Research and Development Administration, 1975.

*_Energy Microthesaurus._ Springfield, Va.: U.S. Department of Commerce, National Technical Information Service, 1976.

*_Energy Planning Report._ Vol. 1, No. 4. Washington: Resources News Service, Inc., 1977.

Energy Policy Project of the Ford Foundation. _A Time to Choose: America's Energy Future._ Cambridge, Mass.: Ballinger Publishing Co., 1974.

Federal Energy Administration. _Project Independence Report._ Washington, 1974.

Fischer, John C. _Energy Crisis in Perspective._ New York: Wiley-Interscience, 1974.

Gay, Larry. _The Complete Book of Heating with Wood._ Charlotte, Vt.: Garden Way, 1974.

Gregory, Derek P. The hydrogen economy. _Scientific American_ **228**:13–21 (January, 1973).

*Hammond, Allen L.; Metz, William D.; and Maugh II, Thomas H. _Energy and the Future._ Second Printing. Washington: American Association for the Advancement of Science, 1973.

*Hampel, Clifford A., and Hawley, Gessner G. _Glossary of Chemical Terms._ New York: Van Nostrand Reinhold Co., 1976.

*Hogben, Lancelot. _The Wonderful World of Energy._ New York: Doubleday and Company, Inc., 1968.

Holdren, John P., and Herrera, Philip. _Energy: A Crisis in Power._ San Francisco: Sierra Club, 1971.

*Hollander, Jack M. (ed.). _Annual Review of Energy._ Vol. 1. Palo Alto, Ca.: Annual Reviews Inc., 1976.

*_Information from ERDA Reference Information._ Special Issue. Washington: Energy Research and Development Administration, Office of Public Affairs, 1977.


*Asterisk indicates source from which a select number of terms and definitions were obtained.

*Information from ERDA Weekly Announcements. Vol. 3, Nos. 8, 9, 10, and 11. Washington: Energy Research and Development Administration, Office of Public Affairs, 1977.

Maddox, John. *Beyond the Energy Crisis: A Global Perspective*. New York: McGraw Hill, 1975.

*Marc Reisner (ed.). *NRDC Newsletter*. Vol. 5, Issues 1 and 2. New York: Natural Resources Defense Council, Inc., 1976.

Martell, Charles L. *Batteries and Storage Systems*. New York: McGraw Hill, 1970.

*Master Plan: ERDA-76/122. Washington: Energy Research and Development Administration, Division of Safeguards and Security, 1976.

*Popular Science. New York: Times Mirror Magazine, Inc. July, Aug., Sept., and Oct. 1975; Jan., Mar., Apr., May, Aug., Sept., Oct., and Dec., 1976; Jan., Feb., and Mar., 1977.

Portola Institute. *Energy Primer*, 1974.

Premis, John (ed.). *Energybook #1*. Philadelphia: Running Press, 1975.

Ridgeway, James. *The Last Play: The Struggle to Monopolize the World's Energy Resources*. New York: E. P. Dutton, 1973.

Rocks, Lawrence and Runyon, Richard P. *The Energy Crisis*. New York: Crown Publishers, 1972.

*Rowse, Arthur E. (ed.). *Help: The Useful Almanac: 1976-77*. Washington: Consumer News Inc.

*Science and Public Policy Program, University of Oklahoma, Norman, Oklahoma. *Energy Alternatives, A Comparative Analysis*. Washington: Council on Environmental Quality, 1975.

*Science Policy Research Division. *Energy Facts II*. Prepared for the Subcommittee on Energy Research, Development, and Demonstration, of the Committee on Science and Technology, U.S. House of Representatives Ninety-Fourth Congress, First Session. Washington: U.S. Government Printing Office, 1975.

*Senator Abraham Ribicoff. *Petropolitics and the American Energy Shortage*. Report to the Committee on Government Operations, United States Senate, 93rd Congress, 1st Session Committee Print. Washington: U.S. Government Printing Office, 1973.

*Statistical Committee of the Edison Electric Institute. *Glossary of Electric Utility Terms*. New York: Edison Electric Institute.

Stoner, Carol Hupping (ed.). *Producing Your Own Power*. Rolling Hills, Ca.: Vintage, 1975.

*Thirring, Hans. *Energy for Man—From Windmills to Nuclear Power*. Second Ed. New York: Harper Colophon Books, Harper & Row Publishers, Inc., 1976.

## ENERGY CONSERVATION

*Arbuckle, et al. *Environmental Law Handbook*. 4th ed. Washington: Government Institutes, Inc., 1976.

Belt, Forrest H. *Easi-Guide to Conserving Energy and Materials*. Indianapolis: Howard W. Sams, 1974.

Citizen's Advisory Committee on Environmental Quality. *Citizen Action Guide to*

*Energy Conservation.* Washington: Citizen's Advisory Committee on Environmental Quality, 1700 Pennsylvania Ave., N.W., Washington, D.C. 20006, 1973.

Clagg, Peter. *New Low-Cost Sources of Energy for the Home.* Charlotte, Vermont: Garden Way, 1975.

Clark, Wilson. *Energy for Survival: The Alternative to Extinction.* Garden City: Anchor Press, 1974.

**Common Environmental Terms, A Glossary.* Washington: U.S. Department of Commerce, 1974.

Davis, Albert J., and Schubert, Robert P. *Alternative Natural Energy Sources in Building Design.* Blacksburg, Va.: Passive Energy Systems, 1974.

Dubin, Fred. Energy for architects. *Architecture Plus* **1** (No. 6):38–49, 74–75 (July 1973).

**Environmental Information Summaries C-3.* Rockville, Md.: U.S. Department of Commerce National Oceanic and Atmospheric Administration, Oct. 1976.

Federal Energy Administration. *Tips for Energy Savers.* Washington: U.S. Government Printing Office.

Fond, K.W., et al. (eds.). *Efficient Use of Energy.* New York: American Institute of Physics, 1975.

*Fowler, John M. *Energy Environment Source Book.* Washington: National Science Teachers Association, 1975.

Griffin, Charles William. *Energy Conservation in Buildings: Techniques for Economical Design.* Washington: Construction Specifications Institute, 1974.

Hirst, Eric. *Energy Consumption for Transportation in the U.S., (ORNL-NSF-EP-*15). Oak Ridge: Oak Ridge National Laboratory.

Hittman Associates. Residential Energy Consumption, Single Family Housing: *Final Report.* Washington: U.S. Department of Housing and Urban Development, 1973.

Large, David B. (ed.). *Hidden Waste.* Washington: Conservation Foundation, 1973.

**Minimum Energy Dwelling.* Washington: Energy Research and Development Administration Office of Conservation Division of Buildings and Community Systems.

*Pennsylvania Department of Education. *The Environmental Impact of Electrical Power Generation: Nuclear and Fossil*; ERDA-69. Prepared for the Energy Research and Development Administration, Division of Biomedical and Environmental Research, Washington, D.C.

*Phillip J. Berardelli (ed.). *Energy, Intelligence & Analysis for Energy Consumers.* Vol. 1, No. 7. Washington: Peter S. Nagan, 1977.

Portola Institute. *Energy Primer: Solar, Water, Wind, and Biofuels.* Menlo Park, California: Portola Institute, 1974.

Smith, Thomas W. *Household Energy Game.* University of Wisconsin: Johns Hopkins Marine Studies Center, 1974.

## SOLAR ENERGY

Adelson, E. H. *Solar Air Conditioning and Refrigeration.* Isotech Research Labs., 1975.

American Institute of Aeronautics and Astronautics. *Solar Energy for Earth.* New York: American Institute of Aeronautics and Astronautics, 1975.

Brinkworth, Brian Joseph. *Solar Energy for Man.* New York: John Wiley, 1973.

Chalmers, Bruce. The photovoltaic generation of electricity. *Scientific American* **235** (4):34–44 (October 1976).

Clark, Wilson. *Energy for Survival: The Alternative to Extinction.* Garden City: Anchor Press, 1977.

*Corliss, William R. *Direct Conversion of Energy.* Oak Ridge: U.S. Atomic Energy Commission, Office of Information Services, 1964.

*Daniels, Farrington. *Direct Use of the Sun's Energy.* New York: Ballantine Books, Division of Random House, Inc., 1964.

Duffie, John A., and Beckman, William A. *Solar Energy Thermal Process.* New York: John Wiley, 1974.

Duffie, John A., and Beckman, William A. Solar heating and cooling. *Science,* **191:** 143–149 (January 16, 1976)

*Eaton, William W. *Solar Energy.* Washington: Energy Research and Development Administration, Office of Public Affairs, 1976.

Energy Research and Development Administration. *Solar Energy.* Oak Ridge: ERDA—Technical Information Center, P.O. Box 62, Oak Ridge, Tenn. 37830.

Federal Energy Administration. *Buying Solar.* Washington: U.S. Government Printing Office, June 1976.

Free, Jon F. Solar cells: when will you plug into electricity from sunshine? *Popular Science* **205:**52–55, 230. (December 1974).

Gervais, Robert L., and Piet, B. Solar thermal electric power. *Astronautics and Aeronautics*:38–45 (November 1975).

Glaser, Peter E. Beyond nuclear power—the large-scale use of solar energy. New York Academy of Sciences. *Transactions*, ser. 2, **31** (No. 8):951–967 (December 1969).

Grey, J. Solar heating and cooling. *Astronautics and Aeronautics*:33–37 (November 1975).

*Halacy D. S. *The Coming Age of Solar Energy.* New York: Avon Books, Harper & Row, Publishers, 1975.

Hoke, John. *Solar Energy.* New York: F. Watts, 1968.

Keyes, John. *Harvesting the Sun to Heat Your House.* New York: Morgan & Morgan, 1975.

*Lucas, Ted. *How to Build a Solar Heater.* 3d ed., Pasadena, Ca.: Ward Ritchie Press, 1975.

McCaull, J. Storing the sun. *Environment* **18:**9–15 (June 1976).

Meinel, Aden B., and Meimel, Marjorie P. Physics looks at solar energy. *Physics Today:* 44–50 (February 1972).

Rankins II, William H., and Wilson, David A. *Practical Sun Power.* Lorien House, 1974.

Rau, Hans. *Solar Energy.* Translated by Maxim Schur. Edited and revised by D. J. Duffin. New York: Macmillan, 1964.

Russell, Charles R. Solar energy. *Elements of Energy Conversion.* Oxford, New York: Pergamon Press, 1967.

*Skurka, Norma, and Naar, Jon. *Living with Natural Energy, Design for a Limited Planet*. 1st ed. New York: Ballantine Books, Division of Random House, Inc., 1976.

**Solar Energy for Space Heating & Hot Water*. Washington: Energy Research and Development Administration, Division of Solar Energy, 1976.

**Solar Energy Update*, SEU 77-1 Oak Ridge: Energy Research and Development Administration, Technical Information Center, 1977.

Stoner, Carol (ed.). *Producing Your Own Power: How to Make Nature's Energy Sources Work for You*. Emmaus, Pa.: Rodale Press, Book Division, 1974.

**Sun Language*. Washington: Solar Energy Institute of America.

Williams, James Richard. *Solar Energy: Technology and Applications*. Ann Arbor: Ann Arbor Science Publishers, 1974.

## GEOTHERMAL ENERGY

Barnea, Joseph. Geothermal power. *Scientific American* **226**:70–77 (January 1972).

Eaton, William W. *Geothermal Energy*. Oak Ridge: USERDA, Technical Information Center, P.O. Box 62, Oak Ridge, Tenn. 37830, 1975.

*Energy and Power*. San Francisco: W. H. Freeman [c1971].

*Geothermal energy—the hot prospect. *EPRI Journal* **2**(3):6–13 (1977).

Goldsmith, M. *Geothermal Resources in California: Potentials and Problems*. Pasadena: Environmental Quality Laboratory, California Institute of Technology, December 1971.

Hammond, Allen L. Geothermal energy: an emerging major resource. *Science* **177**: 978–980 (September 15, 1972).

Henahan, John F. Geothermal energy—the prospects get hotter. *Popular Science* **207**:96–99 (November 1974).

Henahan, John F. Full steam ahead for geothermal energy. *New Scientist* **57** (No. 827):16–17 (January 4, 1973).

Kruger, Paul, and Otte, C. (eds.). *Geothermal Energy—Resources, Production, Stimulation*. Stanford, Stanford University Press, 1973.

Muffler, L. J. P., and White, D. E. Geothermal energy. *The Science Teacher* **39**:40–44 (March 1972).

Rex, Robert. Geothermal energy—the neglected energy option. *Bulletin of the Atomic Scientists* **27** (No. 8):52–56 (October 1971).

U.S. Congress, Senate, Committee on Interior and Insular Affairs, *Geothermal Energy Resources and Research*. Hearings pursuant to S. Res. 45; a national fuels and energy policy study, 92nd Congress, 2d Session. June 15 and 22, 1972. Washington: U.S. Government Printing Office, 1972.

Weaver, Kenneth F. The search for tomorrow's power. *National Geographic* **142** (No. 5):650–681 (November 1972).

## WIND ENERGY

Blackwell, B. F., and Feltz, L. V. *Wind Energy—A Revitalized Pursuit*, (SAND-75-0166). Livermore, California: Sandia Laboratories, March 1975.

Carter, Joe. Wind power for the people. *Environment Action Bulletin* **6** (No. 12): 4–5 (June 14, 1975).

Clark, Wilson. Energy from the winds. *Energy for Survival.* New York: Doubleday, 1974.

*Eldridge, Frank R. *Wind Machines.* Prepared for the National Science Foundation. Washington: U.S. Government Printing Office, 1976.

Golding, Edward. *The Generation of Electricity By Wind Power.* London: E. & F. N. Spon, 1955.

Hackleman, Michael A. *Wind and Windspinners: A Nuts and Bolts Approach to Wind-Electric Systems.* Sangus, California: Earthmind, 1974.

Hamilton, R. Can we harness the wind? *National Geographic* **148**:812–829 (December 1975).

*Handbook of Homemade Power.* By the staff of the Mother Earth News, New York: Bantam Books, 1974.

*Hunt, V. Daniel. *Wind Power—A Handbook on Wind Energy Conversion Systems,* Book Manuscript, Burke, VA. 1978.

Inglis, David R. Wind Power now. *Bulletin of the Atomic Scientists* **31** (8):20–26 (October 1975).

Johnson, C.C. et al. Wind power development and applications. *Power Engineering* **78** (No. 10):50–53 (October 1974).

Portola Institute. *Energy Primer: Solar, Water, Wind and Biofuels.* Menlo Park, Portola Institute, 1974.

Putnam, Palmer Cosslett. *Power from the Wind.* New York: Van Nostrand Reinhold, 1974.

Putnam, Palmer Cosslett. *Energy in the Future.* New York: Van Nostrand, 1953.

Reynolds, John. *Windmills & Watermills.* New York: Praeger, 1970.

Simmons, Daniel M. *Wind Power.* Park Ridge, N.J.: Noyes Data Corp., 1975.

Sorensen, Bent. Energy and resources. *Science* **189** (No. 4199):255–260 (July 25, 1975).

Soucie, Gary. Pulling power out of thin air. *Audubon* **76**:81–88 (May 1974).

Steadman, Philip. *Energy, Environment and Building.* London, New York: Cambridge University Press, 1975.

Stokhuyzen, Frederick. *The Dutch Windmill.* Translated from the Dutch by Carry Dikshourn. New York: Universe Books, 1963.

Stoner, Carol. *Producing Your Own Power; How to Make Nature's Energy Source Work For You.* Emmaus, Pa.: Rodele Press, Book Division, 1974.

Torrey, Volta W. Windmills in the history of technology. *Technology Review* **77** (No. 5):8–10 (March/April 1975).

Wolff, Alfred R. *The Windmill as a Prime Mover.* New York: J. Wiley & Sons, 1885.

## FOSSIL FUEL ENERGY

**A Guide to Federal Power Commission Public Information.* Enclosure No. 020137. Washington: Federal Power Commission, Office of Public Information.

*Annual Review of Energy*, Vol. 1., Palo Alto, California: Annual Reviews, Inc., 1976.

Anthony Sampson. *The Seven Sisters.* New York: Viking Press, 1976.

Bakulev, G. D. *An Economic Analysis of Underground Gasification of Coal.* Washington: U.S. Dept. of the Interior, Bureau of Mines, 1962.

*Bituminous Coal Research, Inc. *Glossary of Surface Mining and Reclamation Technology.* Edited by Council for Surface Mining and Reclamation Research in Appalachia. Washington: National Coal Association, 1974.

**Caterpillar Purchasing Guide*, AEC Q 9019-01. Peoria, Ill.: Caterpillar Tractor Co.

Chase, Victor D. Rusty iron may be the key to cheaper gas from coal. *Popular Science* **210**:91–94 (January 1977).

**Coal, Conversion and Utilization.* Washington: Energy Research and Development Administration, Office of Fossil Energy, 1975.

**Coal Power and Combustion, Quarterly Report, Oct.–Dec., 1975.* Washington: U.S. Energy Research and Development Administration, Office of Fossil Energy, 1975.

**Dynasurge.* Baltimore: Catalyst Research Corporation.

**Glossary for the Gas Industry.* Arlington, Va.: American Gas Association.

*Smith, Craig B. (ed.). *Efficient Electricity Use.* New York: Pergamon Press, Inc.

Hoffman, Edward Jack. *Coal Conversion and the Direct Production of Hydrocarbons from Coal-Steam Systems.* Laramie: College of Engineering, University of Wyoming, 1966.

Metz, William D. Power gas and combined cycles: clean power from fossil fuels. *Science* **179** (No. 4068):56ff (January 5, 1973).

Mudge, L. K., et al. *The Gasification of Coal: A Battelle Energy Program Report.* Richland, Wa.: Battelle Pacific Northwest Laboratories, 1974.

National Economic Research Associates, Inc. *Fuels for the Electric Utility Industry, 1971–1985.* New York: Edison Electric Institute, 1972.

National Research Council. *Committee on Chemical Utilization of Coal.* Chemistry of coal utilization. Supplementary volume. New York: Wiley, 1963.

**Naval Petroleum Reserves.* Report of the Armed Services Investigating Subcommittee of the Committee on Armed Services House of Representatives, 93rd Congress, 1st Session, Washington: U.S. Government Printing Office, 1973.

**News Photo Catalog for the Trans Alaska Pipeline.* Anchorage, Alaska: Alyeska Pipeline Service Company, 1975.

Pyrcioch, E. J., et al. *Production of Pipeline Gas by Hydrogasification of Coal.* Chicago: Institute of Gas Technology, 1972.

Report prepared by the General Accounting Office for the National Fuels and Energy Policy Study Committee on Interior and Insular Affairs, U.S. Senate, 94th Congress, Second Session, Serial No. 94-33. *Status and Obstacles to Commercialization of Coal Liquefaction and Gasification.* Washington: U.S. Government Printing Office, 1976.

Risser, Hubert E. *Gasification and Liquefaction: Their Potential Impact on Various Aspects of the Coal Industry.* Urbana, Ill.: State Geological Survey, 1968.

**Summary Project Description of the Trans Alaska Pipeline System.* Anchorage, Alaska: Alyeska Pipeline Service Company, 1975.

*Tetra Tech, Inc. *Energy from Coal.* Washington: Energy Research and Development Administration, Division of Coal Conversion and Utilization.

**The Dependable Life Support Cell.* Baltimore: Catalyst Research Corporation.

*Thrush, Paul W., and Staff of Bureau of Mines (eds.). A *Dictionary of Mining, Mineral, and Related Terms.* Washington: U.S. Department of the Interior, Bureau of Mines, 1968.

U.S. Energy Research and Development Administration. *Energy From Coal.* Washington: U.S. Government Printing Office, May 1976.

Von Fredersdorff, Claus George, Pyricioch, E.J., and Pettyjohn, E.S. *Gasification of Pulverized Coal in Suspension.* Sponsored by the Gas Production Research Committee of the American Gas Association. Chicago: Institute of Gas Technology, 1957.

*Zabetakis, M. G., and Phillips, L. D. *Coal Mining Safety Manual No. 1.* Washington: U.S. Department of the Interior, Mining Enforcement and Safety Administration.

## NUCLEAR ENERGY

American Nuclear Society. *Nuclear Power and the Environment: Questions and Answers.* Hindsdale, Ill.: American Nuclear Society, 1973.

*American Nuclear Society Standards Committee Subcommittee ANS-9. *American National Standard, Glossary of Terms in Nuclear Science and Technology.* Hinsdale, Ill.: American Nuclear Society, 1976.

**Breeder Backgrounder.* 2d ed. Oak Ridge: Energy Research and Development Administration and Breeder Reactor Corporation, 1976.

Ebbin, Steven. *Citizens Group and the Nuclear Power Controversy: Uses of Scientific and Technological Information.* Cambridge, Ma.: MIT Press, 1974.

Eisenbud, Merril. *Environmental Radioactivity.* New York: McGraw Hill Book Company, Inc. 1963.

*Environmental Survey of the Uranium Fuel Cycle, WASH-1248.* Washington: U.S. Atomic Energy Commission, April 1974.

*Foster, Arthur R., and Wright, Robert L., Jr. *Basic Nuclear Engineering.* 3rd ed. Boston: Allyn and Bacon, Inc., 1977.

Geesey, A. H., and Schultz, M. A. *New Safety System Design for Nuclear Power Reactors.* University Park: Pennsylvania State University, College of Engineering, 1971.

Graeub, Ralph. *The Gentle Killers: Nuclear Power Stations.* Translated from the German by Peter Bostok. London: Abelard-Schuman, 1974.

Graham, John. *Fast Reactor Safety.* New York: Academic Press, 1971.

Heckman, Harry A., and Starring, Paul W. *Nuclear Physics and the Fundamental Particles.* New York: Holt, Reinhart and Winston, Inc., 1963.

Komanoff, Charles; Miller, Holly; and Noyes, Sandy. *The Price of Power: Electric Utilities and the Environment.* New York: Council of Economic Priorities, 1972.

Lapp, Ralph E., and Schubert, Jack. *Radiation: What It Is and How It Affects You.* New York: The Viking Press, 1957.

Leachman, Robert B., and Althoff, Phillip (eds.). *Preventing Nuclear Theft: Guidelines for Industry and Government.* New York: Praeger, 1972.

Lovett, James E. *Nuclear Materials: Accountability, Management, Safeguards.* Hinsdale, Ill.: American Nuclear Society, 1974.

Metz, William D. Laser Fusion: A New Approach to Thermonuclear Power. *Science* **177** (No. 4055):1180 (September 29, 1972).

**Nuclear Terms, A Glossary.* 2d ed. Washington: U.S. Atomic Energy Commission, Office of Information Services, Revised 1967, Reprinted 1974.

Sagain, Leonard A. (ed.). *Human and Ecologic Effects of Nuclear Power Plants.* Springfield, Ill.: Thomas, 1974.

**The Clinch River Breeder Reactor Plant and Its Impact on the Environment.* Oak Ridge: Breeder Reactor Corporation, U.S. Electric Systems Supporting the Clinch River Breeder Reactor Plant Project, 1977.

Willrich, Mason (ed.). *International Safeguards and Nuclear Industry.* Baltimore: Johns Hopkins University Press, 1973.

Willrich, Mason, and Taylor, T. B. *Nuclear Theft: Risks and Safeguards.* Cambridge, Mass.: Ballinger, 1974.

## OCEAN THERMAL ENERGY

Davey, Norman. *Studies in Tidal Power.* London: Constable & Co., Ltd., 1923.

Anderson, Edwin P. *Audels Domestic Water Supply and Sewage Disposal Guide: a Practical Treatise.* New York: T. Audel, 1963.

Clegg, Peter. *New Low-Cost Sources of Energy for the Home; with Complete Illustrated Catalog.* Charlotte, Vt.: Garden Way Pub., 1975.

Creager, William Pitcher, Justin, Joel D., and Hinds, Julian. *Engineering for Dams.* New York: J. Wiley & Sons; London, Chapman & Hall Ltd., 1945.

Davitian, Harry, and McLean, William. Power, fresh water and food from the sea. *Science* **184**:938 (May 31, 1974).

*Energy Primer: Solar, Water, Wind and Biofuels.* Menlo Park, Ca.: Portola Institute, 1974.

Energy from the Sea. *Popular Science* (May, June, July, 1975).

Gillett, Colin Anson. *Can the Tides be Harnessed?* Raetihi, N.Z., 1958.

International Conference on the Utilization of Tidal Power, Nova Scotia Technical College, 1970. *Tidal Power: Proceedings.* Gray, T.J. and Gashus, O.K. (eds.). New York: Plenum Press, 1972.

Macmillian, Donald Henry. *Tides.* New York: American Elsevier Pub. Co., 1966.

*Handbook of Homemade Power.* By the staff of the Mother Earth News. New York: Bantam Books, 1974.

Harris, Carl C., and Rice, Samuel O. *Power Development of Small Streams: A Book for all Persons Seeking Greater Comfort and Higher Efficiency in Country Homes, Towns and Villages.* Orange, Ma.: Rodney Hunt Machine Co., 1920.

Morton, M. Granger (ed.). Ocean thermal energy conversion. *Energy and Man: Technical and Social Aspects of Energy.* New York: IEEE Press, 1975.

**New Water.* Washington: U.S. Department of the Interior, Office of Water Research and Technology, 1970.

Paton, Thomas Angus Lyall; and Brown, J. Guthrie. *Power from Water.* London: L. Hill, 1961.

Rash, Don E. et al. *Energy Under the Ocean: A Technology Assessment.* Norman, Ok.: University of Oklahoma Press, 1973.

Reynolds, John. *Windmills and Watermills.* London: H. Evelyn, 1970.

Vallentine, H. R. *Water in the Service of Man.* Baltimore: Penguin Books, 1967.

Wilson, Paul N. *Water Turbines.* London: H. M. Stationery Off., Palo Alto City, Calif.: obtainable in the U.S.A. from Pendragon House, 1974.

Zener, Clarence. Solar sea power. *Physics Today* **26**:48–53 (January 1973).

## ORGANIC (BIOCONVERSION) FUELS

Blum, S.C. Tapping resources in municipal solid waste. *Science* **191**:669–679 (February 20, 1976).

Bureau of Mines Circular No. 8549. *Energy Potential from Organic Wastes: A Review of the Quantities and Sources.* Washington: U.S. Bureau of Mines, 1972.

Clark, Wilson. The chimerical promise of solar bioconversion. *Energy for Survival.* New York: Anchor Press, 1974.

Clark, Wilson. Solar bioconversion. *Energy for Survival.* Garden City: Doubleday, 1974.

Combustion Engineering, Inc. *Technical-Economic Study of Solid Waste Disposal Needs and Practices.* Rockville, Md.: U.S. Bureau of Solid Waste Management, 1969.

Golueke, Clarence G.; and McGauhey, P.H. Waste materials. *Annual Review of Energy,* Vol. 1., Jack M. Hollander (ed.). Palo Alto: Annual Review, Inc., 1976.

Horner and Shifrin, St. Louis, *Energy Recovery from Waste.* Washington: U.S. Environmental Protection Agency, 1972.

Jackson, Frederick R. *Energy from Solid Waste.* Park Ridge, N.J.: Noyes Data Corp., 1974.

Large, David B. *Hidden Waste; Potentials for Energy Conservation.* Washington: Conservation Foundation, 1973.

Lowe, Robert A. *Energy Recovery from Waste; Solid Waste as Supplementary Fuel in Power Plant Boilers.* Washington: U.S. Environmental Protection Agency, 1973.

National Research Council, Canada. Research Plans and Publications Section. *Wood and Charcoal as Fuel for Vehicles.* 3d ed. Ottawa, 1944.

NSF/NASA Solar Energy Panel. Renewable clean fuel sources. *Solar Energy as a National Energy Resource.* Washington: U.S. Government Printing Office, December 1972.

Proceedings of a Conference, March 10–12, 1976, Washington, D.C. *Capturing the Sun Through Bioconversion.* Washington: Center for Metropolitan Studies, 1976.

Reese, E. T.; Mandels, Mary; and Weiss, Alvin H. Cellulose as a novel energy source. *Advances in Biochemical Engineering.* v. 2, Berlin, New York: Springer-Verlag, 1972.

The Science and Public Policy Program, University of Oklahoma. *The Organic Waste Resource System.* Washington: U.S. Government Printing Office, May 1975.

The Science and Public Policy Program, University of Oklahoma. Organic farms. *Energy Alternatives: A Comparative Analysis.* Washington: U.S. Government Printing Office, May 1975.